GAIA SPECTROSCOPY, SCIENCE AND TECHNOLOGY

COVER ILLUSTRATION:

The cover picture compares the emission line spectrum over the GAIA wavelength region of the X-ray transient XTE J0421+560 dominated by CaII triplet, Paschen head and multiplets #1 and 8 of NI with the absorption line spectrum of the A2 supergiant HD 197345 governed by the same line species.

A SERIES OF BOOKS ON RECENT DEVELOPMENTS IN
ASTRONOMY AND ASTROPHYSICS

Publisher

THE ASTRONOMICAL SOCIETY OF THE PACIFIC

390 Ashton Avenue, San Francisco, California, USA 94112-1722
Phone: (415) 337-1100   E-Mail: orders@astrosociety.org
Fax: (415) 337-5205   Web Site: www.astrosociety.org

ASP CONFERENCE SERIES - EDITORIAL STAFF
Managing Editor: D. H. McNamara
Production Manager: Enid L. Livingston
Production Assistant: Andrea Weaver

PO Box 24463, Room 211 - KMB, Brigham Young University, Provo, Utah, 84602-4463
Phone: (801) 422-2111   Fax: (801) 422-0624   E-Mail: pasp@byu.edu

LaTeX-Computer Consultant: T. J. Mahoney (Spain) – tjm@ll.iac.es

ASP CONFERENCE SERIES PUBLICATION COMMITTEE:
Joss Bland-Hawthorn
George Jacoby
James B. Kaler
J. Davy Kirkpatrick

A listing of all of the ASP Conference Series Volumes and IAU Volumes
published by The ASP may be found at the back of this volume

ASTRONOMICAL SOCIETY OF THE PACIFIC
CONFERENCE SERIES VOLUMES

Volume 298

GAIA SPECTROSCOPY, SCIENCE AND TECHNOLOGY

Proceedings of a conference held at
La Residenza del Sole Congress Center
Gressoney St. Jean, Aosta, Italy
9-12 September 2002

Edited by

Ulisse Munari
INAF Astronomical Observatory of Padova at Asiago, Italy

© 2003 by Astronomical Society of the Pacific. All Rights Reserved

No part of the material protected by this copyright notice may be reproduced or utilized in any form or by any means – graphic, electronic, or mechanical including photocopying, taping, recording or by any information storage and retrieval system, without written permission from The Astronomical Society of the Pacific.

Library of Congress Cataloging in Publication Data
Main entry under title

Card Number: 2003106921
ISBN: 1-58381-145-1

ASP Conference Series - First Edition

Printed in United States of America by Sheridan Books, Ann Arbor, Michigan

Content

Conference photo x

Organization xi

Sponsors xii

Notes xiii

List of participants and authors xiv

Preface xvii

Session 1: OVERVIEW
chair M.G. Lattanzi

The GAIA mission 3
    M.A.C. Perryman
GAIA: the satellite and payload 13
    O. Pace
Considerations about the astrometric accuracy of GAIA 25
    F. Mignard
GAIA photometric performances 41
    V. Vansevičius & A. Bridžius
On science goals of GAIA spectroscopy 51
    U. Munari

Session 2: TECHNOLOGY
chair O.Pace

Design and performances of the GAIA spectrograph 65
    D. Katz
Technical issues for GAIA-RVS 75
    M. Cropper
Absorption cells in wavelength calibration of GAIA spectra 85
    S. Desidera & U. Munari
Bragg gratings in multi-mode fiber optics for wavelength calibration
of GAIA and RAVE spectra 93
    C. Pernechele & U. Munari
The GAIA data access and analysis study 97
    S.G. Ansari, J. Torra, X. Luri, F. Figueras, C. Jordi
    & E. Masana

Telemetric flows on GAIA-RVS estimated from star counts
from the GSC-2.2 catalog 105
    Y.-P. Viala, D. Morin, D. Katz & F. Ochsenbein
Status of the GAIA simulation effort: integration of
spectroscopy simulations 113
    C. Babusiaux & X. Luri
Radial Velocity Spectrometer simulator:
objectives and specifications 119
    D. Katz

Session 3: GALAXY AND STARS
chair F.Mignard

Optimising GAIA: how do we meet the science challenge? 127
    G. Gilmore
Galaxy structure and kinematics towards the NGP 137
    A. Spagna, C. Cacciari, R. Drimmel, T. Kinman,
    M.G. Lattanzi & R.L. Smart
Observational constraints to the mass density in the Galactic plane 147
    O. Bienaymé, A. Siebert & C. Soubiran
Kinematics of the Galactic populations in the GAIA era 153
    G. Bertelli, A. Vallenari, S. Pasetto & C. Chiosi
Methods of relativistic astrometry in space: the static case 159
    F. de Felice, A. Vecchiato, M.T. Crosta,
    B. Bucciarelli & M.G. Lattanzi
GAIA and the planetary nebulae 168
    L. Magrini, M. Perinotto & R.L.M. Corradi

Session 4: STELLAR ATMOSPHERES, CLASSIFICATION AND DATABASES
chair D.Katz

ATLAS model atmospheres 173
    N. Nesvacil, Ch. Stütz & W.W. Weiss
Stellar atmospheres and synthetic spectra for GAIA 179
    P.H. Hauschildt, F. Allard, E. Baron,
    J. Aufdenberg & A. Schweitzer
Cool star atmospheres and spectra for GAIA: MARCS models 189
    B. Plez
On the classification and parametrization of GAIA data
using pattern recognition methods 199
    C.A.L. Bailer-Jones
GAIA broad and medium band photometric performances 209
    C. Jordi, J.M. Carrasco, F. Figueras & J. Torra
Grids of synthetic spectra in planning for the GAIA mission 215
    T. Zwitter, F. Castelli & U. Munari
The Asiago Database of Spectroscopic Databases (ADSD) 221
    R. Sordo & U. Munari
GAIA spectroscopy of peculiar and variable stars 227
    U. Munari

Session 5: RADIAL AND ROTATIONAL VELOCITIES, STELLAR PULSATIONS
chair A.Henden

Cepheids: observational properties, binarity and GAIA 237
    L. Szabados
The Cepheid and RR Lyrae instability strip with GAIA 245
    G. Bono
Miras and other cool variables with GAIA 257
    M. Feast
Observing RV Tauri and SRd variables with GAIA 267
    G.M. Wahlgren
The accuracy of GAIA radial velocities 275
    U. Munari, T. Zwitter, D. Katz & M. Cropper
Stellar rotation from GAIA spectra 285
    A. Gomboc
Chemical abundances from GAIA spectra 291
    F. Thévenin, A. Bijaoui & D. Katz
Wolf-Rayet stars as seen by GAIA 295
    A. Niedzielski

Session 6: DOUBLE AND BINARY STARS
chair M.Cropper

Fundamental stellar parameters from eclipsing binaries 303
    E.F. Milone
Toward optimal processing of large eclipsing binary data sets 313
    R.E. Wilson & S.B. Wyithe
Stellar atmospheres in eclipsing binary models 323
    W. Van Hamme & R.E. Wilson
Observational tests of the GAIA expected harvest on
eclipsing binaries 329
    T. Zwitter
Open questions in binary star statistics and the contribution of
GAIA to solve them 339
    J.-L. Halbwachs, F. Arenou, A. Eggenberger,
    M. Mayor & S. Udry
GAIA and the spectroscopic binaries: what to expect in terms of
orbit determination? 345
    D. Pourbaix & S. Jancart
Distance-effects in the census of binaries with GAIA: principles
and qualitative results 351
    S. Söderhjelm
Binary stars with GAIA and the mass-luminosity relation 357
    O. Malkov & D. Kovaleva

Session 7: GROUND-BASED AND SPACE SURVEYS, CONCLUDING REMARKS
chair T.Zwitter

Current status and future prospects of ground-based and
space photometry 365
    A. A. Henden
Faint and peculiar objects in GAIA: results from GSC-II 375
    D. Carollo, A. Spagna, M.G. Lattanzi, R.L. Smart,
    S.T. Hodgkin & B.J. McLean
RAVE: the RAdial Velocity Experiment 381
    M. Steinmetz
The ING telescopes in the GAIA era 387
    R.L.M. Corradi, D.J. Lennon & R.G.M. Rutten
Conference summary 391
    M.A.C. Perryman

POSTERS

The significance of wavelength coincidence statistics in
abundance determination within the context of GAIA 399
    S.G. Ansari
NLTE line-blanketed CaII calculations for evaluation of
GAIA spectroscopic performances 403
    I. Busà, I. Pagano, M. Rodonò, M.T. Gomez,
    V. Andretta & L. Terranegra
GAIA spectroscopy and cataclysmic variables 407
    M. Cropper & T. Marsh
An extragalactic reference frame for GAIA and SIM using quasars
from the Sloan Digital Sky Survey 411
    E.K. Grebel, M. Odenkirchen & C.A.L Bailer-Jones
AGB stars as tracers of star formation histories: implications for
GAIA photometry and spectroscopy 415
    A. Kučinskas, L. Lindegren, T. Tanabé & V. Vansevičius
Photometric properties of theoretical spectral libraries for
GAIA photometry 419
    T. Lejeune
GAIA spectroscopy of symbiotic binaries 423
    P.M. Marrese & U. Munari
Expanding the Asiago library of real spectra for GAIA 427
    P.M. Marrese, U. Munari, F. Boschi & L.Tomasella
GAIA and the boon to extrasolar planet studies 431
    E.F. Milone & M.D. Williams
Evaluating the GAIA performance on photometry of
near-contact and contact binaries 435
    P.G. Niarchos & V.N. Manimanis
Hydrogen-to-helium ratio in WR stars from GAIA spectroscopy 439
    A. Niedzielski & T. Nugis

The extended tidal tails of Palomar 5: tracers of the
Galactic potential 443
    M. Odenkirchen, E.K. Grebel, H.-W. Rix, W. Dehnen,
    H.J. Newberg, C.M. Rockosi & B. Yanny
Missing spectroscopy of θ Tucanae 447
    M. Paparó
GAIA spectroscopy of Carbon stars 451
    Ya.V. Pavlenko, P.M. Marrese & U. Munari
Coarse estimation of physical parameters of eclipsing binaries
by means of automative scripting 457
    A. Prša
GAIA spectroscopy of active solar-type stars 461
    S. Ragaini, V. Andretta, M.T. Gomez, L. Terranegra,
    I. Busà & I. Pagano
CaII triplet monitoring of stars with the R CrB type variability 465
    A.E. Rosenbush
Spectroscopy with FLAMES: applications to GAIA RVS 469
    F. Royer
An analysis of the precision of stellar radial velocities obtained
using the HERCULES spectrograph 473
    J. Skuljan, J.B. Hearnshaw & S.I. Barnes
Line indices as age, metallicity and abundance ratio indicators 477
    R. Tantalo & C. Chiosi
Comments on atomic data for the GAIA spectral region 481
    G.M. Wahlgren & S. Johansson
Derivation of stellar parameters from DIVA spectral data 485
    P.G. Willemsen & T.A. Kaempf
Crowding in the GAIA spectrograph focal plane 489
    T. Zwitter & A.A. Henden
Information recovery from overlapping GAIA spectra 493
    T. Zwitter

Subject index... 499

Object index.. 511

Author index... 513

Conference photo

Organization

Scientific Organizing Committee

Fiorella	Castelli	*Trieste*	
Cesare	Chiosi	*Padova*	
Mark	Cropper	*London*	
David	Katz	*Paris*	
Mario	Gai	*Torino*	
Gerry	Gilmore	*Cambridge*	
Mario G.	Lattanzi	*Torino*	
Lennart	Lindegren	*Lund*	
Francois	Mignard	*Nice*	
Ulisse	Munari	*Asiago*	(chair)
Oscar	Pace	*ESA*	
Michael A.C.	Perryman	*ESA*	
Catherine	Turon	*Paris*	
Tomaž	Zwitter	*Ljubljana*	

Local Organizing Committee

Federico	Boschi	*Asiago*	(chair)
Ulisse	Munari	*Asiago*	
Rosanna	Sordo	*Asiago*	
Lina	Tomasella	*Asiago*	

Local Support Team

Corrado	Boeche	*Asiago*	
Luciano	di Giorgio	*Asiago*	
Massimo	Fiorucci	*Asiago*	(chair)
Paola	Marrese	*Asiago*	
Alessandro	Siviero	*Asiago*	

Editorial Office

Federico	Boschi	*Asiago*	
Ulisse	Munari	*Asiago*	(editor)
Rosanna	Sordo	*Asiago*	

Sponsors

The main sponsors of this Conference have been:

European Space Agency

Alenia Spazio

Astronomical Observatory of Padova - INAF

Department of Astronomy, University of Padova

Sources of support to the organization of the Conference have also been the GAIA-related research programs financed by the Italian Space Agency (ASI) and the Italian Ministry of Education, University and Technology (via COFIN projects), with P.L. Bernacca, M.G. Lattanzi, C. Chiosi, U. Munari, S. Aiello and I. Porceddu acting as national and/or local coordinators.

Notes

- In the Object and Subject indices no entry is provided for such general items like LMC, SMC, Sun, Galaxy, GAIA, metallicity, chemical elements, etc, because they are mentioned in the majority of papers.

- A significant fraction of the preparatory work for the GAIA mission has resulted in the preparation of technical notes internal to the ESA GAIA community, usually identified by the working group acronym, the author's initials, and a progressive number, for example RVS-DK-003. A large number of information, not published somewhere else, is available from these technical notes only, and several papers in these Proceedings reference them. They are generally maintained, documented and made available via ESTEC *Livelink* at http://astro.estec.esa.nl/llink/livelink. These documents are listed in the references, for ex., as

  Lindegren, L. 2000, SAG-LL-030 (Livelink)

- Another document frequently mentioned in these Proceedings is the *red book* for the GAIA mission, the *GAIA: Composition, Formation and Evolution of the Galaxy. Concept and Technology Study Report* published by ESA in July 2000 as SCI-4. It has been compiled under the supervision of the ESA Study Scientist, M.A.C. Perryman, the ESA Study Manager, O. Pace, and the members of the Science Advisory Group, K.S. de Boer, G. Gilmore, E. Hoeg, M.G. Lattanzi, L. Lindegren, X. Luri, F. Mignard, and P.T. de Zeeuw. In these Proceedings, it is cited within paper text as

  ESA-2000-SCI-4

- R.L.Kurucz's programs and line-lists to compute model stellar atmospheres and spectra, distributed as a series of well known CD-ROMs, are frequently referenced in these Proceedings. They are cited, for ex., as:

  Kurucz, R.L. 1993a, Kurucz's CD-ROM N. 13, CfA Harvard, Cambridge

List of participants and authors

name	affiliation	e-mail
France Allard	CRAL, Lyon, France	fallard@ens-lyon.fr
Vincenzo Andretta	INAF, Capodimonte, Italy	andretta@na.astro.it
Salim G. Ansari	ESA - ESTEC, The Netherlands	salim.ansari@esa.int
Frédéric Arenou	Obs. Paris, France	Frederic.Arenou@obspm.fr
Jason Aufdenberg	CfA, Cambridge, UK	jaufdenberg@cfa.harvard.edu
Carine Babusiaux	IoA, Cambridge, UK	carine@ast.cam.ac.uk
Coryn A.L. Bailer-Jones	MPIA, Heidelberg, Germany	calj@mpia-hd.mpg.de
Stuart I. Barnes	Univ. Canterbury, New Zealand	S.Barnes@phys.canterbury.ac.nz
Edward Baron	Univ. Oklahoma, USA	baron@nhn.ou.edu
Giampaolo Bertelli	IASF-CNR, Roma, Italy	bertelli@pd.astro.it
Olivier Bienaymé	Obs. Astr. Strasbourg, France	bienayme@astro.u-strasbg.fr
Albert Bijaoui	OCA, Nice, France	Albert.BIJAOUI@obs-azur.fr
Corrado Boeche	INAF, Padova-Asiago, Italy	corrado@hati.pd.astro.it
Henri Boffin	Royal Obs. Brussels, Belgium	Henri.Boffin@oma.be
Giuseppe Bono	INAF, Roma, Italy	bono@mporzio.astro.it
Federico Boschi	INAF, Padova-Asiago, Italy	boschi@pd.astro.it
Audrius Bridžius	Inst. Phys., Vilnius, Lithuania	bridzius@astro.lt
Beatrice Bucciarelli	INAF, Torino, Italy	bucciarelli@to.astro.it
Innocenza Busà	INAF, Catania, Italy	ebu@ct.astro.it
Deborah Busonero	INAF, Torino, Italy	busonero@to.astro.it
Carla Cacciari	INAF, Bologna, Italy	cacciari@bo.astro.it
Daniela Carollo	INAF, Torino, Italy	carollo@to.astro.it
Josep Manel Carrasco	Univ. Barcelona, Spain	carrasco@am.ub.es
Fiorella Castelli	CNR, Trieste, Italy	castelli@ts.astro.it
Cesare Chiosi	INAF, Padova, Italy	chiosi@inaf.it
Romano L.M. Corradi	ING, La Palma, Spain	rcorradi@ing.iac.es
Francoise Crifo	Obs. Paris, GEPI, France	francoise.crifo@obspm.fr
Mark Cropper	MSSL, London, UK	msc@mssl.ucl.ac.uk
Maria Teresa Crosta	CISAS, Padova, Italy	crosta@to.astro.it
Jos de Bruijne	ESA - ESTEC, The Netherlands	Jos.de.Bruijne@rssd.esa.int
Fernando de Felice	Univ. Padova, Italy	defelice@pd.infn.it
Walter Dehnen	AIP, Potsdam, Germany	wdehnen@aip.de
Silvano Desidera	INAF, Padova, Italy	desidera@astras.pd.astro.it
Ronald Drimmel	INAF, Torino, Italy	drimmel@to.astro.it
Anne Eggenberger	Obs. Genève, Switzerland	Anne.Eggenberger@obs.unige.ch
Michael Feast	Univ. Cape Town, South Africa	mwf@artemisia.ast.uct.ac.za
Massimo Fiorucci	Univ. Perugia, Italy	Massimo.Fiorucci@pg.infn.it
Francesca Figueras	Univ. Barcelona, Spain	cesca@am.ub.es
Andrea Frigo	Museo Rovereto, Italy	andrea.frigo@tin.it
Mario Gai	INAF, Torino, Italy	gai@to.astro.it
Gerry Gilmore	IoA, Cambridge, UK	gil@ast.cam.ac.uk

name	affiliation	e-mail
Andreja **Gomboc**	Univ. Ljubljana, Slovenia	andreja@fiz.uni-lj.si
Maria Teresa **Gomez**	INAF, Capodimonte, Italy	gomez@na.astro.it
Eva K. **Grebel**	MPIA, Heidelberg, Germany	grebel@mpia-hd.mpg.de
Jean-Louis **Halbwachs**	Astr. Obs. Strasbourg, France	halbwachs@newb6.u-strasbg.fr
Peter **Hauschildt**	Sternwarte, Hamburg, Germany	phauschildt@hs.uni-hamburg.de
John B. **Hearnshaw**	U. Canterbury, New Zealand	J.Hearnshaw@phys.canterbury.ac.nz
Arne A. **Henden**	USRA/USNO, Flagstaff, USA	aah@nofs.navy.mil
Simon T. **Hodgkin**	IoA, Cambridge, UK	sth@star.le.ac.uk
Sylvie **Jancart**	ULB, Bruxelles, Belgium	sylvie.jancart@fundp.ac.be
Urtzi **Jauregi**	Univ. Ljubljana, Slovenia	urtzi@dome.ago.uni-lj.si
Sveneric **Johansson**	Lund Obs., Sweden	sveneric.johansson@astro.lu.se
Carme **Jordi**	Univ. Barcelona, Spain	carme@am.ub.es
Torsten A. **Kaempf**	SdU, Bonn, Germany	tkaempf@astro.uni-bonn.de
David **Katz**	Obs. Meudon, France	david.katz@obspm.fr
Thomas **Kinman**	NOAO, Tucson, USA	kinman@noao.edu
Dana **Kovaleva**	Acad. Sci., Moscow, Russia	dana@inasan.rssi.ru
Jean **Kovalewsky**	OCA, Nice, France	jean.kovalevsky@obs-azur.fr
Arunas **Kučinskas**	Lund Obs., Sweden	arunas@astro.lu.se
Mario G. **Lattanzi**	INAF, Torino, Italy	lattanzi@to.astro.it
Thibault **Lejeune**	Astron. Obs. Coimbra, Portugal	lejeune@mat.uc.pt
Danny J. **Lennon**	ING, La Palma, Spain	djl@ing.iac.es
Giuseppe **Leto**	INAF, Catania, Italy	gle@ct.astro.it
Lennart **Lindegren**	Lund Obs., Sweden	lennart@astro.lu.se
Xavier **Luri**	Univ. Barcelona, Spain	xluri@am.ub.es
Laura **Magrini**	Univ. Florence, Italy	laura@arcetri.astro.it
Oleg **Malkov**	Acad. Sci., Moscow, Russia	malkov@amon.inasan.rssi.ru
Vassilios N. **Manimanis**	Univ. Athens, Greece	vmaniman@cc.uoa.gr
Paola M. **Marrese**	INAF, Padova-Asiago, Italy	marrese@pd.astro.it
Tom **Marsh**	Univ. Southampton, UK	trm@astro.soton.ac.uk
Eduard **Masana**	IEEC, Catalunya, Spain	masana@ieec.fcr.es
Michel **Mayor**	Obs. Genève, Switzerland	Michel.Mayor@obs.unige.ch
Brian J. **McLean**	STScI, Baltimore, USA	mclean@stsci.edu
Francois **Mignard**	CERGA, Grasse, France	francois.mignard@obs-azur.fr
Shan **Mignot**	Observatoire de Paris, France	shan.mignot@mines-paris.org
Eugene F. **Milone**	RAO, Univ. Calgary, Canada	milone@ucalgary.ca
Danielle **Morin**	Obs. Paris, France	danielle.morin@obspm.fr
Ulisse **Munari**	INAF, Padova-Asiago, Italy	munari@astras.pd.astro.it
Emma **Nasi**	INAF, Padova, Italy	nasi@pd.astro.it
Nicole **Nesvacil**	Univ. Wien, Austria	a9702981@unet.univie.ac.at
Heidi Jo **Newberg**	Polytech. Inst., Troy, USA	newbeh@rpi.edu
Panagiotis G. **Niarchos**	Univ. Athens, Greece	pniarcho@cc.uoa.gr
Andrzej **Niedzielski**	Toruń CfA, Poland	aniedzi@astri.uni.torun.pl
Tiit **Nugis**	Tartu Obs., Estonia	nugis@aai.ee
Francois **Ochsenbein**	Ast.Obs.Strasbourg, France	francois.ochsenbein@astro.u-strasbg.fr
Michael **Odenkirchen**	MPIA, Heidelberg, Germany	odenkirchen@mpia.de
Rob **Olling**	USNO, Washington DC, USA	olling@usno.navy.mil
Oscar **Pace**	ESA - ESTEC, The Netherlands	oscar.pace@esa.int
Isabella **Pagano**	INAF, Catania, Italy	ipagano@ct.astro.it
Margit **Paparó**	Konkoly Obs., Hungary	paparo@konkoly.hu
Stefano **Pasetto**	INAF, Padova, Italy	pasetto@pd.astro.it

name		affiliation	e-mail
Yakiv V.	**Pavlenko**	Astron. Obs., Kiev, Ukraine	yp@mao.kiev.ua
Mario	**Perinotto**	Univ. Florence, Italy	mariop@arcetri.astro.it
Claudio	**Pernechele**	INAF, Padova, Italy	pernechele@pd.astro.it
Michael A.C.	**Perryman**	ESA-ESTEC, The Netherlands	mperryma@rssd.esa.int
Bertrand	**Plez**	GRAAL, Montpellier, France	plez@graal.univ-montp2.fr
Piotr	**Popowski**	MPIA, Garching, Germany	popowski@mpa-garching.mpg.de
Dimitri	**Pourbaix**	ULB, Bruxelles, Belgium	pourbaix@astro.ulb.ac.be
Andrej	**Prša**	Univ. Ljubljana, Slovenia	andrej.prsa@fiz.uni-lj.si
Silvia	**Ragaini**	INAF, Capodimonte, Italy	ragaini@na.astro.it
Hans-Walter	**Rix**	MPIA, Heidelberg, Germany	rix@mpia.de
Constance M.	**Rockosi**	Univ. Washington, USA	cmr@astro.washington.edu
Marcello	**Rodonò**	Univ. Catania, Italy	mrodono@astrct.ct.astro.it
Alexander E.	**Rosenbush**	Astron. Obs., Kiev, Ukraine	mijush@mao.kiev.ua
Frédéric	**Royer**	Obs. Genève, Switzerland	frederic.royer@obs.unige.ch
René G.M.	**Rutten**	ING, La Palma, Spain	rgmr@ing.iac.es
Andreas	**Schweitzer**	Sternwarte, Hamburg, Germany	aschweitzer@hs.uni-hamburg.de
Arnaud	**Siebert**	Obs. Astr. Strasbourg, France	siebert@astro.u-strasbg.fr
Alessandro	**Siviero**	INAF, Padova-Asiago, Italy	siviero@mimir.pd.astro.it
Jovan	**Skuljan**	Univ. Canterbury, New Zealand	j.skuljan@phys.canterbury.ac.nz
Richard L.	**Smart**	INAF, Torino, Italy	smart@to.astro.it
Staffan	**Söderhjelm**	Lund Obs., Sweden	staffan@astro.lu.se
Rosanna	**Sordo**	Univ. Padova-Asiago, Italy	sordo@pd.astro.it
Caroline	**Soubiran**	Obs. Bordeaux, France	Caroline.Soubiran@observ.u-bordeaux.fr
Alessandro	**Sozzetti**	Harvard CfA, USA	asozzetti@cfa.harvard.edu
Alessandro	**Spagna**	INAF, Torino, Italy	spagna@to.astro.it
Matthias	**Steinmetz**	AIP, Potsdam, Germany	msteinmetz@aip.de
Christian	**Stütz**	Univ. Wien, Austria	stuetz@tycho.astro.univie.ac.at
Lazlo	**Szabados**	Konkoly Obs., Hungary	szabados@konkoly.hu
Toshihiko	**Tanabé**	Univ. Tokyo, Japan	ttanabe@ioa.s.u-tokyo.ac.jp
Rosaria	**Tantalo**	Univ. Padova , Italy	tantalo@pd.astro.it
Luciano	**Terranegra**	INAF, Capodimonte, Italy	terraneg@na.astro.it
Frédéric	**Thévenin**	OCA, Nice, France	thevenin@obs-nice.fr
Lina	**Tomasella**	INAF, Padova-Asiago, Italy	tomasella@pd.astro.it
Jordi	**Torra**	Univ. Barcelona, Spain	jordi@am.ub.es
Catherine	**Turon**	Obs. Paris, France	catherine.turon@obspm.fr
Stéphane	**Udry**	Obs. Genève, Switzerland	Stephane.Udry@obs.unige.ch
Antonella	**Vallenari**	INAF, Padova, Italy	vallenari@pd.astro.it
Walter	**Van Hamme**	Florida Int. Univ., USA	vanhamme@fiu.edu
Vladas	**Vansevičius**	Inst. Phys., Vilnius, Lithuania	wladas@astro.lt
Alberto	**Vecchiato**	Univ. Padova, Italy	vecchiato@to.astro.it
Yves-Paul	**Viala**	Obs. de Paris, France	yves.viala@obspm.fr
Glenn M.	**Wahlgren**	Lund Obs., Sweden	glenn.wahlgren@astro.lu.se
Werner W.	**Weiss**	Univ. Wien, Austria	weiss@astro.univie.ac.at
Mark	**Wilkison**	IoA, Cambridge, UK	markw@ast.cam.ac.uk
Philip	**Willemsen**	SdU, Bonn, Germany	willemse@astro.uni-bonn.de
Michael D.	**Williams**	RAO, Calgary, Canada	williamd@ucalgary.ca
Robert E.	**Wilson**	Univ. of Florida, USA	wilson@rigel.astro.ufl.edu
Stuart B.	**Wyithe**	Harvard Coll. Obs., USA	swyithe@cfa.harvard.edu
Brian	**Yanny**	Fermilab, Illinois, USA	yanny@fnal.gov
Tomaž	**Zwitter**	Univ. Ljubljana, Slovenia	tomaz.zwitter@uni-lj.si

Preface

The idea of this Conference begun to take shape after the approval by ESA's SPC of the GAIA mission in October 2000, and went into sharp focus during the Les Houches school devoted to GAIA held in May 2001. It was originally planned to take place in Venice, and announced as GAIA 2002 Venice conference, following the Hipparcos Venice-97 celebrative venue. The September 11 events and consequent troubles for the tourist, hotel and airlines sectors vanished the preliminary work done with local operators and, to avoid sky-rocketing costs, suggested a relocation of the Conference. Mrs. Anna Mello Rella, manager of Serenissima Viaggi, has been the key factor in successfully selecting La Residenza del Sole Congress Center in Gressoney St. Jean, facing Monte Rosa and its glaciers. After the 2001 Les Houches Conference on GAIA held in front of the highest Alp mountain (Mont Blanc 4807 m), and this conference facing the second highest (Monte Rosa 4646 m), the trend toward a future meeting next to the third highest peak (Matterhorn 4478 m) looks intriguing indeed. Among the previous major international conferences devoted to GAIA it is worth to note *Future Possibilities for Astrometry in Space* (Cambridge, June 1995, M.A.C. Perryman and F. van Leeuwen ed.s, ESA SP-379), *GAIA* (Leiden November 1998, V. Straižys ed., Baltic Astronomy vol 8, n.1+2), *GAIA: A European Space Project* (Les Houches May 2001, O. Bienaymé and C. Turon ed.s, EAS Pub. Ser. 2), and *Census of the Galaxy* (Vilnius July 2001, V. Vansevičius, A. Kučinskas and J. Sudžius ed.s, Kluwer).

This Conference has been preceded and followed by a series of meetings of the GAIA Radial Velocity Spectrometer (RVS) Working Group:

RVS-1	Meudon	1-2 October 2001
RVS-2	Asiago	7-8 February 2002
RVS-3	Ljubljana	10-11 June 2002
RVS-4	Gressoney	13 September 2002
RVS-5	Paris	28-29 November 2002
RVS-6	London	June 2003

The RVS Working Group (http://wwwhip.obspm.fr/gaia/rvs/objectives.html), lead by David Katz with assistance from the writer, has been in charge to define by the end of 2002 the baseline instrumental configuration for the GAIA spectrograph, evaluate the performances, investigate the data acquisition-telemetry-reduction strategies and assess the scientific return of the spectroscopic observations. The fourth RVS meeting has taken place at the same location of this Conference the day following its conclusion, in parallel with a meeting of the Double Star Working Group lead by F. Arenou (http://wwwhip.obspm.fr/gaia/dms/), both taking advantage of the fact that a significant fraction of the GAIA community was already gathering in Gressoney St. Jean. The RVS Working Group is composed by scientists from all over Europe, in particular France, UK, Italy and Slovenia. A lot of work as been done since the kick-off meeting in Meudon in October 2001, which has been possible thanks to the dedication and full commitment of all its members and in particular to the friendly and highly cooperative atmosphere that the Working Group has enjoyed, with prompt and open circulation of all relevant information, for which gratitude goes in particular to David Katz, Mark Cropper and Tomaž Zwitter.

Editing this Proceedings has profoundly benefitted from expert advises by Enrico Corsini, Michael Perryman, Thibault Lejeune and Enid Livingston. Great help has been received from Rosanna Sordo and Federico Boschi, that patiently assisted in pre-processing the contributions as their arrived, handled e-mail communication with authors and cooperated in typing lists and indices. They however carry no responsibility for any unintentional error possibly introduced during the text editing, layout homogenization, figure re-drawing, table and reference re-styling of the contributions, for which the Editor has the sole responsibility.

My gratitude goes to the SOC (cf. pag. xi) for the assistance in shaping and assembling the conference science program, to the Chairs for governing the smooth course of the sessions, and to the LOC and LST young enthusiastic collaborators (cf. pag. xi) that spent so much time over the last year in preparing for a conference to be held 400 km away from their Institute. Mark Cropper and Francois Mignard made a great work in, respectively, reviewing in details all the posters and in preparing and moderating the concluding general discussion. I want also to thank Giorgio Martorana for his assistance in exploring accommodation and conference facilities in Venice, Sergio Dalle Ave for taking care of conference poster, badges and photographic post-processing, and Maria Antonia Rossi and Federico Boschi for the wood mock-up of the GAIA spacecraft and payload on show on the speakers' desk during the Conference.

Thanks goes to Davide Porro, the manager of the Congress Center, for his commitment to conference organization and for offering the concluding gala and dance dinner, and again to Anna Mello Rella, the manager of Serenissima Viaggi tour operator, for organization of the door-to-door shuttle service with Milan and Turin airports and railway stations, and the social afternoon and dinner in Aosta.

My gratitude is also for Davide Scomazzon and Emma Rigoni for the realization of the hand-made and hand-painted conference mug, personalized with own name for each participant, and to Daniele Mania, manager of Gressoney St. Jean tourist office, for gadgets, souvenirs, guided tours and Sunday's welcome cocktail offered to participants.

The conference would have been much harder (actually impossible) to organize without the support of the sponsors (cf. pag. xii). Special thanks goes to Michael Perryman for his *buoni uffici* in securing the support of ESA. The raised founds allowed to wave the registration fee and assist with hotel and travel expenses a significant fraction of the Participants. The geographical distribution of parent Institutes of participants and authors

Italy	41	Sweden	5	New Zealand	3	Ukraine	2
France	18	The Netherlands	4	Canada	2	Estonia	1
USA	12	Slovenia	4	Greece	2	Japan	1
Germany	11	Switzerland	4	Hungary	2	Poland	1
Spain	9	Austria	3	Lithuania	2	Portugal	1
UK	7	Belgium	3	Russia	2	South Africa	1

is an indication of the interest growing around GAIA within as well as outside the European scientific community.

The pictures displayed in these proceedings have been taken by M. Perryman, A. Siviero and F. Boschi during the Conference venue.

I would like to conclude with a personal *grazie!* to PierLuigi Bernacca, Mario G. Lattanzi e Cesare Chiosi, for their support to the GAIA mission and leading national grant applications in support of the Italian participation to GAIA, to Gianfranco Dezotti, Massimo Calvani and Piero Rafanelli for aid by INAF Astronomical Observatory of Padova and Department of Astronomy of the University of Padova, Lina Tomasella, Dina Moro and Massimo Fiorucci for their help with early efforts on GAIA spectroscopy and photometry, and to Francesco Bertola and Alessandro Bressan for assistance in the organization of a seminal GAIA workshop at Accademia Nazionale dei Lincei in Rome.

Further information, useful links and a picture gallery can be found at the Conference web site

http://ulisse.pd.astro.it/GAIA2002/

Asiago, February 2003   Ulisse Munari (Editor)

session 1
OVERVIEW
chair: M.G. Lattanzi

The valley of Gressoney, ending at the base of Monte Rosa (4633 m), as seen from *Belvedere* view point on the way to *Cima Regina* peak

Gressoney St. Jean: the parish church (XVI c.) on the right

The GAIA mission

Michael A.C. Perryman

Astrophysics Missions Division, ESA-ESTEC, 2200AG Noordwijk, The Netherlands

Abstract. As of June 2002, GAIA is a confirmed mission within the ESA 'Cosmic Vision 2020' science programme, with a target launch date of mid-2010. GAIA will build on the observational principles of Hipparcos to measure detailed properties of the brightest 1 billion stars in the sky. Astrometric accuracies of 10 microarcsec at 15 mag should lead to 20 million stars measured with distance accuracies of better than 1%, and more than 100 million better than 5%. Tangential velocities will be measured astrometrically at better than 1 km s^{-1} for about 100 million stars, while the dedicated radial velocity spectrometer will gather radial velocities to 1–10 km s^{-1} to 16–17 mag, depending on spectral type. GAIA will provide multi-colour (in 11 medium and 5 broad bands), multi-epoch (of order 100 epochs over 5 years) photometry for each object to 20 mag, with great care being invested in devising the photometric bands to maximise their astrophysical diagnostic power. Scientific preparations for the mission involve the participation of some 15 working groups, taking responsibility for (amongst other aspects) the accuracy modelling, the radial velocity instrument optimisation, preparation of simulated data, and the development of a data processing framework to handle the complex and large (of order 1 Petabyte) GAIA data set.

1. Introduction

Following the success of ESA's Hipparcos space astrometry mission, the GAIA project has been approved as an ambitious space experiment to extend highly accurate positional measurements to a very large number of stars throughout our Galaxy. GAIA's contribution to the understanding of the structure and evolution of our Galaxy is based on three complementary observational approaches: (*i*) a census of the contents of a large, representative, part of the Galaxy; (*ii*) quantification of the present spatial structure, from distances; (*iii*) knowledge of the three-dimensional space motions, to determine the gravitational field and the stellar orbits. Astrometric measurements uniquely provide model-independent distances and transverse kinematics, and form the basis of the cosmic distance scale.

Complementary radial velocity and photometric information are required to complete the kinematic and astrophysical information about the individual objects observed. Photometry, with appropriate astrometric and astrophysical calibration, gives a knowledge of extinction, and hence, combined with astrometry, provides intrinsic luminosities, spatial distribution functions, and stellar chemical abundance and age information. Radial velocities complete the kinematic triad, allowing determination of dynamical motions, gravitational forces, and the distribution of invisible mass. The GAIA mission will provide all this information.

GAIA will be a continuously scanning spacecraft, accurately measuring one-dimensional coordinates along great circles in two simultaneous fields of view, separated by a well-known angle. The payload utilises a large CCD focal plane assembly, passive thermal control, natural short-term instrument stability due to the Sun shield and the selected orbit, and a robust payload design. A 'Lissajous' orbit at the L2 Lagrange point of the Sun-Earth system is the proposed operational orbit, from where about 1 Mbit of data per second is returned to the single ground station throughout the 5-year mission. A more detailed description of the project is given elsewhere (Perryman et al. 2001), based on the extensive study conducted between 1998–2000 (ESA-2000-SCI-4). Following a system re-assessment phase during the first half of 2002, certain payload simplifications have been made with respect to the previous design, and the revised system now fits within a Soyuz launch configuration, without the relaxation of the adopted scientific goals.

2. Scientific goals

This section gives a concise introduction to some of the scientific topics that can be addressed by performing astrometric measurements at the microarcsec level for very large numbers of stars. By way of introduction, it may be noted that GAIA is expected to observe, or discover, very large numbers of specific objects, for example: $10^5 - 10^6$ (new) Solar System objects; 30 000 extra-Solar planets; 200 000 disk white dwarfs; 10^7 resolved binaries within 250 pc; $10^6 - 10^7$ resolved galaxies; 10^5 extragalactic supernovae; and 500 000 quasars.

Structure and Dynamics of the Galaxy: One of the primary objectives of the GAIA mission is to observe the physical characteristics, kinematics and distribution of stars over a large fraction of the volume of our Galaxy, with the goal of achieving a detailed understanding of its dynamics and structure, and consequently its formation and history.

The Star Formation History of our Galaxy: One central element the GAIA mission is the determination of the star formation histories, as described by the temporal evolution of the star formation rate, and the cumulative numbers of stars formed, of the bulge, inner disk, Solar neighbourhood, outer disk and halo of our Galaxy. Given such information, together with the kinematic information from GAIA, and complementary chemical abundance information, again primarily from GAIA, the full evolutionary history of the Galaxy is determinable. Determination of the relative rates of formation of the stellar populations in a large spiral, typical of those galaxies which dominate the luminosity in the Universe, will provide for the first time quantitative tests of galaxy formation models. Do large galaxies form from accumulation of many smaller systems which have already initiated star formation? Does star formation begin in a gravitational potential well in which much of the gas is already accumulated? Does the bulge pre-date, post-date, or is it contemporaneous with, the halo and inner disk? Is the thick disk a mix of the early disk and a later major merger? Is there a radial age gradient in the older stars? Is the history of star formation relatively smooth, or highly episodic? Answers to such questions will also provide a template for analysis of data on unresolved stellar systems, where similar data cannot be obtained.

Stellar Astrophysics: GAIA will provide distances of unprecedented accuracy for all types of stars of all stellar populations, even those in the most rapid evolutionary phases which are very sparsely represented in the Solar neighbourhood. All parts of the Hertzsprung–Russell diagram will be comprehensively calibrated, from pre-main sequence stars to white dwarfs and all transient phases; all possible masses, from brown dwarfs to the most massive O stars; all types of variable stars; all possible types of binary systems down to brown dwarf and planetary systems; all standard distance indicators, etc. This extensive amount of data of extreme accuracy will stimulate a revolution in the exploration of stellar and Galactic formation and evolution, and the determination of the cosmic distance scale.

Photometry and Variability: The GAIA large-scale photometric survey will have significant intrinsic scientific value for stellar astrophysics, providing basic stellar data (effective temperatures, surface gravities, metallicities, etc) and also valuable samples of variable stars of nearly all types, including detached eclipsing binaries, contact or semi-contact binaries, and pulsating stars. The pulsating stars include key distance calibrators such as Cepheids and RR Lyrae stars and long-period variables. Existing samples are incomplete already at magnitudes as bright as $V \sim 10$ mag. A complete sample of objects will allow determination of the frequency of variable objects, and will accurately calibrate period-luminosity relationships across a wide range of stellar parameters including metallicity. A systematic variability search will also allow identification of stars in short-lived but key stages of stellar evolution. Prompt processing will identify many targets for follow-up ground-based studies. Estimated numbers are highly uncertain, but suggest some 18 million variable stars in total, including 5 million 'classic' periodic variables, 2–3 million eclipsing binaries, 2000–8000 Cepheids, 60 000–240 000 δ Scuti variables, 70 000 RR Lyrae, and 140 000–170 000 Miras.

Planetary Systems: There are a number of techniques which in principle allow the detection of extra-Solar planetary systems: these include pulsar timing, radial velocity measurements, astrometric techniques, transit measurements, microlensing, and direct methods based on high-angular resolution interferometric imaging. A better understanding of the conditions under which planetary systems form and of their general properties requires sensitivity to low mass planets (down to $\sim 10\,M_\oplus$), characterization of known systems (mass, and orbital elements), and complete samples of planets, with useful upper limits on Jupiter-mass planets out to several AU from the central star. Astrometric measurements good to 2–10 μas will contribute substantially to these goals, and will complement the ongoing radial velocity measurement programmes. GAIA's strength will be its discovery potential, following from the astrometric monitoring of all of the several hundred thousand bright stars out to distances of ~ 200 pc.

Solar System Objects: Solar System objects present a challenge to GAIA because of their significant proper motions, but they promise a rich scientific reward. The minor bodies provide a record of the conditions in the proto-Solar nebula, and their properties therefore shed light on the formation of planetary systems. The relatively small bodies located in the main asteroid belt between Mars and Jupiter should have experienced limited thermal evolution since the early epochs of planetary accretion. Due to the radial extent of the main belt, minor planets provide important information about the gradient of mineralogical composition of the early planetesimals as a function of heliocentric distance. It is therefore important for any study of the origin and evolution of the Solar system to investigate the main physical properties of asteroids including masses, densities, sizes, shapes, and taxonomic classes, all as a function of location in the main belt and in the Trojan clouds. The possibility of determining asteroid masses relies on the capability of measuring the tiny gravitational perturbations that asteroids experience in case of a mutual close approach. GAIA is expected to discover a very large number, of the order of 10^5 or 10^6 new objects, depending on the uncertainties on the extrapolations of the known population. It should be possible to derive precise orbits for many of the newly discovered objects, since each of them will be observed many times during the mission lifetime. These will include a large number of near-Earth asteroids.

Fundamental Physics: The reduction of the Hipparcos data necessitated the inclusion of stellar aberration up to terms in $(v/c)^2$, and the general relativistic treatment of light bending due to the gravitational field of the Sun (and Earth). The GAIA data reduction requires a more accurate and comprehensive inclusion of relativistic effects, at the same time providing the opportunity to test a number of parameters of general relativity in new observational domains, and with much improved precision. The dominant relativistic effect in the GAIA measurements is gravitational light bending, quantified by, and allowing accurate determination of, the parameter γ of the Parametrized Post-Newtonian (PPN) formulation of gravitational theories. While the angular separation to objects observed by GAIA, χ, is never smaller than $35 - 40°$ for the Sun, grazing incidence is possible for the planets. With the astrometric accuracy of a few μas, the magnitude of the expected effects is considerable for the Sun, and also for observations near planets. Detailed analyses indicate that the GAIA measurements will provide a precision of about 5×10^{-7} for γ, based on multiple observations of $\sim 10^7$ stars with $V < 13$ mag at wide angles from the Sun, with individual measurement accuracies better than 10 μas.

3. Overall design considerations

The proposed GAIA design has arisen from requirements on astrometric precision (10 μas at 15 mag), completeness to $V = 20$ mag, the acquisition of radial velocities, the provision of accurate multi-colour photometry for astrophysical diagnostics, and the need for on-board object detection.

3.1. Astrometry

A space astrometry mission has a unique capability to perform global measurements, such that positions, and changes in positions caused by proper motion and parallax, are determined in a reference system consistently defined over the whole sky, for very large numbers of objects. Hipparcos demonstrated that this can be achieved with milliarcsecond accuracy by means of a continuously scanning satellite which observes two directions simultaneously. With current technology this same principle can be applied with a gain of a factor of more than 100 improvement in accuracy, a factor 1000 improvement in limiting magnitude, and a factor of 10 000 in the numbers of stars observed.

Measurements conducted by a continuously scanning satellite are optimally efficient, with each photon acquired during a scan contributing to the precision of the resulting astrometric parameters. The over-riding benefit of global astrometry using a scanning satellite is however not efficiency but reliability: an accurate instrument calibration is performed naturally, while the interconnection of observations over the celestial sphere provides the rigidity and reference system, immediately connected to an extragalactic reference system, and a realistic determination of the standard errors of the astrometric parameters. Two individual viewing directions with a wide separation is the fundamental pre-requisite of the payload, since this leads to the determination of absolute trigonometric parallaxes, and absolute distances, exploiting the method implemented for the first time in the Hipparcos mission. In the revised design, these two viewing directions are combined into a single focal plane.

3.2. Radial velocity measurements

There is one dominant scientific requirement, as well as two additional scientific motivations, for the acquisition of radial velocities with GAIA: (i) astrometric measurements supply only two components of the space motion of the target stars: the third component, radial velocity, is directed along the line of sight, but is nevertheless essential for dynamical studies; (ii) measurement of the radial velocity at a number of epochs is a powerful method for detecting and characterising binary systems; (iii) at the GAIA accuracy levels, 'perspective acceleration' is at the same time both a complication and an important observable quantity. If the distance between an object and observer changes with time due to a radial component of motion, a constant transverse velocity is observed as a varying transverse angular motion, the perspective acceleration. Although the effect is generally small, some hundreds of thousands of high-velocity stars will have systematic distance errors if the radial velocities are unknown.

On-board acquisition of radial velocities with GAIA is not only feasible, but is relatively simple (at least in principle), is scientifically necessary, and cannot be readily provided in any other way. In terms of accuracy requirements, faint and bright magnitude regimes can be distinguished. The faint targets will mostly be distant stars, which will be of interest as tracers of Galactic dynamics. The uncertainty in the tangential component of their space motion will be dominated by the error in the parallax. Hence a radial velocity accuracy of $\simeq 5$ km s^{-1} is sufficient for statistical purposes. Stars with $V < 15$ mag will be of individual interest, and the radial velocity will be useful also as an indicator of multiplicity and for the determination of perspective acceleration. The radial velocities will be determined by digital cross-correlation between an observed spectrum and an appropriate template. The present design allows (for red Population I stars of any luminosity class) determination of radial velocities to $\sigma_v \simeq 5$ km s^{-1} at $V = 18$ mag.

Most stars are intrinsically red, and made even redder by interstellar absorption. Thus, a red spectral region is to be preferred for the GAIA spectrograph.

To maximize the radial velocity signal even for metal-poor stars, strong, saturated lines are desirable. Specific studies, and ground-based experience, show that the Ca II triplet near 8600 Å is optimal for radial velocity determination in the greatest number of stellar types.

Ground-based radial velocity surveys are approaching the one million-object level. That experience shows the cost and complexity of determining some hundreds of millions of radial velocities is daunting. There is also a substantial additional scientific return in acquiring a large number of measurements, and doing so not only well spaced in time but also, preferably, simultaneously with the astrometric measurements (e.g. variables and multiple systems).

3.3. Derivation of astrophysical parameters

The GAIA core science case requires measurement of luminosity, effective temperature, mass, age and composition, in addition to distance and velocity, to optimise understanding of the stellar populations in the Galaxy and its nearest neighbours. The quantities complementary to the kinematics can be derived from the spectral energy distribution of the stars by multi-band photometry and spectroscopy. Acquisition of this astrophysical information is an essential part of the GAIA payload. A broad-band magnitude, and its time dependence, will be obtained from the primary mission data, allowing both astrophysical analyses and the critical corrections for residual system chromaticity. For the brighter stars, the radial velocity spectra will complement the photometric data.

For essentially every application of the GAIA astrometric data, high-quality photometric data will be crucial, in providing the basic tools for classifying stars across the entire HR diagram, as well as in identifying specific and peculiar objects. Photometry must determine (i) temperature and reddening at least for OBA stars and (ii) effective temperatures and abundances for late-type giants and dwarfs. To be able to reconstruct Galactic formation history the distribution function of stellar abundances must be determined to ~ 0.2 dex, while effective temperatures must be determined to ~ 200 K. Separate determination of the abundance of Fe and α-elements (at the same accuracy level) will be desirable for mapping Galactic chemical evolution. These requirements translate into a magnitude accuracy of $\simeq 0.02$ mag for each colour index.

Many photometric systems exist, but none is necessarily optimal for space implementation. For GAIA, photometry will be required for quasar and galaxy photometry, Solar System object classification, etc. Considerable effort has therefore been devoted to the design of an optimum filter system for GAIA. The result of this effort is a baseline system, with 4–5 broad and eleven medium passbands, covering the near ultraviolet to the CCD red limit. The broad-band filters are implemented within the astrometric fields, and therefore yield photometry at the same angular resolution (also relevant for chromatic correction), while the 11 medium-band filters are implemented within the spectrometric telescope. Both target magnitude limits of 20 mag, as for the astrometric measurements.

3.4. On-board detection

Clear definition and understanding of the selection function used to decide which targets to observe is a crucial scientific issue, strongly driving the final scientific output of the mission. The optimum selection function, and that adopted, is to detect every target above some practical signal level on-board as it enters the focal plane. This has the advantage that the detection will be carried out in the same wave-band, and at the same angular resolution, as the final observations. The focal plane data on all objects down to about 20 mag can then be read out and telemetered to ground within system capabilities. All objects, including Solar System objects, variable objects, supernovae, and microlensed sources, are detected using this 'astrometric sky mapper'.

3.5. Payload design principles

The overall design constraints have been investigated in detail in order to optimise the number and optical design of each viewing direction, the choice of wavelength bands, detection systems, detector sampling strategies, basic angle, metrology system, satellite layout, and orbit. The resulting proposed payload design consists of:

(a) two astrometric viewing directions. Each of these astrometric instruments comprises an all-reflective three-mirror telescope with a (revised) aperture of 1.4×0.5 m^2, the two fields separated by a basic angle of 106°. Each astrometric field comprises an astrometric sky mapper, the astrometric field proper, and a broad-band photometer. Each sky mapper system provides an on-board capability for star detection and selection, and for the star position and satellite scan-speed measurement. The main focal plane assembly employs CCD technology, with about 100 CCDs and accompanying video chains per focal plane, a pixel size 10 μm along scan, TDI (time-delayed integration) operation, and an integration time of ~ 3.3 s per CCD;

(b) an integrated radial velocity spectrometer and photometric instrument, comprising an all-reflective three-mirror telescope of (revised) aperture 0.5×0.5 m^2. The field of view is separated into a dedicated sky mapper, the radial velocity spectrometer, and a medium-band photometer. Both instrument focal planes are based on CCD technology operating in TDI mode;

(c) the opto-mechanical-thermal assembly comprising: (i) a single structural torus supporting all mirrors and focal planes, employing SiC for both mirrors and structure. There is a symmetrical configuration for the two astrometric viewing directions, with the spectrometric telescope accommodated within the same structure, between the two astrometric viewing directions; (ii) a deployable Sun shield to avoid direct Sun illumination and rotating shadows on the payload module, combined with the Solar array assembly; (iii) control of the heat injection from the service module into the payload module, and control of the focal plane assembly power dissipation in order to provide an ultra-stable internal thermal environment; (iv) an alignment mechanism on the secondary mirror for each astrometric instrument, with micron-level positional accuracy and 200 μm range, to correct for telescope aberration and mirror misalignment at the beginning of life; (v) a permanent monitoring of the basic angle, but without active control on board.

A mission length of 5 years is adopted for the satellite design lifetime, which starts at launcher separation and includes the transfer phase and all provisions related to system, satellite or ground segment dead time or outage. A lifetime of 6 years has been used for the sizing of all consumables.

3.6. Astrometric accuracy

There are three main components involved in the improved performance of GAIA compared with Hipparcos. The larger optics provide a smaller diffraction pattern and a significantly larger collecting area; the improved quantum efficiency and bandwidth of the detector (CCD rather than photocathode) leads to improved photon statistics; and use of CCDs provides an important multiplexing advantage.

The GAIA astrometric wavelength band G is fixed such that $G \simeq V$ for un-reddened A0V stars. The approximate transformation from $(V, V-I)$ to G can be expressed in the form:

$$G = V + 0.51 - 0.50 \times \sqrt{0.6 + (V-I-0.6)^2} - 0.065 \times (V-I-0.6)^2 \quad (1)$$

which is valid (to ± 0.1 mag) at least for $-0.4 < V\text{--}I < 6$. For $-0.4 < V\text{--}I < 1.4$ we have the convenient relation $G-V = 0.0 \pm 0.1$ mag. Since, by construction, the G magnitude yields a rather uniform accuracy as a function of spectral type, useful mean accuracies can be derived from a straight mean of the detailed calculations, and are given in Table 1.

Table 1. Mean accuracy in parallax (σ_π), position (at mid-epoch, σ_0) and proper motion (σ_μ), versus G magnitude. The values are sky averages.

G (mag)	10	11	12	13	14	15	16	17	18	19	20	21
σ_π (μas)	4	4	4	5	7	11	17	27	45	80	160	500
σ_0 (μas)	3	3	3	4	6	9	15	23	39	70	140	440
σ_μ (μas yr^{-1})	3	3	3	4	5	8	13	20	34	60	120	380

4. Data analysis

The total amount of (compressed) science data generated in the course of the five-year mission is about 2×10^{13} bytes (20 TB). Most of this consists of CCD raw or binned pixel values with associated identification tags. The data analysis aims to 'explain' these values in terms of astronomical objects and their characteristics. In principle the analysis is done by adjusting the object, attitude and instrument models until a satisfactory agreement is found between predicted and observed data. Successful implementation of the data analysis task will require expert knowledge from several different fields of astronomy, mathematics and computer science to be merged in a single, highly efficient system.

The computational complexity of the data analysis arises not just from the amount of data to be processed, but even more from the intricate relationships between the different pieces of information gathered by the various instruments throughout the mission. It is difficult to assess the magnitude of the data analysis problem in terms of processing requirements. Certain basic algorithms that have to be applied to large data sets can be translated into a minimum required number of floating-point operations. Various estimates suggest of order 10^{19} floating-point operations, indicating that very serious attention must be given to the implementation of the data analysis (an effort which is now under way). Observations of each object are distributed throughout the mission, so that calibrations and analysis must be feasible both in the time-domain and in the object domain. Flexibility and interaction is needed to cope with special objects, while calibrations must be protected from unintentional modification. Object Oriented (OO) methodologies for data modeling, storage and processing are ideal for meeting the challenges faced by GAIA.

The global astrometric reductions must be formulated in a fully general relativistic framework, including post-post-Newtonian effects of the spherical Sun at the 1 μas level, as well as including corrections due to oblateness and angular momentum of Solar System bodies.

Processing these vast amounts of data will require highly automated and efficient numerical methods. This is particularly critical for the image centroiding of the elementary astrometric and photometric observations, and the corresponding analysis of spectral data in the spectrometric instrument.

Accurate and efficient estimation of the centroid coordinate based on the noisy CCD samples is crucial for the astrometric performance. Simulations indicate that 6 samples approximately centred on the peak can be read out from the CCD. The centroiding, as well as the magnitude estimation, must be based on these six values. Results of a large number of Monte Carlo experiments, using a maximum-likelihood estimator as the centroiding algorithm, indicate that a rather simple maximum-likelihood algorithm performs extremely well under these idealized conditions, and that six samples is sufficient to determine the centroid accurately. Much work remains to extend the analysis to more complex cases, including in particular overlapping stellar images.

A preliminary photometric analysis, for discovery of variables, supernovae, etc, can be carried out using standard photometric techniques immediately after data delivery to the ground. In addition, more detailed modelling of the local background and structure in the vicinity of each target using all the mission data in all the passbands will be required. A final end-of-mission re-analysis may benefit from the astrometric determination of the image centroids, locating a well-calibrated point spread function for photometric analysis. Studies of these photometric reductions have begun.

The high-resolution (radial velocity) spectrometer will produce spectra for about a hundred million stars, and multi-epoch, multi-band photometry will be obtained for about one billion stars. The analysis of such large numbers of spectra and photometric measurements needs to be performed in a fully automated fashion, with no manual intervention. Automatic determination of (at least) the surface temperature T_{eff}, the metallicity [M/H], and $\log g$ is expected. A fully automated system for the derivation of astrophysical parameters from the large number of spectra and magnitudes collected by GAIA, using all the available information for each star, has been studied, showing the feasibility of an approach based on the use of neural networks. In the classification system foreseen, spectra and photometric measurements will be sent to an 'initial classifier', to sort objects into stellar and non-stellar. Specialist networks then treat each class. For example, stellar data sets are passed to an 'automated stellar parameterization' sub-package.

It is the physical parameters of stars which are really of interest; therefore the proposed system aims to derive physical parameters directly from a stellar spectrum and photometry. Detailed simulations of the automated stellar parameterization system have been completed using a feed-forward neural network operating on the entire set of spectral and photometric measurements. In such a system, the derived values for the stellar parameters are naturally linked to the models used to train the network. Given the extreme rapidity of neural networks, when stellar atmosphere models are improved, re-classification of the entire data set can be done extremely quickly: an archive of 10^8 spectra or photometric measurements could be reclassified in about a day with the present-day computing power of a scientific workstation.

5. Conclusion

GAIA will create an extraordinarily precise three-dimensional map of about one billion stars throughout our Galaxy and the Local group. It will map their space motions, which encode the origin and subsequent evolution of the Galaxy, and the distribution of dark matter. Through on-board photometry, it will provide the detailed physical properties of each star observed: luminosity, temperature, gravity, and elemental composition, which encode the star formation and chemical enrichment history of the Galaxy. Radial velocity measurements on board will complete the kinematic information for a significant fraction of the objects observed. Through continuous sky scanning, the satellite will repeatedly measure positions and colours of all objects down to $V = 20$ mag. On-board object detection ensures a complete census, including variable stars and quasars, supernovae, and minor planets. It also circumvents costly pre-launch target selection activities. Final accuracies of 10 microarcsec at 15 mag will provide distances accurate to 10 per cent as far as the Galactic Centre. Stellar motions will be measured even in the Andromeda galaxy. The GAIA project provides enormous challenges at many levels, but the necessary principles are well established, and the astronomical results will be substantial.

Acknowledgements

This introduction to the GAIA project is a summary of Perryman et al. (2001), modified following the technology re-assessement study during 2001. The efforts of the many people involved in the GAIA project are acknowledged.

References

Perryman M. A. C., de Boer K. S., Gilmore G. et al. 2001, A&A 369, 339

GAIA Spectroscopy, Science and Technology
ASP Conference Series, Vol. 298, 2003
U. Munari ed.

GAIA: the satellite and payload

Oscar Pace

ESA – D/SCI/PF, Keplerlaan 1, 2200 AG Noordwijk, The Netherlands

Abstract. Following the cut of the ESA Science budget, the system level design of GAIA has been reassessed, with the aim to drastically reduce the Cost at Completion (CaC) by at least 20%, keeping the same scientific requirements. This paper will summarise at satellite level the new GAIA Technical Baseline, assumed for the estimation of the CaC, included in the Cosmic Vision, the new ESA Science Programme, approved by the ESA Science Programme Committee (SPC) in May 2002.

The GAIA spacecraft, to be put in a Lissajous-type orbit around the Sun–Earth Lagrangian point L2, has been redesigned to be launched with a Soyuz–Fregat launcher. The satellite, with a wet-mass of 1428 kg and 1331 W of power, consists of a payload and a service module, which are mechanically and thermally decoupled. The sunshield assembly has a span of 11.50 m when deployed, with a fixed solar array of annular type, installed on the bottom side of the service module, like Planck. While the temperature of the service module is $\sim 20°$C, the payload module is at ~ 160 K, with a temperature stability requirement of ~ 50 μK. The payload, thermally insulated, is composed of two astrometric instruments and an integrated radial velocity spectrometer and photometric instrument, with CCDs as detectors and mirrors made all by SiC, mounted on a single structural torus (~ 3 m diameter), also made by SiC. The system, together with the very quiet L2 orbit, provides a stable environment for the payload optical bench. The science data, dumped on ground by an X-band TM link at a rate of 5 Mbps, will be retrieved by a single ground station (Perth), with a minimum visibility of 6 h per day. The nominal stellar data acquisition time is 5 yr, extendable to 6 yr.

To reach GAIA's scientific objectives, comprehensive technology activities have been identified and a GAIA Technology Plan has been established and implemented. This Plan aims at developing the identified technology to a breadboard readiness level, tested in the relevant environment before the start of Phase B. This paper summarizes the current mission technical concept and introduces the technological developments required to make the GAIA mission feasible.

1. Scientific and mission objectives

The GAIA mission objectives are to build a catalogue of $\sim 1\,000\,000\,000$ stars with accurate positions, parallaxes, proper motions, magnitudes and radial velocities. The catalogue will be complete up to $V = 20$ mag. The overall mission can be split in three parts: (a) astrometry, with the measurement of stellar position, parallax and proper motion; (b) photometry, with measurements in different spectral bands; (c) radial velocity measurements up to $V = 17$ mag.

The accuracy targeted for the astrometry mission is 10 μas in positional and parallax accuracy and 10 μas yr^{-1} for proper motion accuracy, for $V = 15$ mag stars. The parallax accuracy is equivalent to an accuracy of 10% on the distance of a star at 10 kpc. The proper-motion accuracy is equivalent to an accuracy of 1 km s^{-1} on the velocity of a star at 20 kpc. The photometry mission must be multi-colour and multi-epoch. Magnitudes will be obtained in 5 wide spectral bands for all stars up to $V = 20$ mag, complemented with measurements in (provisionally) 11 narrow bands for stars up to $V = 17$ mag. The multi-epoch aspect is performed by a regular observation of the stars throughout the mission. The radial velocity measurements will provide a velocity accuracy of 1 km s^{-1} for $V = 15$ mag stars. The radial velocity is presently derived from the Doppler shift of particular spectral lines within the 8480–8740 Å spectral domain.

2. Measurement principle

The main objective of the mission is to perform global or wide field astrometry as opposed to local or narrow field astrometry. In local astrometry, the star position can only be measured with respect to a neighbouring star in the same field. Even with an accurate instrument, the propagation of errors is prohibitive when making a sky survey. The principle of global astrometry is to link stars with large angular distances in a network where each star is connected to a large number of other stars in every direction.

Global astrometry requires the simultaneous observation of two fields of view in which the star positions are measured and compared. Therefore, the payload will provide two lines of sight, obtained with two separate telescopes, but, like HIPPARCOS, the two images will be focalised, slightly spaced, on a unique focal plane assembly. The angle between the instrument lines of sight defines what is called the basic angle.

A dedicated scanning law (Figure 1) does the coverage of the sky to build up the star network. The scanning law for GAIA is similar to the one of HIPPARCOS. A spin motion of the spacecraft with a 6-h period performs the scan of great circles. The lines of sight of both instruments are perpendicular to the spin axis, and the two instruments successively scan the same sky area, which can be viewed as a band which height is equal to that of the field of view. From the instrument standpoint, the stars are crossing each field of view in a regular motion. As for HIPPARCOS, the scan direction is a privileged direction and the star position measurement is only performed in this direction.

A slow precession of the spin axis about the satellite–Sun axis slowly moves the great circle allowing a full coverage of the sky within a few months. For GAIA, the optimisation of the scanning law, having now a value of 60″ s^{-1} (it was 120″ s^{-1}) has led to an orientation of the spin axis at 50° from the Sun direction and a precession of this axis about the Sun direction in 72 days. This scanning law ensures optimal sky coverage and a great rigidity in the data reduction (each star is measured on several great circles — 41 observations on average over 5 yr — with nearly isotropic orientations). The slow precession

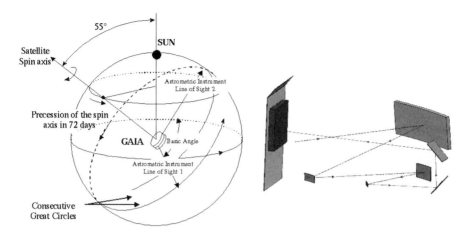

Figure 1. The GAIA scanning law (*left*) and astrometric telescope optical design (*right*).

of the spin axis generates a line of sight motion across scan of ~0.25° over a spin period, while the field of view height is 0.66°. This provides an overlapping between the consecutive bands observed by each field of view, which allows a calibration of some instrument parameters such as the basic angle using the same stars. The data reduction performed on ground starts with a processing of the star data on each great circle (*great circle reduction*): the stars are located relative to each other on the circle. Then, the reduction process has to orient and phase the different circles with respect to each other in what is called the sphere reduction. A more general global approach for the data reduction is also being considered. Large amounts of information for the instruments are derived from the data reduction. In particular, the distortion, the basic angle and the chromaticity can be calibrated as well as their low frequency variations, which make the variations of these parameters less critical. The self-calibration of fixed or slowly variable biases is a crucial advantage of the global measurement concept, which has been unambiguously demonstrated by HIPPARCOS, and without which the GAIA accuracy goal of 10 μas would be difficult to reach. The final catalogue is built from the best estimate of all star positions and of instrument parameters computed from the total amount of data collected during the mission. The final accuracy is only reached at the end of the mission, i.e., after 5 yr of operation, and after complete processing of all the data.

3. Astrometric instrument

3.1. Optical design

The optical design of the astrometric telescopes is represented in Figure 1. Each telescope, separated by a basic angle of 106°, consists of a Three Mirror Anastigmatic (TMA) configuration (M1, M2, and M3), with images combined and

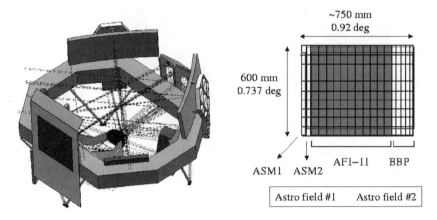

Figure 2. GAIA payload configuration (ASTRO; *left*) and astrometric focal plane (*right*).

focalised by three flat mirrors (M4, M5, and M6), on the same focal plane, slightly staggered in space of one CCD row.

The design features a diffraction limited flat field larger than 0.4 deg^2, a large focal length (46.7 m) in comparison to the inter-mirror distance (less than 3 m), and a low optical distortion compatible with a time delayed integration of the star pattern over $\sim$3.3 s. The entrance pupil is located on the primary mirror and is rectangular: 1.4 m $\times$ 0.5 m. It is elongated in the scan direction, so as to provide the narrowest point spread function in the measurement direction, while being compatible with the optical quality, available volume, field size and radiometric needs. The pupil collecting area is such that the total number of collected photons per star over 5 yr of operation in orbit is compatible with the 10 μas positioning accuracy at magnitude 15. The large focal length of 46.7 m allows a proper sampling of the diffraction pattern (4 pixels) along the scan direction, with a pixel size along scan of 10 μm, which is compatible with the available CCD detector technology. The astrometric instrument consists of two identical telescopes, rotated one with respect to the other by an angle equal to the basic angle, fixed at 106°. The optical design allows a mounting of the mirrors and focal planes of both telescopes on the same torus structure (Figure 2). This provides a high symmetry and dimensional stability of the system and simplifies its overall integration.

3.2. Focal plane design

The astrometric focal plane layout is represented in Figure 2. An observed object follows a nearly horizontal line on this figure, with a speed of $\sim$13.6 mm s^{-1} given by the scan law, and therefore successively crosses all the columns of CCDs. All the CCDs are backside illuminated and work in time-delayed integration mode over their whole width. Therefore, for every observed star, each CCD provides the star pattern corresponding to the optical point spread function sampled by the CCD pixels. All the detectors are physically identical, with a pixel size of

10 μm along scan and 30 μm across scan. The columns of detectors are functionally grouped in three parts: the sky mapper, the astrometric field, and the broad band photometric field. The sky mapper and the astrometric field work in white light, the wavelength band being defined by the CCD quantum efficiency. Although the CCD detectors are physically identical, a different smart analogue binning process is implemented for the three focal plane parts, depending on the data to be extracted. The sky mapper is made of the first column of CCDs (ASM1 for the 1st FOV and ASM2 for the 2nd FOV) and by the first column of the Astrometric Field (AF1) and fulfills several functions:
– Object detection function: no catalogue is implemented on board, and the system will be autonomous for the object detection. Simulations performed by the GAIA Science Team (GST) showed that the detection process is efficient up to $V = 20$ mag, for which ~ 300 e$^-$ per star are detected per second.
– Windowing function: once the object is detected by the first column of the sky mapper, a detection window is defined for the following CCDs for this object, so as to reduce the readout noise and data rate.
– Attitude control function: a centroiding process is applied to two consecutive CCD columns for deriving the star speed measurement in both directions. Note that the speed measurement does not require any catalogue.

The astrometric field is made of the 11 columns of CCDs following the sky mapper. Since only the star position along scan is measured, the pixels are binned across scan in this area, and each CCD provides 6 samples per star representing the star pattern profile along scan. For avoiding any unrecoverable loss of information, these profiles are sent to the ground station without applying any further processing on board, aside from lossless compression techniques. The star dynamical range is managed by an appropriate gating of the CCDs, which allows to perform the star profile measurement up to $V = 3$ mag.

The broad band photometry area is made of the five last columns of CCDs, and provides colour information for all observed objects in five spectral bands of ~ 100 nm width. Such colour data are used for astrophysics science, and also for the calibration of star chromaticity effects.

In summary, the focal plane assembly is constituted of 180 CCDs of the same type, with 10 μm pixels along scan and 30 μm across scan. Out of 180, the 110 CCD of the astrometric field have a physical array size of 49 mm × 60 mm (4500 × 1996 pixels), while the remaining 70 CCDs (20 for the Sky Mapper and 50 for the broad band photometer BBP) have half along-scan size, i.e., 30 mm × 60 mm (2600 × 1996 pixels). The total dimension of the focal plane, encompassing the two FOVs, will be ~ 600 mm × 750 mm.

3.3. Astrometric performance

The final astrometric performance is not only given by the star localisation accuracy for one passage, but is driven by the total number of photons collected by both telescopes over the total observation time in orbit. Table 1 presents the astrometric accuracy versus magnitude for a yellow star (G2V), assuming 5 yr of observation in orbit. Similar results are obtained for blue stars, while the accuracy improves for red stars. The noise floor for bright stars is evaluated at ~ 3 μas r.m.s. and is driven by the CCD full-well capacity estimated at

Table 1. Summary of the astrometric instrument performances.

	Astrometric accuracy (μas), $V = 15$ mag		
	B3V	G2V	M8V
CCD #1	11.3	13.5	13.1
CCD #2	12.7	12.0	12.0
CCD #3	6.9	4.8	5.3

Radiometric performance (CCD #3, across scan speed = 0)			
	B3V	G2V	M8V
Total detected e$^-$, $V = 15$ mag	70 040	86 300	507 050
Detected e$^-$ in sample 6×8 pixels	47 870	61 650	375 580
Maximum e$^-$ signal per pixel, $V = 15$ mag	14 350	14 900	62 345
Saturation magnitude ($Q_{\text{sat}} = 250\,000$, TBC)	11.9	11.95	13.5
Noise accuracy floor (μas)	3.1	2.9	2.6

250 000 e$^-$ (TBC). Even under pessimistic evaluations of the saturation level, the noise figure for bright stars ($V < 12$–13 mag) will be better than 4 μas r.m.s.

4. Spectrophotometric instrument

The spectrometric instrument works up to $V = 17$ mag and is composed of the medium band photometer (MBP) and the stellar radial velocity measurement instrument (RVS). The two functions share the field of a common TMA telescope (Figure 3), including a classical three mirror telescope plus a dioptric spectrometer working at 1:1 magnification. The spectrometric telescope provides a diffraction-limited field of 4.8° across scan by 2° along scan, with a focal length of 2.1 m and a rectangular entrance aperture of dimensions 0.5 m × 0.5 m. The mirrors and the shared focal plane are mounted on the same optical bench, with the telescope deployed in a horizontal configuration (deployment perpendicular to spin axis), while the dispersion direction of the spectrometer is along the scan (Figure 4). The MBP focal plane is constituted by two dedicated detection areas (MBP#1 and MBP#2), separated by the RVS focal plane, having each an optical field of view of 2° (along scan) and 1.6° (across scan; Figure 4). Each field is composed of 2 modules of 8 CCDs with each 5.5 s integration time. The CCDs are constituted of 336 × 3930 pixels, with 10 μm × 15 μm pixel dimensions and an active area of 3.36 mm × 59 mm. The total number of MBP CCDs is 32.

The star measurement is performed in 11 (TBC) different bands (sky mapper excluded) by means of filters located in front of the CCDs, with an integration time of 5.5 s per CCD.

The radial velocity instrument is constituted of a classical configuration assembly collimator/disperser/imager, working at unit magnification. Although in evolution, the RVS focal plane assembly is at present constituted of 6 RVS detectors and a RVSM (RVS Sky Mapper), located in the telescope focal plane,

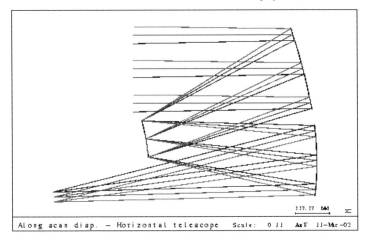

Figure 3.   Spectrometric telescope design.

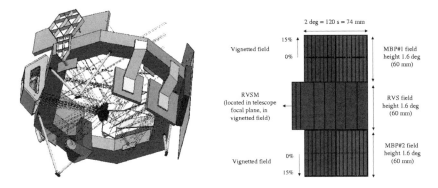

Figure 4.   GAIA payload configuration (SPECTRO; *left*) and SPECTRO focal plane (*right*).

in the vignetted field. The RVSM detects the stars and determines the windows for the RVS detectors. The detection area is formed by 6 CCDs , of the same type of the MBP CCDs (10 μm × 15 μm pixel, 10 mm × 59 mm active area).

The major characteristics of the RVS spectrometer are the following: Wavelength range: 8480–8740 Å; Pixel size: 10 μm × 15 μm; Number of spectral samples per star: 670; Spectral sampling: 0.375 Å pix^{-1}; Available integration time, per star passage: 100 s; Total integration time (5 yr lifetime): 10 200 s.

5. Payload configuration

Although fixed or slowly variable biases are self-calibrated by the great-circle or sphere reduction process, line-of-sight variations at a frequency higher or equal to the spin frequency (6 h) cannot be retrieved. It is therefore mandatory for the astrometric instrument that the design ensures a short-term basic angle stability

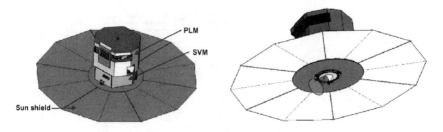

Figure 5. GAIA in orbit: top view (*left*) and bottom view (*right*).

(or at least knowledge of the basic angle variations) significantly below 10 μas r.m.s. over the spin period of 6 h.

The proposed design was shown to actually meet this stringent requirement with a passive thermal control. The major design features for that purpose can be summarised as follows:

– The payload is protected from Sun radiation by means of a sunshield (Figure 5); the sunshield is obtained by connecting deployable booms with thermal foils.
– The payload module, in particular the torus, is also radioactively decoupled from the sunshield, by means of a thermal tent closing the payload completed with baffles on the optical apertures.
– The payload module is radioactively and conductively decoupled from the service module.
– The payload module is also mechanically decoupled from the service module by releasing in orbit two of the three bipods connecting one module to the other.
– A single low-expansion, high-conductivity and homogeneous material, namely Silicon Carbide (SiC), is used for the reflectors, the mounting plates and the torus structure.

The payload temperature is passively stabilised at ~160 K. Although the basic angle stability requirement is passively met, a device was designed by Astrium for continuously monitoring the basic angle variation in-orbit with accuracy better than 1 μas r.m.s. and therefore guaranteeing the mission performance. It basically consists of an artificial star on board (laser source) illuminating simultaneously the two astrometric telescopes.

6. Satellite configuration

The payload module is mounted on the service module (SVM) by means of three bipods in order to have an isostatic interface between the two structures. The SVM structure is built around a stiff and stable primary structure, composed of a cylindrical central tube, 6 shear walls and 6 external panels, supporting the electronic units, and upper panel for payload mechanical interface, all made by CRFP to fulfill the stability requirements. The SVM interfaces on one side with the standard 1194 mm adapter of the Soyuz–Fregat launcher and on the other side with the payload module. All units accommodated into the module are thermally coupled to the lateral panels of the module, which are used as radiators and covered with optical solar reflectors.

The solar array assembly, of an annular shape like Planck, fixed and installed under the Service Module bottom platform, is based on Ga–As cells and has a total surface of ∼10 m². Figure 5 shows the satellite configuration in orbit. The sunshield assembly is deployed with a span of 11.50 m. The optical covers are removed from the instrument entrance apertures in orbit.

7. Science data processing

Because of the very large number of CCDs within each astrometric focal plane, the data rate at the output of these focal planes is huge. A strong filtering of the data flow is mandatory on board in order to minimise the quantity of data to be stored and then transmitted to ground. This on-board filtering process includes several steps described briefly below.

– A star detection/discrimination function, using the sky mapper implemented within each astrometric focal plane, allows to detect all stars up to $V = 20$ mag. The detection process includes a background evaluation and suppression function, a cosmic hit discrimination function and a connectivity test, which allows to discard noise and extended objects. Thanks to this on-board autonomous star detection/discrimination function, the payload can operate without the support of any pre-loaded star catalogue.

– A windowing function propagates in real time the address of each detected star as it crosses the astrometric focal plane. It allows to read only the CCD-pixels that include the star signal.

– Finally, a compression function reduces the data rate, with minimum or no impact on the data quality. Thanks to this on-board filtering process, the data flow at payload output does not exceed an average value of 1 Mbps.

8. Service module

8.1. Orbit acquisition

A simple and secure spin stabilisation approach is selected for this transfer phase, making use of standard Sun sensors and one gyroscope. The satellite is launched by a Soyuz–Fregat, from Baikonour, into a 200 km, circular transfer orbit, by using the Soyuz stage. The satellite transfer from the circular orbit to the final L2 orbit is done by means of the Fregat stage, with transfer time from 70 to 110 days, depending on the day of launch along the year. The selected operational orbit is a Lissajous orbit providing at least 6 yr of observation with no eclipse, and even more than that assuming a limited correction manoeuvre after a few years of observation. A redundant set of Hydrazine 10 N thrusters is used for orbit correction during the transfer phase, for the final insertion into the operational orbit at L2 and in case slew manoeuvre in operations are required. The total ΔV for a 6 months launch window is 326 m s^{-1}, for a 10° Lissajous orbit, while the total fuel mass, including AOCS fuel, is 242 kg.

8.2. Attitude control

Very stringent pointing stability and rate stability requirements have to be met during the observation phase (e.g., 10 mas s^{-1} 3σ, over 1 s, for the absolute rate error) in order to follow adequately the pre-defined satellite scan law without blurring the star image at focal plane level. This is achieved thanks to a set of microthrusters (field emission electric propulsion thrusters - FEEP - with a thrust in the range of 1 mN) controlled in a continuous proportional mode. They are used combined with the wide field star sensors of the astrometric instruments (sky mapper fields) as attitude and rate sensors. Thanks to the high specific impulse of the FEEP microthrusters and because of the very low level of dynamic perturbations to be controlled, only a few kg of Caesium propellant are necessary.

8.3. Power

The satellite power demand, in observation phase, is $\sim$1331 W, including margins. This power is provided by $\sim$10 m^2 of Ga–As cells, with 40° Sun aspect angle, having available power of 1476 W (20% cell efficiency EOL, 0.92 absorptivity at 70°C and 90% packing density). A small Lithium-ION battery provides the required energy during the launch phase. No eclipse is expected during the transfer orbit or in operational orbit.

8.4. Data handling

The satellite electrical architecture may be split in two main parts:
– A dedicated science data chain includes all the units required for the acquisition, discrimination and compression of the payload data. A solid state recorder of 300 Gbits ensures the temporary storage of the compressed science data between two consecutive transmission periods with the ground station.
– A more standard set of equipment, connected to a central computer, provides all the general services to the spacecraft and payload (power, data transmission, attitude control, thermal control, etc.).

8.5. Communications

The communication between the spacecraft and the Earth (the ESA ground station located in New Norcia — Perth, Australia — is today considered for the control of the spacecraft) is done via two sets of equipment:
– A classical telemetry and telecommand (TT&C) link with an omni-directional coverage provides a permanent control of the spacecraft whatever the mission phase and the spacecraft attitude. Both transmission and reception are done in the X-band. A solid state power amplifier provides the 17 W of RF power required to transmit safely the 6 kbps of telemetry data over the $\sim$1.5 million km between spacecraft and Earth.
– A dedicated telemetry link, also in the X-band, provides the transmission to the Earth of the scientific data. The *high* transmission rate (5 Mbps, taking into account a minimum of 6 h of ground station visibility per day), the large distance between the spacecraft and the ground station and the wide antenna field of view induced by the spacecraft spin and orbit motions impose altogether the use of a *high* gain, high power phased array antenna.

9. Satellite interfaces and budgets

9.1. Launcher compatibility

In the launch configuration, the sunshield is stowed against the service module structure. The optical covers are closed against the payload module thermal tent. The satellite envelope, with an overall diameter of 3.80 m and a height of ~3.10 m is then compatible with the Soyuz–Fregat fairing, being developed for the Metop project with a launch in 2005. Use is made of the standard Soyuz–Fregat 1194 mm diameter adapter.

9.2. Lifetime

Five years of total observation are necessary to finally reach the targeted 3–10 μas astrometric accuracy. The satellite (including the consumables) has been designed for 6 yr of operation.

9.3. Mass

The spacecraft dry mass, including margins, is 1136 kg, while the launch mass, including the launcher adaptor (50 kg) and the propellant (242 kg), is 1428 kg, with 212 kg margin with respect to the lift-off capability of the Soyuz–Fregat, version 2-1a (planned to be available in 2005 for the Metop launch).

10. The GAIA technology programme

The main objectives of the GAIA technology programme are to ensure effective technological preparation of the GAIA project by the development of critical technologies, by demonstrating their feasibility and timely availability at flight standard. Such a demonstration is an essential prerequisite enabling the planned mission at an acceptable level of risk in terms of maturity, cost and schedule. The technology activities have been identified in conjunction with the industry, in the context of the system level study performed in the past for GAIA.

The GAIA Technology Programme has been approved by ESA and it is now being implemented. Furthermore, all the activities have to fulfill the requirements identified at system level, i.e., at payload and at spacecraft levels. Furthermore, in order to help ESA to perform these activities, ESA is selecting in open competition two parallel Technology Assistance/Definition Study Contractors, at Prime Contractor level, starting with a Technology Assistance activity of ~1.5 yr duration and followed by a Definition Study activity of 12 months duration within the period 2002–2005. Although the GAIA Technology programme schedule has been established having in mind a GAIA launch date in mid-2010, the planning assumptions would not change in case of GAIA launch date put forward to the 3rd Quarter of 2009, by anticipating selected Phase B tasks during the definition phase.

11. Conclusions

A complete design of the GAIA spacecraft, including the payload instruments, consistent with the scientific goals specified by ESA and demonstrating the feasibility of the 10 μas astrometric accuracy target, has been re-established by ESA with the help of the industry. A comprehensive Technology Programme has also been conceived, approved, and on the way to be implemented, leading to a complete verification of the technology required by GAIA quite in time for the GAIA Project implementation, with a launch in mid-2010 or in the 3rd Quarter of 2009, if required.

Acknowledgments. This paper is an update, reflecting the current GAIA technical approach, of the paper "GAIA: THE STAR MACHINE" presented at the 2001 IAF. I would like to express my thanks and recognition to M.A.C. Perryman, the GAIA Project Scientist at ESA, to G. Colangelo, System Engineer from the ESTEC Technical Support, to the members of ESA's GAIA Science Team, and to C. Koeck, P. Charvet, F. Safa and the other components of the GAIA Astrium Team in Toulouse, having contributed to perform, in a very short time, a complete GAIA technical and managerial reassessment at system level, making possible the re-approval of the GAIA Mission. Finally, I should like to thank J. de Bruijne for his help to produce a LaTeX file and make possible the publication of this paper.

Considerations about the astrometric accuracy of GAIA.

Francois Mignard

OCA/CERGA, av. Copernic, 06130 Grasse, France

Abstract. The unprecedented astrometric accuracy expected from the ESA mission GAIA imposes to revisit afresh some standard concepts of astrometry that are taken for granted. In the first part of this paper I summarise the elements used in the mission study to evaluate the astrometric errors expected from the GAIA measurements, starting from the one dimensional error at each CCD crossing before combining the repeated observations to arrive at the final error in parallaxes. In a second part I discuss several of the classical concepts of positional astrometry whose meaning needs to be refined for the future astrometry missions or whose magnitude, so far neglected, will become relevant with GAIA.

1. Introduction

The Hipparcos astrometry mission has been a landmark in astronomical research, opening for the first time the world of millisecond global astrometry. Although the modeling aiming at a final accuracy of 1 mas had been carefully considered, there was nothing fundamentally new. The standard astrometric model assumed a uniform rectilinear space motion and the proper motion components were simply derived from the rate of change of the angular coordinates. In the model the parallax and proper motions retained their usual and intuitive geometrical meaning. Admittedly the second order aberration had to be introduced but its expression had been formulated by A. Einstein and known for decades before Hipparcos.

The planned missions GAIA and SIM will mark the first step in the realm of microarcsecond astrometry, a big leap compared to Hipparcos implying to consider a fully relativistic modeling of positional observations. This formulation has been derived by Klioner & Kopeikin (1992) and, more recently, a practical version suited to the needs of GAIA has been issued by Klioner (2002, in press) in which he discusses the non trivial interpretation of astrometric parameters at the μas level. A discussion of several of the underlying concepts is presented in this paper together with a simple assessment of the expected astrometric accuracy of GAIA.

2. Astrometric accuracy of GAIA.

The astrometric accuracy of GAIA can be firstly estimated from a straightforward extrapolation of the results obtained with Hipparcos. Knowing the

Table 1. Astometric accuracy of GAIA scaled to Hipparcos

Parameter	Hipparcos	GAIA	Ratio	Accuracy ×
Baseline (cm)	29	140	5	5
Collecting area (m^2)	0.035	0.7	20	4.5
Quantum efficiency	0.008	0.6	80	9
Duration[1] (months)	37×0.65	60	2.5	1.6
FOV size[2] (deg^2)	0.9	0.4	0.45	0.65
Multiplexing[3] (%)	20	100	5	2.5
Grid vs. image	1	3	3	3
Total				*1400*

[1] Effective time available for data acquisition
[2] This determines the number of observations
[3] Time sharing between five targets on the focal plane of Hipparcos.

difference between the instruments and the mission parameters a crude estimate gives already a good idea and was used indeed to set very preliminary design constraints. A more refined approach is then developed relying on the error budget.

2.1. GAIA scaled to Hipparcos

The baseline resulting from the science objectives underlying the mission is 10 μas for the positions and parallaxes of a 15 magnitude star, an achievement just unthinkable 10 years ago. Compared to Hipparcos, GAIA has several components which account for the improved performance :

- A much larger optics provides a smaller diffraction pattern leading to a better astrometric definition of the image center.

- This larger collecting area yields many more photon per unit of time, improving the sensitivity and allowing to observe fainter stars.

- Moving from a photoelectric detector to a CCD has a major impact on the detection efficiency and sensitivity, going from an overall efficiency of 0.008 with Hipparcos to a value larger than 0.5 with GAIA.

- The use of detector arrays instead of a single sensitive surface permits to record many stars at a time compared to a sequential acquisition with Hipparcos.

Each of this factor can be taken individually to determine the statistical improvement in the astrometric accuracy, as shown in Table 1. It appears that the largest single improvement is brought by the use of a high quantum efficiency

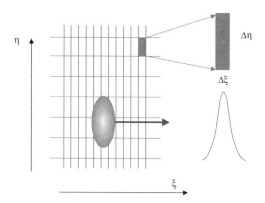

Figure 1. Sketch of the image location procedure with GAIA. The point spread function is sampled over 5 or 6 pixels in the CCD array and a 1-D location of the maximum of light is searched for.

detector made feasible with the modern CCDs. A straight combination of these factors indicates an overall improvement factor as large as 1400 relative to Hipparcos, yielding an accuracy of about 11 μas at $V = 15$, compared to 1 mas at $V = 9$.

A much more refined approach applied during the study phase confirmed this simple-minded estimate. This estimate proceeds along three main steps, each contributing to the final accuracy:

- The accuracy achievable in the CCD local frame resulting from the image centroiding.

- The contribution of the instrument calibrations and attitude error to propagate this local position on the sky.

- The combination of the repeated observations in different scan directions to arrive at the astrometric parameters.

2.2. Centroiding error

The basic measurement on the CCD array consists of locating the center of the diffraction pattern within a pixel. The word *centre* is by itself not very precise, but conveys the idea of associating to an extended and fuzzy image a well defined location linked to the impact point of the chief ray. In fact, no precise definition is required as long as the algorithm does the same thing for every point source and produces the same results for sources that would be located in the same direction.

The analytical model of the 1-D profile (Figure 1) produced by the imaging of a point source has the following form : $I(\xi - \xi_0)$ where ξ_0 is the reference location of the image and $I(\xi)$ is the intensity distribution in the image integrated over the transverse direction. The profile function takes account of the diffraction, the optical distortions, the pixel integration, the smearing due to the motion of the image over the pixel, the magnitude of the source and the overall sensitivity of the detection chain integrated over the spectrum. Therefore the expected number of electrons produced in the jth pixel is given by,

$$n_j = r^2 + b + NI\left(j\Delta\xi - \xi_0\right) \quad (1)$$

where r is the readout noise (rms), b the background contribution and N the total number of electrons expected for this source. The actual number of electrons is then a Poisson random variable $\mathbf{X_j}$ with $E(\mathbf{X}_j) = n_j$. From the observations $n_1, n_2, \cdots, n_j$ one can estimate the reference phase ξ_0 within a pixel. Regarding the accuracy achievable, it depends on the actual procedure used to solve for ξ_0. However it is known that the lower bound of any statistics that estimate ξ_0 is given by the Cramer-Rao limit applied to the Poisson distribution for the model described by Eq. 1 and that this limit can be reached by a maximum likelihood estimator. The computation is rather intricate (cf. ESA-2000-SCI-4) but the result can be conveniently written as,

$$\sigma_\xi \simeq \frac{1}{2\pi} \frac{\alpha}{N^{1/2}} \left[1 + \frac{r^2 + b}{N\Delta\xi}\beta\right]^{1/2} \quad (2)$$

where α and β are angular values related to the width and the shape of the diffraction spot. For the GAIA revised configuration one can adopt $\alpha \sim 200$ mas and $\beta \sim 250$ mas. The other parameters are the pixel width, $\Delta\xi = 44$ mas and the readout noise which should be of the order of 6 e$^-$. The background level is taken conservatively as a star of 22 mag per arcsec^2 yielding ~ 1e$^-$ per pixel. For stars bright enough the second term in the brackets of Eq. 2 is negligible, and one recovers the standard astrometric accuracy for a diffraction limited instrument. When the combination of the background and the RON becomes significant there is a quick degradation of the accuracy as we consider fainter stars. Numerically and for GAIA, $N = \Phi_0 A \tau \rho$, where Φ_0 is the stellar flux in ph m^{-2} s^{-1}, $A = 0.7$ m^2 the collecting area, $\tau = 3.4$ s the integration time over the CCD and $\rho = 0.5$ for the overall quantum and optical efficiency averaged over the visible spectrum. Numerical results are shown in Table 2 for stars of different magnitudes. This gives $\sigma_\xi = 120\,\mu$as for a star of 15 mag, that is to say $1/350^{th}$ of the pixel size. Compared to the best results on the ground in astronomical environment, this appears to be a very tough challenge and could question the feasibility of the mission. This was considered as so critical than a real experiment has been carried out as part of the study phase with parameters comparable to GAIA's. It was found conclusively that the centroid of the image of a point-like object could be determined with that accuracy. Moreover the accuracy found was just 1.2 times larger than the theoretical estimate based on the same arguments as above.

2.3. Propagation to astrometric parameters

The centroiding error of the previous section refers to local measurements on the CCD. This is just the best that one could achieve on the sky if we knew perfectly

Table 2. Elementary centroiding accuracy of GAIA for one CCD crossing in the astometric field and for a G2V star.

V mag	Φ_0 ph m^{-2} s^{-1}	N e$^-$	σ_ξ μas
12	1.1×10^6	1.3×10^6	30
15	7.0×10^4	8.5×10^4	120
18	4.4×10^3	5.3×10^3	500
19	1.8×10^3	2.2×10^3	800
20	7.0×10^2	8.5×10^2	1600
21	2.8×10^2	3.4×10^2	3500

the attitude and the geometry of the CCD arrays which must be calibrated in position, orientation and scale. One expects to map the CCD with an accuracy of 5 to 10 nm, equivalent to 20 to 40 μas.

Table 3. Expected final astrometric errors for GAIA. The second column gives the probability that a star is detected in the skymapper (based on GAIA-1)

G mag	P(G)	n_{obs}	σ_ξ μas	$1.2*\sigma_\pi$ μas
12	1.00	900	30	4
15	1.00	900	120	11
18	0.99	890	500	44
19	0.97	870	800	72
20	0.86	770	1600	150
21	0.40	360	3500	390

Observations are repeated over the 11 CCDs of the astrometric field whenever a star crosses the field of view. Then similar observations are repeated with a very specific pattern over the following months. On the average, the number of crossings of the two astrometric fields is $\sim$ 80 (40 for each of the fields). The ecliptic region is under-observed compared to the average, while the number of field crossings is larger by a factor two at the ecliptic latitude 40 degrees. A typical sequence of astrometric observations will be a crossing in the first field, followed by a crossing of the second field 106 minutes later, and possibly a second pair five hours later. Then one must wait between 2 to 6 weeks for a new sequence of observations to return. One must also take into account the fact that the scan direction is not always the most favorable to probe the longitude, latitude or parallax: a scan along the ecliptic meridian brings no information on the ecliptic longitude and similarly the parallax is not visible if the scan direction is perpendicular to the great circle joining the Sun and the star. The degradation factor is close to 2 for the parallax with the GAIA scanning parameters.

So the final astrometric accuracy takes the following form, where the number of observations should include the detection efficiency, which is less than one at the faint end (faint stars are not detected on-board at each field crossing).

$$\sigma_\pi = g_\pi \left[\frac{1}{n_{\text{obs}}} \left(\sigma_\xi^2 + \sigma_{\text{cal}}^2 \right) \right]^{1/2} \tag{3}$$

The corresponding numbers appear in Table 3 and $g_\pi = 2.2$ has been used.

3. The astrometric modeling

With bright stars being potentially observed with a positional accuracy of 4 μas, one must relate the measurements to be performed by GAIA to the astronomical parameters with an internal accuracy better than 1 μas. As said earlier the basic framework will be the general relativity with the proper modeling of the motion of the observer in the Solar System and consistent relativistic definitions of the parallax and proper motions. This model will be fully compatible with the IAU Resolutions on reference frames and relativistic formulation for celestial mechanics and astrometry. However, apart from the gravitational light deflection, many of the small effects to be considered can be discussed in the newtonian framework, offering thus an easier insight on their meaning.

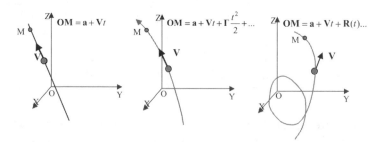

Figure 2. Basic models for the stellar displacement. Uniform rectilinear motion (*left*), uniformly accelerated motion (*middle*), orbital motion superimposed to a linear displacement (*right*).

3.1. The perspective acceleration: simple approach

The simplest space motion of a star is a displacement at constant speed on a straight line, with the position vector at any time given by,

$$\mathbf{R}(t) = \mathbf{R}_0 + \mathbf{V}(t - t_0) \qquad (4)$$

More complex models, as shown in Figure 2 will be used to express a departure

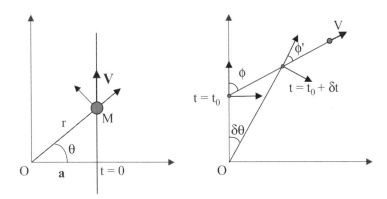

Figure 3. Simplified diagrams illustrating the perspective acceleration. *Left*: a uniform space motion along the straight line leads to a non zero second time derivative of the polar angle θ. *Right*: kinematics in the moving frame with a parallel transport of **V**.

from this situation. Such complexity may appear because the actual motion of the source is not free due to the presence of a another massive body in its vicinity or simply because the curved motion about the centre of the Galaxy becomes noticeable. The objective of the astrometric modeling is to relate the physical description of the star motion, either in the Newtonian or relativistic framework, to the quantities actually measured, primarily directions (with the exception of the radial velocity based on spectroscopic properties).

The concept of celestial sphere provides the convenient manifold to represent directions and it has been used since the very beginning of mathematical astronomy. Therefore the position being reduced to a direction, this is equivalent to using two parameters u, v such that the parametric equation of sphere embedded in the three dimensional Euclidean space is,

$$X = X(u, v) \qquad (5)$$
$$Y = Y(u, v) \qquad (6)$$
$$Z = Z(u, v) \qquad (7)$$

with $X^2 + Y^2 + Z^2 = 1$. Combining these equations and Eq. 4 allows to determine, at least implicitly, $u(t)$ and $v(t)$. In principle (u, v) are pure numbers without direct physical meaning and for the astronomers (u, v) are usually the angular coordinates like right ascension and declination or the longitude and latitude.

The fact that the motion is uniform and rectilinear with $\mathbf{\Gamma} = 0$ tells nothing about the second time derivatives of $u(t)$ and $v(t)$. Consider the very simple model of stellar motion shown in Figure 3. The observer and the velocity vector define a plane in which the star motion takes place. The x-axis is orthogonal to the velocity vector and the angle θ marks the position of the star. With $\mathbf{r}(t) = \mathbf{a} + \mathbf{v}t$ one has $r = a/\cos\theta$ and $\tan\theta = vt/a$. Then $\dot\theta = va/r^2$. Hence,

$$\ddot\theta = -2\frac{\dot r \dot\theta}{r} \qquad (8)$$

showing that even in this very simple case, a uniform space motion leads to a non-zero second derivative of the polar angle, or equivalently to a time variable proper motion. At the same time one sees also that the radial velocity is not constant since $v_r = v\sin\theta$ and $\dot v_r = r\dot\theta^2$.

Using convenient angular and astronomical units one has the following numerical expression,

$$\left.\begin{array}{l}\dot\theta \text{ in mas yr}^{-1} \\ r \text{ in kpc} \\ \pi \text{ in mas} \\ \dot r \text{ in km s}^{-1} \\ \ddot\theta \text{ in mas yr}^{-2}\end{array}\right\} \Rightarrow \gamma = \ddot\theta = -2.0 \times 10^{-9}\, V_r\, \pi\, \mu$$

The angular acceleration depends on the radial velocity and decreases with the distance. Table 4 provides typical values of the acceleration expected for the nearest stars. With $\delta\theta = \gamma\, t^2/2$ one finds that this effect must be accounted for in the GAIA modeling for stars up to 50 - 100 pc.

Table 4.   Typical acceleration expected for the nearest stars observable with GAIA.

d (pc)	V (km s^{-1})			
	10	50	100	
1	0.04	1	4	mas yr^{-2}
10	0.4	10	40	μas yr^{-2}
100	0.004	0.1	0.4	μas yr^{-2}

Let's consider another derivation which helps grasp the true meaning of this acceleration by using the local frame as shown in the right panel of Figure 3. The motion of a star is seen at two epochs separated by small interval of time δt. At $t = t_0$ the two components of the velocity vector on the local frame are $(V\cos\phi,\ V\sin\phi)$ and they are $(V\cos\phi',\ V\sin\phi')$ at $t + \delta t$. By definition the radial velocity is the projection of the speed on the radius vector and its changes follow simply from the rotation of the radial direction. One has,

$$\frac{dv_r}{dt} = \frac{d(V\cos\phi)}{dt} = -V\sin\phi\,\dot\phi = V\sin\phi\,\dot\theta = r\dot\theta^2 \tag{9}$$

This shows primarily that the radial acceleration is not *essential* but is just an effect of the uses of a rotating frame on which this particular component of the velocity vector is projected.

Now for the transverse component one has

$$\frac{dv_\theta}{dt} = \frac{d(V\sin\phi)}{dt} = V\cos\phi\,\dot\phi = -V\cos\phi\,\dot\theta = v_r\,\dot\theta \tag{10}$$

But since $v_\theta = r\dot\theta$, then $dv_\theta/dt = v_r\dot\theta + r\ddot\theta$ which finally yields,

$$\ddot\theta = -2\frac{v_r\dot\theta}{r} \tag{11}$$

similar to Eq. 8. In this derivation one sees more clearly the respective role of the moving frame and that of the choice of a particular coordinates.

3.2. Rigourous approach

With this understanding in hands it is fairly easy to derive a rigourous expression applicable to the three spherical coordinates, simply by considering that a uniform rectilinear motion in the Euclidean space is characterized by the fact that the velocity vector is parallel transported along the straight line. There are many ways to obtain the corresponding expressions, however the most direct consists of expressing that the 3-D acceleration $d^2\mathbf{r}/dt^2 = 0$ and finding the components of the acceleration in spherical coordinates. In curvilinear coordinates one has for the covariant derivative,

$$\gamma^i = \frac{d^2 x^i}{dt^2} + \Gamma^i_{jk} \frac{dx^j}{dt} \frac{dx^k}{dt} \tag{12}$$

where the Christoffel symbols Γ^i_{jk} are computed from the metric of R_3 in spherical coordinates,

$$ds^2 = dr^2 + r^2 \cos^2\delta\, d\alpha^2 + r^2\, d\delta^2 \tag{13}$$

giving in the local frame (normalised with unit vectors) $\mathbf{e_r}, \mathbf{e_\alpha}, \mathbf{e_\delta}$ and with $\gamma^i = 0$,

$$\ddot r = r\cos^2\delta\,\dot\alpha^2 + r\dot\delta^2 \tag{14}$$

$$\ddot\alpha = -\frac{2}{r}\dot r\,\dot\alpha + 2\tan\delta\,\dot\alpha\dot\delta \tag{15}$$

$$\ddot\delta = -\frac{2}{r}\dot r\dot\delta - \sin\delta\cos\delta\,\dot\alpha^2 \tag{16}$$

Eqs. 15-16 are the general expression for the second derivatives of the right ascension and declination of a star moving on a straight line at constant speed in space. One recovers the standard term of Eq. 8 involving the radial velocity

$\dot{r}$ and the parallax as $1/r$. There are supplementary terms of the same order of magnitude which are different from zero even in the case of a purely tangential motion and will have to be allowed for in GAIA. In addition with $\mu^2 = \mu_\alpha^2 + \mu_\delta^2$ one finds with Eqs. 15-16 that $\dot{\mu} = -2\dot{r}/r\mu$, as expected.

The exactness of these relationships has been confirmed with another derivation obtained by differentiating directly,

$$\mathbf{V} = \dot{r}\,\mathbf{e}_r + r\cos\delta\,\dot{\alpha}\,\mathbf{e}_\alpha + r\dot{\delta}\,\mathbf{e}_\delta \tag{17}$$

and using

$$d\mathbf{e}_r = \cos\delta\,\mathbf{e}_\alpha\,d\alpha + \mathbf{e}_\delta\,d\delta \tag{18}$$
$$d\mathbf{e}_\alpha = (-\cos\delta\,\mathbf{e}_r + \sin\delta\,\mathbf{e}_\delta)d\alpha \tag{19}$$
$$d\mathbf{e}_\delta = -\sin\delta\,\mathbf{e}_\alpha\,d\alpha - \mathbf{e}_r\,d\delta \tag{20}$$

Eq. 14 expresses the additional fact that the radial velocity as well is not constant with the time.

While for the declination on has very naturally $\mu_\delta = \dot{\delta}$ and then $\dot{\mu}_\delta = \ddot{\delta}$, one must be very careful for the motion in right-ascension. After Hipparcos it has become customary to use the component of the proper motion in RA as $\mu_\alpha = \dot{\alpha}\cos\delta$, which is not singular toward the poles and is more meaningful than the time derivative of the coordinate itself. Now for the acceleration we have several possibilities : $\ddot{\alpha}$, $\cos\delta\,\ddot{\alpha}$, $\dot{\mu}_\alpha = \cos\delta\,\ddot{\alpha} - \sin\delta\,\dot{\delta}\dot{\alpha}$.

The last two are equally acceptable and I have not found a compelling argument in favor of one in preference of the other. In the following I will use everywhere $\mu_\delta = \dot{\delta}$ and $\mu_\alpha = \dot{\alpha}\cos\delta$ and their derivative $\dot{\mu}_\delta$ and $\dot{\mu}_\alpha$ to express the accelerations. Then Eqs. 14-16 take the following form,

$$\ddot{r} = r(\mu_\alpha^2 + \mu_\delta^2) \tag{21}$$
$$\dot{\mu}_\alpha = -\frac{2}{r}\dot{r}\mu_\alpha + \tan\delta\,\mu_\alpha\mu_\delta \tag{22}$$
$$\dot{\mu}_\delta = -\frac{2}{r}\dot{r}\mu_\delta - \tan\delta\,\mu_\alpha^2 \tag{23}$$

With the standard units (mas, mas yr^{-1}, km s^{-1} and km s^{-1} yr^{-1}) one finds the convenient expressions,

$$\dot{V}_r = 2.30\times 10^{-8}\,\frac{\mu_\delta^2 + \mu_\alpha^2}{\pi} \tag{24}$$
$$\dot{\mu}_\alpha = -2.05\times 10^{-9}\,V_r\pi\mu_\alpha + 4.84\times 10^{-9}\,\mu_\alpha\mu_\delta\tan\delta \tag{25}$$
$$\dot{\mu}_\delta = -2.05\times 10^{-9}\,V_r\pi\mu_\delta - 4.84^{-9}\mu_\alpha^2\tan\delta \tag{26}$$

The terms in radial velocity become negligible at large distance, but this is also the case for the supplementary terms as the magnitude of the proper motion decreases with the distance, until it is dominated by the galactic differential rotation.

4. The parallactic effect

In this section I discuss various possible definitions of the stellar parallax and their relation to the parallactic effect.

4.1. Definition of the parallax

The parallactic effect is the difference in the direction of a distant celestial object as seen from two different viewpoints, that is to say the difference between two unit vectors. In classical astronomy the usual viewpoint was an observing place on the Earth, and the reference point was taken as the centre of the sun or, better, the barycentre of the solar system. Because of the annual motion, the parallactic vector is not constant during the year and the elliptical apparent displacement of the star cannot be incorporated into a linear proper motion. Quite naturally, one adopts as the standard direction of the star that defined in the barycentric frame.

While the parallactic effect and the parallactic vector are well defined concepts in Newtonian astronomy (provided the source and the observer are at rest in the barycentric frame), one must be careful in the choice of the definition of the parallax itself. (Figure 4). Without paying attention to the details the

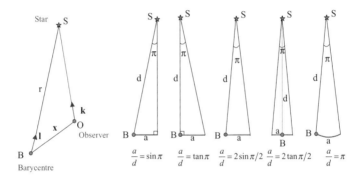

Figure 4. The parallactic effect (at left) and the five possible definitions of the stellar parallax ($a = 1$ AU). The parallactic effect is the vector $\mathbf{k} - \mathbf{l}$.

parallax is understood as the angle subtended by the astronomical unit at the distance of the star. This may lead to at least five more or less equivalent geometric meanings compatible with this definition as illustrated in Figure 4 where a is exactly 1 AU, and not the semi-major axis of the orbit of the Earth. The distance of the star is always measured from the barycentre of the solar system and at the time when the light was emitted (the effect of light propagation will be considered below). The accuracy of classical astronomy made the choice between the five definitions irrelevant and one must even say that not every textbook agrees on the choice. One must note however that the five definitions agree to first order in a/d with $\pi \simeq a/d$.

Fortunately the differences arise only with terms in $(a/d)^3$ which are very very small, even for the new astrometry missions. Considering the nearest star with $\pi \approx 1$ arsec, the third order term is never larger than 10^{-5} μas, much below the modelling accuracy required for GAIA and SIM. This means that one can always, and for many years to come, take the standard definition with $\pi = a/d$ and assimilate the arc to its sine or tangent. One may argue that the barycentric

Table 5. Acceleration in proper motion, variable parallaxes and radial velocity for the nearest stars of the Hipparcos Catalogue.

HIP	Name	π mas	$|\mu|$ mas/yr	V_t km/s	V_r km/s	γ μas/yr^2	$\dot{\pi}$ μas/yr	$\dot{V_r}$ m/s/yr
3829	van Maan.2	226.8	2975	62.2	263.0	-355	-13.5	0.9
24186	Kapteyn	255.1	8661	160.9	245.6	-1085	-15.9	7.3
54035		392.5	4805	58.0	-85.6	323	13.1	1.5
54211		206.9	4509	103.3	67.5	-126	- 2.9	2.5
57939	Groomb. 1830	109.2	7060	306.4	-97.9	151	1.2	11.5
71681	α^2 Cen	742.2	3725	23.8	-21.0	116	11.6	0.5
71683	α^1 Cen	742.2	3711	23.7	-21.6	119	11.9	0.5
87937	Barnard	549.3	10369	89.5	-109.7	1249	33.1	4.9
104214	61 Cyg A	287.2	5282	87.2	-64.7	196	5.3	2.4
104217	61 Cyg B	285.5	5174	85.9	-64.5	190	5.2	2.3
108870	ϵ Ind	275.8	4705	80.9	-40.0	104	3.0	2.0

direction is itself as variable as the geocentric one, because the sun is in motion with respect to the nearest stars at a velocity comparable to that of the orbital motion of the Earth. Therefore there is a parallactic effect of similar magnitude between the barycentric positions at one year interval. This is perfectly true, but the apparent displacement has, to a sufficient approximation, no curvature and is incoporated in the proper motion, which in fact considers the velocity of a star relative to the barycentre of the Solar system. This remarks shows that in accurate astrometry there is no natural separations between proper motion and parallactic effect and this will be done only on the basis of additional conventions.

4.2. The microarcsecond modelling

With the definition of π we can now model the parallactic effect to determine its magnitude as a function of the parallax. Let **k** and **l** be respectively the unit vectors in the direction of the star from the observer frame and from the barycentric frame. The basic equation relating the two directions is given by (Figure 4),

$$\mathbf{OS} = \mathbf{BS} - \mathbf{BO} = \mathbf{BS} - \mathbf{x} \qquad (27)$$

and dividing both members by $|\mathbf{BS}| = r$ (distance at the emission time) one gets,

$$\frac{\mathbf{OS}}{r} = \mathbf{l} - \frac{\mathbf{x}}{r} \qquad (28)$$

With $\mathbf{k} = \mathbf{OS}/|\mathbf{OS}|$ it is easy to write,

$$\mathbf{k} = \left(\mathbf{l} - \frac{\mathbf{x}}{r}\right)\left(1 - 2\frac{\mathbf{l}\cdot\mathbf{x}}{r} + \frac{x^2}{r^2}\right)^{-1/2} \qquad (29)$$

By expressing **x** in astronomical units and with $\pi = 1\mathrm{AU}/r$ one finds to the second order in π, (see also Klioner 2002, in press, for an equivalent expression),

$$\mathbf{k} = \mathbf{l} + \pi\left[(\mathbf{x} \times \mathbf{l}) \times \mathbf{l}\right] + \frac{\pi^2}{2}\left[2(\mathbf{x} \times \mathbf{l}) \times \mathbf{x} - 3|(\mathbf{x} \times \mathbf{l}) \times \mathbf{l}|^2\,\mathbf{l}\right] + \mathcal{O}\left(\pi^3\right) \quad (30)$$

Eq. 30 shows that using $\pi \sim 1/r$ to express the distance of the star leads to a parallactic effect which is not linear in π. The second order term of magnitude π^2 is at most 5 μas for the nearest stars and decreases rapidly with the distance. It is then marginally significant for a handful of stars and should be include in the general modeling of the astrometric observations for GAIA.

4.3. Time variable parallax

When the component of the velocity along the line of sight is non-zero the distance of the star changes with the time implying $d\pi/dt \neq 0$. In principle one should then define the parallax at a specific time as for the position. Setting $v_r = dr/dt$ one gets with the usual units (mas yr^{-1}, mas, km s^{-1}),

$$\dot{\pi} = 10^{-9}\pi^2 v_r \quad (31)$$

giving 50 μas yr^{-1} for a star at 1 pc with $v_r = 50$ km s^{-1}. Hence it is significant for the closest stars and just its measurements would be a nice achievement with GAIA. It is likely that it would be more sensible to allow for the correction instead of adding one more free parameter.

Table 6. Number of Hipparcos stars that would show a value of $d\mu/dt$ larger than a lower boundary.

V_r km s^{-1}	$\gamma >$	100 μas yr^{-2}	10 μas yr^{-2}	5 μas yr^{-2}	2 μas yr^{-2}	1 μas yr^{-2}
10 or V_r		11	70	155	400	800
50 or V_r		12	90	230	650	1500
100 or V_r		14	150	360	1100	3000

5. How many sources

From the Hipparcos Catalogue I have extracted a sample among the nearest stars with large proper motions. Results for the perspective effect, radial acceleration and variable parallax are shown in Table 5. The variable proper motion will be significant for all these stars and therefore for many of the fainter nearby stars that GAIA will detect. On the other hand the variable parallax should not be a serious concern for GAIA. Clearly the accelerated radial motion is irrelevant for

GAIA spectroscopy but may help understand systematic changes between old and recent measurements.

Table 6 gives the number of stars in the Hipparcos Catalogue for which the tangential acceleration is larger than a preassigned value. The radial velocity in the first column is assumed at the value indicated or taken from available measurements. When these numbers are scaled to GAIA it appears that the perspective acceleration will be noticeable or would have to be allowed for in the model for about 1 million stars. But this assumes that the proportion of nearby and distant stars will be the same in the GAIA survey as it is in the Hipparcos Catalogue, which is probably not true. The same evaluation was done for $\dot{\pi}$ in Table 7, indicating that, with the same assumptions, about 10^5 stars would show a variable parallax.

Table 7. Numbers of stars in the Hipparcos Catalogue that would show a variable parallax larger than a preset value.

V_r km s^{-1}	$\dot{\pi} >$	10 μas yr^{-1}	2 μas yr^{-1}	1 μas yr^{-1}
10 or V_r		6	17	33
50 or V_r		7	24	50
100 or V_r		7	50	80

6. Light propagation

In this section I wish to discuss the astrometric effects of the propagation of light between the source and the observer, in particular when the source is moving in the barycentric frame and is observed at different epochs.

6.1. Proper motion and radial velocity

Stellar astrometry has always mapped the apparent sky, that is to say the sky as it would appear to an observer placed at the barycentre of the solar system at a certain epoch t_r tied to the reception of some light signal. The difference between the true (the sky as it is at t_e) and apparent positions is the equivalent of the so-called planetary aberration in the Solar System, and is of the order of $V/c \approx 10 - 50$ arcsec, if V is the barycentric velocity of the star. In principle there is no problem with this choice, even if the difference were much larger, provided it remains constant in magnitude and direction. The change will be seen as an additional proper motion. However, there is a notable exception to this statement in galactic dynamics which should consider the distribution of matter as it is at a specific time and compute the gravitational potential with the positions of the stars at the retarded time.

In microarcsecond astrometry the propagation delay raises interesting questions about the very definitions of the proper motion and the radial velocity. Let $\mathbf{r}(t)$ the position vector of a stellar source in the barycentric frame at time t and $\mathbf{R}(t)$ its position in the apparent sky. Astrometry is characterized by the equation,

$$\mathbf{R}(t_r) = \mathbf{r}(t_e) \qquad (32)$$

where the reception time t_r and the emission time t_e are related by,

$$t_r = t_e + \frac{|\mathbf{r}(t_e)|}{c} \qquad (33)$$

The apparent and true velocity vectors are then,

$$\tilde{\mathbf{V}} = d\mathbf{R}/dt_r \qquad (34)$$
$$\mathbf{V} = d\mathbf{r}/dt_e \qquad (35)$$

and clearly $dt_r \neq dt_e$ in general. The practical procedure applied in the determination of the proper motion and that of the tangent velocity corresponds precisely to Eq. 34. It would be true as well for the radial velocity if it were obtained by comparing apparent distances at two different epochs, which is not the current practice.

By combining the fundamental equation of astrometry, Eq. 32, and Eq. 33 one finds

$$\tilde{\mathbf{V}} = \frac{\mathbf{V}}{1 + \frac{v_r}{c}} \qquad (36)$$

yielding for the proper motion and the tangent velocity,

$$\tilde{\mu} = \mu/(1 + v_r/c) \qquad (37)$$
$$\tilde{v}_t = v_t/(1 + v_r/c) \qquad (38)$$

With $v_r \sim 30$ km s^{-1}, this gives a difference of 10^{-4}, not negligible for a high-proper motion star ($\mu > 100$ mas yr^{-1}). Hence, with GAIA one should indicate accurately the meaning of the proper motion published in the Catalogue and takes care of the definition in the observing equations.

A similar equation holds for the radial velocity considered as a measure of dr/dt, either directly (never done so far) or by using the perspective acceleration with Eq. 3. However this not the method generally employed and spectroscopy remains the only widely used method. In this case one should note that the Doppler effect is precisely the comparison of the period of a periodic signal in the rest frame of the source (period $= dt_e$) to the measured period dt_r, itself compared to a reference on the ground. Therefore spectroscopic v_r are then quite correctly dr/dt_e. One should then distinguish clearly the spectroscopic radial velocity from its kinematical definition, independently of any other systematics affecting the spectroscopic data (Dravins et al. 1999).

Eq. 38 has been widely called upon to explain the super-luminal jets observed in some AGNs. Due to their orientation with respect to the line-of-sight to Earth, jets of gas leaving energetic sources occasionally have the appearance of moving faster than light. Given the speed V of the jet and the orientation along the line of sight θ, Eq. 38 leads to $\tilde{V}_t = V \sin\theta / (1 - (V \cos\theta)/c)$ which is larger than c if $\sqrt{2} V \sin(\theta + \pi/4) > c$.

6.2. Parallax

Even in the Euclidean framework, with absolute space and absolute time a perfect definition of the parallactic effect, and then of the parallax, is easy only when the source and the observer are at rest in the barycentric frame. The parallactic effect is then defined unambiguously by the difference between the direction of the source as seen by the observer and the direction that would be observed at the barycentre, at any time. As soon as the source is in motion the situation is more intricate.

A figure like Figure 4 is misleading if the source is in motion in the barycentric frame, as it does not show contemporaneous positions of points. Worse, there is no clear definition of the parallactic effect, let alone the parallax. The direction $\mathbf{k}$ raises no particular problem (disregarding the possible motion of the observer and the aberration) since it corresponds to a well defined spacetime event occurring at O at the time t_r. There is nothing similar for $\mathbf{l}$ which is not connected to a well identified event. One can imagine an observer located at the barycentre, but at which time should he consider the direction $\mathbf{l}$ as being put in correspondence with $\mathbf{k}$? If it is the same reception time as t_r at O, this leads to a position of the source at a $t'_e \neq t_e$ because the distances that separated O and B from the source are not exactly the same and the solutions in t_e of Eq. 33 are not identical. So $\mathbf{k}$ and $\mathbf{l}$ are unit vectors pointing at the source from two viewpoints, but they are not pointing at the same point in space as a result of the motion of the source. Obviously quantitatively the difference is very small (for the Earth this is at most the motion of the star over 8 minutes of time), but as a matter of principles a clear definition should be provided. The adopted solution which works in a Newtonian framework, is to freeze the position of the source at t_r (observer time) and determine the direction $\mathbf{l}$ for this conventional point. This is also acceptable in special relativity since it is possible to have a set of synchronised clocks within the solar system or, if the gravitational field is included, a common coordinate time in the barycentric frame.

References

Dravins, D., Lindegren, L., & Madsen, S., 1999, A&A 348, 1040
Klioner, S.A., & Kopeikin, S.M., 1992, AJ 104, 897.

GAIA photometric performances

Vladas Vansevičius

Institute of Physics, Goštauto 12, Vilnius 2600, Lithuania

Audrius Bridžius

Institute of Physics, Goštauto 12, Vilnius 2600, Lithuania

Abstract. GAIA present day (end-2002) photometric performances, estimated basing on the stellar populations crucial for understanding the formation and evolution of the Galaxy, are discussed. Performance of the GAIA photometric systems (PSs) is evaluated taking into account their ability to simultaneously determine the main stellar parameters: T_{eff}, $\log g$, [M/H] and E_{B-V}, for a large variety of stars down to $V \sim 20$ mag. A sample of the stars (photometric system test targets, PSTTs) applicable for evaluation of the GAIA photometric systems is presented. Definitions of the 1X PS and its accuracies of stellar parameterization are given. We conclude that there is still no photometric system proposed to date which would allow to achieve the scientific objectives of the GAIA mission at the limiting magnitude $V = 20$.

1. Introduction

Though the primary goal of the GAIA mission is a study of the Galactic dynamics, photometric as well as spectroscopic observations are also of great importance in order to achieve the main GAIA objectives (ESA-2000-SCI-4). Therefore, an optimal multicolour photometric system, adequate to the main mission goals, has to be designed and analysed in detail. One of the most important steps towards such a PS is an elaboration of a method for unbiased evaluation of the PSs proposed for GAIA in terms of the accuracy of stellar parameters derived (Vansevičius, Bridžius, & Drazdys 2002). A tight time schedule for the pre-launch preparation constrains possibilities of thorough testing of the new PS on real sky objects. Therefore, optimization and further choice of the most suitable PS for the GAIA mission must be entirely based on the synthetic and observed spectral energy distributions (SEDs) and the representative set of a large variety of stellar populations and interstellar medium inside the Galaxy.

We would like to note two key points which have been taken into account for evaluating the GAIA PSs in the present study. The same GAIA observation (Deveikis, Bridžius, & Vansevičius 2002) and parameterization (Bridžius & Vansevičius 2002) methods have been applied to all PSs, and the performance of the PSs has been tested on the PSTT stars, which are of the highest importance to the GAIA mission (ESA-2000-SCI-4).

2. Requirements for the GAIA photometric system

The primary objective of the GAIA mission is to obtain data which allows us to study the composition, formation and evolution of the Galaxy. This priority also constrains the requirements for the GAIA PS. Therefore, metallicity and age variations (RGB, AGB and HB), star formation history (SFH) over the last 14 Gyr (A-K MS stars), and dust distribution within the Galaxy are the most important issues to be assessed by GAIA PS.

The variety of the object types and astrophysical circumstances presuppose specific requirements for the GAIA PS performance. It is indisputable that the four main parameters (T_{eff}, $\log g$, [M/H], and E_{B-V}) ought to be derived as precisely as possible at the limiting magnitude of the survey, $V = 20$. However, it is very important to decide which parameters should be determined additionally. Therefore, below we list the most relevant items whose various degree of treatment could significantly influence the final design of the GAIA PS:

- variation of [α/Fe] - should it be determined by PS?
- extinction of the F-K type stars - should it be determined for each individual star by PS or assigned from the 3-D extinction maps?
- binaries - should PS be adapted for their recognition and parameter estimation?
- variable extinction law, R_V - should it be determined by PS?
- chemical peculiarities - should the strong peculiar features (e.g. C, N) be avoided or measured by PS?
- WD, BD, WR, δCep, T Tau, Mira, C, etc. type stars - which parameters should be derived by PS?

Although it is very important to take all mentioned points into account, present day investigations and discussions are devoted mostly to the first two items. Determination of the interstellar extinction of the F-K type stars is fully accounted (Sūdžius et al. 2002) and can be discussed quantitatively. However, the [α/Fe] problem is still pending due to lack of a proper grid of SEDs. Therefore, only these two items are taken into account for selection of the PSTTs.

3. GAIA photometric system test targets

According to ESA-2000-SCI-4, the stars to be observed with GAIA down to $G \sim 20$ can be roughly represented by the following proportions of stellar populations (Ms - stands for millions of stars):

- the total number of stars of all types $\gtrsim$ 1000 Ms;
- disk dwarfs $\sim$ 780 Ms; disk giants $\sim$ 90 Ms;
- thick disk (all types) $\sim$ 100 Ms; halo (all types) $\sim$ 70 Ms;

and spectral types:

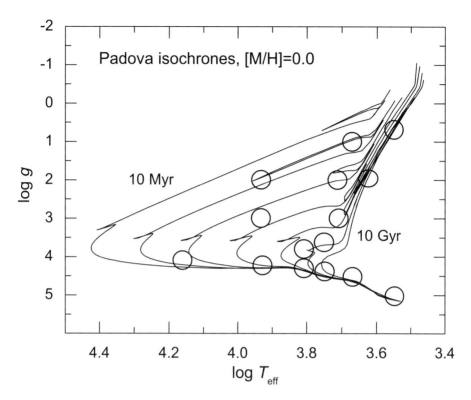

Figure 1. GAIA photometric system test targets (PSTTs).

- O-B ~ 2 Ms; A ~ 50 Ms; F ~ 230 Ms;
- G ~ 400 Ms; K ~ 300 Ms; M ~ 55 Ms.

In order to evaluate any PS proposed for GAIA we need to select some set of standard objects with which to test a PS. PSTTs ought to cover the most important stellar population types mentioned above and the most characteristic variety of astrophysical circumstances expected to be encountered in the Galaxy. A set of stellar types which would make up a minimum number of PSTTs is presented in Table 1 and in Figure 1. We need to keep this list as short as possible in order to quickly test any new improvements to the GAIA PSs. However, it should also be comprehensive in the sense of the main GAIA goals. The same statements are also valid for other parameters listed below. The proper choice of the PSTTs is a very important issue for GAIA PS development and optimization, therefore, the list is open for discussions and suggestions.

The list of PSTTs is complemented by other astrophysical parameters, representing circumstances which could be encountered in the Galaxy (Table 2), as completely as possible. Two variants of a complete set of models based on distance, D, of the PSTTs or their apparent magnitude, V_J, are foreseen. In order to make a grid of PSTTs convenient for computer modelling and analysis, we suggest to use a complete set of various combinations even if one supposes

Table 1. GAIA photometric system test targets (PSTTs).

Code	Sp. Type	T_{eff}	$\log T_{\text{eff}}$	$\log g$	M_V
1	B V	14500	4.16	4.1	−0.9
2	A V	8500	3.93	4.2	1.6
3	F V	6450	3.81	4.3	3.7
4	G V	5600	3.75	4.4	5.1
5	K V	4700	3.67	4.5	6.7
6	M V	3550	3.55	4.7	10.3
7	F IV	6450	3.81	3.8	2.6
8	G IV	5600	3.75	3.6	3.1
9	G III	5150	3.71	3.0	0.9
10	K III	4150	3.62	2.5	0.3
11	M III	3550	3.55	1.2	−0.5
12	BHB	8500	3.93	3.0	0.5
13	RHB	5150	3.71	2.0	0.5
14	A I	8500	3.93	2.0	−5.0
15	G I	4700	3.67	1.0	−4.5

that any particular combination of the parameters is unrealistic and cannot be found in the Galaxy. Therefore, the complete set of models should contain 5625 different combinations of the parameters in total. Such a task can be easily solved even on a PC type computer.

4. The 1X GAIA photometric system

Few different photometric systems have been proposed for the GAIA mission to date: 1F (ESA-2000-SCI-4) and its modification 2F, 2A (Munari 1999), 2G and its modification 3G (Høg, Straižys, & Vansevičius 2000). However, the analysis of their performance has demonstrated that none of them satisfy the

Table 2. Parameters for the PSTTs. V_J is in the UBV Johnson system.

No.	parameter	range				
1	D, kpc	1	2	4	8	20
2	V_J	15	17	18	19	20
3	E_{B-V}	0.05	0.5	1.0	1.5	3.0
4	[M/H]	+0.5	0.0	−0.5	−1.5	−2.5
5	[α/Fe]	0.0	0.3	0.6		

Table 3. Evolution of the GAIA PSs. V_J indicates the limiting magnitude at which PS performance is satisfactory for the GAIA goals.

Proposed, year	Photometric System			Limit, V_J
1999	1F	2A	2G	17
2000			3G	18
2001	2F			18
2002		1X		19

requirements for the GAIA PS (Vansevičius et al. 2002). Therefore, taking all valuable findings in those PSs into account, a new photometric system, 1X, is proposed. Evolution of the GAIA PSs and the limiting magnitudes at which they satisfy GAIA goals are shown in Table 3.

The filter transmission curves of the 1X medium band (MB) PS designed for the Spectro telescope are shown in Figure 2. The central wavelength and width of the bands as well as a number of slots, allocated for each filter in the focal plane, are given in Table 4.

5. Performance of the 1X photometric system

The performance of the 1X PS is evaluated in terms of its ability to derive basic stellar parameters: T_{eff}, log g, [M/H], and E_{B-V}, for the PSTTs. The procedure for simulation of the photometric GAIA observations developed by Deveikis et al. (2002) was used. The simultaneous 4-D stellar parameterization method (Bridžius & Vansevičius 2002) was applied.

Some specific assumptions on the 1X PS application to the GAIA case introduced in the modelling procedure should be noted:
- the design of GAIA-1 is assumed as it is given in ESA-2000-SCI-4;
- only the medium band PS is tested assuming that there is no supplementary information obtained from the GAIA broad band PS (negligible influence of the

Table 4. Definition of the 1X PS. *CWL*: central wavelength (in nm); *FWHM*: full width of the filter transmission band at half maximum (in nm); *NF*: number of slots allocated to the filter in the focal plane.

Filter	x33	x38	x41	x46	x51	x55	x65	x78	x82	x86	x99
CWL	323	382	410	465	510	555	655	785	825	860	992
FWHM	63	29	19	29	19	49	29	29	29	29	56
NF	2	2	2	1	2	1	1	1	1	1	1

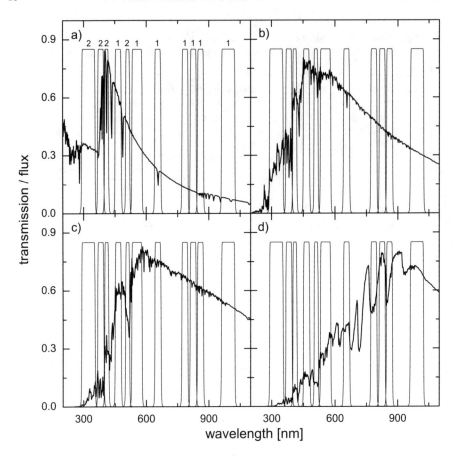

Figure 2. The 1X GAIA PS. SEDs are plotted for main sequence stars with T_{eff} 9000 K (a), 5500 K (b), 4500 K (c), and 3500 K (d). The numbers of slots allocated to each filter in the focal plane are shown in panel a.

BB PS to the performance of MB PS was demonstrated by Vansevičius et al. 2002);
- the total number of slots for MB filters is assumed to be equal to 15 (as it is foreseen in the Radial Velocity Spectrometer (RVS) design for GAIA-2);
- the maximum of filter transmission curves is set conservatively to 85%;
- modelling is performed by employing the SEDs from BaSeL 2.2 (Lejeune, Cuisinier, & Buser 1998) with solar abundance of α-elements, $[\alpha/\text{Fe}]=0.0$;
- the standard ($R_V = 3.1$; Cardelli, Clayton, & Mathis 1989) and invariable extinction law is assumed;
- the aperture photometry procedure is applied for the modelling of star measurement.

The results of the 1X PS performance test are presented in Table 5. A typical value of colour excess, E_{B-V}, was assumed for each particular PSTT star.

All apparent magnitudes were set to $V = 18$, therefore, each PSTT was placed at some distance, D. In order to derive the accuracy of the 1X PS more reliably, 100 independent simulations of each PSTT were performed, and deviations of the stellar parameters from the true value were determined. The r.m.s. estimate was employed, and the σ's of the stellar parameters are tabulated.

We find that performance of the 1X PS at $V = 18$ is excellent and completely satisfies the requirements raised for the GAIA PS by the main GAIA objectives.

Table 5. Performance of the 1X GAIA PS evaluated on PSTTs of solar abundance, [M/H]=0.0, at $V = 18$. SpT: spectral type.

Code	SpT	M_V	E_{B-V}	D,kpc	σT_{eff},%	$\sigma(\log g)$	σ[M/H]	σA_V
1	B V	-0.9	3	0.8	4.3	0.10	1.20	0.02
2	A V	1.6	1	4.6	4.0	0.15	0.30	0.09
3	F V	3.7	1	1.7	2.3	0.35	0.05	0.09
4	G V	5.1	0.5	1.9	2.8	0.30	0.20	0.10
5	K V	6.7	0.5	0.9	1.5	0.30	0.05	0.09
6	M V	10.3	0.05	0.3	0.5	0.10	0.05	0.05
7	F IV	2.6	1	2.9	2.8	0.25	0.05	0.10
8	G IV	3.1	1	2.3	3.3	0.50	0.25	0.11
9	G III	0.9	1	6.3	1.0	0.45	0.05	0.03
10	K III	0.3	1	8.3	0.5	0.05	0.05	0.04
11	M III	-0.5	1	12.	0.5	0.15	0.10	0.09
12	BHB	0.5	1	7.6	2.0	0.10	0.60	0.04
13	RHB	0.5	1	7.6	2.0	0.55	0.15	0.08
14	A I	-5.0	3	5.5	0.8	0.05	0.10	0.01
15	G I	-4.5	3	4.4	1.0	0.15	0.05	0.06

The performance of the 1X PS is significantly better, especially if cases of higher extinction are considered, comparing to the previous PSs, 2F & 3G (Figures 3 – 5; $V = 18$; the r.m.s. estimates of the parameters were derived basing on 100 simulations of each PSTT; for more details see Vansevičius et al. 2002). The 1X PS also performs satisfactory at $V = 19$, however, there is an obvious necessity to make it more accurate at $V = 20$.

6. Conclusions

We have suggested a set of the GAIA Photometry System Test Targets (PSTTs) selected in order to facilitate easy and comprehensive test of the proposed GAIA PSs. We have introduced a new 1X PS which demonstrates superior performance compared to the PSs proposed for GAIA earlier. The performance of the 1X PS was evaluated taking into account its capability to simultaneously determine the main stellar parameters: T_{eff}, $\log g$, [M/H], and E_{B-V}, for the PSTT stars, and assuming that no supplementary information is available, except the data obtained by the GAIA medium band PS.

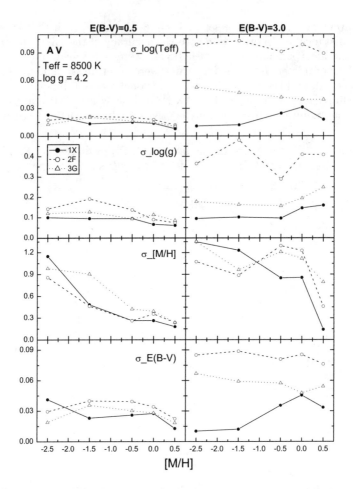

Figure 3. Comparison of the performance of the 1X, 2F, & 3G GAIA PSs, PSTT #2, see Table 1.

GAIA is assumed to measure stars in the Galaxy down to $V \sim 20$, however, even the best photometric system (1X) among the PSs, proposed for GAIA to date (end-2002), performs satisfactory only down to $V \sim 19$. Therefore, we conclude that there is still no optimal PS, in terms of the main GAIA mission goals (ESA-2000-SCI-4), proposed to date. However, excellent accuracy of the stellar parameters determined at $V = 18$ (1X PS) implies fulfilment of the requirements for GAIA PS at $V = 20$ after appropriate fine tuning of each band.

Some gain in performance could be achieved by applying different parameterization schemes for the faintest stars ($V > 18$). Parameterization of these stars should be performed by employing all complementary information obtained by GAIA (parallaxes, spectra, 3-D extinction maps constructed basing on E_{B-V}

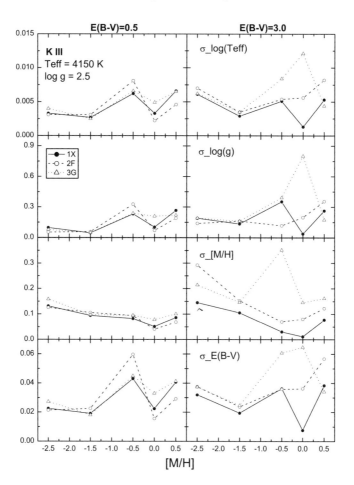

Figure 4. Comparison of the performance of the 1X, 2F, & 3G GAIA PSs, PSTT #10, see Table 1.

derived for the brighter stars, $V < 18$, etc.). However, such an automatic parameterization procedure is not yet developed.

In order to optimize the GAIA PS, the following supplemental information is urgently needed:
- a homogeneous and complete database of theoretical stellar spectra, especially SEDs of various $[\alpha/Fe]$;
- realistic estimates of the limiting magnitudes and accuracies of the stellar parameters derived from the RVS data;
- determination of the limiting distance, resolution and achievable accuracy of the 3-D extinction maps of the Galaxy;
- realistic models of the photometry procedure, taking into account real sky complexity and accuracy of the post-mission calibrations.

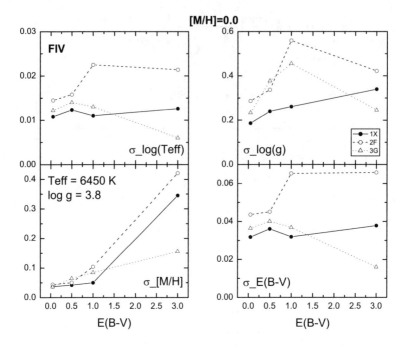

Figure 5. Comparison of the performance of the 1X, 2F, & 3G GAIA PSs, PSTT #7, see Table 1.

Acknowledgments. This work was supported by a Grant of the Lithuanian State Science and Studies Foundation. We are indebted to Valdas Vansevičius for his help on preparation of the article.

References

Bridžius, A., & Vansevičius, V. 2002, Ap&SS 280, 41
Cardelli, J. A., Clayton, G. C., & Mathis, J. S. 1989, ApJ 345, 245
Deveikis, V., Bridžius, A., & Vansevičius, V. 2002, Ap&SS 280, 47
Høg, E., Straižys, V., & Vansevičius, V. 2000, GAIA-CUO-078 (Livelink)
Lejeune, T., Cuisinier, F., & Buser, R. 1998, A&AS 130, 65
Munari, U. 1999, Baltic Astron. 8, 123
Sūdžius, J., Raudeliūnas, S., Kučinskas, A., Bridžius, A., &
    Vansevičius, V. 2002, Ap&SS 280, 109
Vansevičius, V., Bridžius, A., & Drazdys R. 2002, Ap&SS 280, 31

On science goals of GAIA spectroscopy

Ulisse Munari

Astronomical Observatory of Padova – INAF, Asiago Station, 36012 Asiago (VI), Italy

Abstract. The basic aims of GAIA spectroscopy are reviewed. The application of radial velocities to galactic kinematics, perspective acceleration, internal kinematics of stellar aggregates, binaries and pulsations is outlined, and side use of GAIA spectra (for classification, atmospheric analysis and chemical abundances, rotation, mass loss, peculiarities and interstellar medium) considered in the context of the 11 500 resolving power baselined for the GAIA spectrograph.

1. Introduction

The prime goal of GAIA spectroscopy is to provide the radial velocity which combined with astrometric position, distance and tangential motion gives a complete representation of a given star in the phase space. For a small fraction of GAIA targets the radial velocities will be required to correct astrometry for perspective acceleration. The observations will be performed over the region 8480–8740 Å centered on the CaII triplet and head of the Paschen series, at a currently baselined resolving power (R.P.) of 11 500.

The design of the GAIA spectrograph and selection of wavelength range and resolution have been focused to deliver the best possible radial velocities for the largest set of stars in the Galaxy, particularly in those regions like the Halo where low metallicities grossly reduce the number of suitable absorption lines. In this respect, the CaII triplet lines remain core-saturated even at metallicities [Fe/H]=–2.5 (cf. Figure 6 in Munari 2002) when all other lines have disappeared.

The measurement of radial velocities will be performed on the ground, which means that full spectra will be received from GAIA, not only the result of on-board cross-correlation. This will permit many other uses of the spectra other than measurement of radial velocities. Most of them will be covered in details during this conference. The aim of this introduction to the role of GAIA spectroscopy is therefore to provide just an overview. Other general introductions are available from Munari (1999, hereafter M99), ESA-2000-SCI-4, Perryman et al. (2001) and Munari (2002, hereafter M02). A scheme of the general goals of GAIA spectroscopy as discussed in this introduction is given in Figure 1, with the clear division between top priority radial velocities and the rest.

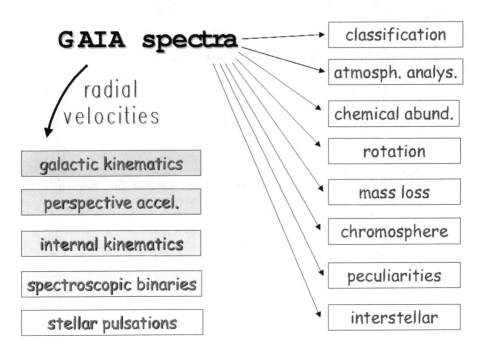

Figure 1. Schematics of the goals for GAIA spectroscopy.

2. Radial velocities

2.1. Galactic kinematics

The requirements on radial velocities for kinematical studies of the Galaxy is, by general consensus formed during meetings of the RVS Working Group and confirmed at this Conference, a mission-averaged error $\lesssim 10$ km sec^{-1} on $V =$ 17.5 mag field single stars (which typical spectral type is G-K). This would allow, for example, to detect and characterize the Halo streams left over by past merging events. The currently baselined 11 500 R.P. can deliver (at least in the uncrowded Halo) such a precision at the faint end, but also providing radial velocities accurate to a few 0.1 km s^{-1} at the bright end where are located the ~40 million stars for which GAIA will provide tangential motion accurate to 0.5 km s^{-1} or better (cf. pag. 275).

The stellar density is so high close to the galactic plane and toward the Bulge that spectra will *all* be in overlap on the GAIA spectrograph focal plane whatever low the R.P. will be in the 20 000–5 000 interval (cf. pag. 71, 105 and 489). At low galactic latitudes, the regions of highest interest are the *holes* through the heavy interstellar extinction, and there the sky density can go up to 10^6 star deg^2 like in the Baade window, far too excessive for GAIA spectrograph slitless operation mode (cf. pag. 76). The crowding will impact the limiting magnitude at which GAIA will deliver radial velocities of a given precision (the higher the crowding, the brighter the limiting magnitude). On the other hand, such limited regions of the sky will be the among the most appealing targets of multi-fibers and integral field spectrograph facilities coming on line in the optical and infrared at the largest Observatories on the ground, equipped with adaptive optics and light-collecting powers far in excess of the 0.5×0.5 m primary mirror of GAIA spectrograph. We can expect great results from the ground on these low latitude, low extinction, limited area windows by the time GAIA will have completed its observations. Where GAIA will remain unsurpassed is in its ability to cover the whole sky, particularly the low density region at high galactic latitudes where the star density is too low to saturate the input capabilities of large telescopes and the area to cover far too large for their small fields of view. Another GAIA uniqueness compared to ground based observations will be the repeated visits to all targets (an average of ~100 each). Even if the determination of the bulk motion properties of galactic ensembles should be little affected by undetected binaries similar to those found in the solar neighborhood (Hargreaves et al. 1996), their discovery and characterization by the ~100 epoch radial velocities provided for each star by GAIA adds to the knowledge of the nature and evolution of the stellar ensemble, and it is mandatory in object-by-object investigations.

Therefore, the galactic kinematical studies where GAIA can be expected to impact the most and still be unique for many years to come are those based on the large area of the sky away from the densest regions of the galactic plane, where Thick Disk and the Halo dominate (with $|b| \gtrsim 6°$ accounting for 90% of the sky).

2.2. Perspective acceleration

GAIA astrometry of nearby, fast moving stars will be affected by their radial motion which will prevent securing accurate data if not corrected for, resulting in misleading proper motions for an *unknown* subset of the target stars (cf. pag. 31, Dravinis et al. 1999). Also parallaxes will be affected, even if quite few in a sizeable manner: for ex. Barnard' star is expected to show a parallax derivative of +34 μas yr^{-1} (ESA-2000-SCI-4).

The perspective acceleration due to a radial velocity V_R is given by $a = -2.0 \times 10^{-9} \pi \mu V_R$ with a in mas yr^{-2}, μ in mas yr^{-1} and V_R in km s^{-1}. The positional error (in mas) integrated over the t years of mission duration become $\epsilon = 1/2 a(t/2)^2$. If ϵ must be less than 1 μas, than the uncertainty in a must be of the order of ~ 0.1 μas yr^{-2} (for 5 yr mission duration). How many stars will be affected by this effect depend on the kinematical properties of the various galactic populations. ESA-2000-SCI-4 estimates in 35 000 the number of stars affected if their V_R >10 km s^{-1} and 200 000 if V_R >100 km s^{-1}.

The accuracy of radial velocities required to account for the effect is not particularly demanding, and ±10 km s^{-1} should suffice, which is well within GAIA spectrograph capabilities, particularly considering that the majority of affected stars will be among those with accurate astrometry, thus bright and close.

2.3. Internal kinematics

The knowledge of radial velocities to complement GAIA proper motions (and stereoscopic view for the closest clusters) for members of stellar clusters and associations would better constraints the internal kinematics and dynamical status of these stellar aggregates, helping to prune the astrometric membership and quantify the energy reservoir kept in binaries. Radial velocities will also assist astrometry in the discovery of new moving groups and in the identification of new types of dynamical aggregations between field stars.

All early type members of the ∼50 known OB associations are well within the magnitude limit of GAIA spectroscopy. The severe extinction caused by the parent clouds in which the associations are usually embedded, and that significantly affects the acquisition of spectra in the classical blue region at the head of the Balmer series, is much less a problem at the relatively long wavelengths of GAIA spectral interval at the head of the Paschen series, for which the I_C magnitude is a more meaningful estimator than V of the collectible flux. Extrapolating to O and B types the accuracy of extensive GAIA radial velocity simulations on A7 stars by Zwitter (2002) and A0 stars by Katz et al. (2002), and integrating with actual cross-correlation data on early B stars by Munari and Tomasella (1998), it can be concluded that for the vast majority of early type stars in stellar aggregates (which are typically brighter than I_C=13 mag) the accuracy of mission averaged radial velocities will be of the order of ± 2–5 km s^{-1}, appropriate to support dynamical characterization of young aggregates.

However, open clusters with ages of 1 billion years or more will be those that best will benefit from GAIA radial velocities. Their accuracy, for members brighter than $V \sim 14.0$ mag, will be of the order of ~ 1.0 km s^{-1} or better. They will therefore be able to revolve the internal kinematics, characterized by dispersion of radial velocities of a few km s^{-1}. Also the number, barycentre kinematics and binding energy of the binaries in such cluster will be accurately mapped (cf. pag. 329).

The impact of GAIA spectroscopy on globular clusters is harder to predict at this stage, and will anyway be limited to the outer regions where stellar crowding is lower. The typical faint magnitudes and low metallicities of globular cluster stars argue against expectation of results as good as those on intermediate age open clusters.

2.4. Binaries

The ∼100 epoch radial velocities that GAIA will on average obtain for each target over the 5 yr mission lifetime are ideally suited to discover and track the orbital motion of binary/multiple stars. Orbital amplitudes of a few km s^{-1} are expected to be detectable for solar type SB1 (single lined) binaries down to V=15.5 mag (cf. pag. 275, Munari et al. 2001a, Zwitter 2002), and their orbits and mass functions will be therefore derivable.

Eclipsing binaries are a gorgeous target for GAIA spectroscopy. They will allow masses, radii and effective temperatures to be derived with great accuracy (masses will be provided by astrometric binaries too). Munari et al. (2001b) estimated that for about 10 000 eclipsing binaries GAIA will deliver masses and radii accurate to 1-2%, with orbital parallaxes accurate to 1–5% for many overcontact systems (cf. pag. 329). Comparison with the extremely accurate GAIA parallaxes will allow to *close the loop* (cf. pag. 309, 316) and refine our understanding of more subtle details (like limb darkening, and atmosphere treatment and modeling, cf. pag. 323) that will allow a better and more general use of eclipsing binaries as a prime distance indicator in general astronomy (next gen-

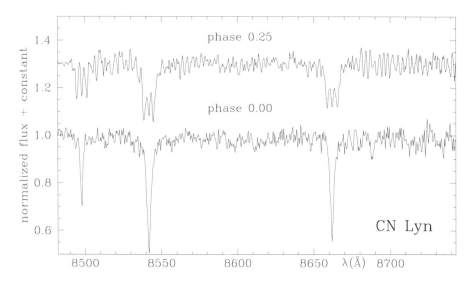

Figure 2. Line splitting and merging in the triple, equal mass eclipsing system CN Lyn in Asiago GAIA-like observations at 11 500 R.P. (spectrum at phase 0.25 displaced by 0.3 in flux for clarity).

eration space and ground telescope will be able to study eclipsing binaries in many external galaxies). The extremely precise $\log g$ and T_{eff} derivable from accurate orbital solutions of eclipsing binaries severely constrain the atmospheric modeling of observed spectra, lifting uncertainties on gravity and temperature and therefore concentrating fitting efforts on chemical abundances alone and other effects like the micro-turbulent velocity or the atmospheric structure.

The 11 500 R.P. baselined for GAIA is large enough to allow detection and study of all solar type SB2 (double lined) eclipsing binaries with an orbital period equal or shorter than the 5 yr mission lifetime. As an example of the GAIA spectroscopic capabilities, Figure 2 reports about discovery on GAIA-like spectra secured in Asiago of a triple nature for the eclipsing system CN Lyn (discovered by Hipparcos photometry), that results to be composed by three equal mass G0 stars (a rare combination), two in a close orbit and the third orbiting at great distance (Marrese et al., in preparation).

2.5. Pulsations

Pulsations affect many areas of the HR diagram. GAIA radial velocities will be able to track the radial ones, in particular the rhythmic expansion and contraction of the outer layers of RR Lyr, Cepheids and Miras. Their amplitudes are large enough for GAIA spectroscopy to map in detail the shape of their pulsation curves. Figure 3 reports an example sequence of pulsation curves for classical Cepheids (for other examples, including RR Lyr and Miras, see Figure 10 of M02).

Defining as 10% of the total amplitude the maximum error of the single epoch radial velocity measurement useful to reconstruct the pulsation curve

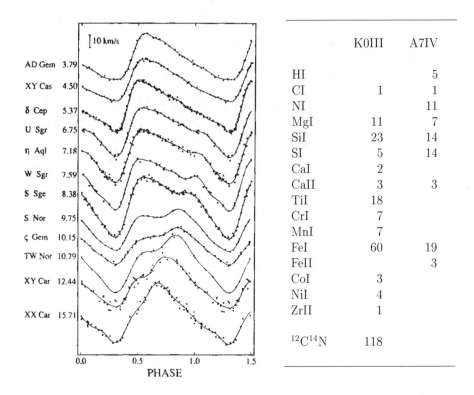

Figure 3. Shape of the radial velocity curve as function of the pulsation period for classical Cepheids (from Bradley 2001). Note the progression of Hertzsprung *bump* for periods between 6 and 15 days.

Table 1. Number of absorption lines of different species which reach a depth of at least 10% of the adjacent continuum in Kurucz high resolution synthetic spectra (R.P. 500 000; from tabular data in Castelli and Munari 2001 and Munari and Castelli 2000).

with ~100 GAIA observations, and taking 75, 35 and 15 km s^{-1} as the typical pulsations amplitudes of, respectively, RR Lyr, Cepheid and Miras variables, the limiting magnitude for proper pulsation curve reconstruction by GAIA radial velocities will be

$V \sim 14.7$ for RRLyr; $V \sim 14.0$ for Cepheids; $V \sim 14.3$ for Miras

(for the current Astrium design of the spectrograph and assuming disturbance from CaII emission lines in 10% of the measurements of Miras. For the Carbon variety of the latter the limiting magnitude improves to $V \gtrsim 15.5$).

3. Other goals

3.1. Classification

The GAIA wavelength interval has a huge classification potential in mapping the MK system and side branches like S- and Carbon stars (M99, M02, Munari & Tomasella 1999, pag. 199, 427, 451).

Paschen hydrogen lines dominate the early spectral types and CaII first appears around B8 and equals Paschen at A3, as illustrated in Figure 4. From mid-A all the way down to mid-M (where it begins to be overwhelmed by TiO absorption, cf. pag. 429), CaII triplet is the dominating feature, with FeI lines becoming prominent with late F types and TiI lines with K stars. The accuracy achievable in MK classification over the GAIA wavelength range is equivalent to that of the classical 3900-4900 Å region for A to M stars, and lower for O and B types due to the weakness of HeI features and absence of significant HeII lines. Considering that the GAIA interval extends for just $1/4$ of the $\Delta\lambda=1000$ Å of the original MK range, the performance appears impressive. Response to gravity (i.e. luminosity class) is equally remarkable, with a marked positive effect on wings of Paschen and negative on those of CaII lines, and powerful intensity ratios built around FeI, TiI, SiI, MgI diagnostic lines (cf. Figure 4 of M02, Boschi et al. 2003).

3.2. Atmospheric analysis and chemical abundances

The automatic atmospheric analysis and chemical abundance determination of GAIA spectra can be performed (with support from GAIA parallaxes and photometry) in basically two ways: modeling each spectrum (which leaves ample freedom in - for example - adjusting individual chemical abundances) and matching it to a grid (built on devoted ground-based observations, or on synthetic spectra, or iteratively defined on GAIA spectra themselves).

The first approach is the one potentially delivering most of information, but will requires large efforts in improving and automation of existing model atmosphere and synthetic spectral codes (cf. pag. 173, 179, 189, 291, 481) and strongly depends on the speed of computers devoted to the analysis of the $\sim 5 \times 10^7$ mission-averaged spectra with a S/N high enough to justify the analysis.

The other approach, comparison with a template grid, is less demanding on computer time and the automation appears much easier than for the previous one. The finite dimensions of the comparison grid calls however in a coarser definition of at least some of the output parameters, in particular the differential chemical abundances, while determination of the overall metallicity (and $[\alpha/\text{Fe}]$) is well feasible. The precise errors to be expected on T_{eff}, $\log g$ and [Fe/H] for various parameter and S/N combinations at the GAIA 11500 R.P. on F-G-K stars have not yet been evaluated. As a guideline, simulations by C.Bailer-Jones at R.P. 5800 (cf. pag. 199) and $30 \leq S/N \leq 170$ over the GAIA wavelength range gives $\Delta T_{\text{eff}} \sim 5\%$, $\Delta \log g \sim 0.5$ and $\Delta[\text{Fe/H}] \sim 0.3$. The errors at 42000 R.P. over the wider wavelength region 3800–6800 Å derived by Katz et al. (1988) are 86 K, 0.28 dex and 0.16 dex, respectively, at S/N=100, and 102 K, 0.29 dex and 0.17 dex at S/N=10.

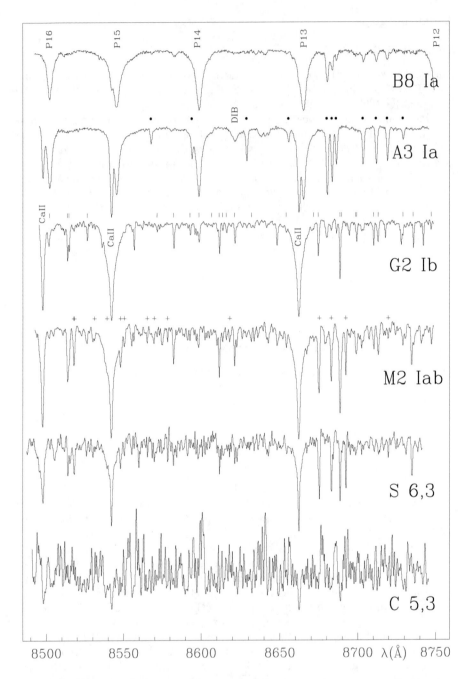

Figure 4. Spectra of super-giants over the GAIA wavelength region. Paschen and CaII lines are indicated, as well as the diffuse interstellar band at 8620 Å. Dots mark NI, dashes FeI and + TiI strongest lines.

3.3. Stellar rotation

The growing importance of rotation in stellar astrophysics is best sensed by the Maeder and Meynet (2000) statement that "stellar evolution is basically a function of mass M, metallicity Z and angular velocity Ω". Rotation has sensible effects on the actual stellar tracks on the HR diagram, lifetime, surface abundances, chemical yields, atmospheric modeling and interplay with magnetic fields and winds.

The distribution in terms of $V_{rot} \sin i$ of the $\sim$12 000 stars in the recent Glebocki et al. (2000) catalog of stellar rotations is presented in Figure 5. The break in the distribution of velocities at late F/early G stars is evident. These and lower mass stars possess a convective envelope that interacting with (differential) rotation sustain a magnetic field which, by gripping the ionized stellar wind, forces loss of angular momentum from the star. The faster the rotation, the stronger the generated field, and consequently a more effective brake and a more active stellar surface. Observations of open clusters suggest that relaxation toward settled rotation for solar-type stars has a typical time scale of 1 billion year. Such a condition has not yet been reached by several of the field M dwarfs in the Glebocki et al. catalog that still rotate appreciably fast (cf. Figure 5), consistent with the time-scale for spin-down increasing with decreasing stellar mass. In is interesting to note that brown dwarfs so far measured all rotates as the fastest rotators of mass $\sim$1 $M_\odot$. In massive O, B, and A stars the fast rotation is connected to such phenomena as the Be stars, the He and N enrichment seen in the spectra of O stars, the decrease in T_{eff} and $\log g$ at equatorial latitudes, or the abundance anomalies in A and early F stars such as the Ap, Am and HgMn stars (all invariably slow rotators).

There are basically two ways to measure the rotation of a star: line broadening and spots transit, both applicable to GAIA.

The accuracy of $V_{rot} \sin i$ from cross-correlation measurements of line broadening on GAIA spectra of non-binary F-G stars as function of resolution, S/N and mismatch between object and template is discussed by A.Gomboc (pag. 285). She found promising results even for solar type stars, the most difficult to measure given the minimal spread in velocity. R=12 000 appears to be a dividing line between accurate and coarse measurements of the population of slower rotators: at R=9 000 the whole range of G stars velocities is mapped into only 2 or 3 bins, while at R=15 000 details of the velocity distribution are revealed.

The periodic transit of inhomogeneities (spots, active areas) on the stellar surface traces the rotation. Particularly extended, long living, isolated and large ΔT_{eff} areas can be detected by epoch GAIA photometry (most effectively by the intermediate bands); however, the small number of observations per star, can confuse interpretation and fail to solve the degeneracy with other types of variability not related to rotation, particularly if more than one single large spot is present on the star at the same time. Such problems are quite alleviated by GAIA spectra, that can trace several different spots at the same time as they cross the projected stellar surface. Emission lines associated with individual active areas split according to projected velocity (cf. BY Dra spectra at pag. 233) and allow accurate rotation period to be measured even with a relatively small number of spectra (of the order of a few tens), in a way similar to the case of period determination in spectroscopic binaries (cf. pag. 332).

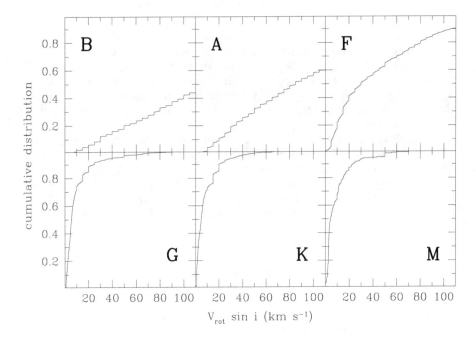

Figure 5. Cumulative distribution of observed $V_{rot}\sin i$ in the catalog of Glebocki et al. (2000).

Among the countless astrophysical applications of the measurement of rotational velocities, we will conclude this section with a couple of examples, to which the GAIA data will provide an obvious fundamental contribution.

The reasonable assumption that rotational axes are roughly aligned with orbital axis in eclipsing binaries allows observations to provide V_{rot} directly. Together with the stellar radii derived from the orbital solution, this gives the spin periods of both components. They can be compared with the orbital period for synchronization, to which the system is driven by tidal energy dissipation. Departure from synchronization points to internal stellar structure, age, formation and evolution history of the binary. For example, *all* symbiotic binaries in the Halo harbor as donor star an F-G giant with [Fe/H]$\simeq$0, [s-elements/Fe]$\sim$1 and $V_{rot}\sin i$ $\sim$50-120 km s^{-1}, which suggested that the progenitor of the now F-G giant was a low mass star spun up by accretion from the AGB precursor of the current white dwarf companion (Smith et al. 2001).

The distribution of $V_{rot}\sin i$ with galactic latitude [b] for a given class of objects relates to the assumption about a random orientation of axes. The similar distribution of rotation in open clusters at high and low b has suggested that stellar rotation does not have its origin linked to the galactic rotation. Recent analysis by De Medeiros et al. (2000) on evolved F-G-K stars suggests a dependence of the mean $V_{rot}\sin i$ with b, in the sense of higher rotations concentrating at lower b, and this on both side of the rotational discontinuity. De Medeiros et al. however note that the correlation does not follow the $\cos b$ relation expected for a perfect axial alignment between stellar and galactic rotation, and other interpretations have to be looked for, like the variation in age, mass and chemical abundance with b.

3.4. Mass loss, active chromospheres and peculiarities

Response of GAIA spectroscopy to peculiarities is very sensitive with CaII and hydrogen lines to be the first to turn into emission in almost every corner of the HR diagram, and displaying P-Cyg profiles even for modest mass loss rates. CaII triplet lines are excellent indicators of chromospheric activity, as much as the celebrated H and K lines of CaII at blue wavelengths. A detailed description in provided at pag. 227.

Detection of peculiarities in GAIA spectra can be used to flag photometric data as "handle with care" cases, de-routing their analysis from the main branch of the reduction pipeline. It can be used also to flag astrometric data, that in objects with for ex. emission lines could mean a moving stellar photo-center that corrupts the accuracy/interpretation of astrometric observations.

3.5. Interstellar medium

In the GAIA wavelength range there are no suitable atomic interstellar absorption lines (for ex. resonant line CsI 8521.10 Å goes undetected in high resolution, high S/N spectra of even the most reddened stars), and weak C_2 lines at 8754, 8764 and 8773 Å are just longward of the red edge of the interval (in any case, at the 11 500 R.P. baselined for GAIA spectrograph, they would be washed in the adjacent continuum).

At least one diffuse interstellar band (DIB) is found within the GAIA wavelength range, at 8620 Å (Jenniskens Désert 1994). The first investigation devoted to this band has been published by Munari (1999) who found a remarkable linear correlation of the equivalent width and reddening (E_{B-V}). The relation has been expanded and confirmed by observations of an enlarged sample of stars by Munari (2000), taking the form $E_{B-V} = 2.69 \times EW(Å)$. It is possible that the slope of the relation depends on galactic 3D coordinates [$E_{B-V} = \alpha(l, b, D) \times EW(Å)$], following the varying local properties of the crossed interstellar medium. Observations to check for this are in progress. Variation with galactic longitude have been already found in the past, for ex. by Herbig (1975 and references therein) that reported DIB 4428 Å to be in Cygnus, for the same E_{B-V}, only 70% of the intensity elsewhere.

GAIA will observe a huge number of hot stars scattered through the whole Galaxy, fixing accurate distances to them. The DIB 8620 Å will be unaffected by stellar absorption lines for all spectral types earlier than $\sim$A3. All those brighter than V=13 mag will deliver mission averaged spectra with $S/N \geq 100$, thus providing opportunity for accurate measurements of the DIB equivalent width (as well as shape, wavelength and central depth). In such a way a detailed 3D map of the DIB 8620 Å intensity through the Galaxy will be obtained, for which important applications could be (a) distance estimate for those stars for which GAIA could not, for whatever reason, lock on the parallax, (b) check the reddening derived from GAIA photometry on problematic objects, and (c) comparison with the 3D map of the properties of dust extinction in the attempt to constrain the nature of the carrier of the DIB 8620 Å absorption, as a possible clue to the identification of the carriers of the other DIBs as well (in spite of massive efforts since discovery in 1922, the origin and carrier(s) of DIBs are still unknown, marking one of the longest standing unsolved problems of astronomical spectroscopy).

A couple of other DIBs have been reported in literature over the GAIA wavelength range, at ∼8648 and ∼8530 Å. Herbig (1995) found uncertain the first and probably spurious the second. The atlas by Jenniskens & Désert (1994) offers only a minimal support to their reality, while Galazutdinov et al. (2000) atlas suggest how DIB 8530 Å could actually be HeI stellar absorption. The synthetic spectra by Castelli and Munari (2001) suggests that both ∼8648 and ∼8530 Å features are quite probably HeI stellar absorption lines.

References

Hargreaves, J.C., Gilmore, G. & Annan, J.D. 1996, MNRAS 279, 108

Boschi, F. et al. 2003, in Symbiotic Stars Probing Stellar Evolution, R.L.M Corradi et al. eds., ASP Conf. Ser., in press (astro-ph/0208319)

Bradley, P.A. 2001, in Enciclopedia of Astronomy and Astrophysics, P.Murdin ed., Nature Publ. Group, pag. 3127

Castelli, F. & Munari, U. 2001, A&A 366, 1003

De Madeiros, J.R., Carvalho, J.C., Soares, B.B. et al. 2000, A&A 358, 113

Dravins, D., Lindegren, L. & Madsen, S. 1999, A&A 348, 1040

Galazutdinov, G.A. et al. 2000, PASP 112, 648

Glebocki, R., Gnacinski, P. & Stawikowski, A. 2000, Acta Astron. 50, 509

Herbig, G.H. 1975, ApJ 196, 129

Herbig, G.H. 1995, ARA&A 33, 19

Jenniskens, P. & Désert, F.-X. 1994, A&AS 106, 39

Katz, D. et al. 1998, A&A 338, 151

Katz, D. et al. 2002, in GAIA: A European Project, O.Bienayme GAIA: A European Project, C.Turon ed.s, EAS Publ. Ser. 2, pag. 63

Maeder, A. & Meynet, G. 2000, ARA&A 38, 143

Munari, U. 1999, Baltic Astron. 8, 73

Munari, U. 2000, in Molecules in Space and in the Laboratory, I.Porceddu and S.Aiello ed.s, Soc. It. Fis. 67, pag. 179 (also astro-ph/0010271)

Munari, U. 2002, in GAIA: A European Project, O.Bienayme GAIA: A European Project, C.Turon ed.s, EAS Publ. Ser. 2, pag. 39

Munari, U. & Tomasella, L. 1998, UM-PWG-005 (Livelink)

Munari, U. & Tomasella, L. 1999, A&AS 137, 521

Munari, U. & Castelli, F. 2000, A&AS 141, 141

Munari, U., Agnolin, P., Tomasella, L. 2001a, Baltic Astron. 10, 613

Munari, U. et al. 2001b, A&A 378, 477

Perryman, M.A.C. et al. 2001, A&A 369, 339

Smith, V.V., Pereira, C.B. and Cunha, K. 2001, ApJL 556, L55

Zwitter, T., 2002, A&A 386, 748

session 2
TECHNOLOGY
chair: O. Pace

The *Residenza del Sole* Congress Center and, above-right, the isolated *Cima Regina* peak (2356 m), target of Sunday 8 excursion

Participants during a Conference session (with M.Cropper and his inseparable laptop at centre)

Design and performances of the GAIA spectrograph

David Katz

Observatoire de Paris, GEPI, 5 place Jules Janssen, 92195 Meudon

Abstract. The Radial Velocity Spectrometer (RVS) is an integral field spectrograph, observing in scan mode a field of view of 2 by 1.6 square degrees. Its spectral interval, 8480–8740 Å, is dominated by the Ca II triplet in late type stars and by the head of the Paschen series in the early type ones. The resolution of the instrument is in definition phase. Configurations with R = 5 000 to 20 000 are currently compared. The resolution, as the other RVS open characteristics, should be frozen by December 2002. The precision of RVS for a solar type star, estimated by Monte-Carlo simulations (assuming R = 10 000), is 1 km s^{-1} per transit at V=12 and 10 km s^{-1} at the end of the mission at V=17.5 mag.

1. Introduction

The idea of a spectrometer on-board GAIA was proposed in 1997, based on earlier evaluations (Perryman 1994; Favata & Perryman 1995). Originally named ARVI for *Auxiliary Radial Velocity Instrument*, its main goal was to complement the astrometric measurements with radial velocities (Favata 1997). At the time GAIA was accepted in October 2000, ARVI was an integral field spectrograph with a 2 by 1 square degrees field of view (FoV), an effective resolution $R \simeq 5750$ and a spectral range of 250 Å located around the ionised Calcium infra-red triplet: 8480–8740 Å (this spectral domain was originally proposed by U. Munari, and noted independently by R. Le Poole). This configuration of ARVI is described in detail in the *GAIA Concept and Technology Study Report* (ESA-2000-SCI-4).

In June 2001, the European Space Agency (ESA) appointed 16 international scientific working groups to cover the various aspects of the preparation and optimisation of the satellite. One of those groups was devoted to the spectrograph, which, since GAIA approval, had changed name to become the *Radial Velocity Spectrometer* (often referred to as RVS). The RVS working group task was to review the spectrograph scientific objectives, priorities and specifications, to assess its performances and to optimise its characteristics (resolution, number and size of the CCD and pixels, etc). The objective was and is still to converge on the RVS characteristics by the end of 2002.

This article presents the status of the RVS instrument after one year of studies by the RVS working group members. Sect. 2 gives a brief overview of the RVS scientific goals (Munari 2003 gives a more detailed description). Sect. 3 describes the RVS design, and sect. 4 discusses the main RVS open characteristics. Sect. 5 gives a summary of the RVS radial velocity (RV) performances (see Munari et al. 2003 for an extensive discussion). The potential systematic errors which could affect the RV accuracies are examined in David (2001, 2002a, 2002b). Many other facets of the RVS performances are presented in these Proceedings. The RVS precision in the determination of the atmospheric parameters using a *minimum distance method* has been assessed by Soubiran 2001.

2. RVS scientific goals

The primary goal of the RVS instrument is the acquisition of radial velocities. The motivation for it is threefold:
(*i*) to correct the astrometric data from the perspective acceleration effect. The motion of a star along the line of sight induces an additional apparent time quadratic displacement on the sky, i.e. the perspective acceleration, proportional to its parallaxe, *true* proper motion and radial velocity. This effect should be corrected to avoid both bias in proper motions and false detections of binary systems. Simulations performed by Arenou & Haywood (1998), have shown that about 100 000 GAIA targets, mostly *bright* stars ($\simeq$98% are brighter than $V = 17$ mag) located in the solar neighbourhood, were concerned. Of course, a complete sky survey is needed, because the affected stars are not known *a priori*;
(*ii*) the astrometric instrument only supplies 2 components of the velocity vector. The third component, the radial velocity, is essential to study the Milky-Way kinematics and dynamics. A key objective of the RVS is to complement the positions and proper motions, in order to decipher the properties, origin and formation mechanism of the Galactic Disk and Halo (see Wilkinson 2002 for a review of the Galactic structure case in the RVS context). In particular the GAIA observations should allow to test the hierarchical formation paradigm and recover a large fraction of the Halo accretion events, if indeed it is the way the Halo has assembled (Helmi & de Zeeuw 2000). The Bulge displays a lower priority level for the RVS, for two reasons. First, the very high stellar density of the Bulge will produce a correspondingly very high level of crowding in the RVS field of view, which will reduce the effective limiting magnitude of the instrument. Second, the Bulge covers a relatively modest surface on the sky and it is most likely that the coming generations of multi-targets ground spectrographs will observe a significant part of it by 2015-17 (at the time the GAIA catalogue will be released);
(*iii*) each star will be observed 100 times on average with the spectrograph. This information will lead to the detection and characterisation of a huge amount of multiple systems: stars and brown dwarfs.

The spectroscopic information will also be used to complement the astrometric and photometric (broad and medium band filters) observations to classify (as star or non star, peculiar or non peculiar objects, etc) and to parameterise (effective temperature, surface gravity, iron or average metal content) the GAIA targets (Bailer-Jones 2003).

The RVS spectral domain, 8480–8740 Å, contains a large number of lines of different elements. According to the resolution that will be adopted next December (the resolution is at the moment an open issue, see Sect. 3.1 and 5), individual stellar abundances of, e.g. Ca, Mg, Ti, Si, N, could be derived from RVS spectra. The RVS wavelength range also contains a *Diffuse Interstellar*

Band (DIB) at 8620 Å. Unlike most of the known DIB, this feature seems to be a reliable tracer of interstellar reddening (Munari 2002), and therefore could be used to derive a 3 dimensional reddening map of the Galaxy.

The RVS spectra will also provide relevant diagnostics (once again depending upon the chosen resolution) for the study of peculiar and variable stars: e.g. line shift, line profile variation. Many future uses of the RVS data to study pulsating stars are presented in these Proceedings.

3. RVS design

3.1. RVS concept

Before going into the details of the Radial Velocity Spectrometer design, let us consider its general concept. The RVS is a 2 by 1.6 square degrees FoV integral field spectrograph, i.e. dispersing the light of the whole field of view. As the other instruments, it operates in scan mode, observing the sources 100 times on average over the 5 years of the mission (the exact number of observations is function of the target ecliptic coordinates).

The RVS wavelength range, 8480–8740 Å, was selected in the near infrared in order to be close to the domain where the spectral energy distribution of the spectrograph principal targets (G and K type stars) picks. In F, G and K type stars this interval displays 3 strong Calcium lines, which allow to derive radial velocities at very low signal to noise ratios. The 8480–8740 Å region is almost free of telluric lines, so that ground-based follow-up in the RVS wavelength range are possible, if needed.

The resolution of the instrument is not frozen. Configurations with resolutions ranging from $R = \lambda/\Delta\lambda = 5\,000$ to $20\,000$ have been compared to establish their merits and drawbacks. The resolution should be chosen by December 2002. This issue is discussed in more detail in Sect. 5.

3.2. RVS optics

The light is collected and reflected toward the spectrograph by a system of three off-axis rectangular mirrors, physically located under the optical bench. The three mirrors system corresponds to an equivalent entrance pupil of 0.5×0.5 m^2. The spectrograph itself is classically made of a collimator (in order that the light rays enter the dispersive element with the same incidence angle), a dispersive element (i.e. a grism) and an imager (to reconstruct the image on the detectors). The focal length of the spectrograph is 2.1 m.

The spectral dispersion orientation is parallel to the scan direction. This configuration presents two advantages with respect to a dispersion perpendicular to the scan direction. First, the spectrum of a star located close to the top or bottom edge of the field of view do not extend outside the detectors boundaries. The second motivation is linked to the pixels dimensions. The pixels are asymmetric: narrower in the along scan direction than in the across scan direction. Therefore, for the same effective resolution (assuming a sampling of 2 pixels per resolution element), the along scan dispersion requires a smaller grism dispersive power than the across scan.

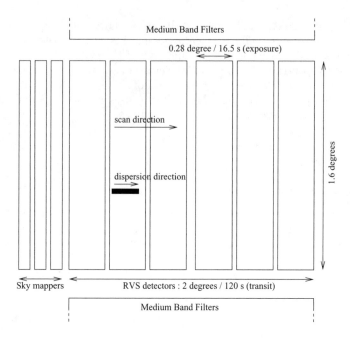

Figure 1. RVS CCD layout.

3.3. RVS focal plane and detectors

The acquisition of the spectra is performed in two steps and each one is associated with a dedicated set of CCD:
(i) unlike the Hipparcos satellite, GAIA has no input catalogue. Therefore the first step of the observation is to detect the sources that will enter the FoV. In the RVS, this task is performed by the 3 CCDs named *sky mappers*. The first one is used to detect the potential targets and the third one to confirm the astrophysical interest of the detection (i.e. to reject cosmic events, false detections, etc). The second CCD is redundant. It is there to replace one of the other two CCDs in case of malfunction. The sky mappers are physically located in the medium band photometer focal plane and are illuminated by undispersed light. Dimension and characteristics of the sky mappers are summarised in Table 1;
(ii) once the sources are detected, the dispersed light is recorded by 6 CCD detectors. The field of view scanned by the detectors is located between the two photometric fields (while not in the same physical plane). The detectors (as the sky mappers) are operating in *time delay integration* (TDI) mode, i.e. the signal is transfered from pixel to pixel, at the same speed the stars are crossing the field of view. The last column is read at each TDI clock. Contrary to the other instruments, where only a constant number of patches are read, in the RVS the whole CCD are read. The dimension and characteristics of the detectors are summarised in Table 1.

Both the sky mappers and detectors are red-enhanced CCDs, with a quantum efficiency greater than 80% over the RVS wavelength range. The overall

Table 1. Dimension and characteristics of the sky mappers and detectors. The detectors pixel width is given for two RVS resolutions: R = 5 000 and 20 000.

	Sky mapper (× 3)	Detector (× 6)
Dimension (mm)	3.4 × 60	10.1 × 60
Dimension (degree)	0.09 × 1.6	0.28 × 1.6
Dimension (pixel)	336 × 3930	1010 × 3930
Exposure time (s)	5.5	16.5
Pixel size (μm)	10 × 15	10 × 15
Pixel size (arcsec)	1 × 1.5	1 × 1.5
Pixel width (Å)	--	0.85 to 0.21
Pixel width (km s^{-1})	--	30 to 7.5
Read-out noise (e$^-$)	TBD	4 (TBC)

efficiency of the RVS instrument, including the optical transmission, is estimated at 35%. The RVS CCD layout is represented on Figure 1.

3.4. RVS spectra

In F, G and K type stars, the strongest feature in the RVS wavelength range is the ionised Calcium triplet (8498.0, 8542.1 and 8662.1 Å). Several other *weak* lines of large astrophysical interest are present, e.g. Si and Mg. In hotter stars, the spectra are dominated by the head of the Paschen series. Figure 2 shows as an example the synthetic spectrum of a solar type stars, in the RVS wavelength range, computed using a Kurucz atmospheric model (Kurucz 1993), the VALD atomic data (Piskunov et al. 1995; Kupka et al. 2000) and the Piskunov's SYNTH program. The RVS is an integral field spectrograph, which disperses all the light entering the field of view and in particular the *background light*, whose main component is the zodiacal light. Nevertheless, the background light (provided it displays a relatively uniform spatial intensity) do not contaminate the stellar signal with any spectral signature but only add white noise. The reason for this is that the background spectral features are averaged in the spatial dimension, i.e. each adjacent sky point source generates the same spectrum, each spectrum being slightly shifted with respect to the CCD pixels. The zodiacal light surface brightness varies from about $V = 21.5$ mag arcsec^{-2} on the ecliptic plane to $V = 23$ mag arcsec^{-2} at 45 degrees ecliptic latitude (Zwitter 2002). At intermediate ecliptic latitudes and assuming a spectrum of 1225 by 3 pixels (R = 10 000 and maximum transverse motion without tilt mechanism), the zodiacal light integrated over the whole spectrum is equivalent to a $V = 14.5$ mag line-free star.

Another consequence of the absence of slit or fibre is that the spectra of neighbouring sources will overlap. Viala et al. (2003) using the GSC2.2 have estimated the RVS field of view *filling* rate as a function of the Galactic coordinates. Figure 3 shows the Galactic directions where the CCD will be full

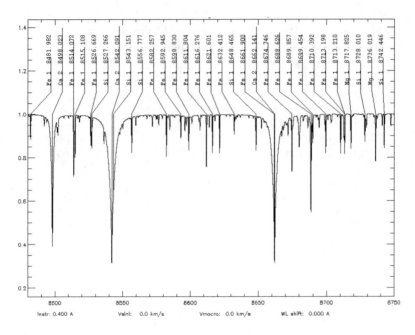

Figure 2. Synthetic spectra of parameters $T_{\text{eff}} = 5750$ K, $\log g = 4.5$ and [Fe/H]=0, in the RVS wavelength range.

of stars of magnitude V = 17 or brighter, i.e. the total area covered by the spectra equals or exceeds the CCD surface. Those areas are mostly located around the Galactic plane. The minimum stellar density to fill the CCD is function of the resolution and spectra width. Assuming 2 pixels wide spectra, the CCD will be filled starting at 14 500 (R = 5 000), 7 200 (R = 10 000) and 3 600 stars deg^{-2} (R = 20 000). Looking at the Milky-Way, the CCD will be *full* over 35% (R = 5 000), 75% (R = 10 000) and 95% (R = 20 000) of the sky for $-10° \leq b \leq +10°$.

Once the *crowded areas* are identified, a second question arises: what is the impact of the crowding on the RVS performances ? In an *exactly full* CCD each star has 87% probability to be contaminated significantly (more than 25%) by one or more stars: one star: 27%, two stars: 27%, three stars: 18% and four stars or more: 15%. Zwitter (2003) has studied the issue of the recovery of the information in dense field and has derived RVS radial velocity precisions as a function of magnitude and stellar density. Except at relatively high densities, the crowding has a modest impact on the performances: at R = 8 500 the radial velocity precision for a V=17 mag K1V star is degraded by a factor of two between 0 and 40 000 stars deg^{-2}, i.e. 6.5 to 13 km s^{-1} (at the end of the mission).

3.5. On-board data processing

The RVS on-board processing effort is mostly devoted to the compression of the data, in order to fit in the 0.25 Mbit s^{-1} telemetry allocated to the spectroscopic

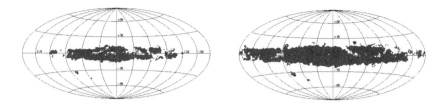

Figure 3. Galactic directions where the CCD will be *full* of stars of magnitude $V = 17$ or brighter, for resolutions $R = 5\,000$ (*left*) and $20\,000$ (*right*).

instrument. The first processing stage will be to select and extract, using the information provided by the sky mappers, the pixels which contain relevant astrophysical data, i.e. to discard the background pixels. In a second stage, the spectra will be *compressed*. Many processes have been proposed and their merits and drawbacks currently assessed: (*i*) sum the successive 6 observations to download a single spectrum per object per transit, (*ii*) sum the pixels in the across dispersion direction keeping a single row per spectrum, (*iii*) extract numerically the three Ca II lines in faint targets (idea proposed by U. Munari), (*iv*) convolve and resample the faint stars (M. Cropper) or (*v*) sum spectra of the same object on successive transits (M. Cropper).

4. RVS open issues

The main RVS open characteristics are the resolution and the presence or absence of a tilt mechanism to rotate the CCD plane. While the resolution will play a critical role on the scientific case of the RVS, the tilt mechanism will have a strong impact on the CCD assembly. The problematics of those two issues are presented below. In both cases, studies are in progress to assess the advantages and disadvantages of the different configurations. The RVS characteristics will be frozen by December 2002.

4.1. Resolution

Because it impacts on most of the RVS aspects, the resolution has been the driver of a very large fraction of the studies performed by the RVS working group during the last year. Configurations with resolutions ranging from $R = 5\,000$ to $20\,000$ have been considered. The advantages of the low resolution are that: (*i*) it allows to push the *limiting* magnitude toward slightly fainter stars, i.e. $\simeq$ 0.25 to 0.5 mag fainter at $R = 5\,000$ than at $R = 20\,000$ (see Sect. 5), (*ii*) the risk of small systematic perturbations *corrupting* the spectra of the faintest targets, because of the very small number of photons collected per spectrum and per transit, is less pronounced, (*iii*) at equivalent stellar density, the CCD will be less crowded and (*iv*) it requires a lower telemetry rate to transmit the data to the earth. The merit of the high resolution is that it carries more spectral information than the low one. This allows to determine more accurately the radial velocities of *bright* to *moderately faint* stars and to determine a broader

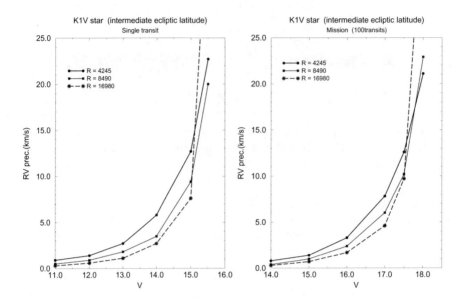

Figure 4. K1V star, single transit (*left*) and mission average (*right*), RV precisions as a function of magnitude and resolution.

range of astrophysical parameters:, e.g. individual element abundances (e.g. Si, Mg), rotational velocities, interstellar redenning via the 8620 Å DIB, line profile variation. The characteristics and performances of the different resolutions are currently compared to the RVS scientific case, in order to choose the configuration showing the best adequation.

4.2. Tilt mechanism

The motion of the GAIA satellite, is the composition of a 6 hours rotation (the sun aspect angle, i.e. the angle between the spin axis and the sun-L2 direction, is 50 degrees), a 70 days precession motion of the rotation axis around the sun-L2 direction and an annual revolution motion around the Sun. The projection of the precession motion on the CCD plane, induces a 6 hours period sine transverse motion of the stars, and therefore of the spectra, with respect to the CCD detectors. The maximum induced transverse velocity is 0.17 arcsec/s. In the 120 s of RVS focal plane crossing, a spectrum would be shifted by up to 20 arcsec ($\simeq$ 14 pixels). Such a blurring of the spectra would increase awfully the background contamination, the read-out noise and the crowding rate. The baseline solution to avoid to spread the spectra over a very large number of pixels, is to split the focal plane crossing in 6 intervals (i.e. 6 CCD detectors). The maximum transverse displacement per CCD is 2.8 arcsec or about 2 pixels. An alternative to the large number of CCD would be to implement a tilt mechanism, that would rotate the CCD detectors, in order that the spectra motion be always parallel to the CCD lines. With respect to the baseline option, this configuration presents the advantage to completely cancel the transverse motion and therefore

Table 2. RV precisions (km/s) as a function of magnitude, obtained for R = 10 000 and K1V, G6I and A8V type stars, for single transits (S) and over the mission (M).

	K1V		G6I		A8V	
	σV_r (S)	σV_r (M)	σV_r (S)	σV_r (M)	σV_r (S)	σV_r (M)
13	2	< 1	1	< 1	5	–
14	4	< 1	3	< 1	14	3
15	10	1	6	< 1	> 40	4
16	> 40	2	27	1	> 40	6
17	> 40	6	> 40	4	> 40	34
18	> 40	23	> 40	12	> 40	> 40

to reduce the contamination by the zodiacal light and read-out noise and to lower the occurrence of spectra overlapping. In the tilt mechanism configuration, the number of CCD detectors could be reduced to 3, with the merit of further decreasing the read-out noise per transit. The drawback of the tilt mechanism is that, as a moving part which should operate continuously during 5 years, it presents a failure risk. This risk is currently quantified and the result will drive the decision of implementing or rejecting the mechanism.

5. RVS radial velocity performances

Several studies have been devoted to the estimation of the RVS RV performances: Katz (2000), Munari et al. (2001), Katz et al. (2002), Munari (2002), Zwitter (2002). They all rely on Monte-Carlo simulations, cross-correlating many times an *RVS-like* object spectrum with a template spectrum. The RV precisions presented here are the synthesis of the above simulations (prepared in collaboration with T. Zwitter and U. Munari), re-run with the new design characteristics. Figure 4 shows the RV precisions as a function of magnitude and resolution derived for a K1V type star. Two regimes can be distinguished. For the *bright* to *moderately faint* objects, high resolution leads to more accurate radial velocities. For the faint stars, *moderate* to *low* resolutions allow to go slightly fainter in magnitude (0.25 to 0.5 mag. fainter) for RV precisions of 15 to 20 km/s. Table 2 summarises the RV precisions as a function of magnitude derived, assuming R = 10 000, for the spectral types K1V, G6I and A8V.

Acknowledgments. I would like to thank all the Radial Velocity Spectrometer Working Group members and the ASTRIUM and ESA RVS team, whose dedicated work and implication in the preparation of the GAIA spectrograph has defined the RVS instrument as presented in this article. I am very grateful to R. Kurucz and the VALD people for making their software packages and atomic and molecular data available to the community.

References

Arenou, F. & Haywood, M. 1998, SWG-OPM-001 (Livelink)
Bailer-Jones, C.A.L. 2003, in GAIA Spectroscopy, Science and Technology, U.Munari ed., ASP Conf. Ser. 298, pag. 199
David, M. 2001, RVS-MD-001 (Livelink)
David, M. 2002a, RVS-MD-002 (Livelink)
David, M. 2002b, RVS-MD-003 (Livelink)
Favata, F. 1997, FF-PWG-002 (Livelink)
Favata, F., & Perryman, M.A.C. 1995, in Future Possibilities for Astrometry in Space, M.A.C. Perryman & F. van Leeuwen ed.s, ESA SP-379, pag. 153
Helmi, A. & de Zeeuw, T. 2000, MNRAS 319, 657
Katz, D. 2000, PhD Thesis, Univ. Paris 7
Katz, D., Viala, Y., Gomez, A. & Morin, D. 2002, in GAIA: A European Space Project, O. Bienaymé and C. Turon ed.s, EAS Pub. Series 2, pag. 63
Kupka, F.G., Ryabchikova, T.A., Piskunov, N.E., Stempels, H.C., & Weiss, W.W. 2000, Baltic Astron. 9, 590
Kurucz, R. 1993, Kurucz' CD-ROM No. 13, CfA Harvard, Cambridge
Munari, U. 2002, in GAIA: A European Space Project, O. Bienaymé and C. Turon ed.s, EAS Pub. Series 2, pag. 39
Munari, U. 2003, in GAIA Spectroscopy, Science and Technology, U.Munari ed., ASP Conf. Ser. 298, pag. 51
Munari, U., Agnolin, P., & Tomasella, L. 2001, Baltic Astron. 10, 613
Munari, U., Zwitter, T., Katz, D., & Cropper M. 2003, in GAIA Spectroscopy, Science and Technology, U.Munari ed., ASP Conf. Ser. 298, pag. 275
Perryman, M. 1994, in Astronomical and Astrophysical Objectives of Sub- Milliarcsecond Optical Astrometry, IAU Symp. 166, E. Høg & P.K. Seidelmann ed.s, Kluwer, pag. 211
Piskunov, N.E., Kupka, F., Ryabchikova, T.A., Weiss, W.W., & Jeffery, C.S. 1995, A&AS 112, 525
Soubiran, C. 2001, RVS-CS-001 (Livelink)
Viala, Y.P., Morin, D., Katz, D., & Ochsenbein, F. 2003, in GAIA Spectroscopy, Science and Technology, U.Munari ed., ASP Conf. Ser. 298, pag. 105
Wilkinson, M. 2002, RVS-MW-001 (Livelink)
Zwitter, T. 2002, A&A 386, 748
Zwitter, T. 2003, in GAIA Spectroscopy, Science and Technology, U.Munari ed., ASP Conf. Ser. 298, pag. 493

Technical issues for GAIA RVS

Mark Cropper

Mullard Space Science Laboratory, University College London, Holmbury St Mary, Dorking, Surrey RH5 6NT, United Kingdom

Abstract. I discuss some of the technical issues facing the RVS on GAIA at the current stage of development. These include data handling and compression (which remains problematic), the inclusion of a scan mechanism, the performance of the detectors at the low exposure levels for most sources, the need for cold electronics within the payload bay and the possible usefulness of L3CCDs.

1. Introduction

The design of RVS is proceeding through a number of studies (see Perryman 2003 and Pace 2003 for an overview of the GAIA design and programmatics). These elucidate a sequence of technical issues, some of which are solved or ameliorated as they are better understood, while others appear to loom larger at any particular instant. This work looks at some of the main issues considered problematic at this phase of the RVS development (autumn 2002). Inevitably, in the future other issues will arise, so this is by nature a snapshot.

I have grouped the discussion in several areas. The bulk of the work concerns the constraints on transmitting the data from RVS to the ground imposed by the current telemetry budget. I discuss windowing/data selection schemes and compression schemes. This has implications on the system design and the requirements on various electronic processing units and on-board mass storage. The remainder of the discussion touches on issues of image confusion (and the need for a scanning mechanism), on the impact of the cold payload environment on the electronics close to the detectors and on CCD performance at low fluxes and the effect of CCD cosmetics. I also suggest that the use of L3CCDs may be beneficial in RVS.

2. Simulated images

At MSSL we have developed an RVS simulator, which generates image data according to RVS parameters, including the RVS (square) point-spread-function and simple spatial broadening from the GAIA scanning law. We use this to obtain impressions of the RVS data, particularly (so far) for testing different data compression schemes – the telemetry bandwidth limitation has serious implications for RVS.

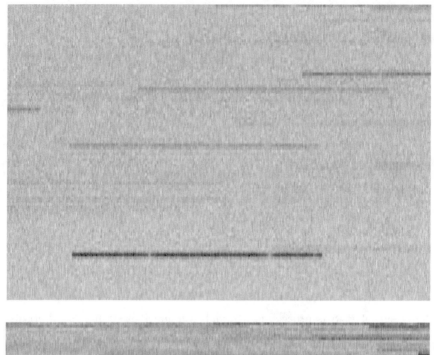
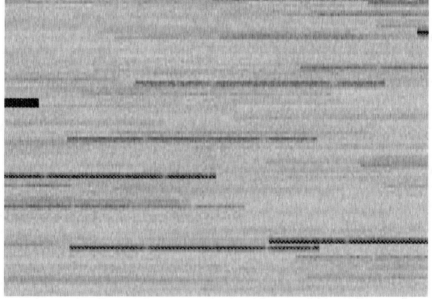

Figure 1. Simulated image data from a combined 99 second scan over 6 RVS CCDs for a high Galactic latitude field (30 − 90°, *top*) and a low Galactic latitude field (5 − 10°, *bottom*). Star densities are taken from ESA-2000-SCI-4.

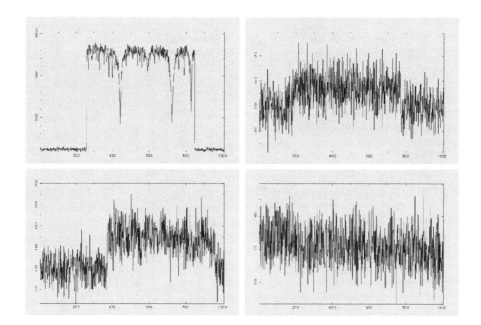

Figure 2. Sample extracted spectra of F stars from the images in Figure 1. Only a single row has been extracted in each case. Anticlockwise from top left the V magnitudes are 10.0, 13.7, 15.0 and 16.4.

Examples of the images created are given in Figure 1. The upper figure shows a high Galactic latitude field, while the lower a low Galactic latitude field. The higher background from faint stars is evident in the low latitude field: indeed, only stars with $V \sim 15$ or brighter are directly visible in these images. Most of the background contribution in the high latitude field is from readout noise, assumed to be $4e^-$.

Figure 2 shows examples of F stars extracted from the images in Figure 1. It is evident that spectra from a star of $V = 10$ mag have excellent signal-to-noise ratios, while those from stars of $V = 16.4$ cannot be discerned in the noise from these single scans. Of course, radial velocities will be extracted from the sum of ~ 100 such scans over the course of the mission.

3. Data handling

3.1. The scale of the problem

The telemetry bandwidth assigned to RVS is a significant constraint for achieving its scientific goals. Most of the stars seen in the RVS will be at the faint limit, where the signal is deeply buried in the noise (Figure 2). Noise is intrinsically incompressible.

The simplest strategy would be to send down full CCD data streams. A CCD format is 1010 by 3930 pixels with 2 bytes pix^{-1} if the signal is digitised

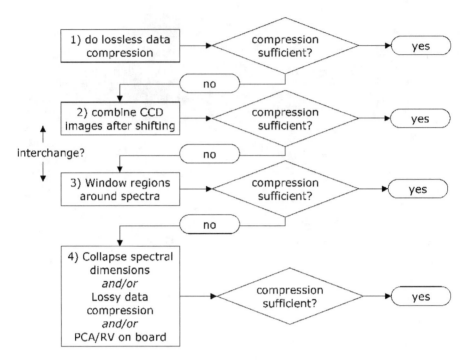

Figure 3. Flowchart indicating priorities for RVS data handling to minimise data loss.

to between 8 and 16 bits. For a CCD transit time of 16.5 sec this results in 18.8 Mbit sec^{-1} for the 6 CCDs in the RVS focal plane. The telemetry allocation is 250 Kbit sec^{-1} so the ratio of resource to allocation is ~ 75.

3.2. Options

The first measure to adopt is to compress the data using a lossless algorithm (see flowchart in Figure 3) . Depending on the nature of the data, a factor of 1.5–3 can be achieved (see Cropper, Smith & Brindle 2002). A second measure is to coadd the 6 CCDs from a single scan. These two measures provide a factor of ~ 12, so that the new ratio of resource to allocation is ~ 6.

Further options include collapsing the spatial information for each source into a 1D spectrum. However, in areas of moderate to high source density, the efficacy of this strategy saturates, once all parts of the CCD have sources brighter than the limiting magnitude. Since Galactic plane information is essential for main GAIA science programmes, it is not scientifically feasible simply to ignore regions of high star density. In addition, the collapse to 1D spectra causes confusion from spectral overlap to become a more serious problem than it is already.

In regions of lower source density, such a strategy may provide a sufficient reduction in the data rate to be useful: then the size of the telemetry buffer becomes important in determining whether data from denser regions can be stored for later telemetry. A second ground station would help significantly here, providing a potential 2-3 times improvement. Of course this has mission cost implications, and it is probably unwise to depend on this as anything other than a backstop at this stage of the programme.

If these options prove to be insufficient, the further options (lossy compression, or data extraction on board) are much less palatable in terms of data loss.

3.3. Lossy compression

Lossy compression can provide significant compression factors. In addition to the likely compression gains, an advantage of lossy compression is that, unlike schemes in which the data are collapsed spatially, the full image is available for analysis, and subsequent reanalysis using techniques which will be learned from the nature of the data itself. However lossy compression introduces a different sort of information loss. If this strategy is to be adopted, it is necessary to quantify its effect.

Many different schemes are available. As a starting point, studies by e.g. Louys et al. (1999) identify the best schemes for astronomical data probably to be *hcompress* (White 1993, developed for HST data) JPEG and the pyramidal median transform (PMT; Starck et al. 1996).

The JPEG standard allows for 8/12 bit greyscale, 24 bit colour. Pseudocolour (24 bit greyscale) cannot be used to compress because colour bleeding introduces spurious greyscale values. Some 16 bit code is however available and needs to be evaluated. PMT is commercial code. We have obtained a demonstration version in MR/1 package for evaluation. *hcompress* is publically available code and we have used it here for an initial evaluation of losses caused by lossy schemes.

Given the linear nature of the spectra from RVS, audio compression schemes (16 bit lossy) may also be worth evaluating.

3.4. Results from *hcompress*

The effects of the *hcompress* algorithm on the spectrum of a $V = 15$ mag F star is shown in Figure 4. It can be seen that compression factors $> 10-20$ modify the spectra significantly, as judged by eye. To provide a more quantitative analysis, extracted spectra from these compressed images at different compression ratios were then cross-correlated with a template spectrum. The results are shown in Figure 5. It should be noted that the extraction scheme was extremely simple, so that the spectra of bright objects were not subtracted before the extraction of fainter objects. This leads to a significant number of outliers, which are an artifact of this simplistic approach.

For compression ratios of ~ 40 and less the radial velocity is recovered adequately (allowing for the outliers) to magnitude $V \sim 15$. However, a more careful extraction procedure is required to examine the effect of the compression in more detail.

As noted earlier, spectra will be summed to make final products. This makes it important to know what the effect is of adding spectra extracted from compressed images. To check, we combined spectra from 20 F stars with $V = 16 - 17$ mag from an image with a compression ratio of 23 and compared the

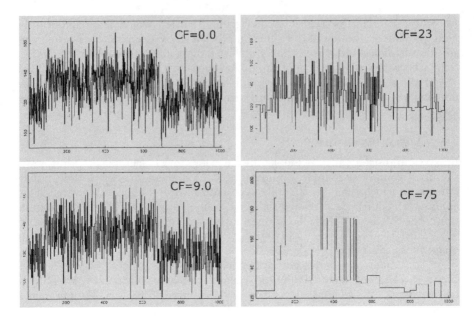

Figure 4. The effect of *hcompress* on a $V = 15.0$ F star spectrum from the high Galactic latitude field in Figure 1, with 1 spatial row extracted. The compression factors (CF) are given on each figure.

resulting spectrum to that of a $V = 15$ mag star from an uncompressed image. The summed spectrum appears broadly similar.

3.5. Are lossy schemes useful?

While the above analysis is encouraging at one level, it must be realised that lossy compression schemes achieve their compression gains by modelling or reducing the noise in some way. Unfortunately in RVS, most of the radial velocity information is held in exposures with extremely few numbers of counts (most pixels will not even have a single count from the source at $V = 17$ mag). It is inevitable therefore that lossy compression schemes have the potential for strongly affecting the information from the large majority of sources that RVS will measure.

The simulations using lossy compression have highlighted the extent to which the signals in RVS are buried in the noise. Our comparison between the *hcompress* and PMT schemes has highlighted the different way in which the noise is modelled. In retrospect, lossy compression schemes may be too dangerous to use, and it may not be useful to pursue them further.

3.6. Where next?

Unfortunately the way forward remains unclear. However, several ideas were discussed at this meeting which may be worth pursuing.

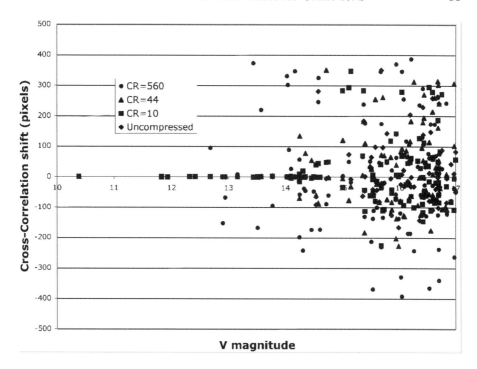

Figure 5. The effect of different compression ratios using *hcompress* on the radial velocity accuracy, as a function of magnitude.

One simple strategy is to send down full data for bright stars while carrying out a spatial chequerboard sampling of the sky for fainter stars. The mark/space ratio can be adjusted. This has the advantage that the selection function is well known and factors up to ~ 10 can be achieved, but with some science loss. If more telemetry is available in practise, perhaps because of a second ground station, the mark-space ratio can be increased to make use of it.

Another strategy could involve the transmission of the full data for bright stars at full resolution, and with faint stars at lower resolution (by binning in the processing unit). This reduces the telemetry requirement but not the effective readout noise which could have been reduced with an intrinsically lower resolution.

A further suggestion (A. Henden) was to use L3CCDs (see below) in photon counting mode. This will reduce the telemetry because of the reduced number of bits per pixel required to describe the noise. At the same time the faint limit would be improved significantly (because $> 75\%$ of the noise is readout noise). Photon-counting mode also avoids the penalty of a factor 2 in the readout noise normally incurred in these devices, as it is simply a detection above a threshold. The TDI mode would then be done in software. The dynamic range will be limited by the readout rate, a guess would suggest $V \sim 11$ but further work is required on this. It might be possible to instigate several thresholds to allow for counting of more than one photon per frame.

A suggestion by U. Munari was to telemeter down only the spectral regions around the Ca lines for stars fainter than some limit (say $V \sim 15$) since this is where almost all velocity information resides. This would improve matters significantly, but would have scientific penalties for non-main sequence stars.

Other suggestions included operating one CCD at full resolution, while binning others on chip, use of different resolution at different Galactic latitudes (for example R=10 000 at high, R=5 000 at low) and combining data from different transits as well (grouping of transits) if there is sufficient on-board storage.

4. Confusion and scan mechanism

A scan mechanism to compensate for the GAIA tracking law adds significantly to complexities and risk for RVS and should be avoided unless absolutely necessary to achieve the scientific goals. However, crowding remains an issue (see Figure 1), despite initial indications that the radial velocities can be extracted from overlapping spectra (Zwitter 2002). Much work remains to be carried out on the extraction of spectra, perhaps with a simultaneous two-dimensional fit, and (or) with an extraction width for spectra optimised as a function of magnitude.

5. Cold electronics

The CCDs and front-end electronics will be running cold. While this is normal for the CCDs, some components in the front-end electronics will require careful consideration.

Analog/Digital (A/D) converters can be identified as one such item of concern. Those parts useful for space use generally contain bipolar elements. Since the junction potential is temperature dependent, such parts may not operate at the temperature of the payload. It may be necessary to adopt the strategy of the transmission of low-level analog signals over a large separation between the CCD detectors and A/D converter. In this case, very careful attention needs to be paid to noise and grounding. One option to increase noise immunity may be to perform a voltage-to-frequency conversion using discrete circuitry. In this case it might even be possible to use a wireless RF (radiofrequency) connection between the immediate detector electronics and warm A/D converter.

The CCD clock driver circuitry may be another challenge (compounded by the large number of video chains): again discrete circuitry may be necessary to effect this at low temperature.

In general, a qualification programme will be required for parts running at below military temperature range. This applies to the sourcing of appropriate parts and technologies, and also a qualification of the packaging.

6. CCD performance at low levels and the effect of cosmetic defects

The RVS CCDs are going to be operated in a regime where most objects contribute an extremely low level of flux in a single transit ($< 1e^-$ pixel^{-1} for an $I_c = 17$ mag object). This is an atypical mode of operation for astronomical

CCDs (this possible concern was first pointed out by D. Katz). The observed spectrum will depend on pixel-to-pixel (and intra-pixel) response, but the normal averaging effects from large number of photons will be absent in RVS. In order to attain nominal performance in the summed spectra, it will be necessary to ensure that CCD performance is understood to a level probably beyond that usually required, particularly as regards operating point shifts, readout noise, the effects of radiation damage and general ageing. It is probably also necessary to ensure that a capability is in place to characterise and monitor the CCD performance in orbit as a continuous process.

The low signals for most stars for each exposure require a sufficiently accurate sampling of the readout noise in order to coadd the spectra from different CCDs. This has implications for dynamic range, for A/D converter accuracy and digitisation (it is not possible to sample the full well capacity of the CCDs with less than 16 bits) and also significant implications for lossy compression (see earlier). The Poisson statistical noise is much larger on brighter objects (although a relatively smaller fraction), so it may be appropriate to consider non-linear A/D conversion (perhaps logarithmic, or square-root, as the latter is easier to implement). Such schemes need to be included *a priori* in the simulations for the compression tradeoffs.

Pixel defects affect the operation of entire rows in RVS. Even low-level defects caused (for example by radiation damage) have large-scale effects. This means that the scheme for combining data from different CCDs needs to be able to exclude particular CCD rows (for example by median filtering as suggested by U. Munari). The need for a monitoring regime has already been noted above: whether this should operate automatically for the flagging of degraded pixels should be considered, together with the implications for the ground segment of any such scheme.

7. L3CCDs

L3CCDs (LLLCCDs) are CCDs with modified readout stages (avalanche gain stages), e.g. CCD 00463-10 from E2V (EEV). These produce reduced or negligible readout noise but with reduced full well capacity and (generally) double the photon noise. They are potentially useful in the case of RVS in that the readout noise is a significant component affecting the faint end of the magnitude range (e.g. Cropper & Mason 2001).

An initial check for radiation tolerance of these devices indicates that L3CCDs are no different in this regard from normal CCD variants from which they are derived (Smith, Holland & Robbins, 2002, in press). However these tests are very preliminary (10 MeV protons; device tested unbiased), so confirmation and further work is required.

Simulations are required to explore the tradeoff in the case of RVS for readout noise gain vs. photon noise and dynamic range loss. The possibility of flexible avalanche gain with star brightness, to increase full well capacity and hence dynamic range should be explored, and the advantages and disadvantages of photon-counting mode investigated.

Acknowledgments. I am grateful to many in the RVS working group and at the Conference itself for very many useful discussions and ideas

References

Cropper, M. & Mason, K. 2001, RVS-MSC-001 (Livelink)
Cropper, M, Smith, P. & Brindle, C. 2002, RVS-MSC-003 (Livelink)
Louys, M., Starck, J.L., Mei, S., Bonnarel, F. & Murtagh, F. 1999, A&AS 136, 579
Pace, O. 2003, in GAIA Spectroscopy, Science and Technology, U.Munari ed., ASP Conf. Ser. 298, pag. 13
Perryman, M.A.C. 2003, in GAIA Spectroscopy, Science and Technology, U.Munari ed., ASP Conf. Ser. 298, pag. 3
Starck, J.L., Murtagh, F., Pirenne, B. & Albrecht, M. 1996, PASP 108, 446
Zwitter, T. 2002, A&A 386, 748
White, R.L. 1993, in Space and Earth Science Data Compression Workshop, J.C. Tilton ed., NASA Conf. Pub. 3173, pag. 117

Absorption cells in wavelength calibration of GAIA spectra

Silvano Desidera & Ulisse Munari

Osservatorio Astronomico di Padova – INAF, vicolo dell'Osservatorio 5, Padova, Italy

Abstract. The wavelength calibration of the GAIA spectra is a crucial point in deriving accurate radial velocity for the program stars. The scanning slit-less mode does not allow to use the usual wavelength calibration techniques. Current plans foresee to calibrate the spectra by the identification of a suitable ensemble of reference stars, that would map the wavelength solution in the different position of the focal plane. We discuss an alternative solution, the use of an absorption cell that imprints its own lines on top of each spectrum, allowing an accurate wavelength calibration. The attanaible precision is discussed in analytical form. Absorption cells were widely used in recent years mainly for the radial velocity searches of extra-solar planets. The major issue for the use of absorption cells within the GAIA spectrograph is represented by the identification of a suitable absorbing medium that produce a few lines in the spectral region of interest with a small optical path and, more crucially, at the satellite operating temperature. Possible media are discussed and tests at the telescope anticipated.

1. Introduction

The GAIA mission will collect spectra with the prime goal of deriving good radial velocities, thus providing the 3rd component of the velocity vector. Accurate wavelength calibration, with a proper control over the whole mission lifetime of possible systematic effects, is therefore mandatory.

The scanning slit-less operation mode for the GAIA spectrograph does not allow the use of conventional wavelength calibration techniques (e.g. emission lines lamps). Current plans foresee to calibrate the spectra by the identification of a suitable ensemble of reference stars, that would map the wavelength solution as they sweep over the focal plane as a consequence of the satellite spin. Such a calibration can be properly performed only at the end of the mission, with iterative preliminary solution building up with the mission advancement. In fact, only radial velocity constancy at later times allows to qualify a star as a standard and use it in setting up a web of standards distributed over the whole sky to calibrate the spectra obtained at earlier times.

An alternative approach to wavelength calibration of GAIA spectra is to place an absorption cell filled with a suitable gas in the optical path of the spectrograph (other possibilities are discussed by Pernechele and Munari 2003).

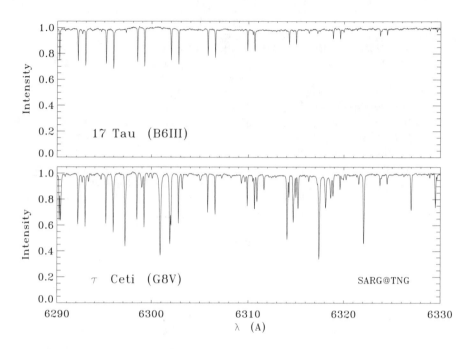

Figure 1. Spectrum of the fast rotating early type star 17 Tau, evidencing the telluric absorptions, and of the slowly rotating solar type star τ Ceti, displaying both the telluric and the stellar lines.

In this way the spectral lines of the absorbing medium are directly superimposed on each stellar spectrum, providing the wavelength calibration looked for. Such calibration should not require continuous refinements along the mission to converge to final values, but should provide immediately end results.

If the intrinsic width of the lines of the selected absorbing medium is small, the absorption cell can also be used to study the spectrograph PSF over the whole focal plane and its evolution during the mission lifetime. Knowing the instantaneous instrumental profile at various wavelengths will help to characterize precisely the spectrograph performances, allowing better measurements of stellar line parameters like the core width and thus $V_{\rm rot} \sin i$ in the slowly rotating G, K and M stars.

2. Absorption cells in astronomy

The simplest realization of the absorption cell approach is to use telluric lines as wavelength reference (Griffin & Griffin 1973). Of course, the wavelength range is limited to the spectral regions with suitable atmospheric features. The latter are abundant longward of Hα (with the notable exception of the GAIA wavelength range, which allows easy access to it from the ground, cf. Munari 1999), with some usable regions also shortward (cf. Figure 1).

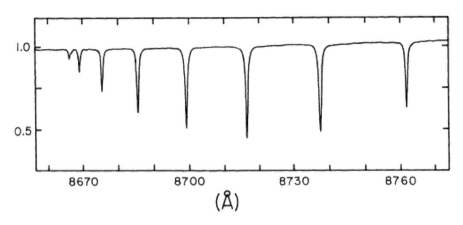

Figure 2. Spectrum of a HF cell (from Campbell & Walker 1979).

To have the reference lines in the desired spectral range it is necessary to build a suitable cell. The first cell used in astronomy was the hydrogen fluoride (HF) cell realized by Campbell & Walker (1979). Such a cell covers part of the GAIA spectral range with a few narrow absorption lines (cf. Figure 2), but it is not suitable for GAIA due to the high operating temperature (100 C), the corrosive nature of the medium and the long optical path (1 m).

Iodine cells (Marcy & Butler 1992) are the absorption cells currently most adopted in astronomy (cf. Figure 3). They are used for very high precision differential radial velocity surveys aiming to the discovery of extra-solar planets. Iodine is not a viable possibility for GAIA, because (a) it concentrates its lines away from the GAIA wavelength range, (b) there are so many lines per wavelength unit to become unresolved at the GAIA foreseen spectral resolution, and (c) to produce absorptions the cell must be properly heated. Experience with existing iodine cells demonstrates that cells can be manufactured to maintain over a 5 year period (the GAIA lifetime) a long term stability better than 2 m s^{-1}.

3. Accuracy of wavelength measurements

The experience of Campbell & Walker (1979) with the HF cell suggests that a suitable absorbing medium with just a few spectral lines over the GAIA wavelength range would allow to achieve the required precision in the wavelength calibration of the spectra. Campbell & Walker where able to derive radial velocities accurate to 0.015 km s^{-1} from $\Delta\lambda$=100 Å wide spectra of 0.35 Å resolution exposed to very high values (S/N=1500) by using five HF reference absorption lines as wavelength calibrators. Campbell & Walker have derived a useful expression for the standard error of the measured position of a single line (in velocity units):

$$\epsilon(RV) = \frac{ck\Delta\lambda}{\lambda d \frac{S}{N}} \tag{1}$$

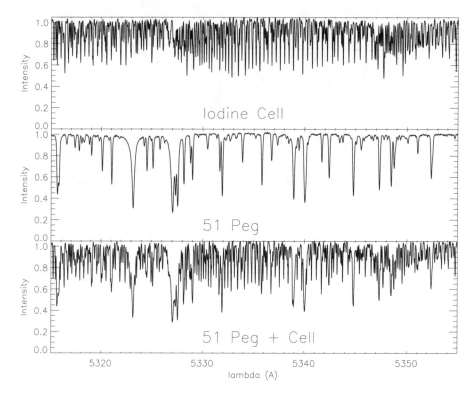

Figure 3. Iodine cell spectra obtained with SARG spectrograph at TNG (Gratton et al. 2001). *Upper panel*: spectrum of the iodine cell illuminated by the flat field lamp. *Central panel*: spectrum of the solar type star 51 Peg without the cell. *Lower panel*: spectrum of 51 Peg recorded through the iodine cell.

where $\Delta\lambda$ is the spectrograph dispersion, k is a constant of order 1 which depends on the method for determining line positions, d is the depth of a line as a fraction of the continuum, and S/N is the signal-to-noise ratio per pixel of the continuum adjacent to the line.

It is interesting to apply the above expression to the fainter stars observed by GAIA, those for which the noise in each epoch spectrum is dominated by the sky background and read-out noise. For this type of spectra, the S/N is linearly proportional to the recorded star light, not its square root. In fact, the general expression for S/N in such a case reduces to

$$\frac{S}{N} = \frac{e^-_{star}}{\sqrt{e^-_{star} + e^-_{bck} + e^-_{dark} + R^2_{readout}}}$$

$$\simeq \frac{e^-_{star}}{\sqrt{e^-_{bck} + e^-_{dark} + R^2_{readout}}} \simeq \frac{e^-_{star}}{\text{constant}}$$

with $\sqrt{e_{bck}^- + e_{dark}^- + R_{readout}^2}$ indendent from spectral dispersion for the GAIA slit-less, integral field, constant exposure time configuration.

Under these conditions, increasing the dispersion means to decrease both the S/N and $\Delta\lambda$ by the same amount in Eq.(1) above. Thus, it seems there should be not net gain or loss. However, increasing the dispersion means to increase the depth of absorption lines (and thus d in the above formula) up to the point when the lines become fully resolved by the spectrograph. Increasing further the dispersion does not bring any further gain in accuracy. Hence, the highest precision in the determination of the wavelength of an absorption line for observations of fixed exposure time is reached when the resolution is high enough to resolve the lines profile.

Let's apply this to GAIA, assuming Gaussian profiles for both the instrumental PSF and the absorption lines, characterized by half widths σ_{PSF} and σ_{line}, respectively. The instrumental PSF for a slit-less spectrograph is the mono-chromatic image of the star (convolved with the pixelization of the detector if under-sampled). Most of the GAIA targets will be G and K stars, that notoriously rotate slowly, with an average around 5 km s^{-1}. For this slow rotators it is:

$$\sigma_{line} = \sqrt{\sigma_{intrinsic}^2 + \sigma_{rotation}^2} \quad (2)$$

The typical intrinsic half width of metallic lines in the stellar atmosphere of G-K stars of intermediate gravity it is 3 km s^{-1}, or 0.09 Å at GAIA wavelengths. Combining with the average $V_{rot}\sin i$=5 km s^{-1}, it gives σ_{line}=5.8 km s^{-1} or 0.17 Å. Thus, the condition requiring the line profiles to be resolved to reach the highest accuracy - for a fixed exposure time - in the measurement of the wavelength or radial velocity of a given line can be expressed as:

$$\sigma_{PSF} \sim \sigma_{line} \quad \longrightarrow \quad \sigma_{PSF} \sim 0.17 \text{ Å} \quad (3)$$

For a Nyquist optimal sampling FWHM(PSF)=2 pixel, this means that for GAIA spectra of faint stars, dominated by the read-out and sky background noises, the optimal dispersion would be 0.17 Å pix^{-1} for G-K giants at typical rotation velocities.

Overlapping stars however increase the noise and reduce d in the above Eq.(1), with overlap probability increasing linearly with increasing spectral dispersion. Estimating from Zwitter and Henden (2003) and Viala et al. (2003) results on crowding at various spectral dispersions and galactic latitudes, away from the galactic plane the best wavelength and radial velocity measurement of an absorption line should be reached by dispersions $\Delta\lambda \sim$0.25 Å pix^{-1}, in agreement with independent findings by Munari et al. (2001, from real observations) and Zwitter (2002, from simulations).

4. Optimal number of reference absorption lines

Let us now estimate the accuracy to which an absorption reference line can be measured. Inserting in Eq.(1) a $\Delta\lambda = 0.25$ Å pix^{-1} dispersion, k=1.6 appropriate for fitting the line with a Gaussian profile, and assuming an average depth d=0.5 of the reference absorption line with respect to the continuum (like in the case of the HF cell in Figures 2), for the GAIA wavelength range it results:

$$\epsilon(RV) = \frac{35}{\frac{S}{N}} \quad \text{km s}^{-1} \qquad (4)$$

Because to calibrate with great accuracy the dispersion function of GAIA spectrograph a small number of constant radial velocity stars would be enough, the main use of the reference absorption lines will be to set the *zero* of the radial velocity scale on each individual spectrum. The accuracy of the wavelength calibration scales with the $\sqrt{N}$ of the number of used lines. Supposing the absorption cell will provide 6 lines of average d=0.5, the above equation, now expressing the calibration accuracy of the whole GAIA spectrum, becomes:

$$\epsilon(RV) = \frac{11}{\frac{S}{N}} \quad \text{km s}^{-1} \qquad (5)$$

Thus, well exposed GAIA spectra ($S/N \sim 100$) would have their wavelength zero-points calibrated to an accuracy of 0.1 km s^{-1}, and the budget error of the radial velocity of the science spectrum will be essentially dominated by the accuracy of its cross-correlation against a proper template. At the faint end, a $S/N \sim 5$ spectrum will have its wavelength zero-point known to ± 2 km s^{-1}, a quantity comparable to the accuracy of its cross-correlation (cf. Munari et al. 2003). Such numbers correspond well to common practice with real spectra obtained from ground-based telescope over a variety of instrumental conditions.

5. Possible absorbing cells suitable for GAIA

We have discussed above how none of the absorbing media already used for cells in astronomy are suitable for GAIA. An alternative has then to be found for GAIA.

The basic characteristics of the absorbing medium are driven by both technical and scientific criteria:

- a few (5-10) narrow lines in the GAIA spectral region, with free transmittance in between;

- no blending with the astrophysically most important stellar lines;

- in gaseous form at GAIA operating temperature at L_2;

- short optical path enough to reach a sufficient optical depth;

- stable and not corrosive (to guarantee long term stability of the cell)

Atomic Nitrogen could be a suitable medium, however it is the strongest absorber after hydrogen and CaII in the spectra of A-type stars (cf. Munari & Tomasella 1999 spectral atlas for the GAIA region). Adopting Nitrogen as the absorber would compromise both the wavelength calibration and the NI chemical abundance analysis of A-type stars. Unfortunately, gaseous Nitrogen

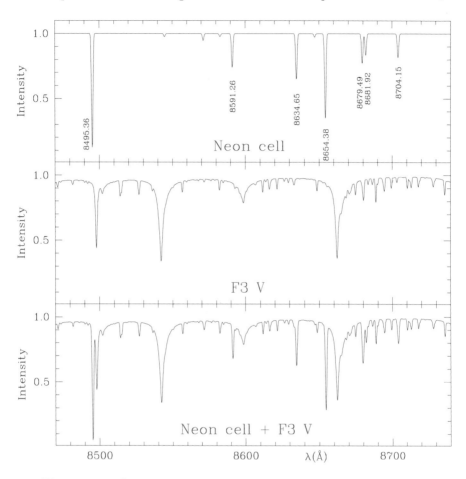

Figure 4. The tentative absorption spectrum of a neon cell plotted using known wavelengths and possible strengths (*upper panel*), the spectrum of a F3 V star (*central panel*), and the same star observed through the tentative neon cell (*bottom panel*).

in an absorption cell operating at GAIA temperatures should be in molecular form (N_2).

There is only one resonant line falling within the GAIA wavelength range, the CsI 8521.10 Å. Unfortunately, the other doublet line at 8943.50 Å fall outside the interval, and thus Cs appears of little interest.

A more promising absorbing medium seems to be Neon. It is a noble gas, thus stable and not corrosive given its very low chemical reactivity, and in atomic gaseous form at the GAIA operating temperatures. It possesses some strong lines in the GAIA wavelength range, which are generally away from the strongest stellar absorption lines (closest proximities are 3 Å with the weaker of the CaII lines, and 1 Å with a strong TiI line in cooler dwarfs). A tentative absorption spectrum of a Ne cell is presented in Figure 4.

Testing of a Neon cell at dome pressure and temperature with the Asiago 1.82m telescope should begin soon. If successful, the tests will have to be rerun at lower temperatures, as much as possible aproximating GAIA conditions.

Acknowledgments. This study has benefitted from fruitful discussions with F.Boschi (Asiago), G.Favero (Padova) and F.Crifo (Paris).

References

Campbell, B., & Walker, G.A.H. 1979, PASP 91, 540
Gratton, R.G. et al. 2001, Exp. Astron, 12, 107
Griffin, R., & Griffin, R. 1973 MNRAS 162, 255
Marcy, G.W., & Butler, R.P. 1992, PASP 104, 270
Munari, U. 1999, Baltic Astron. 8, 73
Munari, U., & Tomasella, L. 1999, A&AS 137, 521
Munari, U., Agnolin, P., & Tomasella, L. 2001, Baltic Astron. 10, 613
Pernechele, C., & Munari, U. 2003, in GAIA Spectroscopy, Science and Technology, U.Munari ed., ASP Conf. Series 298, pag. 93
Munari, U., Zwitter, T., Katz, D., & Cropper, M. 2003, in GAIA Spectroscopy, Science and Technology, U.Munari ed., ASP Conf. Series 298, pag. 275
Viala, Y.P., Morin, D., Katz, D., & Ochsenbein, F. 2003, in GAIA Spectroscopy, Science and Technology, U.Munari ed., ASP Conf. Series 298, pag. 105
Zwitter, T. 2002, A&A 386, 748
Zwitter, T., & Henden, A.A. 2003, in GAIA Spectroscopy, Science and Technology, U.Munari ed., ASP Conf. Series 298, pag. 489

Bragg gratings in multi-mode fiber optics for wavelength calibration of GAIA and RAVE spectra

Claudio Pernechele & Ulisse Munari

Astronomical Observatory of Padova – INAF, vicolo Osservatorio 5, I-35122 - Padova - Italy

Abstract. We propose a new technique, the use of FBGs (fiber Bragg gratings), for accurate, easy and low cost wavelength calibration of GAIA, RAVE and follow-ups spectra at local Observatories. FBGs mark the spectra with absorption lines, freely defined in number and position during the fibers manufacturing. The process goes in parallel with the science exposure and through the same optical train and path, thus ensuring the maximum return in wavelength calibration accuracy. Plans to manufacture and test FBGs for the CaII/Paschen region are underway at the Astronomical Observatory of Padova.

1. Introduction

Modern and demanding application of radial velocities, like those associated with the search for extra-solar planets, have promoted the introduction of new methods of wavelength calibration.

The key to success has been recording of the reference wavelength grid for to the whole duration of the science exposure and through the same optical train and path. Two approaches have been followed: the *parallel* (where Thorium lines going through telescope optics are recoded on the CCD close and parallel to the science spectrum with an exposure lasting as long as the science observation), and the *Iodine-cell* one (the stellar spectrum passes through and is absorbed by a cell filled with Iodine vapors, that mark the stellar spectrum with a huge number of sharp and equally intense absorption lines). Both approaches require delicate instrumentation, careful maintenance and dedicated spectrographs. As such they are not easily manageable on small, all-purpose telescopes.

We propose here a third, independent way to mark with reference wavelengths the science spectrum as it is recorded, through exactly the same optical train and path. It consists of using optical fibers with recorded in them Bragg gratings that allows undisturbed passage of all wavelengths except a selection of few ones that will appear in the spectrum as sharp, deep absorption lines of constant and pre-selected wavelengths. The latter are then used as a reference grid to accurately wavelength calibrate the science spectrum.

The system is easy and fast, allows to decouple the spectrograph from the telescope and operate it in a controlled environment, requires minimal mechanical work to be adapted to existing spectrographs, the marking absorption lines can be placed at any preferred wavelength, and the fibers can be mass-produced at low cost for distribution to interested Observatories. FBGs can be easily installed on spectrographs already designed to operate in multiple fiber mode. The RAVE project (Steinmetz 2003), with its fixed wavelength range, would represent an ideal case for the application of FBGs.

The FBGs (fiber Bragg grating) appear of potential interest also to GAIA. The latter's spectrograph operates in TDI, slit-less mode with thousands of spectra simultaneously crossing the field of view as the satellite spins. No conventional calibration lamp can be used, obviously. The base-lined wavelength calibration procedure foresees using stars of recognized constant radial velocity as they travel across the spectrograph focal plane, by accurately linking their astrometric position on the sky to the wavelength of their absorption lines on the pixels of the CCDs.

A bundle of FBGs properly introduced in the GAIA spectrograph optical train would mark with a set of properly placed reference absorption lines the spectra of all stars entering the field of view. The FBGs could become the prime wavelength calibration mean of GAIA spectra (particularly for preliminary solutions while the mission is still flying), or a backup for the method based on standard radial velocity stars in case this would not satisfactorily operate for whatever reason once the satellite will reach L_2. The FBGs dissipate no power and have no movable parts. They should also stand the low operating temperatures of the satellite in L_2 and should do not degrade with time, and thus looking ideal for use in space.

No FBG has ever been used as proposed in this paper, and thus confirmatory tests are required. At the Astronomical Observatory of Padova we plan to manufacture and test prototype fibers for the GAIA/RAVE wavelength range to check their performance with actual observations at the telescope.

2. Fiber Bragg grating principle

Fiber gratings are made by laterally exposing the core of the fiber to a periodic pattern of intense ultraviolet light. The exposure produces a permanent increase in the refractive index of the fiber's core, creating a fixed index modulation according to the exposure pattern: this fixed pattern is called a grating. At each periodic change of refraction, a small amount of light is reflected (Bragg diffraction). All the reflected light signals combine coherently to one large reflection at a particular wavelength when the grating period is approximately half the input light wavelength. This is referred as the Bragg condition and the wavelength at which this reflection occurs is called the Bragg wavelength (λ_B). Light at wavelengths different from λ_B, which are not phase matched, will pass unabsorbed through the fiber. Thus, light propagates through the grating with negligible attenuation and only those wavelengths that satisfy the Bragg condition are affected and strongly back-reflected.

Application of this technology to astronomical spectra at GAIA wavelengths is attractive for two basic reasons: (a) the technology is already mature with the tele-communication industry that operate it in the near-IR, down to 1 μm, and (b) the limited $\Delta\lambda$ range observed by GAIA and RAVE requires only a limited number of reference lines to be properly calibrated (see discussion in Desidera and Munari 2003), which in turn means an easier FBGs production.

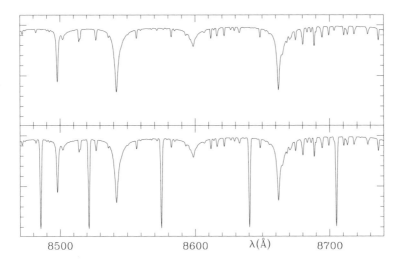

Figure 1. The spectrum of an F3 V star before and after passing through an hypotetical FBG fiber marking on it 5 reference lines.

3. Fiber Bragg grating production

Exposing a photosensitive fiber, usually Germanium-doped (Yuen, 1982), to an intensity pattern of ultraviolet radiation (at wavelengths between 240 and 255 nm, where the Germanium has an absorption band) is used to produce (i.e. write) a fiber Bragg grating. In its basic form the grating selectively reflects light at the Bragg wavelength $\lambda_B = 2nH$, where H is the grating spacing and n is the effective index of refraction. Both H and n depend on fabrication parameters: n depend on the modal dispersion characteristics of the fiber and lies, for each mode, between the core and cladding indices. The functional form of the normalized frequency (ν) parameter is given by:

$$\nu = \frac{\pi d}{\lambda}\sqrt{(n_{co}^2 - n_{cl}^2)} \tag{1}$$

where d is the diameter of the fiber core, λ is the free space wavelength, n_{co} the core index and n_{cl} the clad index. Note that the difference between the two squared index is also the numerical aperture of the fiber. For low enough ν values the fiber is single mode and only one Bragg reflection is observed at λ_B wavelength given above. At a certain value of ν the fiber become two moded with two separated Bragg wavelengths, and for larger ν values many more modes begin to appear in the fiber grating spectrum. The lowest order or fundamental mode has the longest λ_B, and the higher order modes have progressively shorter λ_B with the cladding index giving the limit on the shortest λ_B. The peak reflectivity is an increasing function of the grating length and of the difference between n_{co} and n_{cl} (Lam & Garside, 1981).

Recording FBGs inside a fiber usually involves interference between two coherent UV laser beams that precisely define the spacing between the grating

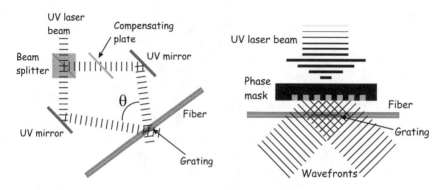

Figure 2. Writing the FBG via bulk interferometry (*left* panel) or phase mask (*right* panel).

planes. For symmetric incidence and an angle θ between the UV beams, the pitch of the recorded grating is $0.5\lambda sin(\theta/2)$, where λ is the illumination wavelength.

The basic interferometric setup splits the laser beam to create an interference pattern on the fiber (Morey et al., 1989) as shown in Figure 2 (left panel). Because the beams travel different paths, the laser must have sufficient temporal coherence to account for path-length mismatch, and good spatial coherence. The sensitivity to mechanical instabilities of this method of FBG production can make difficult to produce identical replicas.

The other way for writing FBGs is to use phase mask to generate multiple beams (Hill et al., 1993). Two of the beams carry approximately 40% of the total energy and closely overlap the phase mask surface to create an interference pattern at the desired Bragg wavelength. The fiber-phase mask assembly can be illuminated either by a large beam to cover the full fiber Bragg grating length or by a small scanning beam. Because the fiber is usually in proximity to the phase mask, the assembly is a very stable mechanical system, suitable for mass-production of identical FBG replicas. The pattern recorded into the fiber is a copy of the phase mask (pitch and chirp) scaled by about 50%.

Typically two types of UV laser are used to manufacture FBGs: continuous-wave frequency-doubled argon-ion lasers at 244 nm, and pulsed excimer lasers operating at either 248 nm (KrF) or 193 nm (ArF).

References

Desidera, S., & Munari, U., 2003, in GAIA Spectroscopy, Science and Technology, U.Munari ed., ASP Conf. Ser. 298, pag. 85
Hill K., Malo B., Bilodeau F. et al. 1993, Appl.Phys.Lett. 62, 1035
Lam, D.K.W. & Garside, B.K. 1981, Appl. Optics, 20, 440
Morey, W.W., Meltz, G., & Glenn, W.H. 1989, Proc. SPIE 1169, 98
Steinmetz M., 2003, in GAIA Spectroscopy, Science and Technology, U.Munari ed., ASP Conf. Ser. 298, pag. 381 (and http://www.aip.de./RAVE/)
Yuen, M.J. 1982, Appl. Optics 21, 136

GAIA Spectroscopy, Science and Technology
ASP Conference Series, Vol. 298, 2003
U. Munari ed.

The GAIA data access and analysis study

Salim G. Ansari

Science Programme Coordination Office, European Space Agency, ESTEC, The Netherlands

Jordi Torra, Xavier Luri, Francesca Figueras, Carme Jordi

Departament d'Astronomia i Meteorologia, Universitat de Barcelona, Spain

Eduard Masana

Institut d'Estudis Espacials de Catalunya, Spain

Abstract. The GAIA Database Access and Analysis Study was initiated by ESA in July 2000 to investigate the feasibility of implementing a Data Reduction System for the mission. In its first phase, the study was limited to a few well-defined algorithms based around the astrometric problematics of data processing. Furthermore, the Study seamlessly integrated the GAIA Simulator, which has in the meantime grown to be a major effort of simulating all the data expected to be produced by GAIA. GDAAS has NOW begun its second phase, where a significant number of algorithms will be implemented and tested for performance and evaluation of the amount of data storage that will be required during the operational phase.

1. Introduction

During its 5-year mission GAIA (ESA-2000-SCI-4) is expected to produce some 150 terabytes of raw data resulting into an expected 1 petabyte archive. The aim of this ongoing study is not only to identify the individual steps and algorithms of data reduction, but to also estimate the processing power and data storage requirements that will need to be procured to handle the daily tasks of data processing and archiving. The aim will also be not to wait until the very end of the mission to produce results on which the GAIA community can work on, but to do this earlier on, thereby producing preliminary results for further analysis.

The initial study began in July 2000, based on a modular architecture that can follow an evolutionary path over a ten-year period prior to launch. The involvement of an industrial partner (GMV, Spain) and an academic partner (University of Barcelona) has allowed the consortium to bring together several experts with a wide range of backgrounds. Furthermore, the responsibility of running the hardware and provision of computational infrastructure has been given to a *flexible* partner, CESCA (Spain), which has adapted very rapidly to the requirements of the study in its initial phase.

Phase I of the GDAAS Study was completed in June 2002 (González 2002) in which two major astrometric algorithms were implemented: Cross-Matching (Lattanzi et al. 2001) and the Global Iterative Solution (Lindegren 2001).In the next phase, which started in September 2002, a number of identified algorithms will be implemented by involving the GAIA community at large. The study will furthermore follow closely the evolutionary path of the mission's ongoing redesign efforts and adjust to them accordingly.

2. The GDAAS architecture

GDAAS has adopted a three-tier architecture (see Figure 1):

Database Management System: an object-oriented architecture was chosen for the reasons of scalability, efficiency and maintainability. At the time of writing this article, Objectivity is being used. The database choice, however, may change in the future, without a major effort in re-implementing the system. In fact, this was one of the major driving requirements of selecting a highly configurable architecture.

Data Manipulation Layer: this layer is composed of two parts: the data model, which describes the bulk of data required by all algorithms and is categorised by individual classes, while the data manipulation itself stores and retrieves the individual *data objects*.

Processing Framework Layer: this layer manages system resources and processes. It is the forefront layer upon which individual algorithms are implemented. This particular layer plays a crucial role in scheduling as well as simultaneously processing more than one task in the system. The number of algorithms may be arbitrarily selected. The limiting factor may be the total processing power at our disposal.

The driving factor of having a distributed architecture is concurrency. This is an advantage, since it allows for processing not only to be distributed across several processors, but also a in time. The system was originally designed having data and processors very close to one another. This meant that the individual computers with large data stores were setup to work on individual processes. This was somewhat limiting, since very large processes would have slowed down the performance. By changing the architecture and centralising the storage array, it became more flexible for a single processor to increase the required storage, whenever the need arose, without being limited by a local disk. The concept of a *node* in GDAAS is virtual, solely defined by an arbitrary processor and the required arbitrary storage. The main advantage of this setup is therefore the uniform distribution of nodes along all physical computers that make up the overall system.

2.1. Configuration control

Due to the expected longevity of the code being currently implemented and the changes that the system will undergo over the next few years, it is highly desir-

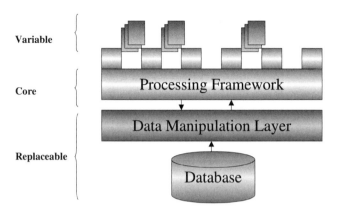

Figure 1. GDAAS System Architecture. A three-tier architecture was adopted to ensure the highest possible flexibility during system design.

able to keep a tight configuration control over various aspects of the algorithm implementation. Not only should the core system be properly and highly detailed and documented, but all changes that are carried out at various stages of system definition need to be explained and noted.

2.2. Algorithms

GDAAS algorithms are classified in three categories:

Critical: algorithms that have a direct impact on the data. These may modify the original data, the result being critical to the overall data reduction. Such algorithms must also be evaluated in terms of optimisation, as they will play a crucial role in the overall data processing performance. Typically cross-matching or the global iterative solution would fall under this category. This code may have to be translated into the internal language of the system for performance and maintenance reasons.

Non-critical: code, which may not have a direct impact on the overall data processing would typically fall into this category. In this case, the algorithm may be run independently of the ongoing processing, and use calibrated data to extract meaningful physical parameters. Photometric measurements or spectra may fall under this category. Such code may be written in any language and a wrapper would then be provided to allow it to communicate with the GDAAS main engine.

Plugins: these algorithms make use of end products and can be considered user-specific. Anyone wishing to make use of the GDAAS database to calculate e.g. the dynamics and evolution of the galaxy, or to extract a set of classified objects, based on given criteria may use this feature. The goal of such an approach at this stage is to ensure that the system can sustain the management of these types of algorithms. This code is physically independent of the GAIA data processing, but access its resulting data.

2.3. Types of users

There are three types of users foreseen for GDAAS:

Administrator: the Administrator has all access privileges to the system and can also allocate disk space and processing power to individual tasks.

Power User: users who require to work on a substantial part of the database and/or require a large amount of processing power can request for an account to be setup on the central node.

Casual User: all other users wishing to extract data to work on individually can get access to the data through a special client being developed.

3. The GDAAS Phase I prototype

During the GDAAS Phase I we developed a model for the satellite, instruments and their operation allowing a description of the raw telemetry data (including needs and rates) and a model of the whole relationship among data, access and processing needs. At present, a user interface provides all the necessary functionality to operate and test the system.

3.1. Satellite model

Several simplifications of the GAIA payload and of the actual observations were considered. Only data coming from the astrometric instruments (two focal planes constituted by 10 x 26 TDI operated CCDs) was considered (the spectro instrument was not taken into account). We assumed that while crossing the focal plane the star follows a single CCD row, with fixed acquisition window size and constant PSF.

The datation of the observations is a critical issue because it determines the astrometric precision. A simplified yet effective model of datation was used where only the time of the ASM detection was transmitted to ground and the datation of the individual observations was inferred from the same algorithm used on-board. This schema allowed a precise datation of the observations while, at the same time, saving bandwidth.

A model of the on-board data flow was developed and, according to it, science data (17 *astrometric patches* plus 5 broad band *photometric patches*) was generated by the GDAAS simulator developed at the UB (Masana et al. 2001). About 1 million stars were used for GDAAS testing and, with the simplifications described above, the prototype managed about 250 GB of telemetry data.

3.2. Definition of the data model

Before any system design can be carried out, a data model describing the structure and relationships of the data has to be defined. This data model essentially contains the definition of the Java classes modelling the different sets of data that enter into the system. In addition to raw science data and source data

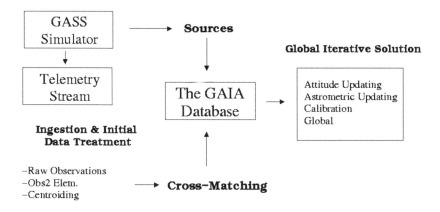

Figure 2. GDAAS Phase I prototype.

(basically astrometry and photometry), several sets of data have been designed to contain attitude, calibration, satellite data as well as auxiliary data. The data model has been designed and is maintained using UML tools.

3.3. Processing prototype

Figure 2 describes the prototype of the processing structure. From the processed telemetry stream the raw observations are stored and enter the initial data treatment, where, *elementary observations*, the ones needed on the GAIA data processing, are derived. Different observations of the same source are linked together during the cross-matching process where the DB *source objects* are created as necessary.

The needs of access in temporal, spatial and instrumental domains were identified, as well as the complex interdependencies between the different types of objects. The DB itself was designed as a scalable system, allowing remote and distributed processing and access. From the early stages of the GAIA project an object oriented DB system has been considered the best suited to handle the complexity of the GAIA structure.

4. The core processing

It is an iterative process requiring access to the DB in the temporal, spatial and CCD domain as well as in the source domain, and it is the most complex process to be executed in the GAIA system. The general concept for the astrometric reduction, is called the Global Iterative Solution (GIS). The GIS aims to compare and minimize the differences between the observed and calculated positions of the stars, provided that a model exists for the satellite attitude, the objects, and the instruments (see Figure 3). Astrometric relativistic treatment is required to reach the μas accuracy.

The practical implementation of GIS, fully described by Lindegren (2001), can be understood as an iterative sequence of four steps, that, through minimiza-

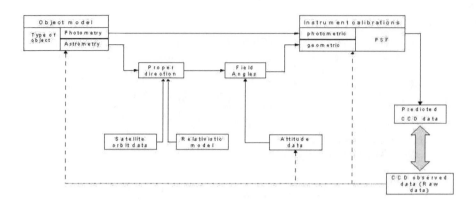

Figure 3.   Core processing

tion, improve individual parts of the data entering in the model: the attitude of the satellite, the source astrometry, the calibration of the instruments and the global parameters.

5. The simulator

The development of a mission simulator is an independent project under the responsibility of the GAIA Simulation Working Group (SWG). However, the provision of simulated data for the GDAAS project is one of the key functions of this group and therefore the development of the GAIA simulator is closely linked to the GDAAS project.

More specifically, a specialised module of the overall simulator, named the GAIA System Simulator (GASS, Masana et al., 2001) was developed to cater for the GDAAS needs. This module generates realistic GAIA telemetry that can be ingested into the GDAAS system to fill its database. During Phase I the GDAAS simulator was adapted to the data model and simplifications assumed in this early stage. GASS is now being updated to take into account the revised GAIA design and to cover the needs of the second phase of GDAAS.

6. Testing the prototype

The tests discussed in this section cover Phase I of GDAAS. This phase included cross matching routines (Lattanzi et al. 2001) and the global iterative solution (Lindegren, 2001). It covers a full range of simulated data acquisition over a five-year period.

The performance of the system has been fully evaluated and the results from several tests have been used to devise possible optimisation techniques. Special attention was paid to the ingestion, cross-matching and GIS processes, which have extensive CPU and DB I/O requirements.

6.1. Daily processing

A basic requirement for the GDAAS system comes from the daily operations of the mission (downloading data from the satellite, storage, processing and DB ingestion). A test was carried out to evaluate the time consumption of the ingestion, initial data treatment and cross-matching of a full day of mission using simulations up to 20th magnitude (realistic conditions). Time consumption for a single processor was about 280 CPU hours using the present design and hardware.

A complementary test for the ingestion, initial data treatment and cross-matching of the full mission telemetry was run by generating and processing a scaled-down simulation up to 13th magnitude. Assuming a scaling factor of 380 from 13th to 20th magnitude (Torra et al. 1999 galaxy model), the final DB size would be around 460 TB. The average ingestion and cross-matching time consumption per single processor is of about 1.5 CPU hours per day of observation at 13th magnitude. The total processing time can be easily reduced using distributed processing.

Therefore, taking into account the increase in hardware performance up to the mission start, the possibility of running a distributed ingestion process and the possible tuning and optimisation of the system, one can safely assume that the daily processing of telemetry is a feasible task.

Finally, the performance of the cross-matching algorithm in crowded regions up to faint magnitudes was also tested. First results indicate that the system is able to cope with high-density areas, although it should be adapted to cope with processing power peaks when certain regions in the galactic plane are observed by the satellite.

6.2. GIS processing

As expected, all tasks of GIS are complex and time consuming. The four GIS steps ran successfully in terms of data retrieval and storage. Because too few observations were processed during the initial GDAAS phase, some tests did not reach convergence. At present, GIS testing of 6 months of mission data considering 1 million of processing sources is underway.

7. Conclusions

The initial phase of GDAAS has provided the necessary experience in initially setting up a complex system, based on a highly flexible architecture on which to further build upon in the coming years prior to launch. It was important to have selected an open architecture, one that is maintainable over a long period of time, and at the same time flexible enough to evolve over time. The choice of technology will certainly change in the coming years, however, the base of the architecture should remain very much stable.

During the next phase of the project, a collaborative environment will be set up allowing the GAIA community to propose solutions to various aspects of the mission. Also, a rigid Configuration Control will be put in place, thereby allowing this effort ample time to evolve into an operational system. The tests have been promising in that the performance issues have not been as critical as one may have thought. However, the technological development in handling this enormous database will be the actual challenge.

References

González, L.M. Serraller, I., Torra, J., et al. 2002, GMV-GDAAS-RP-001 (Livelink)

Lindegren, L. 2001, GAIA-LL-34 (Livelink)

Lattanzi, M., Luri, X., Spagna, A., Torra, J., Jordi, C., Figueras, F., Morbidelli, R., & Volpicelli, A. 2001, GDAAS-TN-005 (Livelink)

Masana, E., Jordi, C., Figueras, F., Torra, J., & Luri, X. 2001, Highlights of Spanish Astrophysics II, pag. 389

Torra, J., Chen, B., Figueras, F., Jordi, C., & Luri, X. 1999, Bal. Astron. 8, 171

Telemetric flows on GAIA-RVS estimated from star counts from the GSC-2.2 catalog

Yves P. Viala, Danielle Morin, David Katz

Observatoire de Paris, GEPI, FRE K2459 CNRS, 5 place Jules Janssen, 92195 Meudon Cedex

Francois Ochsenbein

Observatoire Astronomique de Strasbourg, UMR 7550 CNRS, 11 rue de l'Université, 67000 Strasbourg

Abstract. Using the new (spring 2002) version of the GAIA Nominal Scanning Law (GAIA-NSL) and the distribution of stars over the whole sky determined through star counts from the GSC-2.2 catalog, we have computed mean daily telemetric flows expected from the Radial Velocity Spectrometer (RVS) all along the assumed 5 years (1800 days) GAIA mission. Telemetric flows are determined for several parameter sets (resolution, spectrum width and number of CCD) of the instrument.

1. Introduction

The Radial Velocity Spectrometer (RVS), one of the instruments that will be implemented on the GAIA satellite, is still in the phase of definition of its specifications, a task devoted to the RVS Working Group which has to present its recommandations before the end of 2002. The telemetry budget allocated to RVS, up to now fixed to 0.25 Mbits s^{-1}, is a severe constraint that will impact on the instrument design. The purpose of this paper is to provide predictions of the RVS telemetric flow for various instrument parameters and its variations during the whole mission. Parameters checked here are the spectral resolution, the width (in rows of pixels) of a single star spectrum perpendicular to dispersion and the lowest magnitude limit of objects to be observed. The main objective is to compute telemetric flows for any integration time and at any time during the GAIA mission. For that purpose we need : (*1*) the nominal scanning law of GAIA (GAIA-NSL) which provides the position of the field of view (FOV) of RVS on the sky as a function of time; (*2*) the distribution of star density on the celestial sphere which can be obtained from star counts from any catalogue covering the entire sky; (*3*) the number of bytes by observed object that must be sent to the ground, which depends on the adopted set of RVS parameters.

2. The GAIA nominal scanning law

The GAIA-NSL is described in Lindegren (1998, 2000, 2001). Among the parameters that define it, three of them have recently (spring 2002) been modified to take into account changes in the payload design. The angle of the precession cone of the GAIA rotational z axis around the Sun direction is now $\xi = 50°$; the mean (r.m.s.) speed of the z-axis on the sky, in units of the Sun speed, is $S = 4.095$ leading to a number of loops of the z-axis per year of $K = 5.200$; the rotational velocity is $\omega = 60$ arcsec s^{-1}. The GAIA-NSL allows to compute the matrix attitude of the satellite as a function of time $A(t)$, this matrix relies the components of any direction **u** in the GAIA reference system (GRS): (u_x, u_y, u_z) and in the International Celestial Reference System (ICRS, equatorial coordinates) (u_l, u_m, u_n) through :

$$\mathbf{u}(GRS) = A(t) \times \mathbf{u}(ICRS)$$

In the new design, RVS is located opposite to the ASTRO1 instrument in the XOY scanning plan: XOY in the GRS, so that: $u_{xc} = cos(127°), u_{yc} = sin(127°), u_{zc} = 0$. The size of the RVS-FOV is 2° along scan and 1.6° across scan, so that the FOV transit time along scan is 120 s. Using the attitude parametrization of GAIA-NSL through quaternions formalism by Lindegren (2000) and the Fortran code he provided (Lindegren 2001), we computed the matrix attitude $A(t)$ every 120 s for an assumed 1800 days GAIA mission arbitrarily starting from day J2000.0. At each timestep, from the above relation, we get the celestial coordinates (equatorial, galactic and ecliptic) of the centre and extremities of the RVS-FOV, as well as its orientation on the sky, defined as the angle between the scan direction and the galactic plane. With this chosen timestep, we get successive FOVs nearly juxtaposed (neglecting drift across scan, a fairly good approximation since a drift speed of 0.171″ s^{-1} leads to drift across scan of only 20.5″ between two successive FOVs). One day of scanning covers 720 FOVs and the whole mission corresponds to a total of 1 296 000 FOVs. Each FOV is divided in smaller pavements of equal areas; their number and size are free parameters: here, the FOV was divided in two rectangles of size 1°along scan and 1.6° across scan. The celestial coordinates of each pavement are computed from the attitude relation and the number of stars within each pavement is obtained from counts in the GSC-2.2 catalogue (see next section); the number of stars per FOV is obtained by summation over the pavements.

3. Star counts from the GSC-2.2 catalog

The second generation Guide Star Catalog (GSC) was chosen because it is, up to now, the most complete star catalog covering the whole sky. We used the 2.2 public release set up in the Centre de Données Stellaires (CDS) in Strasbourg which contains positions, classifications and magnitudes for 455 851 237 objects. These data were obtained through 1″ resolution scans of the photographic sky survey plates from the Palomar and UK Schmidt telescopes in F and J bandpasses. More information on the GSC-II project can be found in e.g. McLean et al (2000). The GSC-2.2 release is complete up to magnitude 18.5 in F-band ($\lambda \sim 0.71$ μm) and 19.5 in J-band ($\lambda \sim 0.44$ μm). Magnitudes in V-band ($\lambda \sim 0.55$ μm) are also given, but for a much smaller number of objects. For the moment, the GSC-2.2 release contains only two classes of objects: star and non-star. Apart form extended objects, most of non-star objects consists of unresolved multiple systems. We have performed star counts within boxes of $1° \times 1°$ centered every degree in galactic coordinates ($0° \leq l \leq 359°$ and $-90° \leq b \leq 90°$). Within each box, we performed counts in the three (F, J and V) bands, and, in each band, for stars, non-star objects, and total. Nine ascii files have hence been produced: they give the number of objects by square degree and by 0.5 magnitude interval from 0 (or less) up to magnitude 19, as a function of galactic coordinates for the entire sky. These files are available on the anonymous web site mehipx.obspm.fr. More information can be obtained from yves.viala@obspm.fr. For purpose of comparison, we also performed star counts by using the very simple galactic model presented in the GAIA Concept and Technology Study Report (ESA-2000-SCI-4, Table 6.6, p. 282) to which we applied interpolation between magnitudes 17 and 20 assuming a power law variation of the star number.

4. Results

The number of objects observed per day is obtained by summing over the 720 successive juxtaposed RVS-FOVs scanned per day. Figure 1 displays the number of objects (GSC-2.2 F-band stars and total, and GAIA-CTSR model) up to magnitude limit 17 observed per day for the whole mission. The total number of GSC-2.2 objects observed per day varies within a factor of 10 during the

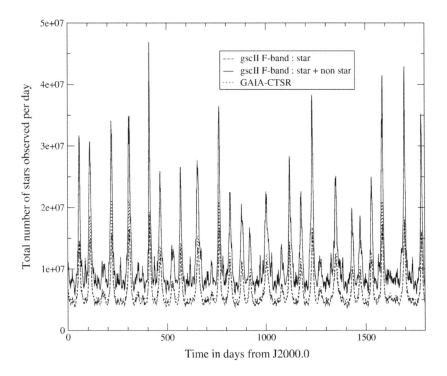

Figure 1. Number of objects observed per day by RVS up to $m_{\text{lim}} = 17$.

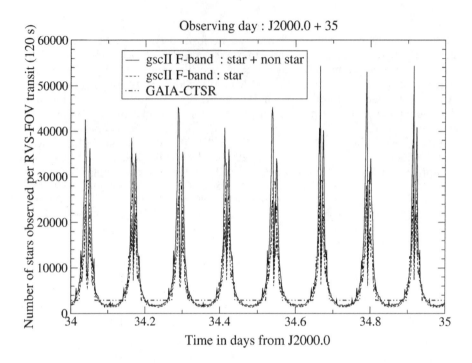

Figure 2. Number of stars observed per RVS-FOV transit up to $m_{\text{lim}} = 17$ on $day = J2000.0 + 35$ during which a minimum of 5.8 millions of objects are observed.

mission. This number depends of course on the position of the satellite. Indeed, the minimum number of scanned objects is reached when the rotational axis of GAIA points towards the galactic plane so that the great circle along which scanning occurs is perpendicular to it. The minimum number of observed GSC-2.2 objects occurs between day 34 and 35. To illustrate this, Figure 2 displays the number of objects up to $m_{\text{lim}} = 17$ observed within each successive FOV during that day; we clearly note the eight crossings through the galactic plane, during which the number of GSC-2.2 objects per FOV lies in the range 35 000-55 000. The number of daily observed objects up to $m_{\text{lim}} = 17$ reaches its maximum value between day 411 and 412 and amounts to nearly 47 millions (compared to 6 millions at minimum). On that day, the GAIA rotation axis points near the South Galactic Pole at latitude ranging from –83° to –81°, so that the instruments scan regions close to the galactic plane, with latitude in the range ±9°. The number of scanned objects per FOV transit is plotted on Figure 3; crossings of the galactic plane are still clearly identified on the figure; the most crowded FOVs observed that day contain up to 300 000 to 400 000 objects. Such a high level of telemetric flow occurs at 23 different epochs during a 5 years mission (see Figure 1).

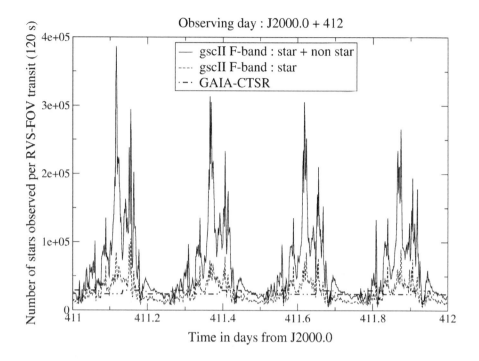

Figure 3. Number of stars observed per RVS-FOV transit up to $mag = 17$ on $day = J2000.0 + 412$ during which a maximum of 46.9 millions of objects are observed.

Once the number of objects observed by RVS in a fixed integration time t_{int} (FOV transit time or day) is known, the telemetric flow within t_{int} is obtained by multiplying this number by the number of pixels per star spectrum and dividing by t_{int}. The number of pixels per star spectrum on a single CCD depends on: (1) the spectrometer resolution which fixes the spectrum extension along scan, and (2) the spectrum width perpendicular to dispersion, i.e. the number of CCD rows over which each star spectrum can extend. The number of pixels per star spectrum on a single CCD is given in Table 1 for three resolutions and three spectrum widths. If the RVS-FOV is covered by 6 CCD, the total number of pixels occupied by each star spectrum over its transit through the FOV is obtained by multiplying by 6 the values of Table 1.

Nevertheless, the possibility to sum the 6 CCD on-board (after readjustement of the images) before sending data to the ground would allow to reduce the effective number of pixels per star spectrum and transit through the FOV to the values listed in Table 1.

At last, when computing telemetric flows, we have to account for CCD filling up in very crowded FOVs. This occurs when all pixels of one single CCD are occupied by at least one star. If $n_{pix/obj}$ is the number of pixels per star spectrum on a single CCD (Table 1), the critical number of objects per square

Table 1. Number of pixels per star spectrum on a single CCD and critical number of objects per square degree above which the CCD is full.

	$n_{pix/obj}$			N_{obj}^{crit}		
number of rows	resolution			resolution		
per star spectrum	20 000	10 000	5 000	20 000	10 000	5 000
3	3750	1875	937	2363	4726	9452
2	2500	1250	625	3544	7089	14178
1	1250	625	313	7089	14178	28306

degree, N_{obj}^{crit}, above which the CCD is full is given by :

$$N_{obj}^{crit} \times S_{CCD}(deg) \times n_{pix/obj} \geq S_{CCD}(pix)$$

where $S_{CCD}(deg) = 0.28 \times 1.6 = 0.45$ deg^2 and $S_{CCD}(pix) = 1010 \times 3930 = 3\,969\,300$ pixels is the surface of a single CCD expressed in deg^2 and pixels, respectively. Values of N_{obj}^{crit} versus spectrometer resolution and width of individual spectrum are listed in the last three columns of Table 1. It is also worth noting that the crowding problem, in terms of spectrum contamination due to overlap between neighbouring objects, becomes serious much before the critical star density leading to saturation occurs.

It is beyond the scope of this paper to present all the results we got on the RVS telemetry budget versus parameters susceptible to impact on it (catalog choice, band of observation, object type, magnitude limit, integration time of observation, RVS parameters, etc): these are available on request to the authors in the form of telemetric files.

We merely present in Table 2, telemetry rates averaged over one day of observation, obtained after CCD summation, taking into account all objects of the GSC-2.2 catalog and for three magnitude limits 17, 18 and 19 in the F-band. For each magnitude, telemetry rates are given for 9 different sets of 2 independent RVS parameters (spectrum width across dispersion and spectrometer resolution). For each magnitude limit and parameters, daily telemetry rates are presented for the two extreme cases corresponding to the minimum and maximum number of objects observed per day, as well as for a mean day of observation (average over the whole mission).

If a telemetry rate of 0.25 Mbits s^{-1} is allocated to RVS, the compression factor to be applied to the data flow can be obtained by multiplying by 4 the values listed in Table 2. If on-board CCD summation is possible, the mean telemetric flow per day range from 0.3 Mbits s^{-1} to the maximum value of 3.8 Mbits s^{-1} corresponding to a full CCD. The needed compression factor to be applied to the data, in the range 1 to 15, appears reasonable in view of the present status of data compression techniques. The situation becomes much more critical if the data from all six CCD must be sent separately to the ground. In that case, numbers listed in Table 2 must be multiplied by 6 ; assuming a magnitude limit of 17, a spectrum width of 2 CCD rows and mean daily telemetry rate over the mission, we need data compression factors of 25, 40 and 56 for resolutions of 5 000, 10 000 and 20 000. For more crowded FOVs, whatever the resolution, RVS observations have to be restricted to stars up to magnitude between 14-15, to stay within the telemetry budget allocated to RVS.

5. Conclusion

Let us conclude with the future work to be done to improve predicted telemetry rates from GAIA-RVS.

In computing telemetric flows, we took into account all objects, including non-star ones, mainly because most of them appear to be multiple star systems supposed to be identified by GAIA. By doing this, we slightly overestimate telemetry rates since we account for extended nebulae and galaxies. The new version of the GSC-II catalog, planned to appear next year, will deal with five classes of objects (instead of two), among which extended objects will be identified. This point, together with the fact that the catalog will achieve completion to fainter magnitudes, will allow a significant improvement of predicted telemetric flows. Improvements can also be obtained by performing star counts with a better spatial resolution on the sky, e.g. 12'x12' or 6'x6'. Conversion of GSC-II F and J photographic magnitudes into GAIA magnitude would also be useful, but not a simple task due to the difficulty (impossibility?) to distinguish lumi-

Table 2. Mean daily telemetry rates (in Mbits s^{-1}) expected from GAIA-RVS (from star counts of GSC-2.2, star + non star objects in the photographic F-band).

CCD	rows	day	resolution 5 000 limiting mag			resolution 10 000 limiting mag			resolution 20 000 limiting mag		
			17	18	19	17	18	19	17	18	19
1	1	35	0.3	0.6	0.7	0.7	1.0	1.2	1.2	1.6	1.8
1	1	mean	0.6	1.0	1.2	1.1	1.5	1.8	1.7	2.2	2.4
1	1	412	2.2	2.9	3.2	3.0	3.6	3.8	3.7	3.8	3.8
1	2	35	0.7	1.0	1.2	1.2	1.6	1.8	1.7	2.2	2.6
1	2	mean	1.1	1.5	1.8	1.7	2.2	2.4	2.3	2.8	3.1
1	2	412	3.0	3.6	3.8	3.7	3.8	3.8	3.8	3.8	3.8
1	3	35	0.9	1.3	1.5	1.5	1.9	2.2	2.2	2.7	3.1
1	3	mean	1.4	1.9	2.1	2.0	2.5	2.8	2.7	3.2	3.5
1	3	412	3.5	3.8	3.8	3.8	3.8	3.8	3.8	3.8	3.8

nosity classes within the catalog. We finally plan to compute telemetry rates by performing star counts from other catalog, such as the point source DENIS catalog in I band, closer to the wavelength band at which RVS works but with the main drawback that the DENIS survey does not cover the entire sky.

Acknowledgments. Star counts have been performed by using the version of the GSC-II catalog settled at the Centre de Données Stellaires de Strasbourg and software of the Vizier database environment.

References

McLean, B.J., Greene, G.R., Lattanzi, M.G., Pirenne, B. 2000, in Astronomical Data Analysis Software and Systems IX, N. Manset, C. Veillet and D. Cratbee ed.s, ASP Conf. Ser. 216, pag. 145

Lindegren, L. 1998, SAG-LL-014 (Livelink)

Lindegren, L. 2000, SAG-LL-030 (Livelink)

Lindegren, L. 2001, SAG-LL-035 (Livelink)

Status of the GAIA simulation effort: integration of spectroscopy simulations

Carine Babusiaux

Institute of Astronomy, Madingley Road, Cambridge, U.K.

Xavier Luri

Departament d'astronomia i Meteorologia, Universitat de Barcelona, Avda. Diagonal 647, 08028 Barcelona

Abstract. The simulation activities for GAIA started early on in the project because simulated data were needed for scientific assessment and mission design. In this first stage the effort was mainly devoted to simulations of the astrometric instrument due to its role as mission driver and to the lack of definition of the spectroscopic instrument. After the mission approval the Simulation Working Group (SWG) aims to coordinate the simulation efforts and to provide a common framework for the development of algorithms. The full inclusion of the spectroscopic instrument is a main goal in the SWG agenda. A review of the coordination structure, common tools and proposed future developments for the GAIA simulator is presented, with special emphasis on the integration of spectroscopy simulations.

1. Introduction

The GAIA Simulation Working Group (SWG) coordinates the mission simulation activities. These activities started early in the project to provide reliable estimates of the mission performances and expected results, thus contributing to the mission design that was approved by the European Space Agency (ESA) as Cornerstone VI.

The simulation activities, however, did not end with the mission approval. Once GAIA was approved its design was revised and improved, allowing a substantial reduction of the mission budget. New simulations are therefore required for mission reassessment and detailed instrument design, thus ensuring that scientific goals are fully covered. Furthermore, GAIA will produce a huge amount of data (about 150 TB of raw telemetry) that has to be carefully reduced to produce microarcsecond-level astrometry. The preparation of this data reduction is a challenging task that requires realistic simulations to develop and test the data handling system before the mission starts.

According to the ESA's Science Programme Committee (SPC) agreements, GAIA will be launched not later than 2012. However, it is likely that the launch will actually take place at an earlier date, with the baseline assumed for technical developments being a launch around mid 2010. This early date sets the schedule for the SWG in the years before launch.

Short term schedule (2002-2005): the SWG should cater for the phase A needs, where the mission design will be consolidated, a detailed instrument design will be produced and a prototype of the data analysis system will be developed. At the end of phase A (around mid 2005) the GAIA design will be frozen, ready to enter the industrial design phase.

Long term schedule (2005-2015): the SWG should first provide inputs for phase B (2005-2006), where a detailed industrial design of GAIA will be produced. At the same time, realistic simulations should be produced to allow the development and testing of the GAIA database and reduction system (2005-2010) that should be fully operational at mission launch.

After that, the need of simulations for the technical aspects of the mission will be reduced, but they will still be required to prepare the scientific exploitation of the GAIA data. Simulated data will allow the scientific community to prepare algorithms and methods to fully take advantage of the unprecedented amount of astrometric, photometric and spectrometric data provided by GAIA.

Although in the early phases of the GAIA design most of the simulation effort has been devoted to the astrometric instruments (these being the mission drivers), it is now time to fully integrate the spectro instrument simulations in the overall effort to cover the SWG duties in the coming years.

2. Overview of the GAIA Simulator

2.1. Structure of the Simulator

In view of the long term effort required from the SWG, a GAIA simulator structure has been set up. This structure is summarized in Figure 1.

All the different individual simulations are gathered into a Common Toolbox, which is divided into three main parts: a universe model, an instrument model and a utility box containing numerical methods and astronomical tools.

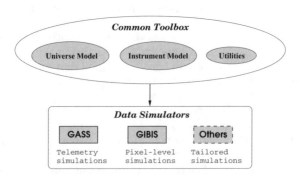

Figure 1.   Structure of the GAIA simulator

Different data simulators can use this common toolbox to generate GAIA simulated data. Two main data generator have already been developed.

GASS (GAia System Simulator) simulates GAIA telemetry, e.g. raw observation and satellite house-keeping data (Masana et al. 2001). It aims at generating quickly a huge amount of realistic data. It has been developed in particular to provide data for the development of the Gaia Data Access and Analysis System (GDAAS) (Ansari & Torra 2003).

GIBIS (Gaia Image and Basic Instrument Simulator) provides pixel-level simulations (Babusiaux et al. 2001). Those very detailed simulations allow tests of the on-board algorithms and the sampling strategy, studies of the effects of the instrument artifacts such as CCD degradation with time, and tests of calibration and data reduction procedures.

Other data generators can be added to this structure if needed. This common structure will allow, when final precise tests will be needed and computation power available, to merge the data simulators to provide simulations of a huge amount of very detailed and realistic data.

2.2. Organization of the development

To coordinate the development of the simulations, maintain the common tool box, ensure its coherence and that the mission needs are covered, a small 'core team' has been created. A 'GASS team' and a 'GIBIS team' are respectively in charge of providing telemetry and pixel-level simulated data. A task list has been distributed for the common toolbox to all the SWG members.

The contributions from the GAIA community will be integrated by the core team into the simulator structure. The resulting system and simulated data will be available for use by the whole GAIA community.

2.3. Working tools

The design of the simulator has been made in UML (Unified Modeling Language). It is a design tool to specify, visualize and document a system under development. Its main advantages for the GAIA simulator are that the visual modeling makes easier the collaboration between different teams and that it allows a modular, iterative and incremental development of the system.

The main structure of the simulator has been implemented in Java. Besides providing a natural implementation of the UML modeling of the system, an object-oriented language allows a high level structure of the code that makes easier the multi-team development and the maintenance of the code. Individual simulations written in other languages (in particular C and Fortran), are integrated and called from the Java core.

The simulator code is maintained using CVS (Concurrent Versions System). This tool allows to maintain a centralized repository of files, ensuring a coordinated development and a controlled code distribution.

To facilitate the integration of contributions, we are currently working on the definition of programming interfaces, conventions and procedures. A clear documentation structure of the simulator is maintained using Javadoc.

Finally the SWG web page (http://gaia.am.ub.es/SWG/) provides access to the UML design, the CVS repository, documentation, convention, simulated data, and contact information.

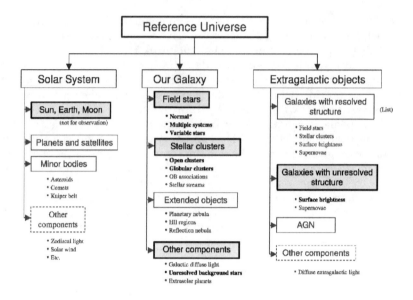

Figure 2. Components of the Universe Model. The modules in grey shade are the ones currently (totally or partially) implemented in the simulations.

3. Universe model

A key component of the simulations is the choice of models to generate synthetic objects to be *observed* by the simulated instruments. We refer to the ensemble of models used here as the 'Universe Model'.

The Universe Model includes many modules representing the different types of objects that GAIA will observe. The availability of realistic simulations of the main types of sky objects during the design phase will allow to optimize the GAIA design for the scientific goals of the mission. Later on, many other types of (more or less exotic) objects will be included allowing the preparation of the scientific exploitation of the mission data.

The development of the Universe Model therefore constitutes a key task in the SWG effort. Its overall structure is depicted in Figure 2.

A module of the Universe Model aims at providing all the relevant information needed to simulate the type of object it represents: spatial distribution, kinematics, photometry and spectra. This information, combined with the instrument models, is used to generate simulated GAIA observations. As seen in the figure, the Universe Model is constituted by three main modules.

The Solar System module describes the objects of the Solar System and contains the ephemeris of the Sun and Earth (needed to describe the GAIA orbit and scan law). It includes a realistic model of the minor planets, including variations in shape and brightness due to the object characteristics and rotation.

The Galactic objects module main components are a Galaxy model and an extinction model; the current versions of these models are rather simplistic

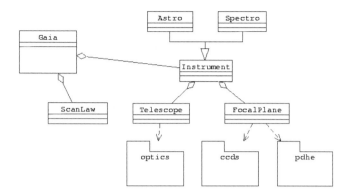

Figure 3. A UML description of the Instrument Model. To illustrate the meaning of the different arrows, here is how to read the right part of this diagram: 'The Spectro class specialize the Instrument class which contains a FocalPlane which is dependent on the package pdhe'.

but they are being refined by the SWG members. These main components are complemented by several other modules describing relevant galactic objects to be observed by GAIA (e.g. variable stars, double stars, extra-Solar planets) or relevant astronomical phenomena (e.g. microlensing).

Finally, the Extragalactic objects module contains models of extragalactic sources to be observed by GAIA, including quasars (whose observations will be crucial to establish the GAIA reference frame and to link it to the ICRS) and external galaxies (both resolved and unresolved). It also includes some extragalactic events observable by GAIA, like supernova explosions.

4. Instrument model

Until recently the simulation efforts have been mainly dedicated to the simulation of astrometric data due to its role as mission driver and to the lack of definition of the spectroscopic instrument. But as illustrated in Figure 3, the Astro instrument model has a lot to share with the spectro instrument. Indeed the Medium Band Photometer (MBP) and Radial Velocity Sky Mapper (RVSM) parts of the Spectro instrument have been included in the Instrument Model, and plans have been deployed to include the RVS simulations in this global model (Katz 2003).

The main characteristics of the instruments need to be simulated, using variable parameters to test different design options, and including possible defaults to access the final accuracies and design the calibration procedure. As for the Universe model, the Intrument model has been divided into several modules.

The scanning law of the satellite allows to determine the pointing direction of each instrument at any time and all the transits of a sky position in the different GAIA instruments.

Optical point spread functions, including realistic aberrations, are combined with the star spectrum properties and the CCD quantum efficiency to provide polychromatic PSFs. Effects of TDI smearing and transverse motion are added.

The main CCD characteristics, including realistic noise, are present. Cosmics, non-linearity, saturation, aging and CTI effects are to be included.

The payload data handling electronics (PDHE) package contains all the on-board data handling algorithms, such as detection, selection, tracking, compression. Prototypes and statistical behaviors are included in the simulator.

5. Roadmap for the spectro simulations

5.1. Image simulations

Several RVS image simulations, presented in these proceedings, have already been developed by different teams and need to be included into the common structure. For this, a detailed RVS instrument model is needed, including realistic PSFs, vignetting model and spectrum simulation methods. Detailed source characteristics should be provided by the universe model to allow an adapted spectrum simulation.

Various simulation modes should be accessible for the different tasks. Indeed the use of different type of approximations, short-cuts or pre-computed data allows to adapt the simulation speed and precision to the individual needs.

5.2. Telemetry simulations

The prototype database of GDAAS will be filled using telemetry simulations coming from GASS and should include spectro instrument data in order to test the corresponding reduction algorithms. Generating such a huge amount of data is a challenging task: the level of realism of these simulations should be tuned so that the CPU time required for its generation does not exceed the available capabilities but allowing at the same time a realistic assessment of the performances of the reduction algorithms in the GDAAS system.

In the coming months, once the RVS design is frozen, these simulation requirements should be carefully addressed by the SWG in order to cope with the demands of the GDAAS implementation schedule.

Acknowledgments. This work has been supported by MCYT under contracts ESP2001-4531-PE and PNAYA-0937.

References

Ansari, S.G. et al. 2003, in GAIA Spectroscopy, Science and Technology, U.Munari ed., ASP Conf. Ser. 298, pag. 97

Babusiaux, C., Arenou, F. & Gilmore, G. 2001, GAIA-CB-01 (Livelink)

Katz, D. 2003, in GAIA Spectroscopy, Science and Technology, U.Munari ed., ASP Conf. Ser. 298, pag. 119

Masana, E., Jordi, C., Figueras, F., Torra, J. & Luri X. 2001, Highlights of Spanish Astrophysics II, pag. 389

Radial Velocity Spectrometer simulator: objectives and specifications

David Katz

Observatoire de Paris, GEPI, 5 place Jules Janssen, 92195 Meudon

Abstract. The Radial Velocity Spectrometer (RVS) simulator is currently in definition phase: definition of the objectives and specifications, coordination with the GAIA simulator development and identification of the interfaces. The RVS simulator is intended to assess the generic performances of the instrument, to be a tool to optimise the instrument design (study of instrumental effects) and to investigate the Galactic or time dependent issues (crowding, multiple systems, etc). Considering the wide scope of the simulator, its main specification is flexibility: on the output data (from single star to field, from one CCD to full mission), on the level of complexity of the background, design and satellite simulation and on the input data (local/simulator or external/user data). The RVS simulator will be developed as part of the more general GAIA simulator, thus using the same interfaces as the other instruments: Galaxy model, general satellite and payload characteristics.

1. Introduction

During the last year and a half, several studies have been conducted to assess the radial velocity spectrometer (RVS) performances and operational characteristics. Dedicated home made simulators were written to reproduce particular aspects of the instrument, e.g. single spectra to assess radial velocity (RV) performances (Katz et al. 2002, Zwitter 2002), spectra plus *contaminating* neighbour stars to investigate the impact of field crowding on the RV performances (Zwitter 2003), images of the RVS field to test compression algorithms (Cropper et al. 2002).

The goal for the coming year is to merge all those simulators in a single *multi-task* simulator in order that, (i) all the people interested in simulating or using simulated RVS data could do so without investing time to code there own simulator, (ii) all the studies performed on RVS spectra or images rely on the same RVS design and modelisation parameters.

As will be developed in the following pages, the RVS simulator will have to be complex enough to fulfill, as much as possible, the variety of expectations from its future users. The first and present step of its development is to define its objectives and specifications. They are respectively reviewed in Sect. 2 and 3. With goals similar to those of the RVS working group, the GAIA Simulation Working Group (http://gaia.am.ub.es/SWG/) develops a simulator of the whole GAIA satellite and instruments (Babusiaux et al. 2001; Babusiaux & Luri 2003), temporary not including the spectrograph. Sect. 4 lists the numerous motivations to merge the RVS simulator into the GAIA simulator and presents an agenda for the simulator development.

2. Objectives

The first objective of the simulator is to produce RVS like spectroscopic data to assess the performances of the RVS instrument: e.g. refine previous studies on the radial and rotational velocities, carry on the estimation of the spectrograph precision in the determination of the atmospheric parameters (T_{eff}, $\log g$, [Fe/H]) or individual abundances (e.g. Ca, Fe, Si, Mg). The description of the RVS will have to be realistic enough, so that design dependent effects be taken into account in the derivation of the accuracy budget: e.g. a systematic radial velocity shift caused by an asymmetric PSF or the decrease of the spectra signal to noise ratio with time due to CCD and optics aging.

During the next 18 months (starting around mid-November 2002), the technical aspects of the RVS instrument will be reviewed, in order to establish the feasibility and optimise the spectrograph optics, detectors and focal plane assembly, mechanics, thermics and proximity electronics. The simulator will be one of the tools used to optimise the different facets of the RVS, allowing to re-derive the instrument performances each time a new option is proposed and to compare with the previous configuration.

Most of the RVS on-board processing is focused on the compression of the data in order to fit in the 0.25 Mbit s^{-1} telemetry stream allocated to the spectrograph: summation of consecutive CCD observations, collapse of the spectrum lines in the across dispersion direction, extraction of the Ca lines regions in faint stars spectra, numerical degradation of the resolution of the faint targets spectra, summation of the spectra over successive transits. The simulator will be used to assess the impact of the on-board processing algorithms on the data and therefore on the performances.

Another application for the RVS simulator is to help to define and test the CCD and wavelength calibration strategies: e.g. derive the accuracy of the calibration of the wavelength dispersion relation as a function of the number of observable RV standards.

Coupling a realistic description of the instrument, with a realistic description of the stellar signal will allow to define and optimise the spectra reduction algorithms. In the case of the derivation of radial velocities, this means for example comparing the advantages and disadvantages of cross-correlation in direct space versus Fourier space, of masks versus synthetic or observed templates, or define the *optimal* mask as a function of spectral type.

The RVS simulator will also have to address Galactic coordinates dependent issues: e.g. performances of the instrument as a function of Galactic coordinates or in high level background areas such as bright nebulae. Along the same line, it will be a tool to investigate time dependent issues, for example test multiple systems orbits reconstruction capabilities or variable stars characterisation.

The simulator will also have to participate to the assessment of the global satellite issue, i.e. problematics relying on the information of several of the satellite instruments. One of the most important of those global questions is the classification and parameterisation of the observed sources which will use astrometric, photometric and spectroscopic information (cf. Bailer-Jones 2003).

3. Specifications

The variety of the simulator objectives implies that it fulfills a large number of specifications, whose key word is flexibility. It starts with the spatial and temporal extent of the data. The simulator will have to be able to generate a single spectrum (e.g. to derive individual target performances), a spectrum plus the signal from *contaminating* neighbours (i.e. to assess impact of crowding) or a full image (e.g. to investigate selection and windowing strategies or to test image processing/compression algorithms). The spectrograph will observe each sky area a large number of times. The simulator must allow to choose the temporal length of the simulated data: a single CCD, a single transit, several transits or a full mission. In this case as for the following specifications, the flexibility gives the possibility of a trade of between complexity/realism and simulation computing time.

Some works will require a very detailed and realistic description of the celestial sphere as seen by the satellite: zodiacal light (in particular small scale variations), stars, galaxies, clouds, solar system objects, cosmic events. But a high level of complexity will not be necessary in all studies. Therefore the simulator will have to provide a precise Galaxy model, but also the option to switch off some aspects of it if not needed or if computing time is favoured over realism.

The simulator will of course integrate its own description of the universe but will also have to allow the users to inject there own model, for example to make test on a real and precise zone of the sky.

The specifications for the simulation of the various facets of the RVS design (optics, mechanics, thermics, detectors) are very similar to the specifications of the universe model. Some applications will require a very realistic description of the instrument: PSF, star line curvature, vignetting rate derived from optical software, CCD read-out noise, dark current, charge transfer inefficiency, aging obtained from laboratory tests. An extensive simulation of the spectrograph design will be extremely time consuming. Therefore, once again, switches will be needed to choose which aspects to include in the computations.

The baseline RVS design will be the configuration considered by default in the simulator. But, the simulator will also be a tool to optimise the spectrograph design. Therefore, it will be necessary that the RVS characteristics could be modified from one run to another.

The simulator will have to be able to take into account the different on-board processing algorithms that are foreseen. Some of them may impact on the format of the output data generated. For example, the on-board summation of the 6 CCD successive observations (proposed to reduce the RVS telemetry stream) has for consequence that a single spectrum is produced per object and per transit instead of six.

The simulator will propose a large choice of observed and/or synthetic spectra. At the same time, users interested in the science case of a particular type of stars (e.g. variables, pre-main sequence, etc) shall have the opportunity to inject their own spectra in the simulator.

Several libraries of spectra covering the RVS wavelength range 8480–8740 Å have been observed or computed in the past few years. Munari & Tomasella (1999), using the Asiago telescope, have mapped the MK system with 131 stars ranging from O4 to M8 types, and Marrese et al. (2003) has extended the mapping to further 92 F-G-K-M stars (including metal rich and poor ones). These libraries will soon be completed by other atlases of Carbon stars, S-type and peculiar stars (cf. Munari 2003). Munari & Castelli (2000) and Castelli & Munari (2001), using the Kurucz software packages (Kurucz 1993a; Kurucz 1993b) and atomic data (Kurucz 1995), have computed a library of synthetic spectra covering a large domain of effective temperatures, surface gravities and metallicities.

Part of the studies will rely on *classical* objects together with the instrument baseline configuration. In such cases and in order to save computing time, the simulator will have to dispose of libraries of pre-processed spectra: e.g. standard type stars spectra computed using accurate optical description.

The definition and implementation of the spectrograph simulator will be a huge task involving several developers. Therefore, it is important, before coding, to develop a clear modelling and management model. This will be achieved using an object oriented modelling tool: the Unified Modelling Language (UML). Because several people will interact and develop the code, it is also necessary that the program be built around a clear structure. This will be obtained by using high level abstraction languages, i.e. object oriented: Java or C++. Some particular subroutines may be very demanding in term of processing power and will then be coded using fast languages such as Fortran, C or C++, with some wrapper to make the interface between the routines and the rest of the simulator.

4. Integration in the GAIA simulator

4.1. Motivations and interfaces

There are numerous motivations to integrate the RVS simulator in the global GAIA simulator. The first one is to avoid work duplication. A huge amount of resources have been developed for the GAIA simulator and could be used by the RVS simulator. The main interfaces are the universe model, the satellite characteristics (e.g. scanning law, attitude stability, etc), the payload characteristics (e.g. CCD properties), the simulation methods (e.g. simulation of the TDI mode) and of course the other GAIA instruments.

Another reason to simulate the astrometric, photometric and spectroscopic instruments with a single program, is that all satellite studies rely on the same assumptions. At the same time, a single simulator for all the GAIA instruments will allow to assess the global satellite issues: e.g. the detection of the sources by the radial velocity sky mappers (located in the photometric focal plane and operating in undispersed light) followed by the selection and windowing of the spectra, the characterisation of the multiple or variable systems and, in general, the parameterisation of the sources based on all the available information (astrometry, broad and medium band photometry and spectroscopy).

4.2. Agenda

The RVS simulator will be a complex software package, which will fulfill many specifications and will be incorporated in the even more complex GAIA simulator. The conception, realisation and integration will take a long time. To fill the gap until the integration of the RVS simulator in the GAIA simulator, some autonomous and more focused versions (with restricted objectives) of the RVS simulator will be distributed. Table 1 lists the planned versions together with their release dates.

Table 1. Release agenda of the planned versions of the RVS simulator.

Version		Release date
v1.0	Single spectra - Gaussian PSF	2002/11
v2.0	Single spectra - Optical PSF	2003/01
v3.0	Image	2003/Q1
	RVS simulator UML model	2003/Q2
	Integration of RVS sim. in GAIA sim.	2003/Q4

The first version of the simulator, to be distributed in November 2002, will generate a single spectrum taking into account the following effects: convolution according to the resolution (assuming a Gaussian PSF profile), normalisation according to the magnitude, source type and RVS design, 2-dimensional sampling according to the pixel size and the across dispersion profile and generation of the photon, zodiacal light and read-out noises.

The second version of the simulator (January 2003) will be similar to the first one, except that it will take into account a realistic description of the spectrograph optics and TDI mode: PSF computed by optical software packages as a function of the wavelength and position in the field of view, star line curvature, spectrum orientation variation, vignetting rate and TDI blurring.

The third version (end of first semester 2003) will be an image simulator relying on a Universe model to produce RVS instrument snapshots as a function of Galactic coordinates.

The development, coding, testing and integration of the spectrograph simulator will be performed in parallel to the development of the RVS simulator versions 1 to 3, with the aim of integrating the Radial Velocity Spectrometer simulator in the GAIA simulator by the end of 2003.

References

Babusiaux, C., Arenou, F., & Gilmore G. 2001, GAIA-CB-001 (Livelink)

Babusiaux, C., & Luri, X. 2003, in GAIA Spectroscopy, Science and Technology, U.Munari ed., ASP Conf. Ser. 298, pag. 113

Bailer-Jones, C.A.L. 2003, in GAIA Spectroscopy, Science and Technology, U.Munari ed., ASP Conf. Ser. 298, pag. 199

Castelli, F., & Munari, U. 2001, A&A 366, 1003

Cropper, M., Smith, P., & Brindle, C. 2002, RVS-MSC-003 (Livelink)

Katz, D., Viala, Y., Gomez, A. & Morin, D. 2002, in GAIA: A European Space Project, O. Bienaymé & C. Turon ed.s, EAS Pub. Ser. 2, pag. 63

Kurucz, R.L. 1993a, Kurucz's CD-ROM No. 13, CfA Harvard, Cambridge

Kurucz, R.L. 1993b, Kurucz's CD-ROM No. 18, CfA Harvard, Cambridge

Kurucz, R.L. 1995, Kurucz's CD-ROM No. 23, CfA Harvard, Cambridge

Marrese, P.M., Munari, U., Boschi, F. & Tomasella, L. 2003, in GAIA Spectroscopy, Science and Technology, U.Munari ed., ASP Conf. Ser. 298, pag. 427

Munari, U., & Tomasella, L. 1999, A&AS 137, 521

Munari, U., & Castelli, F. 2000, A&AS 141, 141

Munari, U. 2003, in GAIA Spectroscopy, Science and Technology, U.Munari ed., ASP Conf. Ser. 298, pag. 227

Zwitter, T. 2002, A&A 386, 748

Zwitter, T. 2003, in GAIA Spectroscopy, Science and Technology, U.Munari ed., ASP Conf. Ser. 298, pag. 493

session 3
GALAXY AND STARS
chair: F. Mignard

Participants on an after-lunch walk, close to the Congress Center

Clockwise from top-left: U. Munari, D. Katz, T. Zwitter, F. Mignard, G. Gilmore and C. Babusiaux.

Optimising GAIA: how do we meet the science challenge?

Gerry Gilmore

Institute of Astronomy, Madingley Road, Cambridge CB3 0HA, UK

Abstract. GAIA has been approved to provide the data needed to quantify the formation and evolution of the Milky Way Galaxy, and its near neighbours. That requires study of all four key Galactic stellar populations: Bulge, Halo, Thick Disk, Thin Disk. The complex analysis methodologies required to model GAIA kinematic data are being developed, and in the interim applied to the relatively simple cases of the satellite dSph galaxies. These methodologies, illustrated here, show that we will be able to interpret the GAIA data. They also quantify what data GAIA must provide. It is very unlikely in the present design that GAIA will be able to provide either radial velocities or worthwhile photometry for study of two of the key science goals: the Galactic Bulge and the (inner) Galactic old disk. The implication is that the radial velocity spectrometer and the medium band photometer should be optimised for study of low density fields – suitable for their low spatial resolution – and the broad band photometry must be optimised for inner galaxy astrophysical studies.

1. Introduction

The GAIA mission has been approved to provide for the first time a clear picture of the formation, structure, evolution, and future of the entire Milky Way. In addition, as secondary goals, GAIA will contribute to many other branches of astrophysics, especially stellar and solar system minor body astrophysics, with a valuable contribution to cosmology and fundamental physics.

This clear scientific prioritisation must drive the design, and all compromises.

Understanding the structure and evolution of the Galaxy requires three complementary observational approaches: (*a*) carrying out a full census of all the objects in a large, representative, part of the Galaxy; (*b*) mapping quantitatively the spatial structure of the Galaxy; (*c*) measuring the motions of objects in three-dimensions to determine the gravitational field and the stellar orbits. In other words, what is required are complementary measurements of distances (astrometry), photometry to determine both extinction and intrinsic stellar properties, and the radial velocities along our line of sight.

Detailed studies of the Local Group are the key specific test of our understanding of the formation and growth of structure in the Universe. The most significant questions here relevant for GAIA include the history of most of the stars in the Galaxy, the nature of the inner Galactic disk and the Galactic Bulge:

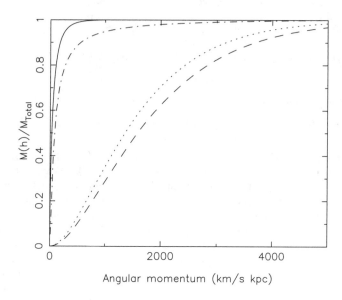

Figure 1. Galactic population angular momentum cumulative distribution functions, showing bulge/halo (solid and dash-dot adjacent curves) and thin/thick disk (dotted and long-dashed adjacent curves) dichotomy. This indicates disparate evolutionary histories, and emphasises the critical need for GAIA to study all four Galactic components.

what are their age, abundance and assembly histories? Stellar studies locally of course are of critical general significance: only locally can we determine the stellar Initial Mass Function directly. This function directly controls the chemical and luminosity evolution of the Universe. Galactic satellite galaxies are proving the most suitable environs to quantify the nature and distribution of dark matter, and to test the small scale predictions of hierarchical galaxy formation models. The Galactic disk itself, of course, is a key test of angular momentum distributions (Figure 1), chemical evolution, and merger histories, and must eventually provide the most robust information on Cold Dark Matter (CDM). Such issues as the number of local dwarf galaxies and the inner CDM profiles of the dSphs are well known demanding challenges for CDM models. Other Local group information is also important: what is the stellar Initial Mass Function, what is the distribution of chemical elements, what is the age range in the Galactic Bulge, when did the last significant disk merger happen, etc ? All these challenge our appreciation of galaxy formation and evolution, and in turn provide the information needed to refine the models. The partnership between local observations and *ab initio* theory is close, and developing well, and must culminate in GAIA.

There are clear similarities and distinctions between fundamental properties of the different Galactic stellar populations, such as age, metallicity, star formation history, angular momentum (Figure 1) which allow study of their individual histories. This is arguably one of the greatest advantages of studies of Local Group galaxies: one is able to disentangle the many different histories which have led to a galaxy typical of those which dominate the luminosity density of the Universe.

2. GAIA and dark matter mapping

Ninety percent of the matter in the Universe is of an unknown nature (to say nothing here of the even larger amount of dark energy). This matter dominates gravitational potential wells in the early Universe on all scales, and everywhere except in the centres of large galaxies and dense stellar clusters today. The smallest scale lengths on which this dark matter is dominant are an important constraint on its nature: for example, if dark matter is concentrated on very small scale lengths it cannot be relativistic. The only known way in which we can study dark matter is through its dynamical effects on test particles orbiting in a potential which it generates. The simplest dark-matter dominated systems known, with scale lengths of interest to constrain the physical nature of the dark matter, and with a dynamical structure simple enough to be amenable to understanding, are the Galactic satellite dwarf spheroidal galaxies. Very extensive studies, using gravitational lensing, galaxy rotation curves, stellar and gas kinematics, X-ray profiles, and so on, have initiated mapping dark matter on larger scales. For the dwarf galaxies, stellar kinematics are uniquely the method of choice.

This dark matter dominates the matter in the Universe, but remains weakly constrained: what is the nature of the dark matter? A variety of theoretical studies suggest that the small scale structure of dark matter is the key test of its nature (Navarro, Frenk & White 1997; Klypin etal 1999; Moore etal 1999). Many extant studies of dark matter on many different scale lengths are available, using techniques such as gravitational lensing, X-ray luminosity and temperature profiles, HI and Hα rotation curves, and stellar dynamics. None has been able to provide a reliable study of dark matter on the crucial small length scales, because the only small dark-matter dominated relatively simple dynamical systems, the Milky Way satellite dwarf galaxies, can be studied in detail only with very accurate kinematics of faint stars.

2.1. The importance of the smallest dark haloes

To be confined on small scales, a system must be cold. For self-gravitating systems the virial velocity, and equivalently virial temperature, of a system in equilibrium is simply related to its mean density, ρ, and characteristic radius, R, as

$$V^2_{virial} = k_B T_{virial}/m_p \sim G\rho R^2.$$

Baryonic material, which can radiate away energy while gaseous, can form systems of a wide range of virial velocities, with the lower limit just set by the cooling law. In particular, baryonic dark matter could form dark haloes of small scalelength. One can relate the 'temperature' of non-baryonic dark matter (assumed in thermal equilibrium in the early Universe) with its velocity dispersion at the epoch in the early universe when it decouples from the ordinary matter (Bond & Szalay 1983). 'Hot' dark matter, for example a massive neutrino, which is relativistic at this epoch, has a large free-streaming length and cannot be confined on small scales (Bond, Szalay & White 1983). In contrast, 'cold' dark matter, which is non-relativistic when it decouples from ordinary matter, can form small-scale structure. One currently popular non-baryonic cold dark matter candidate, the axion, is not produced through processes in thermal equilibrium, has an extremely low 'temperature', and can cluster on scales below that of even a dwarf galaxy.

Thus the study of dark matter in systems with the smallest characteristic radius provides a thermometer of dark matter and can be used to determine its nature. Dwarf spheroidal galaxies have (stellar) scale-lengths of ~ 300 pc (Irwin & Hatzidimitriou 1995), comparable to the scale height of the Galactic thin disk and bulge. Numerical simulations of cosmological structure formation do not have the dynamic range necessary to study such scales. Irrespective of the future developments of computing power and techniques, the simulation of a dwarf galaxy requires the inclusion of the unknown physics of star formation and feedback, expected to be particularly important in these low velocity-dispersion systems (e.g. Dekel & Silk 1986; Wyse & Silk 1985; Lin & Murray 1991). Improved understanding must be led by observations.

What are the smallest systems for which there is evidence of dark matter? Galactic disks are thin, of small scale-height, and we are located in the middle of one. Dark matter confined to the disk (as distinct from the extended, Galactic-scale dark halo) would be 'cold'. However, the local neighbourhood is not a good place to study cold disk dark matter, since it is now accepted there is no evidence that any exists (Kuijken & Gilmore 1989, 1991; Gilmore, Wyse & Kuijken 1989; Flynn & Fuchs 1994; Creze etal. 1998; Holmberg & Flynn 2000). Globular clusters are fairly nearby, and have very small scalelengths, tens of parsecs, but again there is no evidence for dark matter associated with globular clusters (e.g. Meylan & Heggie 1997). Note that in fact in these two cases there is not just lack of evidence for cold dark matter, there is actual evidence for a lack of dark matter, which in itself is significant.

In contrast, the dark matter content of low-luminosity dwarf galaxies, as inferred from analyses of their internal stellar and gas kinematics, makes them the most dark-matter-dominated galaxies (e.g. Mateo 1998; Carignan & Beaulieu 1989). Of these small galaxies, the low-luminosity gas-free dwarf Spheroidals (dSph) are the most extreme. Available stellar kinematic studies provide strong evidence for the presence of dominant dark matter (e.g. review of Mateo 1998), confirming earlier inferences from estimates of the dSph's tidal radii and a model of the Milky Way gravitational potential (Faber & Lin 1983).

2.2. What information is essential?

Substantial analytic methodologies have been developed over the last few years by several groups to interpret colour-magnitude data and stellar kinematics. These studies provide an existence theorem proving that efficient analysis of the GAIA and complementary data is possible.

Adequate kinematic data can determine the potential well locally on a variety of different scales, parsecs up to 1Mpc. The motivation is that the *smallest* scale is set by a fundamental characteristic of the dark matter, its velocity dispersion, analogously its temperature, so that study of the dynamics on these scales provides unique and strong constraints on the nature of dark matter. Such a determination is possible – and is possible only – using full 6-dimensional phase space data, such as will be determinable by GAIA.

3. Dwarf galaxies: an example analysis

The dynamical structure of a dwarf galaxy, while simple, is not trivial. The galaxies are probably triaxial, can sustain complex families of anisotropic orbits, and live in a time-varying Galactic tidal field. Given this complexity, considerable information is required to derive a robust measurement of the 3-dimensional gravitational potential, and hence the generating (dark) mass. Several groups have devised methods to explore the possible dynamical structure of a Galactic satellite, and used these to simulate observational approaches to determination of the mass distribution. These studies show that determination of at least five phase space coordinates – two projected spatial coordinates, and all three components of the velocity – are necessary and sufficient to provide a robust and reliable determination of a dark matter mass distribution. Acquisition of these data are possible only with a combination of radial velocity data and proper motion data for a well-selected sample of stars distributed across the whole face of the target galaxy. In the nearest appropriate target galaxies, Draco and Ursa Minor, the stars of interest are at magnitude $V \sim 19$, and the required proper motion accuracy, corresponding to 1-2 km s^{-1} in velocity, is 3–6 μas yr^{-1}. Thus this specific test of the nature of matter on small scales is better with SIM rather than GAIA. Nonetheless, the analysis principles apply in general, and can be generalised to GAIA studies of the local Galaxy, where SIM cannot contribute.

An important simplification in any dynamical analysis is that the potential be time-independent over an internal dynamical timescale, the tracer stellar distribution be in equilibrium with the potential, and be well-mixed (relaxed). While dSph galaxies show every possible star formation history, it is a defining characteristic that star formation ceased at least a few Gyr ago (e.g. Hernandez, Gilmore & Valls-Gabaud 2000). The internal dynamical times of these galaxies are typically only $t_{\rm dyn} \sim R/\sigma \sim 2 \times 10^7 (R/200 \text{ pc})(10 \text{ km s}^{-1}/\sigma)$ yr; thus star formation indeed ceased many hundreds of crossing times ago and the systems should be well-mixed. This assumption can of course be directly tested by the data.

3.1. The degeneracy between anisotropy and mass

Most studies of galactic dynamics have been impeded by the degeneracy between anisotropy and mass. This degeneracy, that one cannot distinguish between gravitational potential gradients and gradients in orbital anisotropy from measurements of just one component of the velocity ellipsoid (radial velocity), is the fundamental limitation in extant analyses, and the key justification for extending observation to 5 phase space coordinates. The large observed central radial velocity dispersion is compatible with either a massive halo, and a low central density; or no halo, and a large central density. To demonstrate our ability to discern between these possibilities, Wilkinson et al. (2002) constructed a set of models that span the space of halo mass and anisotropy, and perform Monte Carlo recoveries of the input anisotropy and halo mass.

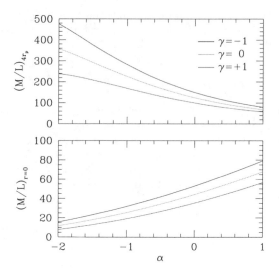

Figure 2. Variation of mass–to–light ratio (in units of $M_\odot/L_\odot$) as a function of α for three different γ values. The *top* panel shows the total M/L within $4r_0$ ($3r_0 \approx$ Draco King tidal radius; Irwin & Hatzidimitrou, 1995). The *bottom* panel shows the central M/L. Upper line: $\gamma=-1$; center line: $\gamma=0$; lower line: $\gamma=+1$.

They assume that the luminosity density of a dSph is given by a Plummer model

$$\rho_\mathrm{p}(r) = \frac{\rho_0}{\left(1+(r/r_0)^2\right)^{5/2}}, \qquad (1)$$

where ρ_0 is determined by the total observed luminosity. Next, assume that the potential of the system has the form

$$\psi(r) = \frac{\psi_0}{\left(1+(r/r_0)^2\right)^{\alpha/2}} \qquad (2)$$

For this dark matter potential, $\alpha = 1$ corresponds to a mass–follows–light Plummer potential, Keplerian at large radii; $\alpha = 0$ yields, for large r, a flat rotation curve; and $\alpha = -2$ gives a harmonic oscillator potential. As the parameter α decreases, the dSph becomes more and more dark matter dominated.

Finally, assume that the distribution function (DF) of the stars in the model potential is a function entirely of the energy E and norm of the angular momentum L. Under this assumption, one can derive a two parameter set of DFs, parameterised by α and an anisotropy parameter γ, namely

$$F(E,L^2) = \frac{\rho_0}{\psi_0^{5/\beta-\gamma/\beta}} \frac{\Gamma(5/\beta-\gamma/\beta+1)}{(2\pi)^{3/2}\Gamma(5/\beta-\gamma/\beta-1/2)} E^{5/\beta-\gamma/\beta-3/2}$$
$$\,_2F_1(\gamma/2; 3/2-5/\beta+\gamma/\beta, 1, L^2/2E) \qquad (3)$$

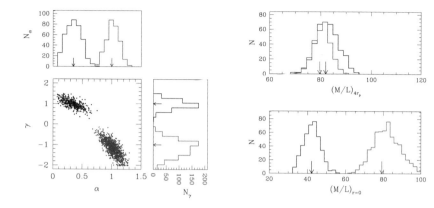

Figure 3. *Left:* recovery of best–fit α, γ for 2 sets of 250 star artificial data ensembles; one set of ensembles was created with $\alpha = 0.2, \gamma = 1$, the other with $\alpha = 1, \gamma = -1$. The centre panel is a scatter plot of the recovered α, γ; the top and right panels are histograms of α and γ, respectively; arrows indicate actual values. We are able to distinguish clearly between these models, and break the mass/anisotropy degeneracy. *Right:* recovered mass to light ratio for simulations in left plot. Top panel shows the total M/L (solar units) within 4 Plummer radii; bottom panel shows the central M/L.

This formula holds for $L^2 < 2E$ and $\alpha < 0$; similar expressions can be derived for all the other cases. These DFs generalise earlier calculations by Dejonghe (1987), which are restricted to the case where there is no dark matter. In terms of Binney's anisotropy parameter β, the radial and tangential velocity dispersions σ_r^2 and σ_θ^2 vary as

$$\beta = 1 - \frac{\sigma_\theta^2}{\sigma_r^2} = \frac{\gamma}{2}\frac{r^2}{1+r^2} \qquad (4)$$

The potential normalisation ψ_0 in Eq.(2) is a well defined function of the observed central radial velocity dispersion, and the assumed α and γ. Figure 2 shows the variation of the central and large–scale mass to light ratio as a function of α and γ.

To simulate our ability to resolve among values of γ and α, we generate 6D phase space coordinates $\{x_i, y_i, v_{xi}, v_{yi}, v_{zi}\}_{i=1...N}$ drawn from the DF $F(x, y, z, v_x, v_y, v_z; \alpha, \gamma)$, discard the z (geocentric radial position) coordinate, and attempt to recover γ and α using a Bayesian likelihood technique (see e.g. Little & Tremaine 1987, Kochanek 1996, Wilkinson & Evans 1999, van der Marel et al. 2000). Specifically, we scan a grid of α, γ, and at each point compute the probability of observing the input data set

$$P(\{x_i, y_i, v_{xi}, v_{yi}, v_{zi}\}_{i=1...N} | \alpha, \gamma) = \prod_{i=1}^{N} \int_{-\infty}^{\infty} dz\, F(x_i, y_i, z, v_{xi}, v_{yi}, v_{zi}; \alpha, \gamma) \qquad (5)$$

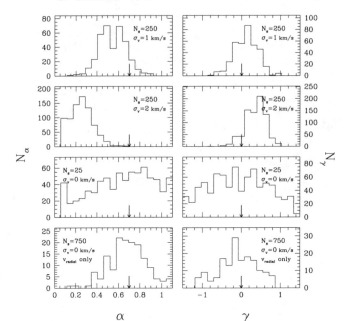

Figure 4. *Top three panels:* Effect of velocity measurement errors of 10%, 20%, and 40% of the system dispersion (=10 km s^{-1} here), respectively, on recovery of halo parameter α (left) and anisotropy parameter γ (right). The input (arrows) α and γ can be recovered with measuring errors of 10% or 20% of the system internal velocity dispersion, albeit with a (calibratable) systematic shift in γ (see text). *Third panel from top:* Recovery of α and γ using only 25 velocities; no meaningful constraints can be placed on α and γ. *Bottom panel:* Recovery of α and γ using 750 radial velocities; α and γ are poorly constrained, demonstrating the necessity of proper motions and radial velocity data.

By Bayes's theorem, and the assumption of uniform prior probabilities of α, γ, the most likely α, γ are given by maximising Eq.(5). Confidence regions are obtained by applying 2D χ^2 statistics to the logarithm of Eq.(5).

3.2. Results

Figure 3 (left) shows the result of a Monte Carlo simulation to recover the input α and β of two model distributions: $\alpha = 0.35, \gamma = 1$, and $\alpha = 0.8, \gamma = -1$, respectively. These values lie along the direction of the usual mass/anisotropy indeterminacy, and thus illustrate our ability to break this degeneracy. Both model fits involve 800 reconstructions, each of which contains an ensemble of 250 simulated stars. The right panel of Figure 3 depicts constraints placed on the central M/L, and the total M/L within four Plummer radii. Although the two models evaluated have a similar total mass within $4r_p$, our method clearly recovers the significant difference in central M/L caused by a halo.

The top two panels of Figure 4 illustrate the effect of anticipated velocity measurement errors which are 10% and 20% of the internal (local) velocity dispersion of the (sub-)system under study, assuming a 250 star sample. These errors cause a systematic displacement of the best–fit α in the small halo ($\alpha = 0.7$) case we examine, because they create a significant increase of the apparent velocity dispersion at the maximum radius ($3r_P$) considered. However, the width of the distribution remains narrow, and the systematic α shift will be rigorously incorporated into the analysis by convolving the velocity errors with the DF. Larger errors would not be acceptable, however, because resolving among α values requires a precise determination of the falloff of the velocity dispersion. The third panel of Figure 4 illustrates the recovery of α, γ using only 25 velocities; the halo and anisotropy are essentially unconstrained by this small sample. The bottom panel shows the effect of using 750 purely radial velocities, with zero errors. Although we have the same number of one–dimensional velocities as in the 250 star cases, we are unable to recover α and γ; this failure indicates that the analysis requires all three velocity components, and this work could not be performed using ground–based radial velocities.

3.3. Triaxial dSph models

A triaxial spheroid seen projected on the plane of sky mimics a rotationally symmetric ellipse. Triaxial systems can support much more complex orbital structures than can 2-D systems; this allows a degeneracy between radially-dependent orbital complexity and system shape, potentially invalidating dynamical analyses. More complex modelling is required in this case, and has been partially developed. The (very obvious!) conclusion from extensive numerical simulations based on those triaxial models is that it is vital to have full kinematics available if we are to investigate triaxial (more generally: not isolated spherical) systems.

4. Implications for GAIA design

The discussion above illustrates the considerable progress in quantitative analyses of stellar kinematics which are currently under development. The introduction recalled that GAIA is funded to study Galactic evolution. What data must GAIA provide so these analysis techniques can deliver this science? The quantitative analysis techniques need distances, 2-D (better 3-D) kinematics, and some astrophysical information on the distribution function of stellar properties. Highly detailed information on a few stars is not what is needed. Can GAIA deliver? Simulations presented at this conference show GAIA radial velocity data will be unable to provide useful scientific data at low Galactic latitudes. Low Galactic latitudes, while *only* a few percent of sky, contain most of the stars, and contain *all* of the inner disk and bulge. It is worth emphasising that these parts of the Milky way are the only parts of high redshift galaxies which we can see. If GAIA fails to provide adequate data on these low latitudes, it fails a primary science goal.

However, as the simulations show, bad data are useless. The lesson for GAIA spectra is clear: do something excellently, not several things badly, optimise for the faintest possible stars in uncrowded regions: leave low latitudes for the other instruments.

What other instruments? The medium band photometr (MBP) unfortunately has been degraded to poor ground-based quality spatial resolution, 1 arcsec. Thus, as long experience has established, MBP will also be of little or no value in the inner Galaxy. Again, the implication is clear: optimise MBP for relatively low extinction higher b regions, with somewhat metal-poor old stars.

Can GAIA meets its primary science goals at all: only if the broad band filters are optimised to deliver the essential minimum astrophysical data. That is, the limited sensitivity of RVS and the poor spatial resolution of MBP require that the broad band filters be optimised for astrophysical analyses, and not other considerations. Without these choices, GAIA will fail to meet its design science goals. With them, it will revolutionise our understanding of galaxy formation and evolution.

References

Bond, J.R. & Szalay, A.S. 1983, ApJ 274, 443
Bond, J.R., Szalay, A.S. & White, S.D.M, 1983, Nature 301, 584
Carignan, C. & Beaulieu, S. 1989, ApJ 347, 760
Creze, M., Chereul, E., Bienayme, O. & Pichon, C. 1998, A&A 329, 920
Dejonghe, H. 1987, MNRAS 224, 13
Dekel, A. & Silk, J. 1986, ApJ 303, 39
Faber, S.M. & Lin, D.N.C. 1983, ApJL 266, L17
Flynn, C. & Fuchs, B. 1994, MNRAS 270, 471
Gilmore, G., Wyse, R.F.G. & Kuijken, K. 1989, ARA&A 27, 555
Hernandez, X., Gilmore, G. & Valls-Gabaud, D. 2000, MNRAS 317, 831
Holmberg, J. & Flynn, C. 2000, MNRAS 313, 209
Irwin, M. & Hatzidimitriou, D. 1995, MNRAS 277, 1354
Klypin, A., Kravtsov, A.V., Valenzuela, O. & Prada, F. 1999, ApJ 522, 82
Kochanek, C.S. 1996, ApJ 457, 228
Kuijken, K. & Gilmore, G. 1989, MNRAS 239, 605
Kuijken, K. & Gilmore, G. 1991, ApJL 367, L9
Lin, D.N.C. & Murray, S. 1991, in The Formation and Evolution of Star Clusters, K. Janes ed., ASP Conf. Ser. 13, pag. 55
Little, B. & Tremaine, S. 1987, ApJL 320, L494
Mateo, M.L. 1998, ARA&A 36, 435
Meylan, G. & Heggie, D.C. 1997, A&A Rev 8, 1
Moore, B., Ghigna, S., Governato, F., Lake, G., Quinn, T., Stadel, J. & Tozzi, P. 1999, ApJL 524, L19
Navarro, J.F., Frenk, C.S. & White, S.D.M. 1997, ApJ 490, 493
van der Marel R.P., Magorrian J., Carlberg R.G., Yee H.K.C. & Ellingson E. 2000, AJ 119, 2038
Wilkinson, M.I. & Evans, N.W. 1999, MNRAS 310, 645
Wilkinson, M., Kleyna, J., Evans, N.W., & Gilmore,G. 2002, MNRAS 330, 778
Wyse, R.F.G. & Silk, J. 1985, ApJL 296, L1

Galaxy structure and kinematics towards the NGP

Alessandro Spagna

INAF – Osservatorio Astronomico di Torino, I-10025 Pino Torinese, Italy

Carla Cacciari

INAF – Osservatorio Astronomico di Bologna, I-40127 Bologna

Ronald Drimmel

INAF – Osservatorio Astronomico di Torino, I-10025 Pino Torinese, Italy

Thomas Kinman

Kitt Peak National Observatory, NOAO, Tucson, AZ 85726-6732, USA

Mario G. Lattanzi

INAF – Osservatorio Astronomico di Torino, I-10025 Pino Torinese, Italy

Richard L. Smart

INAF – Osservatorio Astronomico di Torino, I-10025 Pino Torinese, Italy

Abstract. We present a proper motion survey over about 200 square degrees towards the NGP, based on the material used for the construction of the GSC-II, that we are using to study the vertical structure and kinematics of the Galaxy. In particular, we measured the rotation velocity of the halo up to 10 kpc above the galactic plane traced by a sample of RR Lyræ and BHB giants for which radial velocities were used to recover the complete distribution of the spatial velocities. Finally, the impact of astrometric and spectroscopic GAIA observation are discussed.

1. Introduction

It is generally accepted that the Galaxy is constituted by four discrete main components, the *bulge*, the *thin disk*, the *thick disk* and the *halo*, which are characterized by distinctive stellar populations in terms of spatial distribution, kinematics properties, metallicity and age. A detailed knowledge of such galactic components is essential to achieve a complete description of the Milky Way, as well as of the various processes and evolutionary phases which occurred during the history of our Galaxy (and other galaxies too) and that are responsible for the existence and the properties of those components we observe today.

Ground based surveys providing photometry, proper motions and/or spectroscopic observations have been carried out in the past to study the structure and kinematics of Galactic populations, and these will continue and be extended in the next years thanks to availability of all-sky catalogs such as GSC-II and USNO-B, based on multi-epoch photographic surveys, as well as other current and future photometric and spectroscopic surveys (e.g. 2MASS, DENIS, SDSS, EIS, HES, HK, VST, etc.) that benefit from the availability of dedicated scanning and large field cameras as well as large multi-fiber spectrographs (e.g. 2dF, 6dF, FLAMES).

Clearly, the GAIA mission will provide an enormous contribution to the understanding of the formation and evolution of the Galaxy, thanks to very accurate astrometric parameters complemented by photometric and spectroscopic measurements which will permit the direct determination of the 6D phase space distribution and the chemical abundance of large and complete samples of stellar tracers belonging to the different galactic components.

Here we describe a new project which combines astrometric, photometric and spectroscopic data in order to investigate the kinematics of the outer halo by means of velocities derived from proper motions and radial velocities for a sample of RR-Lyræ and BHB giants towards the NGP.

2. NGP survey: proper motions and radial velocities

At the moment we have surveyed about 200 square degrees towards the NGP, and produced positions, proper motions and photographic photometry for about 500 000 objects down to plate limits ($R_F < 20.5$).

Table 1.   Plate material

Survey	Epoch	Pixel	Band	Emulsion + Filter
POSS-I E	1950-1956	25 μm	E	103a-E + red plexiglass
POSS-I O	1950-1955	25 μm	O	103a-O unfiltered
Quick V	1982-1983	25 μm	V_{12}	IIaD+Wratten 12
POSS-II J	1988-1996	15 μm	B_J	IIIaJ + GG385
POSS-II F	1989-1996	15 μm	R_F	IIIaF + RG610
POSS-II N	1990-1998	15 μm	I_N	IV-N + RG9

Our material consists of $6.4° \times 6.4°$ Schmidt plates from the Northern photographic surveys (POSS-I, Quick V and POSS-II) carried out at the Palomar Observatory (Table 1). All plates were digitized at STScI utilizing modified PDS-type scanning machines with 25 μm square pixels ($1''.7$ pix^{-1}) for the first epoch plates, and 15 μm pixels ($1''$ pix^{-1}) for the second epoch plates. The digital copies of the plates were initially analyzed by means of the standard software pipeline used for the construction of the GSC-II (e.g. Lasker et al.

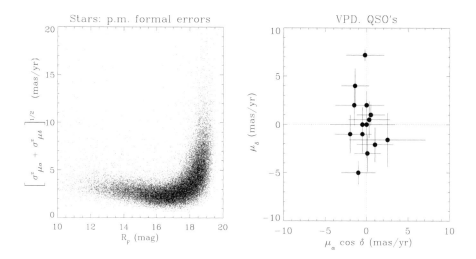

Figure 1. *Left panel*: formal errors of fitted proper motions as a function of the magnitude (all stellar objects). *Right panel*: vector point diagram (VPD) of a sample of 15 QSO's, with 1 σ error bars. Weighted means are $\langle \mu_\alpha \cos\delta \rangle = -0.02 \pm 0.23$ mas yr^{-1} and $\langle \mu_\delta \rangle = +0.33 \pm 0.28$ mas yr^{-1}, respectively. Both plots refer to the POSS-II field no. 442.

1995, McLean et al. 2000). The pipeline performs object detection and computes parameters and features for each identified object. Further, the software provides classification, position, and magnitude for each object by means of astrometric and photometric calibrations which utilized Tycho 2 (Høg et al. 2000) and GSPC-2 (Bucciarelli et al. 2001) as reference catalogs. Accuracies better than 0.1-0.2 arcsec in position and 0.15-0.2 mag in photographic magnitude are generally attained. Relative proper motions were derived by applying the procedure described in Spagna et al. (1996) and afterwards transformed to the absolute reference frame forcing the extended extragalactic sources to have null tangential motion. As shown in Figure 1, the typical precision ($\sigma_\mu \sim 3$ mas yr^{-1} down to $R_F \simeq 18$) has been estimated from the formal errors of the fitted proper motions, while the zero point accuracy of the absolute proper motions have been tested by checking the mean motion of a set of known QSO's that give values smaller than 1 mas yr^{-1} on each component.

Radial velocities and chemical abundances of the sample of RR Lyræ and BHB giants were derived by means of spectroscopic observations carried out with the 4m Mayall telescope at Kitt Peak and with the 3.5m TNG on La Palma. The data were processed with standard procedures and routines (IRAF), and typical errors are $\sigma_{RV} \leq 40$ km s^{-1} and $\sigma_{[Fe/H]} \sim 0.2$ dex.

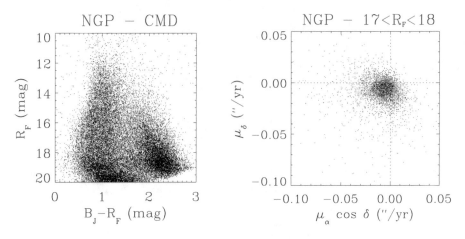

Figure 2. Color magnitude diagram (*left panel*) and vector point diagram (*right panel*) for stars with $17 \leq R_F \leq 18$. Both plots are based on data from the POSS-II field no. 442.

3. The vertical structure

The color magnitude diagram and the vector point diagram observed in one field of our NGP survey are shown in Figure 2. The observed distributions are the result of the complex mixture of the stars belonging to the various populations which are present towards high galactic latitudes. In particular, these may include:

1. the flat and rapidly rotating old thin disk;
2. the extended thick disk, including its metal weak tail;
3. a flattened and slowly rotating inner halo;
4. a spheroidal non-rotating outer halo;
5. satellite debris and kinematics substructures.

Actually, the physical properties of these components are not completely established and various problems are still controversial. For instance: (*a*) the density scale factor, rotation and metallicity distribution of the thick disk; (*b*) the nature of the metal weak thick disk (MWTD) and its relation with the standard thick disk (satellite debris or initial phase of a dissipative formation?); (*c*) the halo velocity ellipsoid, the spatial distribution and axial ratio of the halo as a function of the distance, (*d*) the search of halo streams; (*f*) the determination of the luminosity and mass function of the faintest Pop.II stars (white dwarfs, late M dwarfs and subdwarfs).

Due to the lack of trigonometric parallaxes for our and similar surveys, accurate distances and tangential velocities cannot be directly determined. In such a case it is convenient to analyze these large data-sets by comparing the observations against *ad hoc* Galaxy models which describe both the density and

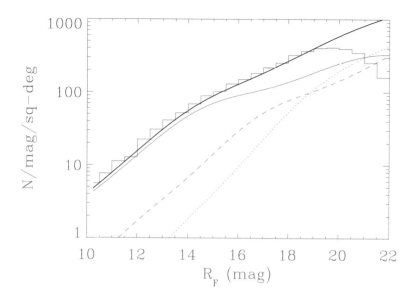

Figure 3. Starcounts derived from plate XP444 (*histogram*) and compared with the distribution predicted by the Mendez's Galaxy model (*thick solid line*), which includes thin disk (*solid line*), thick disk (*dashed line*) and halo (*dotted line*) components.

kinematics, as for instance the Besançon model (Robin & Oblak 1987), the IAS Galaxy model (Bachall, Casertano & Ratnatunga 1987), or the models developed by Mendez & van Altena (1996) and Chen (1997). To this regard, in Figure 3 starcounts are compared against the Mendez's Galaxy model which have been extended to the photographic B_J and R_F magnitudes by A. Spagna (STScI-NGST-R-0013B 2001, http://www.ngst.nasa.gov/public/unconfigured/doc_0422/rev_03/NGST_GS_report5.pdf).

Alternatively, using tracers with known brightness, it is possible to derive distance and space velocity by means of photometry, proper motions and spectroscopic radial velocities. This approach has been adopted to investigate the kinematics of the outer halo traced by a set of 31 RR Lyræ and 65 BHB giants, as will be discussed in the following section. These objects are distributed between $2 \lesssim Z \lesssim 12$ kpc (V=12-16.5 mag) and accurate photometric parallaxes ($\sigma_d/d \lesssim 10\%$) have been computed by means of the M_V vs. $B - V$ relation for BHB giants derived by Preston et al. (1991), while for RR Lyræ stars we adopted the M_V as a function of metallicity or from Fourier components, as following:

$$M_V = 0.23\,[\text{Fe/H}] + 0.92 \qquad (1)$$
$$M_V = -1.876 \log P - 1.158 A_1 + 0.821 A_3 + 0.448 \qquad (2)$$

where Eq.(1) is from Cacciari (2003, in press) and Chaboyer (1999), while Eq.(2) is based on the relation from Kovács & Walker (2001) where P is the period (in days), A_1 and A_3 are the Fourier amplitudes (mag) of the fundamental and second harmonic components, respectively. The zero point has been calibrated by Kinman (2002, in prep.) with respect to the absolute magnitude ($M_V = 0.61^{+0.10}_{-0.11}$) of RR Lyr derived from the HST/FGS parallax measured by Benedict et al. (2002). Finally, the extinction E_{B-V} has been estimated from the maps of Schlegel et al. (1998).

4. Halo rotation

A retrograde rotation of the outer halo has been suggested by Majewski (1992) who measured a mean velocity $\langle V \rangle = -275$ km s^{-1}, which corresponds to a galactocentric retrograde velocity $v_{\rm rot} \simeq -55$ km s^{-1} adopting $V_{\rm LSR} = 220$ km s^{-1}, from the analysis of a pure sample of halo subdwarfs at $Z > 5.5$ kpc towards the NGP. As shown in Table 2, this parameter is still controversial, in fact Carney (1999), after correcting the kinematics bias of his kinematically-selected subdwarf sample, found a net prograde rotation of about $\langle V \rangle = -196$ km s^{-1}. On the contrary, Chiba & Beers (2000) measured a prograde rotating inner halo, $v_{\rm rot} \simeq 20$-60 km s^{-1}, up to about 1 kpc, with a decreasing vertical gradient of $dV/d|Z| = -52 \pm 6$ km s^{-1} kpc^{-1}, while they did not detect any significant rotation above $Z \sim 1.2$ kpc for very low abundance stars ($-2.4 \leq$ [Fe/H] ≤ -1.9), where contamination of thick disk stars should be negligible. In addition, their halo sample at larger distances (212 stars with $4 < Z_{\rm Max} < 20$ kpc and [Fe/H]≤ -1.5) still does not support any significant rotation: $v_{\rm rot} \simeq 0 \pm 8$ km s^{-1}.

However, recently Gilmore et al. (2002), who carried out a spectroscopic survey at intermediate galactic latitudes of about 2000 F/G stars, revealed a significant excess of *retrograde* halo stars in their faintest magnitude bin ($18 < V < 19.5$) corresponding to a vertical distance $|Z| \approx 5$ kpc.

The fact that the velocities of halo stars do not match an exact gaussian distribution is well known (see e.g. Martin & Morrison 1998). How much this depends on the properties of the whole population or on the effects of kinematic substructures, such as satellite debris of ancient accretion events (e.g. Helmi et al. 1999), remains to be established.

As shown in Table 2, a preliminary analysis of our sample of RR Lyræ and BHB giants (Kinman et al. 2002, JENAM, in press) seems to support a retrograde rotation of the outer halo. In fact, we measured a heliocentric velocity $\langle V \rangle = -285 \pm 17$ km s^{-1}, which corresponds to $v_{\rm rot} \simeq -60$ km s^{-1} adopting a solar motion with respect to the LSR of $V_\odot = +5.25 \pm 0.62$ km s^{-1}, from Dehnen & Binney (1998) and assuming $V_{\rm LSR} = 220$ km s^{-1}. This result is confirmed also by the separate analysis of the RR Lyræ and BHB giants, which both provide a retrograde rotation.

How much this value may be affected by a velocity bias can be estimated by the level of systematic errors on proper motions which in practice give the main contribution to the U and V galactic components along this line of sight towards $b \approx 90°$. As reported in Sect. 2, we found systematic errors $\Delta\mu < 1$ mas yr^{-1}, from which velocity bias up to 20-30 km s^{-1} can be expected for such

Table 2. Recent measurements of the halo rotation (*: heliocentric velocities (wrt. the LSR in the other cases.); **: as a function of distance with a gradient $dV/d|Z| = -52 \pm 6$ km s^{-1} kpc^{-1}; ***: preliminary results based on 18 RR Lyræ stars and 35 BHB giants).

Tracers	N. Objects	$\langle V \rangle$ (km s^{-1})	Reference
	INNER HALO		
RR Lyræ	162 ($|Z| < 2$ kpc)	-210 ± 12*	Layden et al. (1996)
RR Lyræ	84 ($|Z| < 2$ kpc)	-219 ± 10*	Martin & Morrison (1998)
RR Lyræ	147 ($|Z| < 2$ kpc)	-217 ± 13*	Gould & Popowski (1998)
Subdwarfs	($|Z| < 1$ kpc)	$-(160 \div 200)$**	Chiba & Beers (2000)
	OUTER HALO		
Subdwarfs	21 ($Z > 4$ kpc)	-275 ± 16	Majewski (1992)
Subdwarfs	30 ($Z_{\text{Max}} > 5$ kpc)	-196 ± 13	Carney (1999)
Subdwarfs	212 ($Z_{\text{Max}} > 4$ kpc)	-220 ± 8	Chiba & Beers (2000)
RR & BHB	53 ($2 < Z < 12$ kpc)	-285 ± 17*,***	this survey

stars located at $Z \approx 5$-6 kpc, on average. Clearly, this is a critical point which need further analysis.

5. Kinematics simulation

A Montecarlo simulation has been developed in order to compare the GAIA capability to recover the halo kinematics with respect to the velocity precision that can be attained by the current and future ground based surveys. For simplicity we considered an uniform spatial distribution with Pop.II-like kinematics, $(\sigma_U, \sigma_V, \sigma_W) = (150, 100, 100)$ km s^{-1} and an observer rotating with a velocity of 220 km s^{-1} at a distance of 8 kpc from the galactic center. Implicitly we assumed to measure bright tracers, such as RR Lyræ, BHB and red giants, having apparent magnitude $V \sim 16$ mag at a distance of about $d \sim 10$ kpc. The following cases were tested:

- Case A (GAIA): $\sigma_\pi = 10$ μas, $\sigma_\mu = 10$ μas yr^{-1} (per component) and $\sigma_{\text{RV}} = 10$ km s^{-1}, from astrometric and spectroscopic observations;

- Case B (ground-based surveys): $\sigma_{m-M} = 0.2$ mag, $\sigma_\mu = 1$ mas yr^{-1} (per component) and $\sigma_{\text{RV}} = 10$ km s^{-1}, from astrometric, photometric and spectroscopic observations.

Note that the two cases differ essentially in a factor 100 on the proper motion accuracy. In fact, at ~ 10 kpc the distance accuracy is the same for both cases. This points out the fact that, although radial velocities and distance moduli

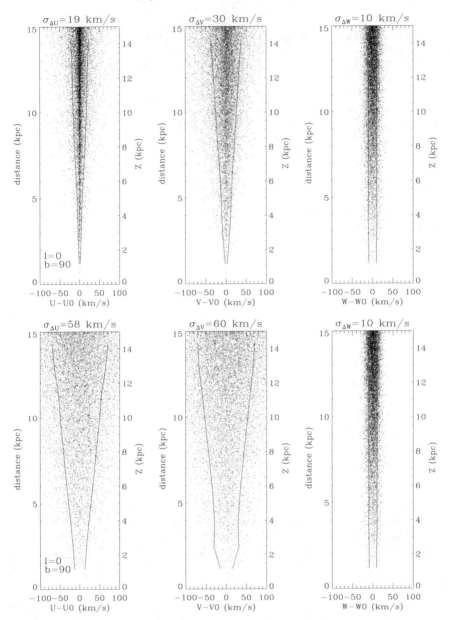

Figure 4. Kinematics simulation towards NGP. Velocity residuals for Case A (*top panels*) and Case B (*bottom panels*). Solid lines show the precision level ($\pm 1\sigma$) as function of the distance up to Z=15 kpc. Above each plot, the rms of the residuals of the whole sample is reported.

derived from ground based spectro-photometric surveys can attain similar accuracy, GAIA astrometry will uniquely provide accurate and reliable tangential velocities up to large distances.

Figure 4 shows the results of Montecarlo simulations with 5 000 stars towards $b = 90°$. Top panels present the velocity errors (i.e. the differences between the observed velocity and the true value) for Case A, and those at the bottom for Case B. As expected, errors increase linearly with distance, and for GAIA the r.m.s. of ΔU and ΔV residuals vary from ~ 10 km s^{-1} at 5 kpc to 20-30 km s^{-1} at 10 kpc. For Case B the errors on U and V are about a factor 2-3 larger, while $\sigma_{\Delta W} = \sigma_{RV} \equiv 10$ km s^{-1} in both the cases. Note that in Case A the error on the tangential velocity is in practice dominated by the distance uncertainty, σ_π, while both the errors on proper motions and photometric parallaxes contribute to the velocity errors of Case B. Similar results are derived in other directions.

These velocity errors should be compared with the typical motion of the halo stars (100-200 km s^{-1}) for which GAIA will provide individual 3D spatial velocities with a significant signal-to-noise, σ_v/v, up to $d \approx 10$-15 kpc, a distance where ground based surveys are not able to measure reliable tangential velocities. GAIA will measure direct distances and velocities for large samples of halo tracers, selected *in situ* without kinematics nor metallicity bias, from which it will possible to determine accurately the halo velocity ellipsoid and its orientation. In particular, the GAIA μ-arcsec level accuracy will be essential in order to take advantage of the $\sqrt{N}$ statistical factor and avoid the risk of velocity biases due to the presence of systematics errors affecting proper motions and parallaxes as discussed in Sect. 4.

Finally, accuracy of the order of 15-20 km s^{-1}, such as that attained by GAIA for $d \lesssim 10$ kpc, is the value requested to resolve kinematics substructures in the halo, as the satellite debris predicted by the hierarchical scenarios of galaxy formation. To this regards, Helmi (2002) estimated that GAIA will be able to recover 2/3 of the accretion events with a velocity accuracy better than 20 km s^{-1} and distance $\sigma_d/d < 20\%$ down to $V \simeq 18.5$, or 1/2 of the events down to $V \simeq 15$ mag.

6. Summary

Large field proper motion surveys ($\sigma_\mu \approx$ 1-10 mas yr^{-1}) based on photographic surveys digitized with fast measuring machines (APS, MAMA, PDS, PMM, SuperCOSMOS) are still useful tools for the study of the structure and kinematics of the galactic stellar populations, especially when combined with radial velocities and chemical abundances from spectro-photometric observations. In particular, we have shown how bright halo tracers, such as BHB giants and RR Lyræ stars with $V \lesssim 16$-17 mag, can be used to investigate *in situ* the halo kinematics up to several kiloparsecs via photometric parallaxes if proper motions with $\sigma_\mu \sim 1$ mas yr^{-1} accuracy and radial velocities with precision of about $\sigma_{RV} \sim 10$ km s^{-1} are available. For statistical analysis of large samples, systematic errors on the proper motions (and M_V too) are critical and can result in biased mean velocities and dispersions. These problems will be dramatically reduced by the astrometric parameters and spectroscopic radial velocities provided by GAIA, which will permit to determine (*a*) direct distances from parallaxes,

(b) radial and tangential velocities with an accuracy of a few tens of km s^{-1} to a larger distance ($d \sim$ 10-15 kpc) for (c) unbiased and larger sets of tracers identified by means of its spectro-photometric and astrometric data.

Acknowledgments. The GSC II is a joint project of the Space Telescope Science Institute and the Osservatorio Astronomico di Torino. Space Telescope Science Institute is operated by AURA for NASA under contract NAS5-26555. Partial financial support to this research comes from the Italian CNAA and the Italian Ministry of Research (MIUR) through the COFIN-2001 program.

References

Bachall, J.N., Casertano, S. & Ratnatunga K.U. 1987, ApJ 320, 515
Benedict, G.F., McArthur, B.E., Fredrick, L.W. et al. 2002, AJ 123, 473
Bucciarelli, B., Garcia Yus, J., Casalegno, R. et al. 2001, A&A 368, 335
Carney, B.W. 1999, Proc. of 3rd Stromlo Symposium, B.K. Gibson, T. Axelrod & M. Putnam ed.s, ASP Conf. Ser. 165, pag. 230
Chaboyer, B. 1999, in Post-Hipparcos Cosmic Candles, A.Heck and F.Caputo ed.s, Kluwer, pag. 111
Chen, B. 1997, ApJ 491, 181
Chiba, M., Beers, T.C. 2000, AJ 119, 2843
Dehnen, W. & Binney, J.J. 1998, MNRAS, 298, 387
Gould, A. & Popowski, P. 1998, ApJ 508, 844
Gilmore, G., Wyse, R.F.G. & Norris, J.E. 2002, ApJL 574, L39
Helmi, A., White, S.M., de Zeeuw, P.T. & Zhao H. 1999, Nature 402, 53
Helmi, A. 2002, RVS-CoCo-003 (Livelink)
Høg, E., Fabricius, C., Makarov, V.V. et al. 2000, A&A 355, L27
Kovács, G. & Walker, A.R. 2001, A&A 374, 264
Lasker, B.M., McLean, B.J., Jenkner, H., Lattanzi, M.G., Spagna, A. 1995, in Future Possibilities for Astrometry in Space, M.A.C. Perryman, F. van Leeuwen & T.-D. Guyenne ed.s, ESA SP-379, pag. 137
Layden, A.C., Hansen, R.B., Hawley, S.L. et al. 1996, AJ 112, 2110
Majewski, S.R. 1992, ApJS 78, 87
Martin, J.C. & Morrison, H.L. 1998, AJ 116, 1724
McLean, B.J., Greene, G.R., Lattanzi, M.G., Pirenne, B. 2000, in Astronomical Data Analysis Sotware and Systems IX, N. Manset, C. Veillet & D. Crabtree ed.s, ASP Conf. Ser. 216, pag. 145
Mendez, R.A. & van Altena, W.F. 1996, AJ 112, 655
Preston, G.W., Shectman, S.A, & Beers, T.C. 1991, ApJ 375, 121
Robin, A. & Oblak, E. 1987, in 10th IAU European Astronomy Meeting, J. Palous ed., Prague, 4, 323
Schlegel, D.J., Finkbeiner Douglas, P. & Davis, M. 1998, ApJ 500, 525
Spagna, A. Lattanzi, M.G., Lasker, B.M., McLean, B.J., Massone, G. & Lanteri, L. 1996, A&A 311, 758

GAIA Spectroscopy, Science and Technology
ASP Conference Series, Vol. 298, 2003
U. Munari ed.

Observational constraints to the mass density in the Galactic plane

Olivier Bienaymé, Arnaud Siebert

Strasbourg Observatory, France

Caroline Soubiran

Bordeaux Observatory, France

Abstract. Nearly 400 Tycho-2 stars have been observed in a 720 deg^2 field in the direction of the North Galactic Pole with the high resolution echelle spectrograph ELODIE. Their absolute magnitudes, T_{eff}, $\log g$, [Fe/H] have been estimated as well as their distances and 3D velocities. We obtain new constraints on classical questions concerning the vertical force perpendicular to the galactic plane, the dynamical estimate of the galactic surface mass density, and also concerning the disk star formation rate. This work is a simple illustration of some results that can be obtained in combining the kinematics, the dynamic and the stellar evolution as it could done on a really much larger scale with GAIA.

1. Introduction

We have selected stars out of the Tycho-2 catalogue (Høg et al 2000) within the $0.9 \leq B - V \leq 1.1$ colour interval, towards the North Galactic Pole, to increase the number of red clump stars with respect to dwarfs, subgiants and AGB stars. A total of 387 stars were observed down to magnitude $V_J = 10.1$ with the echelle spectrograph ELODIE of the 193 cm-telescope at the Observatoire de Haute Provence (France), covering the full range 3900–6800 Å at a resolution of 42 000. Exposure time were short and the mean S/N of the spectra at 5500 Å is 23.

Atmospheric parameters and absolute magnitudes of observed stars were obtained with the TGMET software (Katz et al. 1998). TGMET is a minimum distance method (reduced χ^2 minimisation) which meaures similarities between the observed spectra and a library, the TGMET library, built from the ELODIE database (Prugniel & Soubiran 2001) including high S/N spectra of reference stars having well known and measured atmospheric parameters from published detailed analysis listed in the catalogue of [Fe/H] determinations (Cayrel de Strobel et al. 2001).

According to our selection criteria, $B - V \sim 1$ Tycho-2 stars towards the NGP, our sample is dominated by clump stars (Figure 1). 117 reference spectra from TGMET were effectively selected to calibrate our sample of NGP targets.

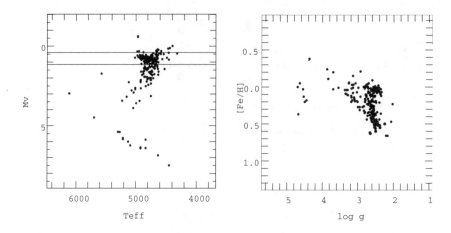

Figure 1. Properties of observed stars. Horizontal lines mark our selection to extract the red giant clump subsample used for the Oort limit analysis.

The resulting errors on determined absolute magnitude is ~ 0.36, relative distance $\sim 18\%$ and [Fe/H] ~ 0.16

2. Kinematics and metallicity

Our NGP sample was used to investigate the kinematics and metallicity of the stellar disk up to 800 pc from the galactic plane with a typical error of 18% on distances, 3 km s^{-1} in U and V velocities, and less than 1 km s^{-1} in W, at the mean distance of the sample (400 pc; Soubiran et al. 2002, submitted).

Our colour-magnitude cuts were efficient to select K giants since less than 6% of the sample consists in dwarfs or subgiants. We notice a lack of metal-poor stars ([Fe/H] < −0.65) which is only partially explained by the method used to estimate the metallicities and by the colour cut. Velocity and metallicity distributions were compared to simulations of the Besançon model (Robin & Crézé 1986) using 2 models of Thick Disk: a metal-poor kinematically hot one and another one with parameters closer to the Thin Disk. The NGP sample is in a better agreement with the moderate Thick Disk model. Alternatively, using a non-informative method (SEM, Celeux & Diebolt 1986) in the case of a mixture of gaussian populations, we find our sample to be consistent with a superposition of two disks (see Table 1). The first one is associated with the Thin Disk in agreement with previous studies from local samples. No vertical gradient is observed in the parameters of the Thin Disk.

The parameters of the Thick Disk were also estimated (Table 1). Compared to previous studies, these values are on the hot side for U, but on the cold, metal-rich side for V, W and [Fe/H]. The proportion of Thick Disk stars is found to be higher than expected with a local normalisation of 15±7%. Inconsistent determinations of the Thick Disk parameters in the recent literature may be reconcilied if 2 components are involved. In the context of a Thick Disk formation by a significant merger, we may have observed, in our NGP sample, the early heated Thin Disk, whereas other authors may have observed the relics of a shredded satellite (Chiba & Beers 2000; Gilmore et al. 2002).

The vertex deviation of the velocity ellipsoid has been studied with respect to distance above the plane and metallicity. For this task, the NGP sample was completed by a local Hipparcos sample for which metallicity and radial velocities were retrieved from the literature. A null vertex deviation is found for the low metallicity subsample whereas the high metallicity sample exhibit a positive one. This suggests no strong deviation from axisymmetry of the galactic disk at solar radius.

Table 1. Adopted parameters for the Thin and Thick Disks from the deconvolution of the NGP sample. The relative densities are given for $z=0$. The error bars were estimated from the dispersion of the SEM solutions on real and simulated data.

disk	%	[Fe/H]	V_{lag} (km s^{-1})	σ_U (km s^{-1})	σ_V (km s^{-1})	σ_W (km s^{-1})
Thin	85±7	−0.17±0.03	−12±2	39±2	20±2	20±1
Thick	15±7	−0.45±0.05	−51±5	63±6	39±4	39±4

3. The Oort limit and the surface mass density

We extract from our NGP observations the subsample of the giant clump stars (200 pc to 800 pc distances above the galactic plane). It is combined with the sample of analogous clump giants observed with Hipparcos (the complete sample inside a 125 pc radius sphere). All these stars are used as a tracer of the galactic potential perpendicular to the galactic plane (Siebert et al. 2002, submitted).

We model the vertical gravitational potential through the galactic disk according to the 3-parameter representation proposed by Kuijken & Gilmore (1989):

$$\Phi(z) = 2\pi G \left(\Sigma_0 (\sqrt{z^2 + D^2} - D) + \rho_{\text{eff}} z^2 \right) \quad (1)$$

where Σ_0 is the total surface mass density of the galactic disk at the Sun galactic radius, D the scale height of the disk mass distribution and ρ_{eff} is the dark halo contribution (assumed here to be 0.01 M$_\odot$ pc^{-3}).

The star counts versus the vertical distance and the vertical velocities allow to constrain the vertical potential. The modelling of the samples consists in solving the Boltzmann and Poisson equations and in reproducing the distribution of the observed apparent and absolute magnitudes, distances and vertical velocities by adjusting the vertical potential parameters. Solutions found for Σ_0 and D parameters, within 1, 2 and 3–σ error, are plotted on Figure 2 (left).

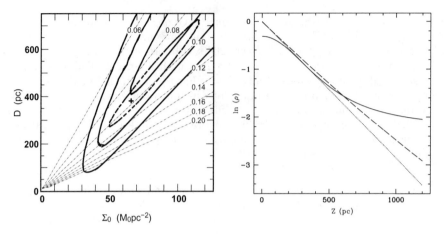

Figure 2. *Left*: 1, 2 and 3-σ error contours for Σ_0 and D solutions (continuous curves) from our clump giant sample. Using also the recent Oort limit determination from Hipparcos data gives better constraints (1-σ error level: dashed curve). *Right*: vertical distribution of the volume mass density in the galactic disk deduced from our dynamical determination (continuous line). An exponential disk with a 350 pc scale height (dotted line), and with a Thick Disk, scale height 750 pc, and relative density of 15 % per cent (dashed line).

We determine, from our NGP sample and from the Hipparcos local counterpart sample, the Oort limit (i.e. the total local mass density at $z=0$, given here by $\rho_{\text{total}}(z = 0) = \Sigma_0/(2D) + \rho_{\text{eff}}$). It ranges between 0.07 and 0.11 $M_\odot$ pc^{-3} (dotted lines on Figure 2 left) in agreement with recent and independent determinations based on Hipparcos data. This total mass determination is also compatible with the local mass density of known, stellar and gas, matter.

Stronger constraints are obtained including the recent determinations of $\rho_{\text{total}}(z = 0) = 0.10 \pm 0.010\, M_\odot$ pc^{-3} obtained with Hipparcos data (Crézé et al. 1998; Holmberg & Flynn 2000). Using this constraint allows to determine more accurately Σ_0 and D (dashed curve on Figure 2, left, indicating 1-σ error): it gives a large scale height $D = 392^{+328}_{-121}$ pc, and for the total surface mass density $\Sigma_0 = 67^{+47}_{-18}\, M_\odot \text{pc}^{-2}$.

The vertical potential $\Phi(z)$ that we determine at high z ($\sim$ 400–800 pc) is compatible with previous estimates. We are also able to measure for the first time, the thickness D of the total disk mass distribution.

This scale height D corresponds to a 350 pc scale height of a vertically exponential density law in the range of z-height 200 to 500 pc. Figure 2 (right) is a plot of the resulting total mass density $\rho_{total}(z)$ and of the stellar density (two exponential disks, old and Thick Disks respectively, with 350 pc and 750 pc scale heights). This mass scale height estimate is in agreement with existing determinations of the stellar disk scale height, $h_z \sim 330$ pc.

We conclude that the local volume mass density and the surface mass density of the disk are in agreement with our current knowledge of the known volume

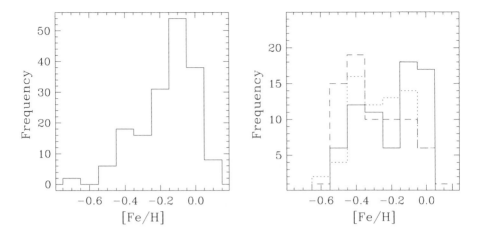

Figure 3. *Left:* metallicity histogram of nearby Hipparcos clump giants with measured Fe/H (a third of nearby clump stars). *Right:* metallicity histogram of NGP clump stars in the 200–350 pc (continuous line) range, in the 350–500 pc range (dotted line), beyond 500 pc (dashed line). The mean metallicity is decreasing with the distance from the galactic plane.

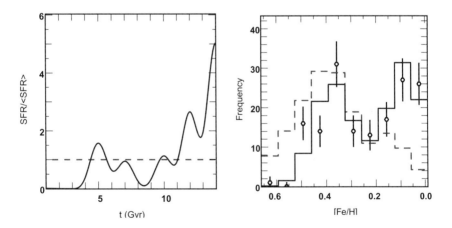

Figure 4. *Left:* two models of Star Formation Rate (the dark line is a non-parametric inversion of data) and for comparison a constant SFR model (dashed line). *Right:* The corresponding predicted metallicity distributions and the observed metallicity histogram (open symbols) for our NGP sample and of clump giants.

and surface density from gas and stellar components. Moreover, the thickness of
the disk mass density distribution is compatible with the thickness of the stellar
old disk.

4. Star formation history

Each stellar disk component may be descibed by its vertical density distribution,
its kinematics and also by its metallicity distribution, the younger component
being the thinner and having the lowest vertical velocity dispersion. The mean
[Fe/H] abundance is also decreasing with the distance above the galactic plane.
This is clearly seen in our sample (Figure 3), and has been modelled and described here with a simple model of stellar evolution.

We model not only the kinematics and the vertical density distribution,
but also the metallicity distribution, using isochrones up to the core helium
burning phase (Girardi et al. 2000) and a Salpeter IMF (constant with time) to
reproduce and explain all the observed distributions ($\rho_*(z)$, $\sigma_w(z)$, [Fe/H]). For
that purpose we assume an age-metallicity relation and age-velocity dispersion
relation (Rocha–Pinto et al. 2000). Different star formation rates (SFR) have
been tried and also a non-parametric solution (Figure 4).

We identify a recent star burst (t=14–15 Gyr) and no star older than 11 Gyr
(t=0–4 Gyr). The lack of old star just mean that we do not find stars with
metallicity [Fe/H] less than −0.6.

References

Cayrel de Strobel, G., Soubiran, C. & Ralite, N. 2001, A&A 373, 159
Celeux, G. & Diebolt, J. 1986, Rev. Statistique Appliquée 35, 36
Chiba, M. & Beers, T.C. 2000, AJ 119, 2843
Crézé, M., Chereul, E., Bienaymé, O. & Pichon, C. 1998, A&A 329, 920
Gilmore, G., Wyse, R.F.G. & Norris, J.E. 2002, ApJ 574, L39
Girardi, L., Bressan, A., Bertelli, G. & Chiosi, C. 2000, A&AS 141, 371
Høg, E., Fabricius, C., Makarov, V.V. et al. 2000, A&A 355, L27
Holmberg, J. & Flynn, C. 2000, MNRAS 313, 209
Katz, D., Soubiran, C., Cayrel, R. et al. 1998, A&A 338, 151
Kuijken, K. & Gilmore, G. 1989 MNRAS 239, 605
Prugniel, P. & Soubiran, C. 2001, A&A 369, 1048
Robin, A. & Crézé, M. 1986, A&A 157, 71
Rocha–Pinto, H.J., Scalo, J., Maciel, W.J. & Flynn, C. 2000, A&A 358, 869

Kinematics of the Galactic populations in the GAIA era

Giampaolo Bertelli[1], Antonella Vallenari[1], Stefano Pasetto[1,2] & Cesare Chiosi[1]

[1] INAF – Osservatorio Astronomico di Padova, Vicolo Osservatorio 5, 35122 Padova

[2] Dipartimento di Astronomia dell'Universitá di Padova, Vicolo Osservatorio 3, 35122 Padova

Abstract. GAIA data will create a precise 3-dimensional map of the Galaxy, providing positional information, radial velocities, luminosity, temperature and chemical composition of a representative sample of stars. Here we present a new implementation of the Padova Galaxy Model, where kinematics simulations are included. A few examples of application to GAIA science are discussed, in particular concerning radial velocities determinations of the Thin and Thick Disk.

1. Introduction

GAIA will provide detailed phase space coordinates for about 1 billion stars within a sphere of 20 Kpc. In addition to this on board multicolour photometry and spectroscopy will give complete chemical measurements (including α, s, r process elements) down to V=17-19 mag. Owing to the precision of the data, the problem of the formation of the Milky Way will be addressed. In the standard cold dark matter scenario the Galaxy should have formed by merges of small substructures. However this scenario is at odds with several observational constraints: i.e. a small scale length of the disk of the order of 300 pc is expected instead of 2000-3000 pc (Steinmetz & Navarro 1999); the formation process predicts that the Milky Way should have a thousand of satellites which are missing in the Local Group. A substantial revision of the model is necessary, by means of a detailed comparison with high quality data. GAIA will bring into evidence fossil remnants of the galaxy formation tracing back the star formation history and revealing gradients in the star formation intensity, chemical abundances, and kinematics across the disk. Chemical and age gradients are not expected inside the Thin Disk. N-body simulations show that radial mixing can wash out gradients close to the Galactic plane after a few Gyr. Concerning the Thick Disk, if it was formed by an heating event (merge with a satellite) traces of it should still be detectable (Freeman & Bland-Hawthorn 2002). In fact vertical gradients in metallicity and velocity dispersions which are still uncertain from the present data are possibly not washed out by the radial mixing far from the Galactic plane and can be used to trace back the formation history. In order to simulate Galactic data we update the Padova Galaxy Model including velocity space and we apply it to the GAIA science.

2. Padova Galaxy Model

The Galaxy is modeled with the code already described by Bertelli et al. (1995) and revised as in Vallenari et al. (2000) where more detail can be found. The Padova model has been newly updated including: (*1*) the usage of new stellar tracks from Z=0.0001 till Z=0.03 with low mass stars down to 0.15 (Girardi et al. 2000). The 0.1 $M_\odot$ track is taken from Baraffe et al. (1998); (*2*) the use of the carbon star models taking into account the effect of the variation of the stellar molecular opacities during the evolution as a result of the dredge-up (Marigo 2002); (*3*) the extinction along the line of sight derived following Drimmel & Spergel (2001) model obtained from COBE-DIRBE infrared data; (*4*) the velocity space simulated in a consistent way as described in the following Section.

3. The kinematic model

The velocity distribution for the whole disk has been computed using the velocity ellipsoids formulation by Schwarzschild. Concerning the Thin Disk, the assumed local values of velocity dispersions are taken as in Mendez et al. (2000). The V_{lag} is derived following Binney & Tremaine (1994, pag. 190). The diagonal terms of the dispersion velocity tensor are from Lewis & Freeman (1989):

$$\sigma_{RR}^2 = \sigma_{RR,0}^2 exp(-(R-R_0)/H_R) \qquad (1)$$
$$\sigma_{\Phi\Phi}^2 = 1/2(1 + (dlnV_{LSR}(R))/(dlnR)) \times \sigma_{RR}^2 \qquad (2)$$
$$\sigma_{ZZ}^2 = \sigma_{ZZ,0}^2 exp(-(R-R_0)/H_R) \qquad (3)$$

The vertical gradient in the velocity dispersions is taken from Fuchs & Wielen (1987) for σ_{RR}^2 and $\sigma_{\Phi\Phi}^2$ and the non vertical isothermality of the Thin Disk from Amendt & Cutterford (1991) which is in good agreement with the observations till 1 Kpc for σ_{ZZ}^2. The off diagonal term σ_{RZ} is derived from Amend & Cutterford (1991):

$$\sigma_{RZ}^2(R,Z) = \sigma_{RZ}^2(R,0) + z\partial/\partial z \sigma_{R,Z}^2(R,0) \qquad (4)$$

where

$$\partial/\partial z \sigma_{RZ}^2(R,0) = \lambda(R)(\sigma_{RR}^2 - \sigma_{ZZ}^2/R)(R,0) \qquad (5)$$

and

$$\lambda(R) = (R^2 \Phi_{Rzz}/3\Phi_R + R\Phi_{RR} - 4R\Phi_{ZZ})(R,0) \qquad (6)$$

where $\Phi_{RZZ}, \Phi_{RR}, \Phi_{ZZ}$ are the derivatives of the Galactic potential Φ obtained from the density model by Dehnen & Binney (1998) by inverting the Poisson equation with the Bessel integrals (Quinn & Goodman 1986). λ is an approximate expression of the vertical tilt of the velocity ellipsoid close to the Galactic plane (at z=0). λ is 1 in the case of a spherical potential, when the ellipsoid is pointing towards the Galactic center, while λ is 0 for a cylindrical potential when the ellipsoid is always parallel to the Galactic plane. Amendt & Cutterford (1991) show that λ is related to the mass gradient in the Galactic plane. No vertex deviation is included in the model although a deviation decreasing from 25° for young stars to near zero for an old population has been found by various authors (Dehnen & Binney 1998, Bienaymé 1999, Soubiran et al. 2002 submitted). The Thick Disk is isothermal with $(\sigma_{RR}, \sigma_{\Phi\Phi}, \sigma_{zz}) = (70, 50, 45)$ km s^{-1} as starting values. The V_{lag} is assumed to have a canonical value of 35 km s^{-1}. The possibility of simulating a vertical gradient in the rotational velocity of the Thick Disk is included as suggested by Chiba & Beers (2000). The halo has $(\sigma_{RR}, \sigma_{\Phi\Phi}, \sigma_{zz}) = (130, 95, 95)$ km s^{-1}. The projection matrixes of the space velocities are derived from Mendez et al. (2000). In the following Sections we show a few examples of applications to the GAIA science. In particular the Thin and Thick Disk are discussed.

4. Disentangling various stellar populations

Spectrograph aboard on GAIA operating in the near-IR will measure radial velocities with an accuracy better than 2 km s^{-1} if an high dispersion of 0.25 Å pix^{-1} is chosen for stars brighter than $V=16$ mag. The expected accuracy will still be better than 8 km s^{-1} for magnitudes as faint as $V=17$, but will drop to 30 km s^{-1} at $V=18$ (Munari et al. 2003) for a G2V stars. This amount of unprecedented good quality data will allow to disentangle Thin Disk, Thick Disk and halo populations, deriving ages, metal content and kinematics. Figure 1 presents a simulation of the expected radial velocities at $(l,b)=(270,-45)$ for a 2.5×2.5 deg^2 field for $V < 17.0$, $V < 17.5$ and $17.5 \leq V \leq 18$ stars at heliocentric distances of 2000-4000 pc where the contribution of the Thick Disk is relevant. Expected observational uncertainties are included. The Thick Disk population begins to contribute significantly at magnitudes fainter than $V=17$. In order to disentangle Thin and Thick Disk populations, a fainter magnitude limit would be more effective, unless kinematic information is coupled with chemical abundances determinations.

5. The Thick Disk velocity gradient

The presence of a gradient in the Thick Disk velocity dispersion or a multicomponent structure is indicative of the formation process as described in Sect. 1. From the observational point of view the situation is far from being clear. Soubiran et al. (2002, submitted) find a Thick Disk with a moderate rotational and vertical kinematics, but no vertical gradient suggesting a quick heating of the precursor Thin Disk. Chiba & Beers (2000) suggest that far from the plane the Thick Disk has lower rotational velocity and higher velocity dispersion than close to the Galactic plane. These data are interpreted as a vertical gradient of about 30 km s^{-1} kpc^{-1}. Gilmore et al. (2002) propose a different interpretation of the data: the Thick Disk has a composite form, indicating the presence of relics of disrupted satellites. Figure 2 shows a simulation of the Thick Disk population in a column of 0.09×0.09 deg^2 at $(l,b)=(270,-45)$ for stars brighter than $V=17.5$. Even including the expected GAIA accuracy for an intermediate dispersion of 0.5 Å pix^{-1}, still the effect of a vertical gradient of 10 km s^{-1} kpc^{-1} is visible.

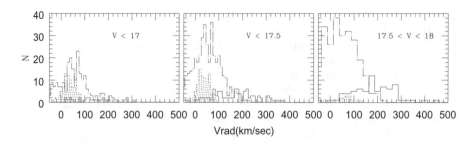

Figure 1. Simulation of a field at $(l,b)=(270,-45)$ including Thin Disk (shaded histogram), Thick Disk (dashed-dotted line) and Halo (solid line) at an heliocentric distance d=2000–4000 pc in various magnitude ranges. The bin size is related to the expected accuracy.

6. The vertical tilt of the Thin Disk velocity ellipsoid

Up to now the vertical tilt parameter $\lambda(R)$ of the Thin Disk velocity ellipsoids is ill-determined. λ is in fact strongly related to the coupling of the U and W velocities and is better constrained by 3-dimensional velocities. It is found to vary from 0.4 to 0.6 at the solar circle (Amendt & Cutterford 1991, Bienaymé 1999). Figure 3 shows the expected proper motions and radial velocities of the Thin Disk population under different assumptions for $\lambda(R)$, namely 0, 1 and following Amendt & Cutterford (1991) in the direction $(l,b)=(26,6)$. The simulations take into account the uncertainties on GAIA determinations for stars brighter than $V=17$. In comparison to the models having $\lambda(R) = 0$ or 1, the model with $\lambda(R)$ related to the potential predicts differences of the mean value of the radial velocities going from 5 to 15 km s^{-1} depending on the distance. On the proper motion μ_l the effect is at maximum 0.2-0.4 mas yr^{-1}, while it is definitely less than 0.08 mas yr^{-1} on μ_b. GAIA expected precision on radial velocities and proper motions will be able to put strong constraints on the determination of the tilt of the velocity ellipsoid.

7. Conclusions

GAIA data will create a precise 3-dimensional map of the Galaxy, providing positional information, radial velocities, luminosity, temperature and chemical composition of a representative sample of stars. Here we present a new implementation of the Padova Galaxy Model, where kinematics simulations are included. This model includes a description of the vertical tilt of the ellipsoids of the velocity of the Thin Disk following Amendt & Cutterford (1991), in addition to isothermal Thick Disk and halo. This formulation might be specially useful to simulate the kinematics of the Thin Disk till a vertical height of 1 Kpc. A few examples of application to the GAIA science are presented, in particular concerning radial velocities determinations. The main conclusions are:

(1) the expected accuracy on radial velocities will put strong constraints on the determination of the vertical tilt of the velocity ellipsoids of the Thin Disk;

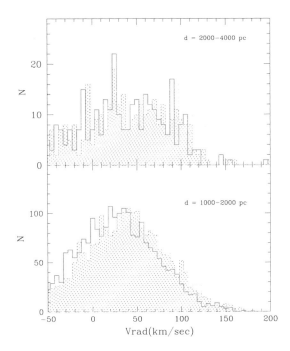

Figure 2. The effect of the Thick Disk velocity gradient. Dashed line shows a model with no vertical gradient; heavy solid line presents a vertical gradient of 10 km s^{-1} kpc^{-1}, thin solid line is the analogous for a vertical gradient of 30 km s^{-1} kpc^{-1}. d is the heliocentric distance

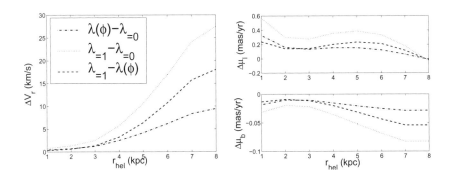

Figure 3. The differences of the mean values of V$_{rad}$ and μ at varying $\lambda(R)$ are plotted as functions of the Galactic radius for magnitudes brighter than $V=17$. Dotted lines show the difference between the models using $\lambda = 0$ and 1; the dashed-dotted lines are the analogous for $\lambda = \lambda(R,\Phi)$ and 0; and finally the dashed lines are the analogous for $\lambda = 1, \lambda(R,\Phi)$.

(*2*) since the Thick Disk begins to contribute significantly at magnitudes fainter than V=17, to disentangle Thin and Thick Disk populations at high latitude, a fainter magnitude limit would be more effective, unless kinematic information are coupled with chemical abundances determinations;

(*3*) finally, the expected accuracy on radial velocities at intermediate dispersion will allow the determination of a possible vertical velocity gradient in the Thick Disk of at least 10 km s^{-1}.

Acknowledgments. The authors wish to thank U. Munari and the Organizing Committee for the excellent organization of this extremely stimulating meeting.

References

Amendt, P. & Cudderford, P. 1991, ApJ 368, 79

Bertelli, G., Bressan, A., Chiosi, C., Ng, Y.K. & Ortolani, S. 1995, A&A 301, 381

Baraffe, I., Chabrier, G., Allard F. & Hauschildt P. H. 1998, A&A 337, 403

Binney, J.J. & Tremaine, S. 1994, Galactic Dynamics, Princeton Univ. Press

Bienaymé, O. 1999, A&A 341, 86

Chiba, M., & Beers T.C. 2000, AJ 119, 2843

Dehnen , W., & Binney, J.J. 1998, MNRAS 298, 387

Drimmel, R. & Spergel D.N. 2001, ApJ 556, 181

Freeman, K. & Bland-Hawthorn, J. 2002, ARA&A 40, 487

Fuchs B., & Wielen R. 1987, in The Galaxy, G. Gilmore & B. Carwell ed.s, NATO-ASI Ser., Reidel, pag. 375

Gilmore, G., Wyse R.F.G., & Norris J.E. 2002, ApJL 574, L39

Girardi,L., Bressan, A., Bertelli, G., & Chiosi C. 2000, A&AS 141, 371

Lewis, J. R., & Freeman, K. C. 1989, AJ 97, 139

Marigo, P. 2002 A&A 387, 507

Mendez, R.A., Platais, I., Girard, T.M. et al. 2000, AJ 120, 1161

Munari, U., Zwitter, T., Katz, D. & Cropper, M. 2003, in GAIA Spectroscopy, Science and Technology, U.Munari ed., ASP Conf Ser. 298, pag. 275

Quinn, P.J.,& Goodman J. 1986 ApJ 309, 472

Steinmetz M., & Navarro J.F. 1999, ApJ 513, 555

Vallenari A., Bertelli G.,& Schmidtobreick L. 2000, A&A 361, 73

Methods of relativistic astrometry in space: the static case

Ferdinando de Felice
Department of Physics, University of Padova, via Marzolo 35131, Padova, Italy

Alberto Vecchiato
Department of Physics, University of Padova, via Marzolo 35131, Padova, Italy

Maria Teresa Crosta
Department of Physics, University of Padova, via Marzolo 35131, Padova, Italy, and
INAF – Astronomical Observatory of Torino, strada Osservatorio 20, Pino Torinese (TO), Italy

Beatrice Bucciarelli
INAF – Astronomical Observatory of Torino, strada Osservatorio 20, Pino Torinese (TO), Italy

Mario G. Lattanzi
INAF – Astronomical Observatory of Torino, strada Osservatorio 20, Pino Torinese (TO), Italy

Abstract. We present a general relativistic model of astrometric data reduction having in mind applications to the GAIA mission. The accuracy of GAIA's observations is expected to be of the micro-arcsec hence we numerically implemented a model which takes into account the effects by all bodies of the Solar System. In this paper we limit ourselves to the order of $(v/c)^2$ since this is a basic step for extension to higher orders.

1. Introduction

The purpose of this paper is to construct a model of the celestial sphere using the algorithms of the General Theory of Relativity, in order to treat the relativistic effects concerning the light propagation in the gravitational field of the Solar System. All mathematical and physical assumptions about the model are essentially tailored for applications to the Astrometric Missions GAIA (ESA) which will be launched not later than 2012 as Cornerstone 6. The accuracy of the satellite GAIA will be pushed to the micro-arcsec level (μas) and, with such expectation, we have to implement a model of the celestial sphere and of the observables which is more accurate than the one applied to the Hipparcos mission (milli-arcsec). In fact, to the μas level of accuracy, it has been shown (Klioner & Kopeikin 1992; Kopeikin & Mashhoon 2002) that the light deflection effects are generated not only by the masses of the Sun and Planets, but also by their gravitational quadrupole moments together with their translational and rotational motion. In what follows, greek indices run from 0 to 3 and latin indices run from 1 to 3.

2. The geometry of the celestial sphere

To identify the background geometry we make four main assumptions. The first is that the Solar System is isolated; this means that there are no perturbing bodies intervening between the emitting stars and the Solar System boundaries. The second assumption is that the Solar System generates a weak gravitational field. This means that the space-time metric can be written as a linear perturbation of the Minkowski metric $\eta_{\alpha\beta}$ as $g_{\alpha\beta} = \eta_{\alpha\beta} + h_{\alpha\beta} + O(h^2)$ where the $h_{\alpha\beta}$'s describe effects generated by the bodies of the Solar System and are *small* in the sense that $|h_{\alpha\beta}| \ll 1$ and $|\delta h_{\alpha\beta}/h_{\alpha\beta}| \sim 1$. The orders of magnitude of each term entering the $h_{\alpha\beta}$'s, are expressed in terms of powers of (v/c), where v is the average relative velocity within the perturbing system and c is the velocity of light in vacuum. In this paper we shall only describe a model accurate to $(v/c)^2$ in order to have a basic touchstone for higher order extensions. Our third assumption is that all time variations of the metric coefficients can be neglected together with terms proportional to h_{0i}. A time derivative of the metric coefficients, in fact, generates terms of the order of at least $(v/c)^3$, and terms as h_{0i} are themselves of the order of $(v/c)^3$ at least. Our fourth and crucial assumption is that space-time can be *foliated*. This means that, given a general coordinate system $(x'^i, t')_{i=1,2,3}$, there exists a family of three-dimensional hypersurfaces described by the equation $\tau(x', t') = $ constant where $\tau(x', t')$ is a real, smooth and differentiable function of the coordinates. We shall denote each slice as $S(\tau)$. Let us now introduce a new coordinate system $(\xi^0 = \tau(\mathbf{x}', t'), \xi^i = \xi^i(\mathbf{x}', t'))$, where ξ^i are space-like coordinates defined on each slice. It is always possible to choose the spatial coordinates in such a way that they remain constant along the normals to the slices. In this case the transformed metric $\tilde{g}_{\alpha\beta}$ satisfies the condition $\tilde{g}^{0i} = \tilde{g}_{0i} = 0$ and the transformed vector field tangent to the normals to $S(\tau)$ have components $\tilde{u}^\alpha(\tilde{\xi}^0, \tilde{\xi}^i) = \frac{d\tilde{\xi}^\alpha}{d\sigma} = e^{\Phi(\tilde{\xi})}\delta_0^\alpha$, $\tilde{u}_\alpha(\tilde{\xi}^0, \tilde{\xi}^i) = e^{\Phi}\tilde{g}_{0\alpha} = -e^{-\Phi}\frac{\partial \tau}{\partial \tilde{\xi}^\alpha}$ where $e^\Phi = (-\tilde{g}_{00})^{-1/2}$ and σ is a parameter which makes the vector field $\tilde{\mathbf{u}}$ unitary.

Here we must bear in mind that the condition $\tilde{g}_{0i} = 0$ is preserved under gauge transformations only to the order of $(v/c)^2$, hence, since we neglect terms as $O[(v/c)^3]$, the constraint $\tilde{g}_{0i} = 0$ is gauge invariant.

3. The light trajectories

A photon traveling from a distant star to the astrometric satellite within the Solar System, would see the space-time as a time development of $\tau = $ constant slices. Since the spatial coordinates $\tilde{\xi}^i$ are constant along the unique normals ũ going through the point with those coordinates, the parameter σ along them will be function of τ only; i.e. $\sigma = \sigma_{\tilde{\xi}^i}(\tau)$. Let us now consider a null geodesic Υ with tangent vector field $k^\alpha \equiv d\tilde{\xi}^\alpha/d\lambda$ which satisfies the following equations:

$$k^\alpha k_\alpha = 0 \qquad \frac{dk^\alpha}{d\lambda} + \tilde{\Gamma}^\alpha_{\rho\sigma} k^\rho k^\sigma = 0 \qquad (1)$$

here λ is a real parameter on Υ and $\tilde{\Gamma}^\alpha_{\rho\sigma}$ are the connection coefficients of the given metric. Assume that the trajectory starts at a point P_* on a slice $S(\tau_*)$ (say) and with spatial coordinates $\tilde{\xi}^i_*$. The light trajectory will *end* at the satellite on a slice $S(\tau_0)$ and at a point with spatial coordinates $\tilde{\xi}^i_{(0)}$. The origin of the coordinate system is meant to be the barycenter of the Solar System. The purpose of our model is to determine $\tilde{\xi}^i_*$, namely the coordinates of the star, from a prescribed set of observables and eventually their variations with time.

Applying a suitable projection technique whose details we omit here, the light trajectory is mapped into a *spatial* path $\bar{\Upsilon}$ which lies entirely in the slice $S(\tau_0)$ which is also the *rest-space* of the baricentric observer at the time of observation τ_0. The vector field ℓ tangent to this curve has components which are equal to the projection of **k** into the slices $S(\tau)$ at the point where the light ray crosses them. It is possible to choose a parametrization on this curve which makes the tangent field ℓ unitary; if we define $d\sigma = -(\tilde{u}_\alpha k^\alpha)d\lambda$, then the tangent vector fields **k** and ℓ change their components into $\bar{k}^\alpha \equiv -\frac{k^\alpha}{(\tilde{u}_\beta k^\beta)}$, $\bar{\ell}^\alpha \equiv -\frac{\ell^\alpha}{(\tilde{u}_\beta k^\beta)}$ and satisfy the following relations $\bar{k}^\alpha = \bar{\ell}^\alpha + \tilde{u}^\alpha$ which implies $\bar{\ell}^\alpha \bar{\ell}_\alpha = 1$. Moreover it is $\bar{\ell}^0 = 0$ showing that $\bar{\Upsilon}$ lies everywhere on the slice $\tau_0 =$ constant as expected. From the previous relations and neglecting time variations of the metric as well as terms proportional to h_{0i}, the second of Eq. (1) writes:

$$\begin{aligned}\frac{d\bar{\ell}^\alpha}{d\sigma} &+ \frac{1}{2}\left(\bar{\ell}^i \partial_i h_{00}\right)\delta^\alpha_0 + \frac{1}{2}\left(\bar{\ell}^i \partial_i h_{00}\right)\left(\bar{\ell}^\alpha + \delta^\alpha_0\right) \\ &+ \eta^{\alpha\sigma}\left(\partial_i h_{\sigma j} - \frac{1}{2}\partial_\sigma h_{ij}\right)\bar{\ell}^i \bar{\ell}^j + \eta^{\alpha\sigma}\left(\partial_i h_{0\sigma}\right)\bar{\ell}^i \\ &- \frac{1}{2}\eta^{\alpha\sigma}\partial_\sigma h_{00} = 0\end{aligned} \qquad (2)$$

If $\alpha = 0$, Eq. (2) leads to $\frac{d\bar{\ell}^0}{d\sigma} = 0$ assuring again that condition $\bar{\ell}^0 = 0$ holds true all along the curve $\bar{\Upsilon}$; if $\alpha = k$ we obtain the spatial trajectory equation that we need to integrate, namely:

$$\frac{d\bar{\ell}^k}{d\sigma} + \bar{\ell}^k\left(\frac{1}{2}\bar{\ell}^i \partial_i h_{00}\right) + \eta^{k\lambda}\left(\partial_i h_{\lambda j} - \frac{1}{2}\partial_\lambda h_{ij}\right)\bar{\ell}^i \bar{\ell}^j - \frac{1}{2}\eta^{k\lambda}\partial_\lambda h_{00} = 0 \qquad (3)$$

where $i, j, k = 1, 2, 3$.

4. Observables and boundary conditions

Each measurement takes place at some time τ_0, say, which fixes the *space* $S(\tau_0)$ where the light trajectory is mapped on. We then consider as observables, at time τ_0, the angles Θ's that the direction of the incoming photon forms with the three spatial directions of a frame adapted to the satellite. These three angles provide the three boundary conditions which are needed to solve Eq. (3) for $\bar{\ell}^i$. Let $\mathbf{u}'$ be the vector field tangent to the satellite's world-line and let $\{\lambda_{\hat{a}}\}$ (where $\hat{a} = 1, 2, 3$) be a space-like triad carried by the satellite. The angles $\Theta_{(\lambda_{\hat{a}}, \bar{\ell})}$ are given by de Felice & Clarke (1990):

$$\cos \Theta_{(\lambda_{\hat{a}}, \bar{\ell})} \equiv e'_{\hat{a}} = \frac{\gamma'_{\alpha\beta} \bar{k}^\alpha \lambda_{\hat{a}}^\beta}{(\gamma'_{\alpha\beta} \bar{k}^\alpha \bar{k}^\beta)^{1/2} (\gamma'_{\alpha\beta} \lambda_{\hat{a}}^\alpha \lambda_{\hat{a}}^\beta)^{1/2}} \quad (4)$$

where no sum is meant over $\hat{a}$ and $\gamma'_{\alpha\beta}$ is the operator which projects into the satellite's rest-frame. From the relation between $\bar{\mathbf{k}}$, $\bar{\ell}$ and $\tilde{\mathbf{u}}$, the above equation can be written more conveniently as

$$e'_{\hat{a}} = \frac{\bar{\ell}^i_{(o)} \lambda_{i\hat{a}} + \tilde{u}^\beta \lambda_{\beta\hat{a}}}{u'_j \bar{\ell}^j_{(o)} + u'^\rho \tilde{u}_\rho} \quad (5)$$

where $\bar{\ell}_{(o)} \equiv \bar{\ell}(\tau_o)$ and recalling that $\gamma'_{\alpha\beta} \lambda_{\hat{a}}^\alpha \lambda_{\hat{a}}^\beta = 1$.

The direction of the light ray, as it is *observed* from within the satellite, depends on the motion of the latter relative to the center of mass which we take as origin of the spatial coordinates on each slice. This dependence gives rise to the stellar *aberration*. Eq. (3) and the three conditions (5) together with the coordinate positions $\xi^i_{(o)} = \xi^i(\sigma(\tau_o))$ of the satellite at the time of observation, form a closed system of equations whose solutions are the coordinate positions of the observed star provided one is able to identify the value of the parameter σ at the emission. The solution we obtain, in fact, is of the type $\xi^i(\sigma(\tau), \bar{\ell}^k_{(o)}, \xi^k_{(o)})$, hence $\sigma_* \equiv \sigma(\tau_*)$ corresponding to the event of the photon emission is an implicit unknown. The way how to determine σ_* will be discussed in the following section while a general solution of (5) is given in de Felice et al. (2002, in prep.).

5. Testing the model

Eq. (3) can only be integrated numerically hence we shall pursue a test campaign to check the accuracy of the results.

5.1. The unitary test

In our approximation it is $(\eta_{\alpha\beta} + h_{\alpha\beta}) \bar{\ell}^\alpha \bar{\ell}^\beta = 1 + O((v/c)^3)$, so the first check was whether the spatial vectors $\bar{\ell}$ obtained at each step of integration satisfy this unitarity condition up to the correct order of accuracy (i.e. up to $O(10^{-12})$). The results show that this happens always on the integration path.

5.2. Identifying the emission time

Let us assume that the star has no proper motion relative to the center of mass of the Solar System. In this case the spatial coordinates $\xi^i_* \equiv \xi^i(\sigma_*)$ of the star remain fixed with τ. If the first photon is emitted at τ_*, and observed at τ_0 then

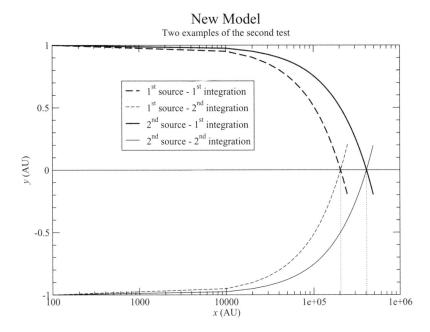

Figure 1. Two examples of a test that deduces the distance of a star from two observations of the same objects taken at opposite positions on the orbit of the observer. The first star is approximately at 1 pc, the second 2 pc. Each couple of observation gives two geodesic that cross at the distance of the object.

its trajectory is mapped into a spatial path on the slice $S(\tau_o)$. Given boundary conditions as explained before, integration along this path leads to a solution $\xi^i = \xi^i(\sigma(\tau), \bar{\ell}^k_{(0)}, \xi^k_{(0)})$. The same star is now observed at a subsequent time $\tau'_o = \tau_o + \Delta\tau$ collecting a photon emitted at a coordinate time $\tau'_* = \tau_* + \Delta\tau$; the light trajectory is now mapped into a slice $S(\tau'_o)$. Evidently, this second observation implies a different set of boundary conditions hence the solution will be a new function $\xi'^i = \xi'^i(\sigma(\tau), \bar{\ell}'^k_{(0)}, \xi'^k_{(0)})$. Since the spatial coordinates of the star are preserved under mapping into the slices of constant τ, they will be uniquely identified by the value of the parameter $\sigma(\tau_*)$ such that $\xi'^i(\sigma(\tau_* + \Delta\tau)) = \xi^i(\sigma(\tau_*))$. A check of the procedure is shown in Figure 1; in this case we have assumed that on the light path acted only the Sun. Two stars are considered; one at a coordinate distance of 1 pc from the observer and the second at 2 pc. Fixing the boundary conditions corresponding to observations of the stars in two symmetrically opposite directions with respect to the Sun and assuming $\Delta\tau$ equal to six months, we find, in logarithmic units, the points of intersections of the two integrated spatial paths. The actual distances are slightly less that 1 and 2 pc respectively because we calculate proper rather than coordinate distances with respect to the baricenter.

5.3. Recovering spherical symmetry

An important self-consistency test is whether, taking the Sun as the only source of gravity, the model recognizes a fully spherically symmetric situation. We have considered an observer at $r_\odot = 1$ AU and a set of stars placed at different angular distances ψ from the Sun. For each ψ, we have taken four stars symmetrically positioned with respect to the Sun-observer direction. The results are reported in Table 1, where the $\delta\psi_i$ are the deflections calculated in the four cases. They have been calculated to the 0.1 μas level and, as expected, are the same for a given angular distance. The way how we calculated the deflection angles requires a detailed analysis which is contained in de Felice et al. (2002, in prep.).

Table 1. Results for the self-consistency tests. ψ is the angular distance from the Sun, $\delta\psi_1 \ldots \delta\psi_4$ the deflection obtained in four symmetric cases at the same ψ.

ψ			$\delta\psi_1$	$\delta\psi_2$	$\delta\psi_3$	$\delta\psi_4$
0°	16′	5″.1428100	1″.7406216	1″.7406216	1″.7406216	1″.7406216
1°	0′	0″.0	0″.4666385	0″.4666385	0″.4666385	0″.4666385
2°	0′	0″.0	0″.2333012	0″.2333012	0″.2333012	0″.2333012
5°	0′	0″.0	0″.0932707	0″.0932707	0″.0932707	0″.0932707
10°	0′	0″.0	0″.0465464	0″.0465464	0″.0465464	0″.0465464
45°	0′	0″.0	0″.0098314	0″.0098314	0″.0098314	0″.0098314
60°	0′	0″.0	0″.0070534	0″.0070534	0″.0070534	0″.0070534
75°	0′	0″.0	0″.0053071	0″.0053071	0″.0053071	0″.0053071
85°	0′	0″.0	0″.0044441	0″.0044441	0″.0044441	0″.0044441
95°	0′	0″.0	0″.0037316	0″.0037316	0″.0037316	0″.0037316
105°	0′	0″.0	0″.0031248	0″.0031248	0″.0031248	0″.0031248
120°	0′	0″.0	0″.0023511	0″.0023511	0″.0023511	0″.0023511
135°	0′	0″.0	0″.0016868	0″.0016868	0″.0016868	0″.0016868
170°	0′	0″.0	0″.0003563	0″.0003563	0″.0003563	0″.0003563
175°	0′	0″.0	0″.0001778	0″.0001778	0″.0001778	0″.0001778
179°	43′	54″.8571900	0″.0000095	0″.0000095	0″.0000095	0″.0000095

5.4. Light deflection tests

A formula for the light deflection as a function of the angular displacement (ψ) of a star from the Sun, is given in Misner et al. (1973) in the case of the exact Schwarzschild solution, namely:

$$\Delta_m = \frac{2M_\odot}{r_o}\sqrt{\frac{1+\cos\psi}{1-\cos\psi}}, \qquad (6)$$

where $M_\odot$ is the solar mass in geometrized units, r_o is the distance of the observer from the Sun. Since (6) is an exact formula, we expect that its predictions coincide with those of the present model at least up to the order of $(v/c)^2$, that is 10^{-8} rad.

Taking the same stars and computing the light deflection using the methods adopted in our model, the tests have shown that the difference between the two predictions is $\lesssim 15$ μas. In particular this happens for limb grazing light rays and the difference becomes rapidly less than 0.1 μas for $\psi > 5°$ (see Table 2).

Table 2. The deflection Δ_m is computed with an exact formula in the Schwarzschild metric for increasing angular displacement ψ of the light source from the Sun. This value is compared with the deflection deduced from our model, namely Δ. The difference $\Delta - \Delta_m$ goes rapidly under the 0.1 μas level for $\psi > 5°$, and the magnitude of its maximum value is comparable to that expected for a $(v/c)^2$ approximation.

ψ			Δ	Δ_m	$\Delta - \Delta_m (\mu as)$
0°	16'	5″.1428100	1″.7406216	1″.7406073	14.3
1°	0'	0″.0	0″.4666385	0″.4666375	1.0
2°	0'	0″.0	0″.2333012	0″.2333010	0.2
5°	0'	0″.0	0″.0932707	0″.0932706	0.1
10°	0'	0″.0	0″.0465464	0″.0465464	0.0
45°	0'	0″.0	0″.0098314	0″.0098314	0.0
60°	0'	0″.0	0″.0070534	0″.0070534	0.0
75°	0'	0″.0	0″.0053071	0″.0053071	0.0
85°	0'	0″.0	0″.0044441	0″.0044441	0.0
95°	0'	0″.0	0″.0037316	0″.0037316	0.0
105°	0'	0″.0	0″.0031248	0″.0031248	0.0
120°	0'	0″.0	0″.0023511	0″.0023511	0.0
135°	0'	0″.0	0″.0016868	0″.0016868	0.0
170°	0'	0″.0	0″.0003563	0″.0003563	0.0
175°	0'	0″.0	0″.0001778	0″.0001778	0.0
179°	43'	54″.8571900	0″.0000095	0″.0000095	0.0

5.5. Deducing stellar positions: consistency with the Schwarzschild model

In recent years a model for astrometric observations was developed assuming as background the umperturbed Schwarzschild metric (see de Felice et al. 1998; de Felice et al. 2001; Vecchiato 1996). In that model the components of the vector k^α tangent to a light trajectory relative to the spatial axes of a *phase-locked* tetrad (de Felice & Usseglio-Tomasset 1992) adapted to an observer moving in a circular orbit around the Sun, were expressed in terms of the distance r_o of the observer from the Sun, its orbital angular velocity ω and the two angular constants of motion of the null geodesic. The latters moreover, can be expressed as functions of the impact parameter r_c of the light trajectory with respect to the Sun and of the angular coordinates of the star (θ, ϕ). Finally, r_c can be implicitly expressed as a function of all three stellar coordinates (r, θ, ϕ). This means that, given the position of a star and that of an observer (in Schwarzschild coordinates), we are able to deduce the cartesian components of the tangent to the null geodesic at the position of the observer by means of an almost completely analytical procedure. These components allow us to define the observables as in Eq. (5).

In the present model we can reproduce the same physical situation by taking the $h_{\alpha\beta}$ terms as approximations of the Schwarzschild metric and using the tetrad adapted to an observer on a circular orbit found in de Felice et al. (2002, in prep.). Then once we have fixed the observables and the position of the observer for a given stellar position, we use them as the boundary conditions needed to integrate backwards the set of differential Eq. (3). The test consists to see whether our model reproduces the stellar positions of the Schwarzschild model. We have considered five cases corresponding to different preassigned stellar positions and calculated the parallexes p_e^* from the exact Schwarzschild model, then compared these with the parallaxes p_a^* reconstructed with our model.

Table 3. Difference (in μas) between the parallax p_e^* calculated from the exact Schwarzschild model and that reconstructed using the present $(v/c)^2$ model (p_a^*). The first column shows the exact distance in pc, the remaining four give $(p_e^* - p_a^*)$ as obtained for each of the corresponding ε_{r_c}, namely the level of accuracy for the impact parameter r_c expressed in meters.

r_e^* (pc)	$(p_e^* - p_a^*)$ (μas)			
	$\varepsilon_{r_c} = 4.8$	$\varepsilon_{r_c} = 9.6 \cdot 10^{-2}$	$\varepsilon_{r_c} = 1.9 \cdot 10^{-3}$	$\varepsilon_{r_c} = 4 \cdot 10^{-5}$
1	-262.2	-12.5	-17.8	-17.8
10	124.6	-19.1	-17.8	-17.8
100	150.9	-23.1	-17.8	-17.8
1000	229.1	-20.6	-17.7	-17.8
10000	85.6	-12.8	-17.8	-17.8

The results are shown in Table 3, where we can see that the minimum difference between those two values is of about 18 μas, as one could expect from the level of approximation attained in this model. This table shows also that the determination of the impact parameter of the null geodesic (r_c) with an accuracy of $\sim 10^{-3}$ m, at least, is necessary to reach the theoretical floor for $(p_e^* - p_a^*)$ caused by the approximation of the model, since with this precision on r_c the numerical contribution to the error is reduced to the 0.1 μas level. We stress that in our case, where $r_c \simeq r_o \simeq 1.5 \cdot 10^{11}$ m, this means that the numerical accuracy needed is $\varepsilon_{r_c}/r_c \sim 10^{-14}$, pushing to the edge the typical accuracy for a double precision number.

6. Conclusions

The relativisitc model of astrometric data reduction in the gravitational field of the Solar System has been completed and successfully tested to the order of $(v/c)^2$. The extension to the order of $(v/c)^3$ together with the definition of the algorithm for deducing the stellar motions is now under close investigation.

References

de Felice, F. & Clarke, J.C.S. 1990, Relativity on curved manifolds, Cambridge University Press

de Felice, F. & Usseglio-Tomasset, S. 1992, Gen. Rel. Grav. 24, 1091

de Felice, F., Lattanzi, M. G., Vecchiato, A. & Bernacca P. L. 1998, A&A 332, 1133

de Felice, F., Bucciarelli, B., Lattanzi, M.G. & Vecchiato, A. 2001, A&A 373, 336

Everhart, E. 1985, in Dynamics of Comets: Their Origin and Evolution, A.Carusi & G.B.Valsecchi ed.s, Proc. IAU Coll. 83, Ap&SSL 115, pag. 185

Klioner, S.A. & Kopeikin, S.M. 1992, AJ 104, 897

Kopeikin, S. M., & Mashhoon, B. 2002, Phys.Rev.D 65, 4025

Misner, C.W., Thorne, K.S. & Wheeler, J.A. 1973, Gravitation, Freeman

Vecchiato, A. 1996, Relativistic astrometry and application to GAIA, degree dissertation (in Italian), Univ. of Padova

GAIA and the planetary nebulae

Laura Magrini, Mario Perinotto

Dipartimento di Astronomia e Scienza dello Spazio, Università di Firenze, L.go E. Fermi 2, 50125 Firenze, Italy

Romano L. M. Corradi

Isaac Newton Group of Telescopes, Apartado de Correos 321, 38700 Santa Cruz de La Palma, Canarias, Spain

Abstract.

GAIA appears to be an excellent opportunity to provide fundamental contribution to knowledge of three dimensional distribution of Planetary Nebulae (PNe) and consequently of low and intermediate mass stars in our Galaxy through the determination of their distances and of their kinematical behaviour. Its highly accurate astrometry will allow us to measure to unprecedent deep levels the distances to Galactic PNe. The GAIA spectral range (8480-8740 Å) will permit the measurement of accurate radial velocities from the Paschen emission lines, which are expected to be well observed in at least 200 PNe. Proper motions will be measured thought repeated imaging observations of the same objects. In total, GAIA imaging and spectroscopic observations will give a very accurate and comprehensive three-dimensional description of the Galactic PNe, allowing to build their luminosity function which will become one of the better standard candle in extragalactic distance measurements and to obtain a quite more accurate understanding of the properties of individual objects.

1. Distances to Planetary Nebulae

Planetary nebulae represent a late stage of evolution of low- and intermediate-mass stars ($0.8\ M_\odot < M < 8\ M_\odot$), approximately 90% of the total. About 1 600 PNe are known in our Galaxy (Acker et al. 1992, 2002 in press) while their total expected population amounts to approximately 15 000 PNe (cf. Kwok 2000). One of their most important parameters, the distance, is generally very poorly known, much worse than that of normal stars. The reason is that none of the physical and geometrical parameters of PNe, like the absolute luminosity of the nebula and of the central star, the diameter, the mass, is constant for the whole PNe population. Some of these parameters, as the mass of the nebulae and that of the central star are different in different PNe whereas others, as the diameter and the electron density, also vary with time.

So far, distances to Galactic PNe have been determined using either individual or statistical methods. The first include trigonometric parallaxes, spectral distance of the companion (if any), expansion parallaxes, and interstellar extinction. Statistical methods assume the constancy of one of the nebular or stellar parameters as the ionized mass of the nebula (Shklovsky 1956), the radio flux (Milne & Aller 1975) or the absolute magnitude of the nucleus. The individual methods set the zero point of the distance scale, while the statistical methods are used when none of the individual methods can be applied.

Except for few cases of nearby planetary nebulae where trigonometric parallaxes have been measured (of the order of a dozen, Kwok 2000, pag. 183), of some PNe with a known binary companion whose spectral type can be determined (cf. Pottash 1984, pag. 101) or the cases in which the intestellar extinction method provided useful results (cf. Pottash 1984, pag. 103), most of the distances to PNe had to be estimated using statistical methods from their nebular properties with somewhat uncertain results. In practice distances to the vast majority of galactic PNe whose distances could be determined so far, are believed to be uncertain by a factor of 2-3. Such uncertainty in distances reflects into a significant uncertainty in our knowledlge of the absolute luminosity of the central stars and thus of their positions on the HR diagram. It is therefore difficult to compare real objects with the predictions of stellar evolution. From the above it is clear how much better distances than those presently available are essential for a real understanding of the evolutionary stage represented by PNe. Distances are also important to the knowledge of the space density and then of the total number of PNe in our Galaxy.

2. The GAIA solution

One of the GAIA's goals is to create an extraordinary precise three-dimensional map of approximately one billion stars in our Galaxy and in closests galaxies. The accuracy of GAIA astrometry will be: (i) 3-4 μas for stars with M_V<10-11; (ii) 10 μas for stars with M_V=15; (iii) 20-40 μas for stars with M_V=17-18; (iv) 100-200 μas for stars with M_V=19-20. That means that to the distance of 1 kpc the distance to a star of M_V<10-11, M_V=15, M_V=17-18, M_V=19-20 will be measured respectively with an accuracy better than 1%, of 1%, 2-4%, and 10-20%; to the distance of 10 kpc the accuracy will be 3-4%, 10%, 20-40%, and 100-200%. Typical distances of central stars of galactic PNe are between 1 and 10 kpc. From the ESO-Strasbourg catalog of Planetary Nebulae (Acker et al. 1992), we found approximately 300 PNe whose central stars have M_V brighter than M_V=18. Their (uncertain) distances are d<10 kpc (see Figure 1). That means that we will have with GAIA the extraordinary possibility to determine with an accuracy between 1% and 20-40% the distance of about 300 Galactic PNe.

GAIA results will thus drastically change our knowledge of PNe and consequently of low- and intermediate-mass stars evolution.

The GAIA spectral range (8480-8740 Å) will include important lines of the PNe spectrum, i.e. the Paschen emission lines. Their observation will convey precious information on the PNe radial velocities, whereas the proper motions will be measured with repeated images of the same sky region. From the the Hβ fluxes presented by the ESO-Strasbourg catalog, we estimate that Paschen lines will be well observed in about 200 Galactic PNe.

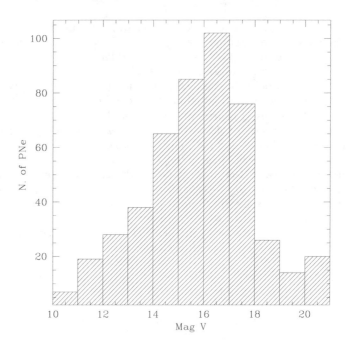

Figure 1. Distribution of the V magnitude of the central stars of the Galactic PNe (data from Acker et al. 1992).

The GAIA information will then allow a complete three-dimensional picture of the Galactic PNe population: distances, proper motions and radial velocities. This will enormously increase our knowledge of late stages of stellar evolution, strongly improving our understanding of low- and intermediate-mass stars. In addition, the precise knowledge of the Galactic planetary nebulae luminosity function (PNLF) will allow to calibrate its bright cutoff directly in our Galaxy. The PNLF, which is used to measure extragalactic distances (cf. Jacoby 1989), will constitute a primary standard candle usable with 10-m telescopes up to a distance of 40 Mpc.

References

Acker A., Ochsenbein F., Stenholm B., Tylenda R., Marcout J. & Schohn C. 1992, *Strasbourg-ESO catalog of Galactic Planetary Nebulae*, ESO
Kwok S. 2000, *The Origin and Evolution of Planetary Nebulae*, Cambridge Univ. Press
Jacoby G.M., 1989, ApJ 339, 39
Milne D.K. & Aller L.H. 1975, A&A 38, 183
Pottash S.R. 1984, *Planetary Nebulae*, Reidel
Shklovsky I.S. 1956, AZh 33, 222

session 4
STELLAR ATMOSPHERES, CLASSIFICATION AND DATABASES

chair: D. Katz

On the way to *Alpenzu Grande* and *Alpenzu Piccolo* on Saturday 14 for a great view of Monte Rosa

B. Plez, N. Nesvacil and P. Hauschildt

ATLAS model atmospheres

Nicole Nesvacil, Christian Stütz & Werner W. Weiss

Institute for Astronomy, University of Vienna, Türkenschanzstrasse 17, 1180 Vienna, Austria

Abstract. We present a review of the basic characteristics of ATLAS model atmospheres and synthetic spectra. The tools used for detailed spectra analysis in Vienna are described. We also give an overview of our new ATLAS9 model grids with different convection models.

1. Introduction

The model atmosphere program ATLAS9 which we describe in this paper was published by R.L.Kurucz (1993b). Apart from this program which is based on grids of Opacity Distribution Functions (ODF) of scaled solar models, another version, ATLAS12 (Kurucz 1993c), exists. It is an opacity sampling code that allows to compute models with individual abundances. It was originally intended to be used for modelling abundance stratification as well, but this part has not been implemented so far. Together with the ATLAS model atmosphere package, Kurucz introduced other software, such as SYNTHE, a spectrum analysis program and WIDTH, which is a spectrum analysis program based on equivalent widths.

2. Basic concepts

An ATLAS model atmosphere is based on the following physical concepts. First of all, a static atmosphere is assumed, so none of the parameters vary with time. That means, for example, pulsation cannot be modelled with ATLAS. Furthermore, we assume the consistancy of flux throughout the whole atmosphere, meaning that our source of energy lies deeper in the star and there are no secondary sources within the atmosphere. An ATLAS model atmosphere is very thin, compared to the stellar radius, so one can assume homogeneous, plane parallel layers. The homogeneity of these layers implies that parameters only change discontinuously with depth. There are no horizontal inhomogeneities, e.g. granules. Throughout the atmosphere, element abundances are constant. And finally, the star is assumed to be in hydrostatic equilibrium. In the original ATLAS9 code and some adaptions by Castelli, Gratton, & Kurucz (1997), the convection treatment is based on the Mixing Length Theory (MLT, Böhm Vitense 1958). The mixing length L is the distance a cell can travel before it dissolves into surrounding material. It is defined as the local pressure scale height H times a mixing length parameter α, which for the Sun was found to be 0.5 in case of no overshooting. In the other case, where convective cells can penetrate into convectivley stable regions, α was determined to be 1.25. In general, the models with no overshooting have turned out to represent observations more accurately. There are also some other convection models implemented for ATLAS9. Two so called FST (full spectrum of turbulence) models are available: the CM (Canuto & Mazzitelli 1991) and the closely related CGM (Canuto, Goldmann, & Mazzitelli 1996). Other than the MLT model, which is more or less a one-eddy approximation, FST uses a turbulence model to compute the full spectrum of a convective flow for a given efficiency. None of these models are actually realistic, because they are all *local* models. Some non-local models are being implemented at present. An impression of the influence of different convection models on our spectra is given in Figure 1. These plots show observed hydrogen Hα and Hβ line profiles for the Sun, Procyon and β Ari, in comparison with synthetic profiles calculated with different convection models (Heiter et al. 2002). For the Sun $T_{\text{eff}} = 5777\ K$ and $\log g$=4.44 were assumed, for Procyon the authors chose $T_{\text{eff}} = 6480\ K$ and $\log g$=3.90, while β Ari was modelled with $T_{\text{eff}} = 8000\ K$ and $\log g$=4.00. The original Kurucz grids (1993a) contain solar models for 13 metallicities from -5 to +1 dex relative to the solar values. They cover a range in $\log g$ from 0.0 to 5.0 in steps of 0.5, and effective temperatures from 3 500 K to 50 000 K. More recently, some models where α-process elements are enhanced by +0.4 and +1.0 dex were released and are available on Kurucz' website (http://kurucz.harvard.edu/). There are also some grids based on ODF computed with new opacities and without overshooting by F.Castelli available for free download.

3. Tools

To calculate a synthetic spectrum which can be compared to observations, other software and databases are needed. Atomic line data fot calculation of synthetic spectra can be taken from various databases, for example Kurucz' line database (Kurucz 1995) or VALD (Vienna Atomic Line Database, Piskunov et al. 1995). So far, VALD does not include molecular data while the Kurucz database already provides a number of diatomic and triatomic molecules. Both databases are constantly being updated. After model computation and selection of all lines which contribute most to the line opacity for a given set of parameters, one can compute a synthetic spectrum using, for example, SYNTHE (Kurucz 1993b), SYNTH (Piskunov 1992) or SYNTHMAG (Piskunov 1998). The basic physical assumptions made for SYNTH (LTE, static atmospheres and plane parallel layers) are consistent with the ATLAS atmospheres. Rotational broadening is not included in this code, but it is part of a program for comparison with observations called ROTATE, which was created by the same author (Piskunov 1998). SYNTH makes use of an adaptive wavelength grid, which means there are more points in places where lines appear, than between the lines. SYNTHMAG is the followup of SYNTH, based on magnetic radiative transfer.

4. Applicability range and automatic analysis

With ATLAS9 it is possible to model mainly main sequence stars from spectral types B to K, and up to luminosity class II, which is the limit for plane par-

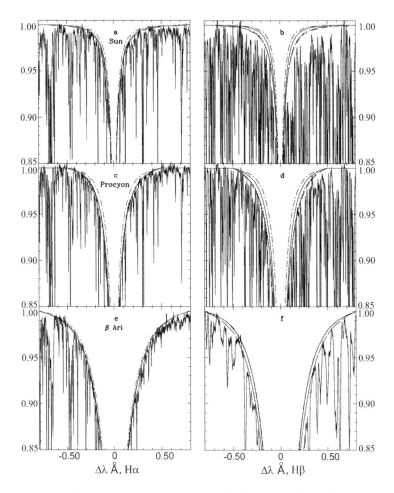

Figure 1. Observed and synthetic Hα (*left*) and Hβ (*right*) profiles of the Sun (T=5777 K, $\log g$=4.44), Procyon (T=6480 K, $\log g$=3.90) and β Ari (T=8000 K, $\log g$=4.00). Linestyles: *line*: MLT, $\alpha = 1.25$; *dash-dotted*: MLT+overshooting, $\alpha = 1.25$; *dotted*: MLT, $\alpha = 0.5$; *dashes*: CGM; *long dashes*: CM. All models with less efficient convection (MLT with $\alpha = 0.5$, CM and CGM) represent observed profiles best (Heiter et al. 2002).

allel treatment. Hot stars cannot be modelled because of spherical symmetry, winds and NLTE effects. For cooler stars, convection becomes very important and most tri-atomic molecules are still missing. All of these shortcomings are presently being improved. Taking all this into account, it should be possible to model a vast number of stars observed by GAIA automatically. Currently, hardly any satisfying all-automatic analysis software exists. In Vienna, we make use of some semi-automatic procedures which, due to the number of free parameters, still require some user interference. Examples for such procedures are

Table 1. Parameters for the Vienna grid of model atmospheres.

T_{eff} (K)	from 4000 to 10000, step 200
$\log g$	from 2.0 to 5.0, step 0.2
[M/H]	−2.0, −1.5, −1.0, −0.5, −0.3, −0.2, −0.1,
	0.0, +0.1, +0.2, +03, +0.5, +1.0
ξ (v_{micro}, km s^{-1})	0, 1, 2, 4
convection	MLT CGM CGM CM
number of layers	72 72 288 288

SME-Spectroscopy Made Easy (Valenti, & Piskunov 1996), VWAutofit (Bruntt 2002), which is based on equivalent widths, and AAP-Abundance Analysis Procedure (Gelbmann et. al 1997), a script to compute model atmospheres, extract lines from VALD and calculate synthetic spectra. Basically, what an automatic procedure could do, is to take the starting values of T_{eff}, $\log g$ and [M/H] from photometry, calculate a small grid of spectra and find the right model and abundances by performing e.g. a least squares fit to the observations. Unfortunately, this would only be that easy if one did not have to deal with some general problems.

4.1. Problems

In any abundance analysis, regardless of whether it is automatic or not, we are probably dealing with inhomogeneous atmospheres, spots, stratification etc. Magnetic fields have to be taken into account, because if we ignore them, we will be getting different values for microturbulent velocity for different lines, and hence, different abundances for different lines of the same element. Rotational broadening has to be considered, as well as the problem of handling noise. Another important error source are the damping constants and inaccurate or even missing gf values in all our databases. Attention should be paid to the fact that we do not know accurate abundances of any star, not even the Sun. As new and better techniques become available, values derived for our best studied star, even for the most abundant elements, change significantly. Comparison of the published solar iron abundances by Grevesse & Sauval (1998) and Holweger (2001) illustrates that within 3 years the value for iron has changed by 0.05 dex!

For all these reasons, a lot of improvements still have to be made to allow the GAIA spectra to be analyzed with an accuracy of a few tenths of a dex in abundances, and a few hundreds K for temperatures.

5. Recent developments

To give an idea of the improvements recently made in the field of stellar atmospheres, we will outline the most important projects made in Vienna so far. First of all, we have calculated new grids of scaled solar model atmospheres with parameters as listed in Table 1. In addition to these *classical* solar scaled

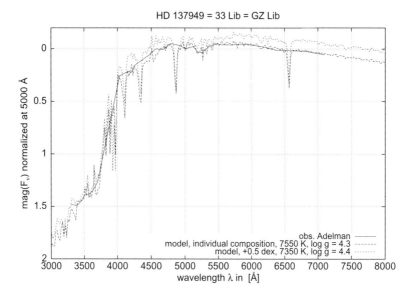

Figure 2. Comparison of spectrophotometry of 33 Lib with best scaled solar fit and individual ODF model.

grids another important development has dramatically influenced our work on chemically peculiar A stars. Piskunov & Kupka (2001) developed a fast method to calculate ODF with individualized abundances. Before we could use this technique we were facing an inevitable discrepancy between the atmospheric parameters derived from comparison with synthetic spectra and the ones obtained by comparison with observed energy distributions. As an example, Figure 2 shows the extremely peculiar Ap star 33 Lib.

The thick line represents spectrophotometry obtained by Adelman (1989). The upper dashed line denotes a scaled solar model which represents observed line intensities best but obviously fails to reproduce the observed fluxes ($T_{\rm eff} = 7350$ K, $\log g = 4.4$). The lower dashed line shows a slightly hotter individual ODF model with $T_{\rm eff} = 7550$ K and $\log g = 4.3$. Evidently, only the latter manages to reproduce the spectrophotometry, while both models can be used for spectrum synthesis. Accurate reproduction of line intensities with the scaled solar model, of course, requires that abundances from the original model are altered before spectrum synthesis - bearing in mind that this method is quite inconsistent. In the future, we will continue to model all our Ap stars based on the more accurate individualized ODF.

6. Future improvements

Presently, F.Kupka is working on the implementation of a non-local convection model, the Reynolds Stress Model (Canuto, Minotti, & Schilling 1994), into our modified version of ATLAS9. These and other new models will have to be tested carefully, ideally by comparison with spectrophotometry. New observations of this kind will be provided by the ASTRA project, maintained by Adelman et al. (http://www.citadel.edu/physics/astra). This project aims to set up an automated spectrophotometric telescope, which is planned to become operational in 2004. It will provide a revision of all standard sources, such as Vega, as well as high accuracy spectrophotometric observations from the near UV to the near IR for a large number of bright stars.

Other projects planned for the future by our team in Vienna include: (*1*) computation of new ODF grids for interesting regions accross the HR diagram, where scaled solar models fail, (*2*) inclusion of opacity sampling into our code, to allow faster computation of individual models in combination with modelling stratification, and of course (*3*) our line database VALD will soon be updated and will include molecular data as well.

References

Adelman, S. J. 1981, A&AS 43, 183
Böhm-Vitense, E. 1958, ZAp 46, 108
Bruntt, H. et al. 2002, A&A 389, 345
Canuto, V.M., Goldman, I. & Mazzitelli, I. 1996, ApJ 473, 550
Canuto, V.M. & Mazzitelli, I. 1991, ApJ 370, 295
Canuto, V.M., Minotti, F.O. & Schilling, O. 1994, ApJ 425, 303
Castelli, F., Gratton, R.G. & Kurucz, R. L. 1997, A&A 318, 841
    (erratum: 1997, A&A 324, 432)
Gelbmann, M. et al., 1997, A&A 319, 630
Grevesse, N. & Sauval, A. J. 1998, Space Sci.Rev. 85, 161
Heiter, U. et al. 2002 A&A 392, 619
Holweger, H. 2001, in Solar and Galactic Composition,
    R.F. Wimmer-Schweingruber ed., AIP Conf. Proc. 598, pag. 23
Kurucz, R. L. 1993a, Kurucz's CD-ROM No. 1–12, CfA Harvard, Cambridge
Kurucz, R. L. 1993b, Kurucz's CD-ROM No. 13, CfA Harvard, Cambridge
Kurucz, R. L. 1993c, in Peculiar Versus Normal Phenomena in A-Type and
    Related Stars, M.M. Dworetsky, F. Castelli & R. Faraggiana ed.s, ASP
    Conf. Ser. 44, pag. 87
Kurucz, R. L., 1995, in Workshop on Laboratory and Astronomical High Resolution Spectra, A. J. Sauval, R. Blomme & N. Grevesse ed.s, ASP Conf. Ser. 81, pag. 583
Piksunov, N. 1992, in Proceedings of the Int. Meeting on Stellar Magnetism, Nauka, St. Petersburg, pag. 92
Piskunov, N. 1998, Cont. Astron. Obs. Skalnate Pleso 27, 374
Piskunov, N. & Kupka, F. 2001, ApJ 547, 1040
Piskunov, N. et al. 1995, A&AS 112, 525
Valenti, J.A. & Piskunov, N. 1996, A&AS 118, 595

Stellar atmospheres and synthetic spectra for GAIA

Peter H. Hauschildt

Hamburger Sternwarte, Gojenbergsweg 112, 21029 Hamburg, Germany

France Allard

Centre de Recherche Astronomique de Lyon (CRAL), Ecole Normale Supérieure de Lyon, Lyon, Cedex 07, 69364 France

Eddie Baron

Dept. of Physics and Astronomy, University of Oklahoma, 440 W. Brooks, Rm 131, Norman, OK 73019-0225

Jason Aufdenberg

Center for Astrophysics, 60 Garden Street, Cambridge, MA 02138

Andreas Schweitzer

Hamburger Sternwarte, Gojenbergsweg 112, 21029 Hamburg, Germany

Abstract. In this paper we discuss our new model atmosphere grids for objects relevant to the GAIA mission. The grids are computed using version 13 of our general-purpose model atmosphere code PHOENIX and included the most recent updated to the micro-physics and line opacities added to the code. We will make the synthetic spectra of the new grid available to support the preparation of the GAIA mission. By the time results from GAIA become available, we expect that advances in modeling stellar atmospheres will result in vastly improved models, but the current grids will be perfectly adequate for the preparation of the mission and initial data analyses.

1. Introduction

Stellar atmosphere modeling has experienced a renaissance in the past decade with the advent of better algorithms and faster computers. This has allowed research groups to remove or relax many of the *standard* assumptions that were made in the 70s through 80s and that had become accepted wisdom over the years. Not surprisingly the new calculations show that many of these assumptions are actually quite bad and can lead to spurious results or incorrect interpretations of observed spectra. The intricate connection between geometry (plane parallel or spherical), line blanketing (atomic and/or molecular) and non-LTE effects (using small to extremely large model atoms and molecules) began

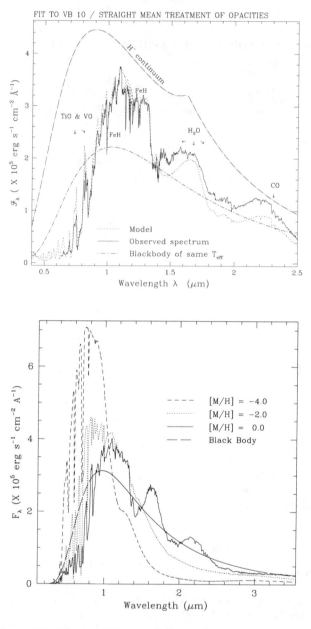

Figure 1. The effects of line blanketing on the emitted spectrum of M dwarfs is illustrated in the *top* panel, the *bottom* panel demonstrates the effects of metallicity changes on the emitted spectra. From Allard et al. (1997).

to emerge slowly as crucial ingredients for physically correct and meaningful interpretations and analyses of stellar spectra. Unfortunately, easy and simple solutions do not really work for stellar atmospheres (although everybody likes the easy way out and some of them are useful for teaching purposes) and have actually hindered progress and reduced the reliability of results.

Our group has developed the very general non-LTE (NLTE) stellar atmosphere computer code PHOENIX (Hauschildt 1992, 1993, Hauschildt et al. 1995, Allard & Hauschildt 1995, Hauschildt et al. 1996, Baron et al. 1996a, Hauschildt et al. 1997, Baron & Hauschildt 1998, Hauschildt & Baron 1999, Allard et al. 2001) which can handle extremely large model atoms as well as line blanketing by hundreds of millions of atomic and molecular lines. This code is designed to be both portable and flexible: it is used to compute model atmospheres and synthetic spectra for, e.g., novae, supernovae, M, L, and T dwarfs, irradiated atmospheres of extrasolar giant planets, O to M giants, white dwarfs and accretion disks in active galactic nuclei (AGN). The radiative transfer in PHOENIX is solved in spherical geometry and includes the effects of special relativity (including advection and aberration) in the modeling.

The PHOENIX code allows us to include a large number of NLTE and LTE background spectral lines and solves the radiative transfer equation for each of them *without* using simple approximations like the Sobolev approximation. Therefore, the profiles of spectral lines must be resolved in the co-moving (Lagrangian) frame. This requires many wavelength points (we typically use 150,000 to 300,000 points). Since the CPU time scales linearly with the number of wavelength points, the CPU time requirements of such calculations are large. In addition, (NLTE) radiative rates for both line and continuum transitions must be calculated and stored at every spatial grid point for each transition, which requires large amounts of storage and can cause significant performance degradation if the corresponding routines are not optimally coded.

An important problem in stellar atmosphere calculations is to find a consistent solution of the very diverse equations that describe the various physical processes. We have developed a scheme of nested iterations that enables us to separate many of the variables (e.g., separating the temperature correction procedure from the calculation of the NLTE occupation numbers). This allows us to compute far more detailed stellar atmosphere models than was previously possible (see references for details).

In order to take advantage of the enormous computing power and vast aggregate memory sizes of modern parallel supercomputers, both potentially allowing much faster model construction as well as more sophisticated models, we have developed a parallel version of PHOENIX which is now the default production version for all our model grids.

2. Methods and models

For our model calculations, we use our multi-purpose stellar atmosphere code PHOENIX (version 13). Details of the numerical methods are given in the above references, so we do not repeat the description here.

One of the most important recent improvements of cool stellar atmosphere models is that new molecular line data have become available and that they have improved the fits to observed spectra significantly. Our combined molecular line list includes about 700 million molecular lines. The lines are selected for every model from the master line list at the beginning of each model iteration to account for changes in the model structure (see below). Both atomic and molecular lines are treated with a direct opacity sampling method (dOS). We do *not* use pre-computed opacity sampling tables, but instead dynamically select the relevant LTE background lines from master line lists at the beginning of each iteration for every model and sum the contribution of every line within a search window to compute the total line opacity at *arbitrary* wavelength points. The latter feature is crucial in NLTE calculations in which the wavelength grid is both irregular and variable from iteration to iteration due to changes in the physical conditions. This approach also allows detailed and depth dependent line profiles to be used during the iterations. This is important in situations where line blanketing and broadening are crucial for the model structure calculations and for the computation of the synthetic spectra. It is highlighted in Figure 1 which shows the dramatic effects of line blanketing on M dwarf spectra and the correspondingly large effects of metallicity on the emergent spectrum (and model structure).

Although the direct line treatment seems at first glance computationally prohibitive, it leads to more accurate models. This is due to the fact that the line forming regions in cool stars and planets span a huge range in pressure and temperature so that the line wings form in very different layers than the line cores. Therefore, the physics of line formation is best modeled by an approach that treats the variation of the line profile and the level excitation as accurately as possible. To make this method computationally more efficient, we employ modern numerical techniques, e.g., vectorized and parallelized block algorithms with high data locality Hauschildt et al. 1997, and we use high-end workstations or parallel supercomputers for the model calculations. With these techniques, a complete model calculation (about 20 iterations) takes less than 20 minutes on 24 CPUs of an IBM pSeries 690 parallel supercomputer.

In the calculations presented in this contribution, we have included a constant statistical velocity field, $\xi = 2$ km s^{-1}, which is treated like a microturbulence. The choice of lines is dictated by whether they are stronger than a threshold $\Gamma \equiv \chi_l/\kappa_c = 10^{-4}$, where χ_l is the extinction coefficient of the line at the line center and κ_c is the local *b-f* absorption coefficient (see Hauschildt et al. 1999 for details of the line selection process). This typically leads to about $10 - 250 \times 10^6$ lines which are selected from the master line lists. The profiles of these lines are assumed to be depth-dependent Voigt or Doppler profiles (for very weak lines). Details of the computation of the damping constants and the line profiles are given in Schweitzer et al. (1996). We have verified in test calculations that the details of the line profiles and the threshold Γ do not have a significant effect on either the model structure or the synthetic spectra. In addition, we include about 2000 photo-ionization cross sections for atoms and ions (Mathisen 1984, Verner & Yakovlev 1995).

The equation of state (EOS) is an enlarged and enhanced version of the EOS used in Allard & Hauschildt (1995). We include about 1000 species (atoms, ions and molecules) in the EOS. The EOS calculations themselves follow the method discussed in Allard & Hauschildt (1995). For effective temperatures,

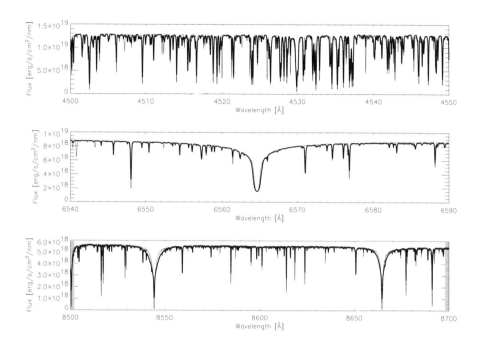

Figure 2. Detailed comparison of the solar spectrum from the Kitt Peak Atlas to a simple LTE PHOENIX model with $T_{\rm eff} = 5770\,{\rm K}$, $\log g = 4.4$ and solar abundances. No attempts of fine tuning were made.

$T_{\rm eff} < 2500\,{\rm K}$, the formation of dust particles has to be considered in the EOS. In our models we allow for the formation (and dissolution) of a variety of grain species. For details of the EOS and the opacity treatment see Allard et al. (2001).

The NLTE treatment of large model atoms or molecules such as H_2O and TiO which have several million transitions is a formidable problem which requires an efficient method for the numerical solution of the multi-level NLTE radiative transfer problem. Classical techniques, such as the complete linearization or the Equivalent Two Level Atom method, are computationally prohibitive for large model atoms and molecules. Currently, the operator splitting or approximate Λ-operator iteration (ALI) method (e.g. Cannon 1973, Rybicki 1972, Rybicki 1984, Scharmer 1984) seems to be the most effective way of treating complex NLTE radiative transfer and rate equation problems. Variants of the ALI method have been developed to handle complex model atoms, e.g., Anderson's multi-group scheme (Anderson 1987, 1989) or extensions of the opacity distribution function method (Hubeny & Lanz 1995). However, these methods have problems if line overlaps are complex or if the line opacity changes rapidly with optical depth, a situation which occurs in cool stellar atmospheres. The ALI rate operator formalism, on the other hand, has been used successfully to treat very large model atoms such as Fe directly and efficiently (Hauschildt &

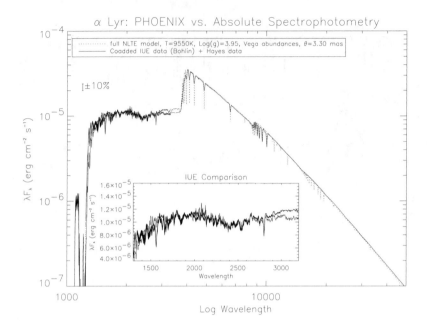

Figure 3. Detailed comparison of the spectrum of Vega to a PHOENIX model with $T_{\rm eff} = 9550\,\rm K$, $\log g = 3.95$ and Vega abundances.

Baron 1995, Hauschildt et al. 1996, Baron et al. 1996b). It allows us to currently treat a large number of species and atomic levels in direct NLTE. NLTE models are *absolutely necessary* for effective temperatures larger than about 10 000 K as the structure of the atmosphere starts to depend significantly on a proper treatment of NLTE effects. Using LTE models will cause misleading results and should be avoided. In general, models with low gravities will require a full NLTE treatment (with spherical geometry) for lower $T_{\rm eff}$ than high gravity models.

3. Some results

In the following paragraphs we will give a few representative results that highlight the type of results that can be obtained today.

With our current model code we can fit the spectra of such fundamental standards like the Sun (Figure 2) and Vega (Figure 3) quite well. For cool giants the models can reproduce high-resolution spectra (e.g., Figure 4) and fit simultaneously the interferometrically observed size as function of wavelength. This demonstrates that current models can reproduce the detailed radiation field of giants very well, which strongly supports of the model approaches that were taken since the early 1990's.

For cooler objects, e.g., M, L & T dwarfs, progress in modeling has been substantial over the last few years. This is shown in Figure 5 where we compare

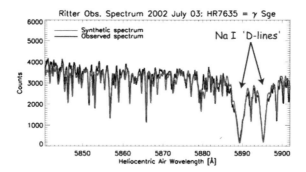

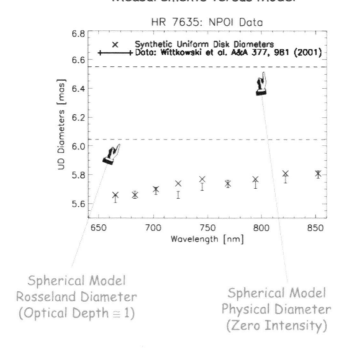

Figure 4. Comparison of the spectrum of γ Sge to a PHOENIX model with $T_{\text{eff}} = 3900\,\text{K}$, $\log g = 0.55$, $M = 1.0\,\text{M}_\odot$, and solar abundances (*top*) and the measured wavelength dependent size distribution (*bottom*). From Aufdenberg & Hauschildt (2002).

models for an L3.5 dwarf with model spectra that were computed under the assumptions (*a*) dust forms and remains at the depth that it formed (*Dusty*), (*b*) dust forms and immediately and completely rains out below the photosphere of the object (*Cond*) to (*c*) the physically better model in which the amount

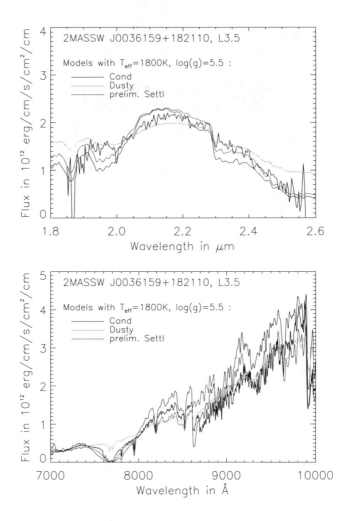

Figure 5. The spectrum of an L3.5 dwarf compared to *Dusty*, *Cond* and preliminary dust *Settling* calculations.

of dust in a given layer is calculated by balancing depletion by gravitational settling with mixing by convection and other global motions in the atmosphere (*Settling*; Allard et al, in preparation).

4. The GAIA grid

The current state of the art in model atmosphere and synthetic spectra calculations will be able to reliably aid the development of the filter and resolution system and also help the design of analysis packages for the GAIA mission. With modern algorithms and fast parallel computers it is possible to construct a fairly detailed grid of synthetic spectra for this purpose. While it is clear that this grid will be obsolete by the time GAIA will be launched, however, the basic features of the models will stay the same even in 10-15 years (after all, current models reproduce observed spectra quite well !). We have thus started to compute such a PHOENIX grid that will cover the parameter range of interest to the GAIA community. We will make the grid of synthetic spectra publicly available as models are completed. The grid will use simplifying assumptions, e.g., LTE, for the effective temperature range of $T_{\text{eff}} = 2500\,\text{K}$ to $10000\,\text{K}$, but will feature the latest set of line lists and be calculated in spherical symmetry for a fixed mass of $1\,M_\odot$.

We use a setup of the microphysics that gives the currently best fits to observed spectra of M, L, and T dwarfs for the low T_{eff} regime and that also updates the microphysics used in the NextGen (Hauschildt et al. 1999a, 1999b) model grid. The water lines are taken from the AMES calculations Partridge & Schwenke (1997), this list gives the best overall fit to the water bands over a wide temperature range. TiO lines are taken from Schwenke (1998) for similar reasons. The overall setup is similar to the one described in more detail in Allard et al. (2001).

5. Conclusions

In this paper we have discussed a few new results of stellar atmosphere modeling that have helped to resolve some outstanding problems understanding and interpreting observed stellar spectra. During the last decade, progress was made by breakthroughs in both methodology and computer technology, which has led to substantially improved models and synthetic spectra. This is in good part due to the pioneering work of R. Kurucz who has provided superb atomic line lists that are crucial to obtain reasonable fits. Although there is still a lot of work to be done, advances over the last decade or so have enabled us to deliver far better models than ever before. In many cases, even our current 'best effort' models cannot reproduce observed spectra satisfactorily, this is in particular the case for L and T dwarfs. However, this is due to physical effects that we 'know' but we cannot currently describe well enough (e.g., incomplete line lists for key molecules or dust and cloud formation processes). Another area that requires much more work is our detailed understanding of winds from both hot and cool stars. There is currently a lot of effort being put into the solution of these key problems, although it is clear that once they are solved, others will pop up in unexpected places.

Acknowledgments. This work was supported in part by NSF grants AST-9720704 and AST-0086246, NASA grants NAG5-8425, NAG5-9222, as well as NASA/JPL grant 961582 to the University of Georgia and in part by NSF grants AST-97314508, by NASA grant NAG5-3505 and by NASA grant and NAG5-12127, NSF grant AST-0204771, and an IBM SUR grant to the University of Oklahoma. This work was supported in part by the Pôle Scientifique de Modélisation Numérique at ENS-Lyon. Some of the calculations presented in this paper were performed on the IBM pSeries 690 of the Norddeutscher Verbund für Hoch- und Höchstleistungsrechnen (HLRN), on the IBM SP 'Blue Horizon' of the San Diego Supercomputer Center (SDSC), with support from the National

Science Foundation, and on the IBM SP and the Cray T3E of the NERSC with support from the DoE. We thank all these institutions for a generous allocation of computer time.

References

Allard, F., & Hauschildt, P.H. 1995, ApJ 445, 433
Allard, F., Hauschildt, P.H., Alexander, D.R., et al. 1997, ARA&A 35, 137
Allard, F., Hauschildt, P.H., Alexander, D.R., et al. 2001, ApJ 556, 357
Anderson, L.S. 1987, in Numerical Radiative Transfer, W. Kalkofen ed., Cambridge Univ. Press, pag. 163
Anderson, L.S. 1989, ApJ 339, 558
Aufdenberg, J.P., & Hauschildt, P.H. 2002, in Interferometry in Optical Astronomy II, W. Traub ed., Proc. SPIE 4838, in press (http://cfa-www.harvard.edu/~jaufdenb/INTERFEROMETRY/spie.pdf)
Baron, E., & Hauschildt, P.H. 1998, ApJ 495, 370
Baron, E., Hauschildt, P.H., Nugent, P., & Branch, D. 1996a, MNRAS 283, 297
Cannon, C.J. 1973, JQSRT 13, 627
Hauschildt, P.H. 1992, JQSRT 47, 433
Hauschildt, P.H. 1993, JQSRT 50, 301
Hauschildt, P.H., & Baron, E. 1995, JQSRT 54, 987
Hauschildt, P.H., & Baron, E. 1999, J. of Comp. and Appl. Math. 102, 41
Hauschildt, P.H., Starrfield, S., Shore, S.N., et al. 1995, ApJ 447, 829
Hauschildt, P.H, Baron, E., Starrfield, S., & Allard, F. 1996, ApJ 462, 386
Hauschildt, P.H., Baron, E., & Allard, F. 1997, ApJ 483, 390
Hauschildt, P.H., Allard, F., & Baron, E. 1999a, ApJ 512, 377
Hauschildt, P.H, Allard, F., Ferguson, J., Baron, E., et al. 1999b, ApJ 525, 871
Hubeny, I., & Lanz, T. 1995, ApJ 439, 875
Mathisen, R. 1984, Photo cross-sections for stellar atmosphere calculations, Inst. of Theoret. Astrophys. Univ. of Oslo, Publ. Series No. 1
Partridge, H., & Schwenke, D.W. 1997, J. Chem. Phys. 106, 4618
Rybicki, G.B. 1972, in Line Formation in the Presence of Magnetic Fields, G.Athay, L.L. House, and G.Newkirk Jr. ed.s, High Altitude Observatory, Boulder, pag.145
Rybicki, G.B. 1984, in Methods in Radiative Transfer, W. Kalkofen ed., Cambridge Univ. Press, pag. 21
Scharmer, G. B. 1984, in Methods in Radiative Transfer, W. Kalkofen ed., Cambridge Univ. Press, pag. 173
Schweitzer, A., Hauschildt, P.H., Allard, F., & Basri, G. 1996, MNRAS 283, 821
Schwenke, D.W. 1998, Chemistry and Physics of Molecules and Grains in Space, Faraday Discussion 109, 321
Verner, D.A., & Yakovlev, D.G. 1995, A&AS 109, 125

Cool star atmospheres and spectra for GAIA: MARCS models

Bertrand Plez

GRAAL, cc72, Université Montpellier II, F-34095 Montpellier cedex 05, France

Abstract. After general considerations on the use of model atmospheres and spectra, with special attention to the expected GAIA spectral harvest, I expose the general characteristics of classical line-blanketed model atmospheres (1D, LTE, static). The history and current status of MARCS model atmospheres, initially developed by Gustafsson et al. (1975), are then briefly described, with a few recent achievements in photometry and spectroscopy of cool stars. It seems that the MARCS models have reached their ultimate level of sophistication, within their set of limiting hypotheses. Further progress require more elaborate modeling, including non-LTE, and hydrodynamics. Synthetic spectra of M and S stars calculated for the GAIA spectral range are shown and discussed. I demonstrate that important work is necessary on the existing line lists before we can reliably extract stellar parameters from GAIA spectra.

1. On the use of model spectra

GAIA will provide us with millions of stellar spectra, as well as with broad- and narrow-band photometry, for stars of all spectral types. This wealth of data will have to be understood, and stellar parameters extracted from it (T_{eff}, $\log g$, metallicity, and if possible more detailed abundance patterns). In the actual phase of instrument design, it is necessary to assess what resolution of the spectrograph will allow a given accuracy on the determination of stellar parameters. Huge efforts have thus been devoted to build libraries of observed stellar spectra in the GAIA range (Munari & Tomasella, 1999; Marrese et al. 2003; Munari 2003). These are extremely helpful, but lack completeness for obvious reasons. Model spectra can be used to complete these libraries with, e.g., rare spectral types, or extremely metal poor stars. Model spectra have the additional advantage that they can be built at any spectral resolution, for any spectral range, and at any special chemical composition, for any combination of T_{eff}, $\log g$, etc, in principle at least. Various problems may be encountered: necessary physics not implemented in the code (highly ionized species must be included for hot stars, molecules have to be included in very cool stars, etc), approximations not appropriate (to compute an O star spectrum in LTE is futile), inadequate line lists (missing lines, incorrect gf values, etc), or even numerical problems. With model spectra, we can also check the effect of a change of the C abundance on the spectrum of a carbon star, or of the Zr abundance on an S star. We can determine the amount of absorption due to a haze of lines, or to some given continuous opacity. For these exercises to be useful there is however a necessary condition: that our models are sufficiently good that we trust them to reproduce real star spectra.

We know real star atmospheres are not simple spherically symmetric layers, at hydrostatic equilibrium and LTE. We know from detailed observations and modeling of the Sun that hydrodynamics, magnetic fields, non-LTE, all must enter the picture if we have the ambition of reproducing the solar spectrum. We know that our line lists are incomplete, or suffer from uncertainties in line strength or broadening parameters. This latter point is however improving, thanks to the enormous work done by our colleagues in atomic and molecular spectroscopy (see Wahlgren & Johansson 2003, for problems directly related to the planed GAIA spectral range). We have detailed models of the solar atmosphere (e.g. Asplund et al. 2000), that accurately reproduce most features of the solar spectrum (line strengths, shifts, and asymmetries, granulation distribution of shapes and intensities, etc). These 3D radiative-hydrodynamics models are extremely costly, and cannot yet be used for routine analysis of a large number of stars. They also still need to be tested in detail for stars different from the Sun (much lower metallicities, and lower temperatures and gravities). Recently, the first 3D model of the atmosphere of the supergiant Betelgeuse was computed by B. Freytag (Freytag & Mizuno-Wiedner 2003; see also http://www.astro.uu.se/~bf), showing enormous convection cells, moving at supersonic velocities, on time-scales of months. This looks extremely promising, and such models will eventually be used to carry detailed analyses of giant and supergiant stars, but they need to be refined and tested extensively. No doubt such models will be available at the time we get the GAIA spectra, and we should bear that in mind when the time comes for their detailed analysis. Note also the recent full non-LTE models of cool stars of Hauschildt et al. (1999), and the new pulsation models for long-period variables of Höfner (1999), that will be indispensable to process mira spectra.

In the mean time, we must use what we have at hand, that can be used easily on a large scale: we want thousands of model spectra spread all over the HR diagram, to carry various tests on instrument design, spectra extraction, photometric system design, and stellar parameter determination. I will show that, the existing classical models have reached a high level of sophistication, within their own limited hypotheses (in the following I call classical models, models built assuming a 1D stratification of the atmosphere, i.e. without horizontal structure, in hydrostatic equilibrium, at LTE, and with conservation of the radiative and convective luminosity). Actually, it is probably not useful to further develop classical models. Instead, one must focus efforts towards the development of models that relax the classical hypotheses.

2. Classical model atmospheres and spectra

We have seen that model spectra are a complement to observed spectra, when building libraries, and that they allow extensive experiments on the impact of the variation of stellar parameters. Their basic use, the one they were initially developed for, is to interpret observed spectra. The extraction of chemical abundances requires the comparison of the observations to a model spectrum. Adjustment of the line profiles (or equivalent widths, when it is possible), provides the abundances, if the model atmosphere used reflects the actual stellar atmosphere thermodynamical conditions sufficiently well. At this stage it is important to clearly make the difference between a model atmosphere and a model spectrum. A classical model atmosphere provides the run of temperature, pressure of gas, electrons, and radiation, convective velocity and flux, and more generally of all relevant quantities, as a function of some depth variable (geometrical, or optical depth at some special frequency, or column mass). Such a model is computed assuming conservation of the luminosity (no energy production nor sink in the atmosphere), hydrostatic equilibrium (radiative and gas and turbulent pressure balance gravity), with radiative transfer treated in LTE, using some theory for the convective transport (usually mixing-length theory, or some refinement of it). As these hypotheses, combined with the 1D (plane-parallel or spherical) geometry, are relatively computationally cheap to deal with (despite strong non-linearities, and numerical problems to take care of), it is possible to treat the radiative field in a much elaborate way. While 3D radiative-hydrodynamics models treat the radiative field with a handful of frequencies, current classical models evaluate intensities at 10^5 wavelengths taking into account 10^8 spectral lines, in minutes, on workstations.

Once we have built a model atmosphere, we can use it to generate synthetic spectra, like the Calcium IR triplet region for GAIA, or larger chunks that may be used to compute synthetic colors. The synthetic spectrum quality will be affected by the quality and adequacy of the atmosphere model, but also by the details of the line list adopted, its accuracy, and completeness. To build a good model atmosphere requires an extensive account of the opacity sources, but the exact position of lines is not so critical, as long as they appear approximately a the right place and strength. The situation is different when computing a synthetic spectrum at a resolution of 100 000: line positions really matter. It is that latter situation that may not be fully satisfactory for molecular lines in the GAIA spectral range, even if the opacities are sufficiently good and complete to construct model atmospheres (for the atomic line situation see Wahlgren & Johansson 2003).

3. MARCS models

The MARCS models were initially developed in Uppsala by Gustafsson et al. (1975), as plane-parallel, convective and radiative flux conservative, LTE, line-blanketed models. The line opacity was treated through opacity distribution functions (ODF). The first grids were for solar-type dwarfs and giants of various metallicities. Carbon star models were produced by Eriksson et al. (1984), after inclusion of additional carbon-bearing molecular opacities, and extension of the chemical equilibrium to a larger number of species. A major update was done by Plez, Brett, & Nordlund (1992), who included spherical symmetry, the opacity sampling (OS) treatment of line opacities, further extended the chemical equilibrium to a few hundreds molecular species, and added opacity for TiO and H_2O, to allow the modeling of M type stars. In parallel, Edvardsson et al. (1993) introduced better continuous and atomic line opacities. Later, continuous opacities were further refined and completed, to build H-deficient and R CrB type models. The current version of the MARCS code is a final evolution merging all these changes with newer updates of the opacities, and a new chemical equilibrium including all the neutral and first ions, and some second ions of the atoms, and about 600 molecular species. The atomic lines are based on VALD (Kupka et al. 1999), with collisional broadening by hydrogen following Anstee, Barklem and O'Mara (see Barklem, Piskunov, & O'Mara 2000). The molecular opacities comprise extensive line lists for CH, NH, OH, C_2, CN, CO, CaH, FeH, MgH, SiH, SiO, TiO, VO, ZrO, H_2O, HCN, C_3, C_2H_2. The data comes primarily from calculations, supported by laboratory data when available (Jørgensen 1997, Jørgensen et al. 2001, Plez 1998). The opacities are sampled with a resolution of 20 000, between 91 and 20 000 nm (more than 150 000 points).

The new models were recently presented by Gustafsson et al. (2003a, b), and Plez et al. (2003), where more details can be found, as well as discussions of the thermal structure of the models. No future extensive updates are foreseen for these models. They are, and will be, used for systematic, detailed analyses of cool stars, and will also provide a reference for more elaborate 3-D, non-LTE modeling.

4. Recent results

The updated MARCS models have been used in recent years for a variety of purposes: from abundance analyses of cool stars, to the analysis and calibration of ISO-SWS spectra, or calibration of photometric systems. This utilization of the models in various conditions, shows that they provide a remarkably consistent description of cool star atmospheres. In some cases, however, the models may still be improved: in very cool carbon stars, failure to accurately reproduce the whole IR spectrum is probably due to persistent uncertainties in polyatomic opacities, esp. C_2H_2. In very cool stars in general (T_{eff} <2500 K), the absence of dust condensation and opacities is a limitation. The ambition of the new grid of model is therefore to cover all spectral types with 2500< T_{eff} <8000 K, with any metallicity, or chemical peculiarity (esp. CNO abundances). Among recent uses of the MARCS models, Bessell, Castelli, & Plez (1998), established a calibration of the Johnson-Cousin broad-band photometric system (also for the Kurucz's ATLAS models, see Nesvacil et al. 2003). The models reproduce well most of the color-color relations, and the temperature-color relations. Bolometric corrections are also provided. Figure 1 shows one example of the match of a model spectrum to a flux calibrated observation.

In a series of papers, Decin et al. (2000; 2002a,b,c in press) studied in detail ISO-SWS spectra of A to M stars, contributing to the final calibration of the instrument. A nice consistency check of the models and their calculated emergent spectra is provided by the fact that temperatures derived from the photometry of the stars using the T_{eff}-color relations established with the models, luminosities derived from bolometric corrections from the models, distances from Hipparcos, and angular diameters from interferometry or from the ratio of ISO absolute flux and model flux, are all consistent.

Recently, Plez et al. (2003) computed a grid of models with parameters appropriate for M and S giants (3000 < T_{eff} < 4000 K, 0.5<C/O<0.99). They studied the effects of varying C/O, Fe/H, and the s-element abundances on

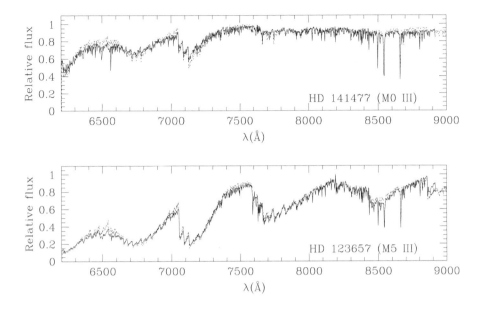

Figure 1. Observed spectra of M giants from Serote Roos, Boisson, & Joly 1996 (full line), and MARCS model spectra (dotted line), for $T_{\text{eff}}=3800$ K, $\log g=1.5$ and $T_{\text{eff}}=3500$ K, $\log g=0.9$. The model spectra were computed with a solar chemical composition. (From Alvarez & Plez 1998)

the thermal structure of the models and on their spectra. They also computed narrow-band TiO and ZrO indices, as well as broad-band colors. They could predict, from the model calculations, that the T_{eff}-scale of S stars is different of that of M stars, but that a combination of broad-band colors and narrow-band indices allows to estimate simultaneously T_{eff}, C/O and Zr/Fe of M and S stars. This is confirmed by the first comparisons to observations. Further work is in progress.

5. The GAIA spectral window, and a few lessons we can learn from models

It is interesting to use some specific models to calculate theoretical spectra in the GAIA window, to check e.g. the impact of spectral resolution, or how different opacity contributors react to changes of T_{eff}. I made this exercise for models of the Sun, of Arcturus, and a few M and S giants and dwarfs. The idea was to take the appropriate MARCS model, an atomic line list from VALD, as extracted from their web site, and the molecular line lists used in our MARCS models. I then computed the spectrum between 848 and 874 nm, at a resolution of more than 800 000. It was then degraded to the required resolution, if needed.

5.1. The Sun

A comparison of the Solar spectrum to the full resolution absolute flux atlas of Neckel (1999) shows that the model performs well on average (see Figure 2), but that mismatches appear at closer examination: (*i*) the continuum flux of the model is slightly below the observations. This must be looked at, as it could be due to a calibration problem of the observations, or to some problem with the continuous opacity of the model. The effect is small however (less than 2 percent). This stresses the importance of comparing the models to well calibrated (absolute flux) observations, as this would not appear in comparisons with observations normalized to the continuum; (*ii*) some lines appear in the observations that are absent in the model spectrum, and conversely. This shows that work needs to be done on the line lists (line identifications, positions, *gf-* values). Wahlgren & Johansson (2003) provide a more extensive discussion of this aspect, and propose solutions; (*iii*) further, a comparison of model and observed spectra degraded to a resolution of about 20 000, shows that as lines blend with their neighbors, it becomes more difficult to recover the information on what lines are missing, misplaced, or of erroneous strength. In addition faint lines tend to behave like a pseudo-continuum (see the right wing of the Ca II line on Figure 2); and finally (*iv*) at low resolution, line profiles become unrecoverable, and most information on velocity fields is lost. This may impact the radial velocity (supergiants show velocity gradients of up to 20 km s^{-1} in their atmospheres; Josselin, Plez, & Mauron 2003), and abundance determinations.

5.2. Cool M and S stars

Similar comparisons for Arcturus ($T_{\rm eff}$= 4300 K, log g=1.5) lead to the same conclusions as for the Sun, with more complications due to the greater crowding, also by molecular lines. This crowding is best illustrated by series of calculated spectra for $T_{\rm eff}$= 3600 and 3200 K (Figure 3). I have separated the contributions of the main species acting in that spectral range (TiO, CN, FeH, and atomic lines). FeH only appears redwards of the Ca II lines, and becomes more prominent in dwarfs. TiO totally dominates the spectrum, except for C/O ratios close to one, where CN takes over. The haze of molecular lines at these low temperatures forms a pseudo-continuum, relative to which the strength of atomic line absorptions must be gauged. Inspection of Figure 3 shows also that the slope of the spectrum carries information, even in the restricted GAIA range. The influence of $T_{\rm eff}$, C/O, Ti/H, and log g on the appearance of the spectrum is strong in cool stars. Stellar parameters cannot probably be recovered from GAIA spectra only, for the coolest stars.

6. Conclusions

In conclusion, it is crucial to carefully carry out detailed comparisons of model and observed spectra, at very high resolution, before we can be sure that our model spectra have the required quality that enables the recovery of stellar parameters from low resolution spectra. This will imply hard work on line lists. The model atmospheres themselves must be tested on a much broader set of data (photometry, spectra at other wavelengths, etc), to ensure that they adequately

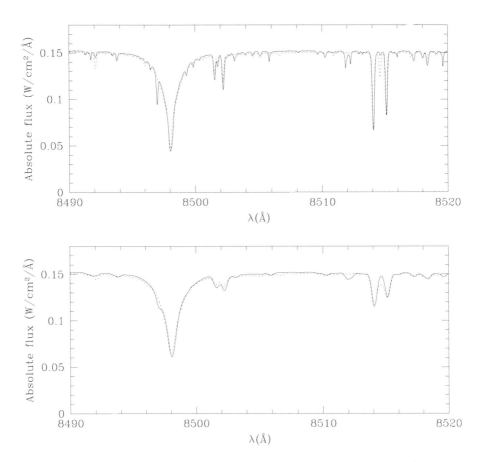

Figure 2. *Upper panel*: detail of the full resolution spectrum of the Sun, around the $\lambda\lambda 8498$ Ca II line. Full line: observed absolute flux from Neckel (1999), dotted line: computed spectrum using a MARCS model and existing line lists. Note the mismatches in line positions and strengths, and the missing lines. *Lower panel*: same spectra degraded to R≈20 000. Note how lines get smoothed out to a point where faint lines totally disappear into a pseudo-continuous opacity. In particular the red wing of the Ca line seems perfectly matched, whereas inspection of the high resolution spectrum shows it is not the case (this seems to be due to a slight difference in absolute flux level).

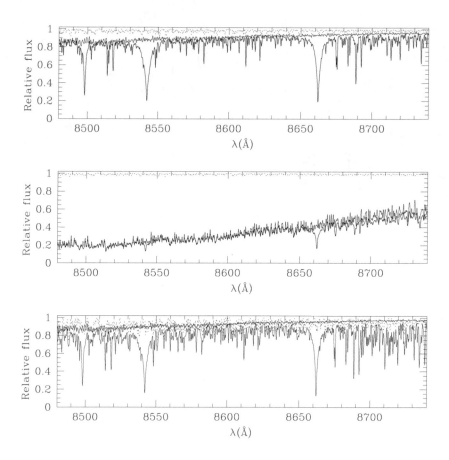

Figure 3. Synthetic spectra for M and S giants, at R≈20 000, computed using MARCS models (see text), and normalized to the continuum. The full spectrum is the thick line. Contribution from TiO alone is the full thin line, and CN alone produces the dotted line spectrum. *Upper panel*: $T_{\rm eff}=$ 3600 K, $\log g$=1.0, C/O=0.5, corresponding to an early-type M giant. Note the lowering of the apparent continuum due to TiO. Many atomic lines are still easily seen. *Central panel*: $T_{\rm eff}=$ 3200 K, $\log g$=0.35, C/O=0.5, a late-type M giant. TiO dominates the spectrum. The flux level is between 20% and 60% of the continuum only! Very few atomic lines, among which the Ca II lines, are detected. *Lower panel*: $T_{\rm eff}=$ 3200 K, $\log g$=0.35, C/O=0.99, a late-type S giant. The molecular veiling is much decreased compared to the M giant of identical temperature. CN and TiO contribute equally. Atomic lines are more prominent than in the early-type M giant model. These changes are due to the displacement of the chemical equilibrium caused by the disappearance of almost all oxygen into CO, and the associated decrease of the TiO an H_2O opacity. Many lines above 8690 Å are from FeH.

represent the conditions in the real star atmosphere. Preliminary tests of the new generation MARCS models shown above are comforting in that sense.

Finally, we should keep in mind that the actual GAIA data will probably be processed using more elaborate models, accounting for non-LTE, if not for 1- 2- or 3D hydrodynamics.

References

Alvarez, R., & Plez, B. 1998, A&A 330, 1109
Asplund, M., Nordlund, Å., Trampedach, R., Allende Prieto, C., & Stein, R.F. 2000, A&A 359, 729
Barklem, P.S., Piskunov, N., O'Mara, B.J. 2000, A&AS 142, 467
Bessell, M.S., Castelli, F. & Plez, B. 1998, A&A 333, 231
Decin, L., Waelkens, C., Eriksson, K., Gustafsson, B., Plez, B., Sauval, A.J., Van Assche, W., & Vandenbussche, B. 2000, A&A 364, 137
Edvardsson, B., Andersen, J., Gustafsson, B., Lambert, D.L., Nissen, P.E., & Tomkin, J. 1993, A&A 275, 101
Eriksson, K., Gustafsson, B., Jørgensen, U.G., & Nordlund, Å. 1984, A&A 132, 37
Freytag, B., & Mizuno-Wiedner, M. 2003, in Modelling of Stellar Atmospheres, IAU Symp 210, N.Piskunov et al. ed.s, ASP Conf. Ser., in press
Gustafsson, B., Bell, R. A., Eriksson, K., & Nordlund, Å. 1975, A&A 42, 407
Gustafsson, B., Edvardsson, B., Eriksson, K., Mizuno-Wiedner, M.,Jørgensen, U.G, & Plez, B. 2003a, in Modelling of Stellar Atmospheres, IAU Symp 210, N.Piskunov et al. ed.s, ASP Conf. Ser., in press
Gustafsson, B., Edvardsson, B., Eriksson, K., Mizuno-Wiedner, M., Jørgensen, U.G, & Plez, B. 2003b, in Modelling of Stellar Atmospheres, IAU Symp 210, N.Piskunov et al. ed.s, ASP Conf. Ser., in press
Hauschildt, P.,H., Allard, F., Ferguson, J., Baron, E., & Alexander, D.R. 1999, ApJ 525, 871
Höfner, S. 1999, A&A 346, L9
Josselin, E., Plez, B., & Mauron, N. 2003, in Modelling of Stellar Atmospheres, IAU Symp 210, N.Piskunov et al. ed.s, ASP Conf. Ser., in press
Jørgensen, U.G. 1997, in IAU Symposium 178, 441
Jørgensen, U.G., Jensen, P., Sørensen G.O., & Aringer, B. 2001, A&A 372, 249
Kupka, F., Piskunov, N., Ryabchikova, T.A., Stempels, H.C., & Weiss, W.W. 1999, A&AS 138, 119
Marrese, P.M., Munari, U., Boschi, F. & Tomasella, L.2003, in GAIA Spectroscopy, Science and Technology, U.Munari ed., ASP Conf. Ser. 298, pag. 427
Munari, U., & Tomasella, L. 1999, A&AS 137, 521
Munari, U. 2003, in GAIA Spectroscopy, Science and Technology, U.Munari ed., ASP Conf. Ser. 298, pag. 227
Neckel, H. 1999, Sol. Phys. 184, 421

Nesvacil, N., Stütz, Ch. & Weiss, W.W. 2003, in GAIA Spectroscopy, Science and Technology, U.Munari ed., ASP Conf. Ser. 298, pag. 173

Plez, B., Brett, J.M., & Nordlund, Å. 1992, A&A 256, 551

Plez, B. 1998, A&A 337, 495

Plez, B., Van Eck, S., Jorissen, A., Edvardsson, B., Eriksson, K., & Gustafsson, B. 2003, in Modelling of Stellar Atmospheres, IAU Symp 210, N.Piskunov et al. ed.s, ASP Conf. Ser., in press

Serote Roos, M., Boisson, C., & Joly, M. 1996, A&AS 117, 93

Wahlgren, G.M., & Johansson, S. 2003, in GAIA Spectroscopy, Science and Technology, U.Munari ed., ASP Conf. Ser. 298, pag. 481

GAIA Spectroscopy, Science and Technology
ASP Conference Series, Vol. 298, 2003
U. Munari ed.

On the classification and parametrization of GAIA data using pattern recognition methods

Coryn A.L. Bailer-Jones

Max-Planck-Institut für Astronomie, Königstuhl 17, 69117 Heidelberg, Germany

Abstract. I discuss various aspects of source classification and physical parametrization using data from the future Galactic survey mission GAIA. Due to the heterogeneity of the data, the large variety of objects observed and problems of data degeneracy (amongst other things), efficiently extracting physical information from these data will be challenging. I discuss the global and local nature of commonly used pattern recognition algorithms and outline two alternative frameworks for classification – parallel and hierarchical – and describe some aspects of each. A method for calibrating the classification algorithms is proposed which requires only a limited amount of additional (ground-based) data. By way of illustration, an example of stellar parametrization using GAIA-like RVS data is presented.

1. Introduction

The primary scientific goal of GAIA is a detailed study of the composition, structure and formation of our Galaxy. The major contribution which GAIA will make in this area is high precision astrometry of around one billion stars, providing accurate positions, parallaxes and proper motions. Of course, to be able to use this astrometric data for Galactic structure studies, it is essential that the intrinsic properties of the stars so observed are known. For this reason, GAIA will also employ multiband photometry and high resolution spectroscopy (see Sect. 2).

The classification[1] requirements for GAIA have been outlined in Bailer-Jones (2002), but include: (*1*) discrete classification of GAIA sources as star, galaxy, quasar, solar system object etc.; (*2*) determination of stellar astrophysical parameters (APs: T_{eff}, [Fe/H], [α/Fe], $\log g$, V_{rot}, mass, age, activity, etc.); (*3*) accurate determination of interstellar extinction (which is unique for a given star so can effectively be assumed to be a stellar parameter); (*4*) detection and

---

[1]I draw a distinction between *classification* and *parametrization*. Classification refers to the allocation of *discrete* classes, such as (*1*) star or nonstar, (*2*) star, galaxy, quasar, asteroid or other, or even (*3*) hot star or cool star. Parametrization, on the other hand, refers to the placing of sources on a *continuous* scale, such as T_{eff}, [Fe/H], star formation rate, albedo etc. The distinction is important in terms of the way in which algorithms are used. However, where the distinction is not important, I will use the term *classification* to refer collectively to the process of assigning attributes (classes or parameters) to sources.

description of stellar multiplicity; (5) identification of new types of objects. The goal of GAIA is not *just* to produce a catalogue of astrometric parameters and associated photometry, but also detailed information on source classification and APs.

2. GAIA data

GAIA is an all sky magnitude-limited survey (to $V \sim 20$). Due to telemetry limitations from its orbit around the Earth–Sun L_2 point, not all data will be transmitted to ground. Instead there will be real time on board detection of all sources above a magnitude limit ($V \sim 20$) and only the CCD pixels in patches around each object will be transmitted.

The primary information for classification purposes will come from the Medium Band Photometer (MBP) a set of 10–20 (the system is not yet fixed) medium band filters over the wavelength region 200–1100 nm. This must obtain information on every object down to the GAIA magnitude limit. This will be supplemented by about five broad bands from the astrometric instrument (which are primarily intended to give a chromatic correction to PSF centroiding).

Data relevant to stellar parametrization will also be provided by the Radial Velocity Spectrograph (RVS), the capabilities and optimization of which are the subject of this meeting. The RVS will obtain slitless spectra over the whole sky over the wavelength range 8480–8740 Å, with a resolution (to be decided) of between 5 000 and 20 000 (Katz 2003). Due primarily to signal-to-noise considerations, RVS data will only be obtained down to $V \sim 17$ mag. For the brighter stars, it will also provide an information on stellar activity (via emission lines), rotational velocities (via line broadening), individual element abundances and permit an independent determination of $T_{\rm eff}$, [Fe/H], [α/Fe] and $\log g$. The RVS instrument has significant implications for parametrization of GAIA data, as it provides different amounts of information (and perhaps even different formats if lossy compression schemes are used) for each object, depending on magnitude (and crowding).

Each point on the sky is observed about 100 times over the course of five years, providing variability information relevant for identifying some types of stars and quasars. Astrometric information will also be very useful for classification, e.g. parallaxes for determining stellar luminosity and radius, (zero) proper motions for identifying quasars. However, we should not use kinematic information to parametrize stars, as this would require a Galaxy model and hence introduce classification biases based on our *current* and limited understanding of Galactic structure.

In summary, classification with GAIA is characterised by the *heterogeneity* of the data: photometry (from two separate instruments), spectroscopy (only for some targets, varying formats) and astrometry. Making full interdependent use of these data is a challenge, and, as discussed by Bailer-Jones (2002), is not something which classification methods used to date in astronomy have had to deal with on this scale. This is further complicated by the wide range objects and astrophysical parameter scales which will be encountered.

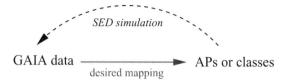

Figure 1. Classification or parametrization is the process of determining the mapping from a data domain to a class or astrophysical parameter (AP) domain. The opposite mapping is equivalent to the simulation of the data, e.g. the emergent stellar spectral energy distribution (SED).

3. Classification principles

Classification is the problem of assigning object classes or APs and generally involves determining some kind of mapping from the data space to the parameter space (Figure 1).[2] A frequently used approach is the *supervised* or pattern matching approach, in which pre-classified data (*templates*) are used to infer the desired mapping. This mapping is then applied to new data to establish their classes or APs. Perhaps the most familiar such technique is the minimum distance method (MDM), shown schematically in Figure 2. This is a *local* template matching method, in which only the properties of the local neighbours in the data space influence the APs of the new object. This is in contrast to a neural network, which attempts to do a *global* interpolation of the function APs=f'(data) over the whole data space.

Classification, then, is the process of mapping from the data space to the AP or class space. By contrast, simulation of source SEDs is the opposite mapping, i.e. from the AP space to the data space. This is generally a many-to-one mapping: a given set of APs provides a unique SED but because of photon noise and degeneracies, two sets of APs could produce the same SED (within the noise). Thus the inverse mapping (i.e. classification) is generally one-to-many and not unique.

This is illustrated schematically in Figure 3. In the left panel we see that there are four templates (those lying within the noise bounds) which give rise to data consistent with the new observation. Confronted with this degeneracy we must decide what to do. Do we quote all results? Do we average the APs? There are in fact whole ranges of the AP which are consistent with the data, so an unweighted average will be biased by the distribution of the nearest templates. Moreover, at large AP, there is actually another solution which we have completely failed to recognise due to the low density of templates in that region. The problem is worse with a lower density template grid (right panel of Figure 3), or, equivalently, lower noise data. Clearly, a MDM which just assigns the APs of the nearest neighbour or even averages over nearby neighbours will give

---

[2]The *data space* refers to the data acquired from GAIA, such as fluxes in different filters or the RVS spectrum. The *parameter space* refers to those properties of the sources we wish to determine, such as T_{eff} or extinction, but could also refer to discrete classes (e.g. star, galaxy, quasar etc.).

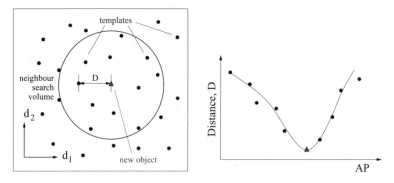

Figure 2. Schematic illustration of the generic minimum distance method (MDM). *Left*: a two-dimensional data space populated with pre-classified templates. Assigning parameters to a new object involves looking at the APs of the nearest neighbours (with the data dimensions suitably scaled). APs are assigned either by interpolating in the data space (i.e. solving the function APs=f'(data) locally at the new object – in the simplest case this is just an average of one or more neighbours) or in the parameter space (i.e. minimising the function D=g(APs), shown for one AP in the *right* panel).

biased results. We might want to get around this by having a *very* dense template space, but this will probably become prohibitive if we have a large number of APs. But even this will give rise to biases in the classifications where the mapping function is nonlinear, as one particular neighbour will be preferentially selected. Although the errors would be small for a single star, it could produce a significant systematic error in the average classification of many similar stars.

Thus, any sensible implementation of MDM or other pattern matching method will do some kind of interpolation to provide solutions between templates, i.e. provide us with an approximation to the curve shown in Figure 3. But this will only be a single-valued function if done in the reverse manner, i.e. data=f(APs). This is the inverse of the function which we would like to have, APs=f'(data), which would enable us to deduce APs given new data. This is important, because it means global interpolation methods for determining the function APs=f'(data), such as neural networks, will give poor interpolants in the presence of data degeneracies. (Think of rotating the left panel of Figure 3 by 90° and trying to fit a single-valued function through the templates.)

Appropriate design of the GAIA photometric system is a prerequisite to avoiding such degeneracies, but given the relatively few photometric bands and large variety of objects observed, they cannot be avoided entirely. The issue, then, is how to recognise degeneracies and appropriately report multiple solutions without biasing the parameter determinations. This is of course simple in the one dimensional example in Figure 3 which we can visualise, but it is more complex with 5–10 APs and tens or even hundreds of data dimensions, as will be the case with GAIA. An appropriate solution to this problem is the subject of ongoing work.

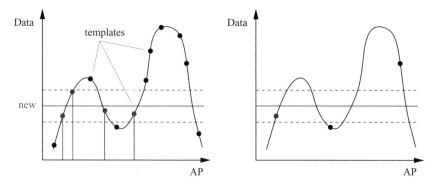

Figure 3. Schematic diagram of the functional relationship between a one-dimensional data and a one-dimensional AP space showing degeneracies, i.e. several AP solutions for a single given data measurement (shown by the horizontal line). The dashed lines show the noise level, so that (in the *left* diagram) any of the four templates with vertical lines are consistent with the new measurement.

4. Classification schemes

It seems unlikely that a single classification algorithm will be able to deal with the large variance of astrophysical sources which GAIA will observe (many results in the literature indicate improved performance when a classification problem is broken down into subsets covering smaller ranges of APs). This can be dealt with in one of two broadly different approaches. The first, which I describe as the *hierarchical scheme*, uses a Global Classification Model (GCM; a model which can deal with the entire range of sources) to produce a coarse classification. Based on this, one of several refined classifiers (each of which I call a Local Classification Model, or LCM) is used to produce a more precise classification or set of APs. Each of these LCMs only 'knows about' (i.e. is capable of producing good results in) a limited part of the parameter space, e.g. a restricted range of effective temperature.[3] This approach is shown schematically in the left part of Figure 4. An example of such a scheme was presented in Bailer-Jones (2002; Sect. 5 and Figure 1).

The alternative approach, shown in the right part of Figure 4, is to pass the data to each of many *local* classifiers right from the start. I refer to this as the *parallel scheme*.[4] The key difference is that every single LCM is given the chance to say something about the data, and, crucially, to provide a probability that this source corresponds to a source of its class (or the range of APs which it deals with). A decision regarding which of these classes the source belongs

---

[3] Generally, these spaces should overlap between the LCMs to obviate the problem of small classification errors from the GCM occuring at the boundaries between the LCMs, which would result in entirely the wrong LCM being chosen.

[4] In Figure 4 I show coarse and refined levels in the parallel scheme, but this could be reduced to a single refined level.

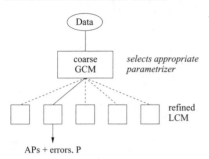
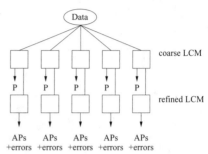

Figure 4. Two alternative classification philosophies. AP = astrophysical parameter, P = probability (that the source conforms to that class or set of APs), GCM = global classification model (one acting on a very wide range of the data space or of source types), LCM = local classification model (one acting on a narrow range of the data space or source types).

to (i.e. which LCM) is then made independently, for example by taking the LCM which yields the largest probability. This approach can be seen from the perspective of Bayesian inference: the determination of the APs within each LCM is the process of determining the posterior probability distribution over the APs, assuming the LCM to be the correct one. The higher level of inference – model comparison – comes about by assessing the *evidence* for each LCM, regardless of the optimal or marginalised APs provided by each LCM, and is provided in this case by the probability, P.

5. A classification framework for GAIA

In the context of the GAIA classification problem, the parallel scheme described in the previous section offers a number of advantages over the hierarchical one.

One of the key advantages relates to the *inhomogeneity of source models*. At some level, classification or parametrization involves matching the observed data to template data of sources with known APs or classifications. A single model which must classify all types of astrophysical sources (even if only coarsely) demands a strong degree of homogeneity of the template data. For example, if a single classifier is to classify both main sequence and pre-main sequence (PMS) stars based on synthetic spectra of such stars, then these synthetic spectra will have to show smooth and self-consistent variations as a function of the APs. In practice, however, different models for different types of stars may be produced independently and according to different assumptions (opacities, treatment of convection and so forth). Thus it may be unrealistic to expect modellers to produce a single, homogeneous grid of synthetic spectra across the full range of APs of stars which GAIA will observe. In the parallel classification scheme, homogeneity is not required. Here, each LCM need only known about a limited set of self-consistent and homogeneous models (e.g. only PMS stars, and perhaps then only over a small mass range). Each LCM then attempts to classify a data vector according to its own source models, and thus provides the most probable APs, along with error estimates and a probability (P) that this particular data vector is described at all by this ensemble of source models (independently of the specific APs it came up with). So if the data source really is a PMS star, of all the LCMs we would expect the 'PMS LCM' to give the highest probability in the parallel scheme, whereas the 'evolved giant' and 'quasar' LCMs (for example) would give very low probabilities.

The key requirement of the parallel scheme is a robust means of determining the probability (P in Figure 4) that the data vector is described by that LCM. This could be determined from the distance (in the data space) compared to the data uncertainties between the source position and the different templates in that LCM, with due regard for interpolation errors between the templates. A specific algorithm for this is presently under investigation.

A classification approach using LCMs (rather than a single GCM) not only permits inhomogeneous source models to be used. It also allows very different approaches to classifying different types of astrophysical sources, possibly using different parts of the GAIA data in each case. For example, a 'Cepheid LCM' may want to do a light curve analysis to look for characteristic variability, something which would not be relevant for old G dwarfs.

Another advantage of the parallel scheme is that it naturally provides for multiple solutions in the presence of AP degeneracy (see Sect. 3). This would be evident from several LCMs yielding high probabilities, whereas in the hierarchical scheme only ever one LCM is selected per source (although this condition could of course be relaxed). Thus if more than one LCM provided output probabilities above some threshold, all sets of parameters from these LCMs could be reported.

6. Classification example with GAIA/RVS-like data

Optimization of the GAIA instruments and pre-launch estimates of the AP precision which can be achieved with these must be undertaken using existing real or synthetic data. By way of illustration, I show the results of a stellar parametrization procedure using a neural network applied to RVS-like data obtained and parametrized by Cenarro et al. (2001). The selected dataset consists of 611 spectra covering the wavelength range 8480–8740 Å, near-critically sampled at a resolution of 5800 (1.5 Å FWHM). The median SNR per resolution is 85, but with a large range (30–170 for 90% of spectra). These spectra were randomly assigned to two nearly equal sized subsets, and a neural network trained on one subset to determine $T_{\rm eff}$, $\log g$ and [Fe/H]. Once training is completed (according to some optimization criterion), the network parameters are frozen and used to determine the APs on the other data subset, from which the performance of the network can be verified. The distribution of the APs in the training and verification sets is shown in Figure 5.

Figures 6 and 7 summarise the results. $T_{\rm eff}$ can generally be determined to within 5% and shows little trend with $\log g$, although is better determined for near solar metallicity stars. The larger scatter around $\log g=2.5$ may indicate $T_{\rm eff}$ errors in the assignments of Cenarro et al. Gravity ($\log g$) can be deter-

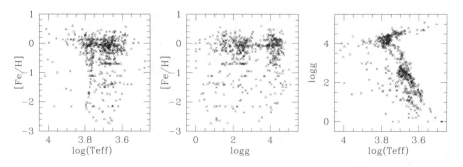

Figure 5. Distribution of the data used from Cenarro et al. (2001) for the training set (crosses, 300 spectra) and the verification set (triangles, 311 spectra).

mined to within 0.5 dex at solar metallicity, degrading to 1.0 dex or more at lower metallicities. We see that $\log g$ is generally harder to determine at lower metallicities, which we would expect as there the line profiles which the $\log g$ depends upon are weaker. We also see that the $\log g$ of the cooler stars is systematically underestimated for the cooler subset (lower left panel of Figure 6) across a range of $\log g$. This may indicate a limitation of the algorithm. [Fe/H] precision is 0.3 dex across all temperatures considered (ignoring low number statistics at the extremes), but shows a trend to poorer performance at low metallicity, particularly for evolved stars (lower right panel of Figure 7). This is expected because the metallicity signature is weaker and small differences are harder to distinguish.

It should be emphasised that these results are based on real data, and therefore include all sources of cosmic scatter. Moreover, the performances assume that the Cenarro et al. parametrizations are true, so only assess the ability of the neural network to reproduce these. Any inconsistencies in that calibration will be reflected by the network. Finally, no attempt has been made to optimize the neural network implementation for this purpose, which furthermore is prone to the degeneracy problem described in Sect. 3.

7. Calibration

The example in the previous section raises the issue of how the parametrization algorithms for GAIA will be calibrated, or, in other words, how the training/template data set(s) will be defined and parametrized. Unfortunately, a homogeneous database of real spectra to serve as the GAIA templates – covering the required wavelengths and APs – does not presently exist. Even if it did, it would presumably consist primarily of ground-based spectra which would need to be processed to remove telluric features (and still the UV data would be lacking), and APs would still have to be assigned to those spectra. Obtaining stellar parameters ultimately requires some kind of stellar model. The emergent spectral energy distributions derived from such models can be used directly in the parametrization process by training pattern recognition methods on such

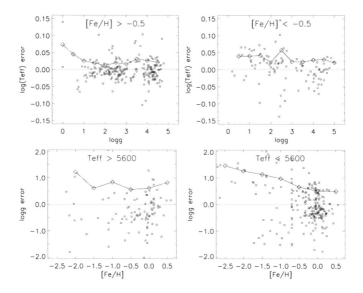

Figure 6. $T_{\rm eff}$ (*top* row) and $\log g$ (*bottom* row) parametrization errors on the verification data set. Each point corresponds to a single spectrum and gives the network determination minus the "true" value (as established by Cenarro et al.). The diamonds joined by a line show the r.m.s. error for all spectra in a bin centered on that point. The $T_{\rm eff}$ errors are plotted as a function of $\log g$ for two metallicity ranges. The $\log g$ errors are plotted as a function of [Fe/H] for two temperature ranges.

spectra, after suitably processing them with the instrument model to look like the data GAIA will obtain (e.g. Bailer-Jones et al. 1997). However, synthetic spectra differ from real spectra in two significant ways. First, they may show systematic differences due to modelling uncertainties (e.g. missing opacities). Second, real spectra show increased cosmic scatter due to unaccounted-for APs (e.g. abundance variations, chromospheres, etc.). So training models on synthetic spectra to apply to real spectra is not ideal.

Fortunately, there is a way around these problems. To assign parameters to GAIA observed objects we need to know: (1) how these objects will appear in the GAIA multidimensional data space; (2) what the *required* APs for these objects are (where *required* means just those APs which can be derived in principle from the GAIA data). However, we do not have to determine the APs of the templates from the same data that we use in the training. We may define a grid of real star (*calibration stars*) on the sky which GAIA will observe, covering the full range of APs at some suitable AP density. Ground-based spectra of each calibration star is obtained with whatever resolution and wavelength coverage is required to determine its APs (probably from detailed line fitting) to limits set by our physical knowledge of stars and the quality of data we can obtain. In some cases existing data or catalogues can be used for this purpose. The GAIA observations form the data for these calibration stars, so at the end of the

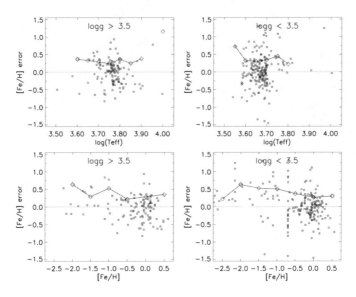

Figure 7. [Fe/H] errors on the Cenarro et al. (2001) data (see Figure 6).

GAIA mission we have both the data and APs on a set of stars which can serve as templates to train our classification algorithms, unaffected by real-synthetic data mismatch.

Prior to and during the mission, GAIA data can be simulated using synthetic spectra, permitting approximate parametrizations. If it turns out that the sky grid of calibration stars is not dense enough in some regions of the AP space, it can be supplemented with synthetic spectra, with broad corrections applied to them to account for systematic differences, in a manner similar to that described by Lejeune et al. (1997). This calibration method would require a ground-based observing program. But it would be on a modest scale, requiring of order 1000 high resolution spectra. Accurate flux calibration is not required, and depending on the wavelength coverage required, might be obtainable with a multi-object fibre or slit spectrograph. Of order 10 nights on 4m and 8m class telescopes would probably suffice.

References

Bailer-Jones, C.A.L. 2002, Ap&SS 280, 21

Bailer-Jones, C.A.L., Irwin, M., Gilmore, G. & von Hippel, T. 1997, MNRAS 292, 157

Cenarro, A.J., Cardiel, N., Gorgas, J., et al. 2001, MNRAS 326, 959

Katz, D. 2003, in GAIA Spectroscopy, Science and Technology, U. Munari ed., ASP Conf. Ser. 298, pag. 65

Lejeune, T., Cuisinier, F. & Buser, R. 1997, A&AS 125, 229

GAIA broad and medium band photometric performances

Carme Jordi, Josep M. Carrasco, Francesca Figueras & Jordi Torra

Dept. Astronomia i Meteorologia, Universitat de Barcelona, Avda. Diagonal 647, E-08028 Barcelona, Spain

Institut d'Estudis Espacials de Catalunya, Gran Capità 4, E-08034 Barcelona, Spain

Abstract. The current design of GAIA in what concerns the broad and medium band photometry is reviewed and estimations of attainable accuracies in the stellar parameters are presented.

1. Introduction

The scientific goals of the GAIA mission require complementary astrometry, photometry and radial velocity data. The main goal of the photometry is astrophysical parameterization of the stars (temperature, luminosity, chemical composition and age) with an accuracy sufficient for the quantitative description of the chemical and dynamical evolution of the Galaxy over all galactocentric distances. Photometry is also necessary to account for the chromatic aberrations in the astrometric focal plane to achieve microarcsec accuracy level.

After five years scanning the entire sky, GAIA will have performed measurements with broad (BBP) and medium (MBP) bands, and white light (G magnitude) providing multi-color and multi-epoch photometry up to $G_{\text{lim}} \sim 20$, i.e. $V_{\text{lim}} \sim 20-25$.

This paper reviews the photometric capabilities of the revised design of GAIA and the estimated accuracies of the derived stellar parameters.

2. The G magnitude

Astrometric observations will be performed in white light (corresponding to a very broad band $\sim 300-1050$ nm). The estimated precisions of the associated G magnitude measurements are given in Figure 1 (left) for single transit and at the end of the mission. The white light measurements yield the best S/N for variability detection among all instruments on board GAIA. As Hipparcos did, GAIA will discover eclipsing binaries, Cepheids, RR Lyrae, etc. The precision of the light curves at the end of the mission (with an average of 82 observations) for variable stars with $V \sim 20$ is similar to those of Hipparcos at $V \sim 9$. The Cepheids in the LMC ($V \sim 13.5-16.5$) will be measured with single transit precision ranging from 0.0006 to 0.003 mag and those in M31 ($V \sim 19.5-22.5$) with precision from 0.015 to 0.10 mag.

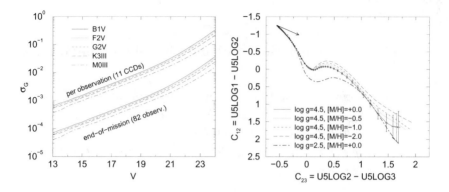

Figure 1. *Left*: estimated precision of the G magnitude per passage and at the end of the mission. *Right*: a color-color diagram for the U5LOG BBP proposal (Lindegren 2001) similar to the $(U-B)-(B-V)$ diagram. The bars represent the end-of-mission errors at $V = 18$ and the arrow corresponds to the reddening associated to $A_V = 1$.

3. Broad band system: BBP

Five columns of CCDs will be allocated in the focal plane common to both astrometric telescopes allowing photometry through 4-5 bands (4 in case of devoting double exposure time to the UV band). Several BBP systems were proposed by the Asiago team, the Vilnius team, the Geneve-Barcelona team and by Lindegren (2001). Among the proposals, several BBP sets by Lindegren have very good performances for the correction of the chromatic effects (r.m.s. residuals < 5 μas), which is the main goal of the BBP. As an example of the attainable precisions, Figure 2 shows the estimated error of magnitudes as a function of V and spectral type for the U5LOG proposal. This proposal (5 bands with almost triangular shape) behaves similarly to the Johnson system in terms of derivation of astrophysical quantities. Figure 1 (right) shows the capabilities of determining reddening and temperatures for early type stars and, for cool stars, temperatures and [M/H] if luminosity class is known. The three redder bands provide a determination of temperature almost independent on metallicity.

With an angular resolution larger than that of the MBP, the BBP will be crucial for astrophysics in case of close pairs and crowded fields. Thus, the PWG works to optimize Lindegren's proposals for the maximum astrophysical return.

4. Medium band system: MBP

Fifteen CCDs for medium band photometry plus one CCD for the detection and selection of sources will be placed up and down of the RVS CCDs in the SPECTRO focal plane, providing groups of 3-4 measurements separated by 6^h each. The MBP is aimed to classifying the observed objects (star, solar system body, QSO, etc) and to characterizing them in terms of astrophysical parameters (i.e. T_{eff}, luminosity or $\log g$, [M/H], anomalies, emission, etc).

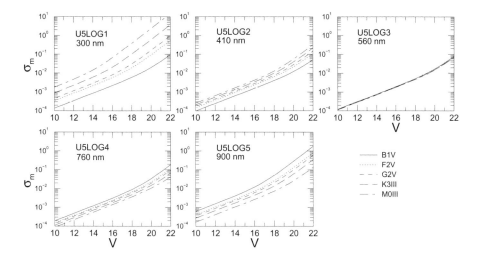

Figure 2. Estimation of the end-of-mission precisions for the five bands in the U5LOG proposal by Lindegren (2001) as a function of V magnitude and spectral type (the errors include a 20% margin for unknown sources of errors). Central wavelength is indicated.

MBP systems have been proposed by the Asiago team, the Vilnius team and the Geneve-Barcelona team. The adopted baseline MBP system (10 bands with $\sim 35-80$ nm width plus a band centered in $H\alpha$) is described in the ESA mission study report (ESA-2000-SCI-4) and Grenon et al. (1999). Although the determination of $T_{\rm eff}$, log g, [M/H], etc, will be performed by automatic algorithms (minimum distance method, neural networks, or others), a quick look to the estimated precision (Figure 3) of several colors and some reddening free indices provides a first estimation of the precision in the derived parameters.

Photometry allows the determination of the individual reddening for main sequence early type stars. However, for cool stars, the A_v and $T_{\rm eff}$ determinations are degenerated, so a 3-D map of the interstellar absorption is of capital importance. Knude (2002) reviews the galactic dust on various scales and several approaches to derive the interstellar absorption by using optical and infrared photometry, star counts, parallaxes, data from other space missions, etc. Estimations by the author show that reddening can be derived to a precision of $\sigma_{E(B-V)} \sim 0.03$.

Figure 4 shows the uncertainties in $T_{\rm eff}$ determination for a G2 V star and an M0 V star as a function of V. The estimated precisions for giants are similar. The main limitation for an accurate determination of $T_{\rm eff}$ for the brightest stars ($V \leq$17-18) is the uncertainty in the reddening correction while for the rest is dominated by the photon statistics. A $\sigma_{E(B-V)} \sim 0.03$ translates into $\sigma_{T_{\rm eff}} \sim 150$ K for a G2 V star (i.e. 2.6%) and 50 K for a M0 V star (i.e. 1.4%).

Photometry will be crucial for the determination of the absolute luminosities (or gravities) in the case of large relative error in the parallax ($> 10-20\%$), as it is the case of giants in the bulge, the outer halo, the Magellanic Clouds,

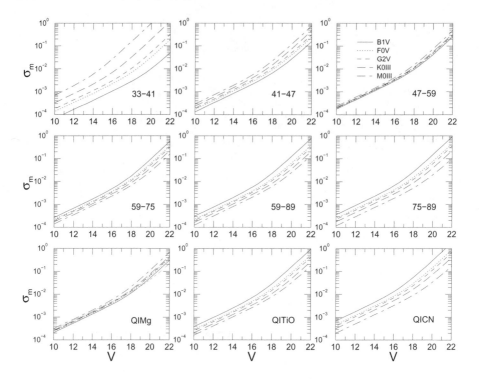

Figure 3. Estimation of the end-of-mission precisions for several colors related to temperature, chemical composition and gravity (see ESA-2000-SCI-4 for explanation of the Q indices).

etc. Absolute magnitude calibrations could be established from the stars with good parallax and photometric distances derived for more distant stars in the traditional way, assuming that they are intrinsically similar to the closest ones, which might not be the case. Thus, luminosities at the level of $\sigma_{M_V} \sim 0.2 - 0.4$ (or $\sigma_{\log g} \sim 0.1 - 0.2$) from photometry would be desirable to match the accuracy of the good trigonometric parallaxes. Figure 5 (left) shows a reddening-free parameter related to luminosity as a function of temperature. Solid and dashed lines represent colors computed from SEDs by ATLAS9 (Kurucz 1997) and from the empirical spectrum library by Pickles (1998), respectively. From this figure one can see that $\sigma_{\log g} \sim 0.7$ dex are derived from the synthetic colors while 0.4 dex are obtained from the empirical spectra for a $V = 19$ star at about 5000 K. This means that (a) the synthetic spectra are not realistic enough, and (b) the estimated precision is SED dependent.

The [Fe/H] catalogue by Cayrel de Strobel, Soubiran & Ralite (2001) compiles high-resolution, high S/N ratio spectroscopic determinations of [Fe/H] for F, G and K stars. The typical $\sigma_{\rm [Fe/H]}$ for the stars with several measurements is 0.10-0.15 dex (someones reaching 0.4 dex). The photometric metallicity calibrations based on the spectroscopic values yield uncertainties of about 0.20-0.30 dex. Low metallicities are worse determined than solar or higher metallicities.

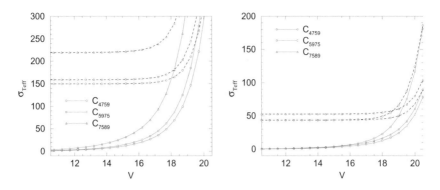

Figure 4. End-of-mission estimated precision of T_{eff} as a function of V for a G2 V star (*left*) and an M0 V star (*right*) derived from several colors. Solid lines assume perfect knowledge of the reddening and dashed lines consider an uncertainty of 0.03 in E_{B-V}.

M.Grenon has given a detailed discussion of the problems involved in the chemical composition determination from photometry with emphasis on the biases involved.

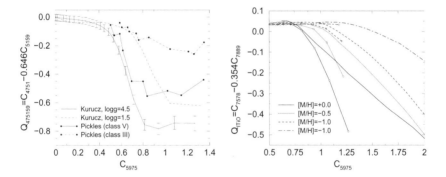

Figure 5. *Left*: Q-index related to luminosity for cool stars. Empirical spectra predict larger Q-changes than synthetic spectra. *Right*: TiO reddening-free index showing the dependence on metallicity for cool dwarfs. Kurucz (open circles) and NextGen (no symbol) SEDs predict different variations for a given Δ[M/H].

Simulations to date show that a precision of 0.10-0.15 dex in [M/H] can be achieved for unreddened Pop. I G-K giants of $V \sim 18-19$ (Figure 8.13 in ESA-2000-SCI-4) when only photon statistics (+20% margin) is taken into account, meaning that giants in the bulge are easily accessible. For Pop. II giants the corresponding error is about twice as large. The spectra of M-type stars are dominated by absorption bands of TiO and only [Ti/H] will be accessible (with slightly better precision than [M/H] for Pop. I G-K giants). On the other hand, a metallicity determination for early type stars (up to $\sim$ A3) is not achievable from photometry. Figure 5 (right) shows the TiO index as a function of the temperature. The ATLAS9 differential effect due to Δ[M/H]=0.5 dex for an M dwarf is about 60% larger than the corresponding change predicted by the NextGen spectra (Hauschildt, Allard, & Baron 1999). Again, the estimated uncertainties are dependent on the SED used.

We refer to Tautišienė & Edvardsson (2002) for the discussion of [α/Fe] determination. The authors point out the inverse behaviour of the Ca II and Mg I spectral features when changing [M/H] and [α/Fe]. Taking advantage of this, [α/Fe] could be determined using narrow bands, but it is unlikely that photometry can provide abundances for the faintest giants and for the dwarfs. Moreover, this may imply the suppression of other bands with the corresponding impact on the determination of other astrophysical quantities. This impact should be investigated as well as the biases due to unsolved binarity.

For ages around 10-14 Gyr, an age variation of 2 Gyr yields a variation of the log T_{eff} turn-off of about 0.01 dex for a given chemical composition. The same variation is derived when an uncertainty of 0.3 dex in [Fe/H] is considered for a fixed age. The variation of the α/Fe abundances has a smaller impact on the age determination than the variation of Fe/H. In summary, an uncertainty of about 4-5 Gyr is estimated for the individual ages at $V \sim 18 - 19$ from the turn-off region of halo and thick disk stars. As usual, subsets of each galactic population (as globular clusters, open clusters, OB associations, a given halo stream, an identified merger, etc.) can be treated statistically and of ages and metallicities of the subset can be obtained with much better precisions than of the individual members.

Intensive work is being carried out within the PWG to optimize the present MBP proposals and to obtain the maximum precision for those stars crucial to the goals of GAIA. In addition, more realistic synthetic spectra are needed to better estimate the final errors and unsolved duplicity, rotational velocity, etc, have to be evaluated.

Acknowledgments. This work has been supported by MCYT under contract ESP2001-4531-PE.

References

Cayrel de Strobel, G., Soubiran, C. & Ralite, N. 2001 A&A 373, 159
Grenon, M., Jordi, C., Figueras, F. & Torra, J. 1999, MGUB-PWG-002 (Livelink)
Hauschildt, P.H., Allard, F. & Baron, E. 1999, ApJ 512, 377
Knude, J. 2002, Ap&SS 280, 97
Kurucz, R.L. 1997, Kurucz's CD-ROM No. 13, CfA Harvard, Cambridge
Lindegren, L. 2001, GAIA-LL-39 (Livelink)
Pickles, A.J. 1998, PASP 110, 863
Tautvaišienė, G. & Edvardsson, B. 2002, Ap&SS 280, 143

GAIA Spectroscopy, Science and Technology
ASP Conference Series, Vol. 298, 2003
U. Munari ed.

Grids of synthetic spectra in planning for the GAIA mission

Tomaž Zwitter

University of Ljubljana, Dept. of Physics, Jadranska 19, 1000 Ljubljana, Slovenia

Fiorella Castelli

Osservatorio Astronomico di Trieste, via Tiepolo 11, 34131 Trieste, Italy

Ulisse Munari

Osservatorio Astronomico di Padova – INAF, Sede di Asiago, I-36012 Asiago(VI), Italy

Abstract. We review the status of the computation of two new grids of Kurucz ATLAS 9 model spectra. The first one covers the GAIA spectral interval, and it is aimed to support mission preparation studies and simulations. The second, extending from 2500 to 10500 Å, is intended to support investigation of the GAIA photometric system, calibration of photometric systems censed in the ADPS and cross-correlation in radial velocity determination of binaries. Beside usual $T_{\rm eff}$, $\log g$ and $[Z/Z_\odot]$ parameters, the grids map also in rotational velocity, resolution and $[\alpha/{\rm Fe}]$.

1. Introduction

GAIA mission will observe an unprecedented number of stars and will set a new standard reference in stellar photometry and spectroscopy. Preparations for such a mission need to build on a vast body of observed and synthetic spectra.

The synthetic spectra required in the preparation phase aims to support evaluation of the potential of the 8480–8740 Å GAIA spectroscopic window, to provide inputs for simulation of RVS operation and results, to assist the shaping of the photometric system and training of classification and automatic data analysis tools, and to provide support to ground-based efforts aiming to better constrain evaluation of GAIA impact on astrophysics. In this respect, such synthetic spectra are not required to be the *ultimate* answer from the community, but it is enough they offer a reasonable aproximation to real spectra. Advancements in stellar amosphere studies and improvment of atomic constants and line lists will rend obsolete even the currently most sophisticated synthetic spectra by the time GAIA spectra will eventually be transmitted down to Earth.

Atlases of observed spectra of normal stars in the GAIA spectral interval were presented by Munari & Tomasella (1999), Cenarro et al. (2001), Marrese et al. (2003) and references therein. Our efforts in computing homogeneous grids of synthetic spectra in support of the preparation activity for the GAIA mission begun with Munari & Castelli (2000) and Castelli & Munari (2001), and have expanded since then. In this note a status report is provided of current efforts and plans.

2. Spectra in the GAIA wavelength range

Synthetic spectra are needed as templates for tests of GAIA performances in the measurement of radial velocities by cross-correlation techniques (Zwitter 2002). Another use, may be less obvious, will be equally important: virtually all spectra obtained by GAIA will suffer from overlaps of faint spectral tracings of neighboring stars (Zwitter & Henden 2003). Most of the overlappers will be too faint to recover their spectrum from GAIA observations. So one will have to rely on photometric classification of the overlapping stars and a proper synthetic spectra database to generate a combined spectrum of the overlapping background stars, subtract it from the studied spectrum and finally analyze it (Zwitter 2003). One could therefore refer to the synthetic spectra database as a critical part of the data reduction code, as no information could be extracted from the spectrograph without subtraction of the calculated background signal.

Most of the GAIA stars will be normal stars of spectral types G and K (Zwitter & Henden 2003). Thus Kurucz ATLAS 9 stellar atmosphere models can be used as an initial approach to synthetic grid calculation (Nesvacil et al. 2003). They are also adequate for instrument and reduction procedure planning, while we note that non-LTE effects, sphericity, dust condensation, convection, etc, need to be taken into account moving toward the extrema of the HR diagram.

The primary goal of the GAIA spectrograph is to measure radial velocities. Moreover, the spectra can be used to calculate or confirm the effective temperature, gravity and metallicity of observed stars and so supplement the photometric results. Finally, in the case of bright targets and provided that spectral resolution is high-enough, the GAIA spectra can also be used to obtain abundances of individual elements, measure stellar rotation and explore spectral peculiarities (Thévenin et al. 2003; Gomboc 2003).

Table 1. Grid of spectra in the GAIA wavelength range. Some models were calculated also for an α-enhanced composition.

parameter	min	max	step	comment
$T_{\rm eff}$	3 500 K	50 000 K	250 K	larger steps for $T_{\rm eff} > 10\,000$ K
$\log g$	0.0	5.0	0.5	till 4.0 for hot stars
$[Z/Z_\odot]$	-3.5	$+0.5$	0.5	
$V_{\rm rot}\sin i$	0 km s^{-1}	500 km s^{-1}		14 values for O-F stars
				11 values ≤ 100 km s^{-1} for G-M
R	5 000	20 000		5 000, 10 000, 20 000

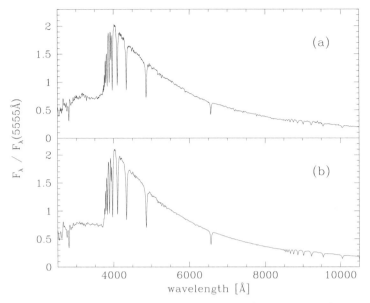

Figure 1. *Top*: observed spectrum of an A5 V star at $R = 500$ resolution, 5 Å sampling (adapted from Pickles 1998). *Bottom*: synthetic spectrum of a non-rotating star with $T_{\rm eff} = 8\,250$ K, $\log g = 4.0$ and solar composition resampled to the same resolution and sampling.

With these goals in mind we generated a grid of synthetic spectra based on the Kurucz ATLAS 9 models computed for a microturbulent velocity of 2 km s^{-1}, a mixing-length convection parameter $l/H = 1.25$, and no overshooting. The spectra were computed with the SYNTHE code from Kurucz and cover the spectral range 8480–8740 Å of the GAIA RVS spectrograph. Details on the database computation will be published elsewhere (Zwitter et al. 2003, in preparation). Table 1 gives the ranges of the spectral grid parameters. Apart from the basic quantities ($T_{\rm eff}$, $\log g$ and [Z/Z$_\odot$]) we include additional dimensions, i.e. rotational velocity ($V_{\rm rot} \sin i$) and spectral resolution (R). Thus the spectral tracings are ready to be included in realistic GAIA simulations. Altogether the database now consists of $\sim 2 \times 10^5$ spectra.

3. Spectra from the near ultraviolet to the near infrared

Recent advances in storage space and computing power permit a calculation of a database of spectra covering the 2 500 Å to 10 500 Å wavelength range. It is computed with Kurucz's codes at a resolution of $R = 500\,000$ and degraded to $R = 20\,000$, a value typical for the Asiago Echelle spectrograph. It is intended to support Asiago cross-correlation works over the whole optical range and be of assistance (after degradation of the resolution) in the optimization of GAIA photometric bands and calibration of photometric systems censed by the ADPS project (Moro & Munari 2000, Fiorucci and Munari 2003 in press). Table 2

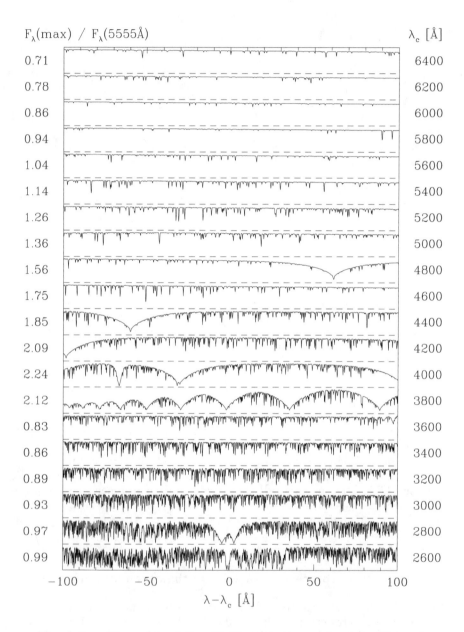

Figure 2. Blue part of the A5 V synthetic spectrum in Figure 1 plotted at the original $R = 20\,000$ resolution. Numbers on the right give the central wavelength and those on the left the maximum flux in each window. The lower limit of the ordinate is always zero.

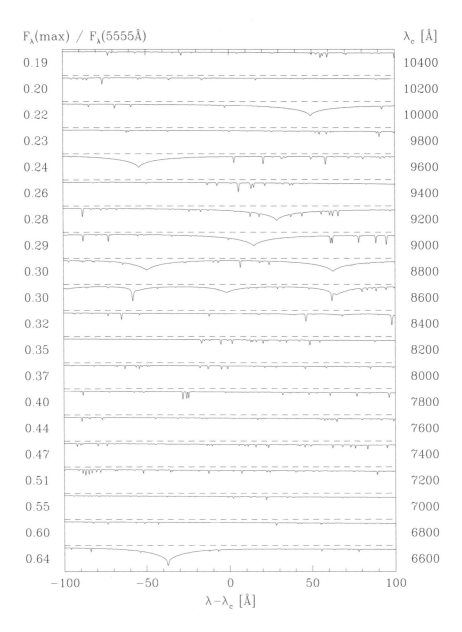

Figure 3. Red part of the A5 V synthetic spectrum in Figure 1 plotted at the original $R = 20\,000$ resolution.

Table 2. Grid of $R = 20\,000$, $2\,500 < \lambda < 10\,500$ Å spectra.

parameter	min	max	step	comment
T_{eff}	5 000 K	15 000 K	250 K	
$\log g$	0.0	5.0	0.5	
$[Z/Z_\odot]$	−3.5	+0.5	0.5	
$V_{\text{rot}} \sin i$	0 km s^{-1}	500 km s^{-1}		14 values for O-F stars
				11 values ≤ 100 km s^{-1} for G-M

summarizes the grid calculated so far (a publication is foreseen for 2003 at the end of the computation phase).

Figure 1 compares an observed and synthetic spectrum of an A5 V star. The observed spectrum was obtained by combining several observed standard star spectra (Pickles 1998) and has a resolution of 500 sampled at 5 Å wavelength bins. The calculated spectrum was resampled to the same resolution and sampling. It was not optimized to match the observed one, still the differences are small. Figures 2 and 3 show the same calculated spectrum at full database resolution ($R = 20\,000$).

References

Castelli, F. & Munari, U. 2001, A&A 366, 1003

Cenarro, A.J., Cardiel, N., Gorgas, J., Peletier, R.F., Vazdekis, A. & Prada, F. 2001, MNRAS 326, 959

Gomboc, A. 2003, in GAIA Spectroscopy, Science and Technology, U.Munari ed., ASP Conf. Ser. 298, pag. 285

Marrese, P.M., Munari, U., Boschi, F. & Tomasella, L. 2003, in GAIA Spectroscopy, Science and Technology, U.Munari ed., ASP Conf. Ser. 298, pag. 427

Moro, D. & Munari, U. 2000, A&AS 147, 361

Munari, U. & Tomasella, L. 1999, A&AS 137, 521

Munari, U. & Castelli, F. 2000, A&AS 141, 141

Nesvacil, N., Stütz, Ch. & Weiss W.W. 2003, in GAIA Spectroscopy, Science and Technology, U.Munari ed., ASP Conf. Ser. 298, pag. 173

Pickles, A.J. 1998, PASP 110, 863

Thévenin F., Bijaoui, A. & Katz, D. 2003, in GAIA Spectroscopy, Science and Technology, U.Munari ed., ASP Conf. Ser. 298, pag. 291

Zwitter, T. 2002, A&A 386, 748

Zwitter, T. 2003, in GAIA Spectroscopy, Science and Technology, U.Munari ed., ASP Conf. Ser. 298, pag. 493

Zwitter, T. & Henden, A.A. 2003, in GAIA Spectroscopy, Science and Technology, U.Munari ed., ASP Conf. Ser. 298, pag. 489

The Asiago Database of Spectroscopic Databases (ADSD)

Rosanna Sordo[1,2] & Ulisse Munari[1]

[1] *Astronomical Observatory of Padova – INAF, Asiago Station, 36012 Asiago (VI), Italy*

[2] *University of Padova, Asiago Astrophysical Observatory, 36012 Asiago (VI), Italy*

Abstract. The Asiago Database of Spectroscopic Databases (ADSD) provides a census and an homogeneous documentation for all the catalogs of real spectra published in literature or available via the *web*, currently complete to March 2002. A total of 262 individual catalogs have been so far censed, divided according to the wavelength coverage (UV, Optical, IR) and format (electronic, tabular or graphic). ADSD applications to the GAIA planning phase are discussed.

1. Introduction

The knowledge of the spectral energy distribution (SED) for a large sample of stars is essential for many astrophysical issues: classification and parameterization, validation of model atmospheres, inputs for synthetic photometry, etc.

The use of synthetic spectra is generally preferred over real ones, because they allow to uniformly explore the parameters space (e.g. T_eff, $\log g$, [Fe/H], [α/Fe], V_{rot}, micro-turbulence velocity). Some of the main families of synthetic spectra are reviewed in this volume (Nesvacil et al. 2003, Plez 2003, Hauschildt et al. 2003). However, the comparison and fine tuning of synthetic spectra against real ones is necessary to improve the underlying atmosphere and atomic physics as well as to expand confidence in their use. Such a comparison needs both real spectra accurately calibrated over the widest wavelength range (even at modest resolution) to check the match to the overall energy distribution, and high resolution data to inspect individual line physics and fine details of atmospheric structure and dynamics.

Securing appropriate real spectra requires huge amounts of telescope time, great care in the observations and excellent sky conditions. With retracing number of small and medium telescopes and increasing over-booking at larger ones, going for vast new observational efforts is becoming more and more a frightening challenge. Thus, assessing what already available in literature is quite worthwhile, and the Asiago Database on Spectroscopic Databases (ADSD) aims precisely to this.

Table 1. Distribution of the 262 spectral catalogs censed by ADSD different sections, according to wavelength range and data type.

name	$\lambda\lambda$ range	total	electronic	tabular	graphic
UV	70–3000 Å	41	14	11	16
Optical	3000–10000 Å	174	64	54	56
IR	1–25 μm	47	18	1	28

2. The ADSD

The ADSD provides a census, documentation and homogeneous access to all (to the best of our knowledge) the catalogs and atlases of real spectra available in literature. ADSD presently contains 262 catalogs: 171 optical, 41 ultraviolet and 48 infrared (the wavelength subdivision is defined in Table 1). The catalogs are also divided according to the data format: electronic, tabular and graphic. The ADSD is currently complete for literature search up to March 2002. The publication of ADSD is foreseen for 2003 (Sordo & Munari, in preparation; hereafter SM03).

2.1. The cards

Documentation *cards* collect, homogenize and integrate the information in support of each individual spectral catalog and atlas. The card for the Jacoby et al. (1984) atlas is shown in Figure 1 as an example. The set of 262 cards constitutes the main body of ADSD. A card is generally made of two pages:

first page: it contains (*a*) a brief description to the original aims of the catalog; (*b*) basic parameters of the atlas (like λ–range, resolution, etc); (*c*) the main bibliographic reference; (*d*) a link to download the data (for catalogs available in electronic form); (*e*) a graphic example of the spectra.

second page: three histograms show the distribution of the stars in spectral type, luminosity class and metallicity, based on current SIMBAD data.

2.2. Synoptical table

An homogeneous set of information for each catalog is collected into synoptical tables. This format provides an easy way to survey the catalogs and to compare them. Figure 2 presents an example of the content of the synoptical tables. The information included are: (*i*) wavelength coverage, dispersion, resolution (or resolving power) and sampling of the spectra; (*ii*) number of stars, their range in spectral type and luminosity class; (*iii*) the instrument or the kind of detector used (i.e. CCD, Reticon, etc); (*iv*) data type (absolute or relative fluxes, norma-

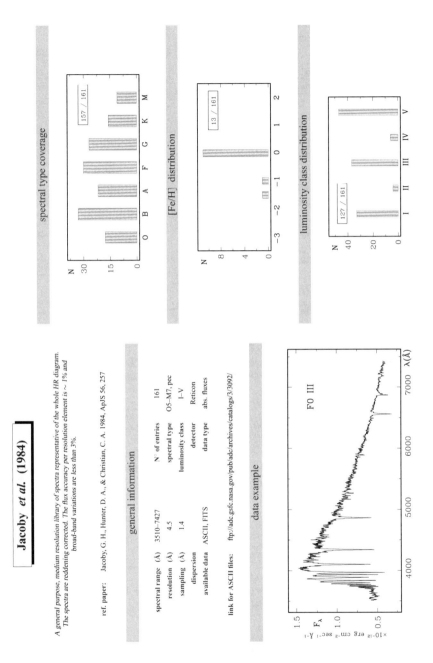

Figure 1. Example of ADSD documentation of a typical catalog. A card has 2 pages: the first reports basic informations on the data type contained into the catalog plus a plot of a sample spectrum, the second the distribution of the stars in $T_{\rm eff}$, $\log g$, [Fe/H].

N.	authors and year of publ.	range (Å)	D (Å/pix) (Å/mm)	R or R_P (Å)	sampl. (Å)	N. stars	spectrum	l. c.	detector/ instr.	data
42	Adelman+ 1989	3300–10800		4–50		207	O9–K4, pec	I–V	Sp.Sc.	rel
43	Alekseeva+ 1997	3200–7500		50	25	602	O5–M4	I–V	Sp.Sc.	abs
		3200–10800		100	25	278	O5–M4	I–V	Sp.Sc.	abs
44	Allen, Strom 1995	5600–9600	1.9	6	1.97	102	F1–M4	V	CCD	counts

Adelman+ (1989): set of previously published energy distributions obtained with rotating scanners. The magnitudes are scaled to zero at 5000 Å: for 33 stars the absolute value at this wavelenght is given. The bandpasses are centered on different wavelenghts depending on the original investigation. AC to Vega like in HL75

Figure 2. Example of ADSD synoptical tables and connected notes. The content is briefly described in Sect. 2.2 and to a full extent in SM03.

lized intensities or raw counts). Individual notes dealing for example with the zero point or the primary standard(s) are provided - where appropriate - for all catalogs.

2.3. The web page

A web interface (http://ulisse.pd.astro.it/ADSD/) is provided to ADSD with two goals: browsing the ADSD in search of catalogs containing (*1*) a given type of data (type of stars, spectral coverage, resolution, etc.) or (*2*) a given star. A direct link is provided to the original data repositories.

Table 2. *a*: example of some of the catalogs covering the GAIA spectral range (8480–8740 Å) with spectral resolution better than 1 Å, useful for RVS planning activities; *b*: example of some of catalogs covering the whole $\lambda\lambda$ 3 000–10 000 Å range, of interest in the optimization of GAIA photometric system. Further catalogs in both categories are censed in the ADSD.

	catalog	R	N. stars	Sp.Type
(a)	Montes & Martin (1998)	0.16	48	F5–M8
	Montes et al. (1999)	0.71	132	F0–M8
	Munari & Tomasella (1999)	0.43	130	O4–M8
	Tinney & Reid (1998)	0.46	4	M8–M9
(b)	Alekseeva et al. (1997)	50, 100	609	O5–M4, WR
	Glushneva et al. (1998)	50	223	O9–M5
	Gunn & Stryker (1983)	20, 40	175	O5–M8, C
	Pickles (1985)	3, 12	48	O–M6
	Pickles (1998)	5	131	O5–M10

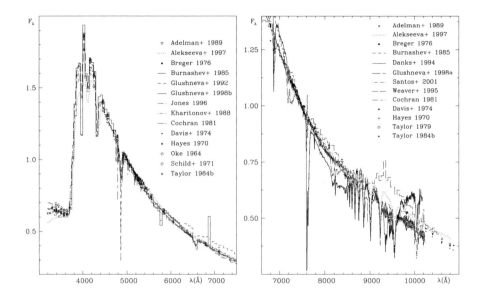

Figure 3. Comparison between flux measurements for γ Gem in the $\lambda\lambda$ 3000–7500 Å region (14 catalogs scaled to 1 at 5000 Å) and in the $\lambda\lambda$ 7000–11000 Å region (13 catalogs scaled to 1 at 7500 Å).

3. ADSD and the GAIA planning phase

Simulation are under way concerning both photometry and spectroscopy, in order to converge on a final photometric system for GAIA and to asses the capabilities of the spectral range adopted for RVS. The ADSD can be of assistance in both efforts.

3.1. Spectroscopy

ADSD at present includes 30 catalogs covering the GAIA spectral range (8480–8740 Å). Only a fraction of them however have a resolution better than 1 Å, thus comparable or better than currently foreseen for RVS (resolving power 11500, equivalent to a resolution of 0.75 Å); some are reported in Table 2. The MK system appear well covered by the higher resolution atlases and catalogs, with stars of high and low metallicity as well as peculiar and variable stars included in significant proportion. ADSD suggests that the currently available material is enough to get a coherent picture of how GAIA spectra will look like over most of the HR diagram. Further observational work should therefore be aimed to specific goals, like building more systematic grids of template stars for automatic classification/analysis of GAIA spectra, or to validate equivalent grids of synthetic spectra.

3.2. Photometry

A significant fraction of the 262 spectroscopic catalogs currently censed by ADSD are of potential interest in the test, optimization and calibration of the GAIA photometric system: all those containing fluxed data extending over large enough wavelength ranges. About flux accuracy, the energy distribution of the same star as given in different spectrophotometric catalogs censed by ADSD is compared in Figure 3, separately for the *blue* and *red* region of the optical spectrum. Understandably, the larger differences affect the far red portion of the spectrum, even if those in the bluer part appear non negligible. About wavelength coverage, less than 10 catalogs among those censed in ADSD cover uniformly the *whole* 3000–10000 Å range over which the GAIA photometric system stretches (some are listed in Table 2). A much larger number of catalogs can be used to investigate subsets of the GAIA photometric bands, like the medium ones, that do not need coverage of the whole wavelength range to be properly tested.

References

Alekseeva, G.A. et al. 1997, BalA 6, 481

Fluks, M.A. et al. 1994, A&ApS, 105, 311

Glushneva, I.N. et al. (1998) VizieR Online Data Catalog, 3208, 0

Gunn, J.E. & Stryker, L.L., 1983, ApJS 52, 121

Hauschildt, P.H., Allard, F., Baron, E., Aufdenberg, J., & Schweitzer, A., 2003, in GAIA Spectroscopy, Science and Technology, U. Munari ed., ASP Conf. Ser. 298, pag. 179

Jacoby, G. H., Hunter, D. A., & Christian, C. A., 1984, ApJS 56, 257

Montes, D. & Martin, E.L., 1998, A&AS, 128, 485

Montes, D. et al. 1999, ApJS, 123, 283

Munari, U., & Tomasella, L., 1999, A&AS 137, 521

Nesvacil, N., Stütz, Ch. & Weiss, W.W., 2003, in GAIA Spectroscopy, Science and Technology, U. Munari ed., ASP Conf. Ser. 298, pag. 173

Pickles, A.J. 1985, ApJS 59, 33

Pickles, A.J. 1998, PASP 110, 863

Plez, B., 2003, in GAIA Spectroscopy, Science and Technology, U. Munari ed., ASP Conf. Ser. 298, pag. 189

Tinney, C.G. & Reid, I.N., 1998, MNRAS 301, 1031

GAIA spectroscopy of peculiar and variable stars

Ulisse Munari

Astronomical Observatory of Padova – INAF, Asiago Station, 36012 Asiago (VI), Italy

Abstract. A long term project to investigate the spectral appearance over the GAIA wavelength range of a large sample of peculiar and variable stars has been undertaken. The project current status is described and sample spectra are presented and discussed. The diagnostic potential toward peculiarities of the GAIA spectral region turned out to be enormous, thanks to dominating emission-line species covering a wide range ionization and excitation potentials (Hydrogen, CaII, HeI, NI, FeII, FeIII and SI). To properly exploit the diagnostic potential of the resolved emission line profiles and to identify emission line cores that would spoil the measure of radial velocities, a resolving power of at least 10 000 is recommended.

1. Introduction

The spectral appearance of normal stars over the GAIA wavelength range at a spectral resolution $\sim$20 000, the upper limit of the range of values considered for the GAIA spectrograph, is already fairly well documented by the atlas of Munari and Tomasella (1999), and its extension by Marrese et al. (2003). At the same resolution, Pavlenko et al. (2003) have explored the Carbon stars, and S-stars are currently under study.

No large scale, systematic attempt has been however undertaken so far to investigate the spectra of peculiar stars as recorded by GAIA. Therefore, we have decided to carry out, in collaboration with L.Tomasella, P.M.Marrese and F.Boschi from Asiago and T.Zwitter from Ljubljana, a large scale documentation project on peculiar stars with the same instrumentation and resolution adopted for the *normal* stars. This paper is a status report on the way to the publication (foreseen for 2003) of the whole atlas of spectra of peculiar stars.

As for the normal stars, the spectral survey of peculiar stars has been conduced with the Asiago + 1.82 m + Echelle + CCD spectrograph. The current status of observed targets is summarized in Table 1. A deeper look to the symbiotic stars spectra is given by Marrese and Munari (2003). All observations have been conducted at $\sim$20 000 resolution. In such a way, beside commonality with the normal stars, a better look to the intrinsic object properties is obtained, the potential of GAIA spectroscopy is clearer and the appearance at any lower resolutions can be easy derived via convolution with a properly degraded instrumental PSF. The spectra have been exposed to reach a S/N generally in excess of 200 and are presented with their continuum normalized to unity.

Table 1. Number of peculiar stars so far observed according to class.

Pulsating stars:		Interacting binaries:	
3	RR Lyr	35	classical symbiotics
4	classical Cepheids	4	VV Cep
4	pop. II Cepheids	2	CVs
4	δ Sct	3	classical Novae
6	Miras	2	recurrent Novae
4	SR (a,b,c,d)	5	symbiotic novae
3	RV Tau	2	MWC 560
3	β Cep	1	LMXRB
		2	MXRB
Chemically peculiar:		2	SECS
5	Ap, Am, SrCrEu	3	β Lyr
1	λ Boo		
2	H-deficient	Young emission-line objects:	
2	J stars	4	pre-ZAMS
2	Ba stars	3	Herbig Ae/Be
2	CN stars	1	Of
2	CH stars	2	WR
Active surfaces, rotating:		Others:	
5	RS CVn	2	R CrB
2	flare stars	4	post-AGB
3	FK Com	116	DIBs, UIPs, ISM
2	BY Dra	5	Be
3	magnetic	4	O, B, A with shell

Table 1 groups the observed peculiar stars into some major families:
pulsating stars. These have been observed essentially to investigate the possible appearance of emission lines (mainly CaII) during the pulsation cycle that could interfere with the measure of the radial velocities. Other matter of interest was to evaluate the effect of low metallicities in the spectra of some of them (like the RR Lyr and W Vir stars) and the ability to support determination of accurate radial velocities (both barycentric and along the pulsation cycle). Both effects cause no special concern. Example of GAIA spectra of pulsating stars have been already presented by Munari (1999);
chemically peculiar stars. These types of objects are classified as such on the base of spectra secured in the classical optical region. It was therefore interesting to see if the peculiarities show up and can be recognized in the GAIA region too. Even if the analysis is still in progress, a parallel classification based on the GAIA spectra is possible at least for part of the targets. However it is already evident that with this type of stars a high spectral resolution is the key to discovery and proper handling;

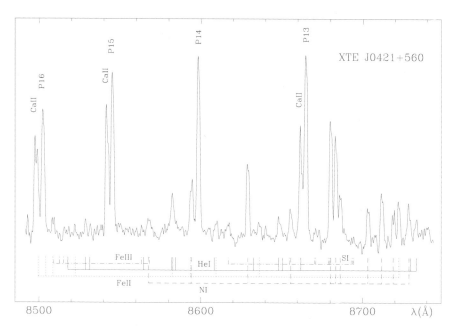

Figure 1. Spectrum of the X-ray transient XTE J0421+560 displaying a significant fraction of the emission lines observed in the GAIA wavelength range. The strongest ones are identified here and listed in Table 2.

Table 2. Wavelength (in Å) of the emission lines identified in Figure 1.

CaII 8498.018	HeI 8518.033	FeII 8499.606	NI 8567.735	SI 8617.090
8542.389	8529.025	8504.033	8594.000	8655.173
8662.140	8531.508	8508.659	8629.235	8670.589
	8564.763	8522.599	8655.878	8671.281
P13 8666.421	8581.856	8582.723	8680.282	8678.927
P14 8599.794	8582.670	8593.842	8683.403	8679.620
P15 8545.984	8584.369	8609.506	8686.149	8680.411
P16 8502.987	8608.312	8636.587	8703.247	8693.137
	8632.770	8722.459	8711.703	8693.931
FeIII 8509.061	8648.258		8718.837	8694.626
8512.713	8650.811	AlII 8640.699	8728.901	
8515.301	8662.171			
8563.493	8729.893	MnI 8740.930		
8568.418	8733.434			

active surface, rotating and magnetic stars. These types of objects are abundant among field F,G,K,M dwarfs. The number of known cases is probably a small fraction of the total, because spectra of proper resolution are required to identify them (frequently obtained as follow-up of satellite detection for those bright in the X-rays). The concern here was with emission cores appearing within the absorption lines of low excitation potential and thus corrupting the measure of radial velocities;

interacting binaries. A whole zoo of quite different beasts is grouped here. Their emission spectrum is rich in strong lines with highly structured profiles. The general absence of a well developed absorption spectrum between the emission lines (with the exception of classical symbiotic stars) suggest that radial velocities from GAIA spectra could be obtained during the automatic data treatment only via special processing, de-routed from the main pipeline.

The spectra of cataclysmic variables over the GAIA wavelength range is discussed by Cropper & Marsh (2003), those of R CrB stars by Rosenbush (2003), Wolf-Rayet stars by Niedzielski (2003) and active solar-type stars by Ragaini et al. (2003).

2. Discussion

The X-ray transient XTE J0421+560 offers one of the richest emission-line inventory known. Its spectrum over the GAIA wavelength range is shown in Figure 1, with major emission lines of Hydrogen, CaII, HeI, NI, FeII, FeIII and SI identified. Their wavelengths are listed in Table 2. Particularly noteworthy in this object are the emissions lines of multiplet #1 and #8 of NI, rivalling in intensity with CaII and Paschen lines.

Spectra of a sample of the objects listed in Table 1 are presented in Figures 2, 3 and 4. Their emission lines are invariably dominated by CaII and Paschen ones, with weaker emissions mainly provided by HeI and NI. A great range of shapes and fine details is presented by the line profiles, and considering the long time over which GAIA could monitor their evolution and the staggering large number of objects it will survey, this would mean the possibility to aim to a comprehensive physical modeling of stellar abnormal behavior and circumstellar material.

The spectrum of XX Oph in Figure 3 offers a quite interesting example of the wide diagnostic potential of the GAIA wavelength range. Paschen 14 shows a strong absorption to the left of the emission line with flat continuum in between, which indicates that the absorption is not forming in a classical wind or expanding atmosphere, but instead in a dynamically de-coupled blob of material (like the erratic jet ejection of MWC 560) with a terminal velocity of -430 km s^{-1} with respect to the emission component. Looking now to the CaII lines, they display instead a classical P-Cyg wind profile (with -270 km s^{-1} terminal velocity) contained within the flat continuum between the emission and absorption components of the Paschen 13, 15 and 16 lines. The CaII and Paschen lines evidently trace different regions and regimes within the interacting binary, which orbital period is comparable to the duration of the GAIA mission.

The spectra of stars with enhanced chromospheric activity presented in Figure 4 confirm the great potential of the GAIA spectra. This type of stars

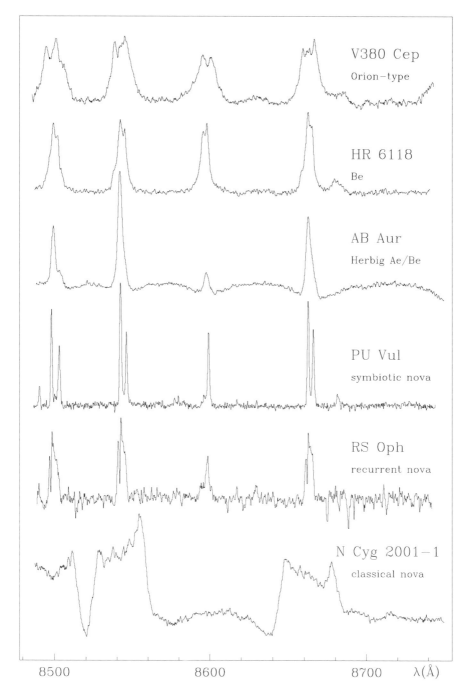

Figure 2. Examples of emission line stars of the early types and of different classes of novae.

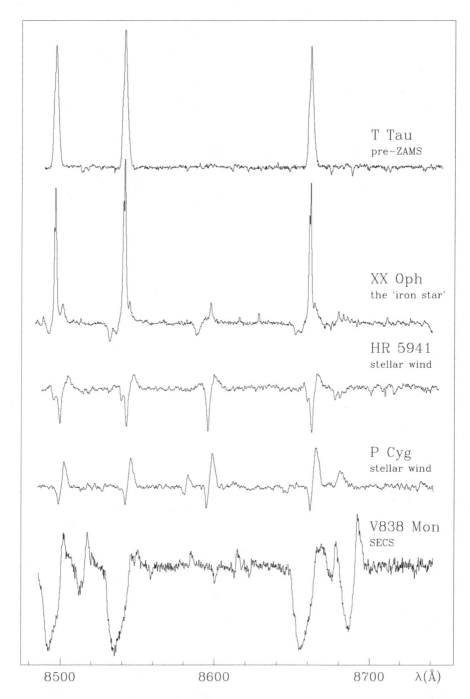

Figure 3. Sample of pre-ZAMS, mass ejecting, stellar-wind and SECS objects.

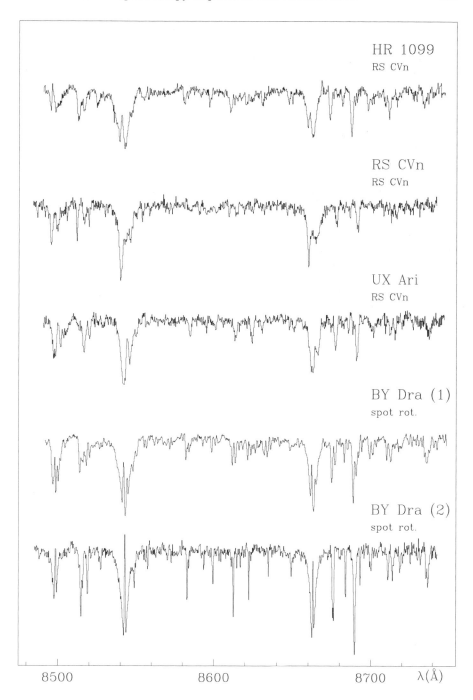

Figure 4. Spectra of a sample of stars with enhanced chromospheric activity. The two spectra of BY Dra have been secured a few months apart to show the typical variability of this type of objects.

displays emission cores in the CaII lines, and sometimes in other metallic lines too depending from the temperature of the surface active regions. Such profiles are highly variable, both for the temporal evolution of the spots and for the rotation that brings the spots in and out of view and shifts forth and back their wavelength. The two spectra of BY Dra in Figure 4, which have been obtained a few months apart, demonstrate the point: in the first case the emission components associated with two major spots (one on the approaching side and the other on the receding one) are readily visible, while in the second spectrum the active areas are passing the central meridian of the star. For the case of spotted stars, resolution is everything. If the latter drops below 10 000, the emission core would not be recognized as such, and the radial velocity measured from the blended absorption+emission profiles would be highly scattered by the continuously changing line photo-center (and in danger to be confused with orbital motion).

Spectral resolution is also the key factor for the other types of peculiarities. Chemically peculiar stars can be identified only if the absorption lines are individually recognized. Objects with P-Cyg profiles can have their mass-loss properly modeled only if the velocity field within the absorption component is properly resolved. Stars dominated by emission lines offer a great deal of details in the line profiles that trace the relative brightness, kinematics and physical conditions of the individual circumstellar nebular regions: if the spectral resolution is too low, anything but the integrated line flux is measurable, preventing access to a lot of exiting physics.

References

Cropper, M. & Marsh, T. 2003, in GAIA Spectroscopy, Science and Technology, U.Munari ed., ASP Conf. Ser. 298, pag. 407

Marrese, P.M. & Munari, U., 2003, in GAIA Spectroscopy, Science and Technology, U.Munari ed., ASP Conf. Ser. 298, pag. 423

Marrese, P.M., Munari, U., Boschi, F. & Tomasella, L. 2003, in GAIA Spectroscopy, Science and Technology, U.Munari ed., ASP Conf. Ser. 298, pag. 427

Munari, U., 1999, Baltic Astron. 8, 73

Munari, U. & Tomasella, L., 1999, A&AS 137, 521

Niedzielski, A. 2003, in GAIA Spectroscopy, Science and Technology, U.Munari ed., ASP Conf. Ser. 298, pag. 295

Pavlenko, Ya., Marrese, P.M. & Munari, U., 2003, in GAIA Spectroscopy, Science and Technology, U.Munari ed., ASP Conf. Ser. 298, pag. 451

Ragaini, S., Andretta, V., Gomez, M.T., Terranegra, L., Busà, I. & Pagano, I. 2003, in GAIA Spectroscopy, Science and Technology, U.Munari ed., ASP Conf. Ser. 298, pag. 461

Rosenbush, A.E. 2003, in GAIA Spectroscopy, Science and Technology, U.Munari ed., ASP Conf. Ser. 298, pag. 465

session 5

RADIAL AND ROTATIONAL VELOCITIES, STELLAR PULSATIONS

chair: A.A. Henden

LOC and LST members A. Siviero, C. Boeche, M. Fiorucci, P. Marrese, R. Sordo and F. Boschi

Clock-wise from top-left: M. Feast, G. Bono, L. Szabados, A. Gomboc, A. Niedzielski and D. Kovaleva

Cepheids: observational properties, binarity and GAIA

Lazlo Szabados

Konkoly Observatory of the Hungarian Academy of Sciences, P.O. Box 67, H-1525 Budapest XII, Hungary

Abstract. Cepheids are primary calibrators of the cosmic distance scale. When deriving luminosity of individual stars from the observed brightness, however, various effects (binarity, crossing number, metallicity, pulsation mode, reddening) have to be taken into account. For this reason, deviations from the regular behaviour are briefly reviewed as determined from the known sample of Galactic Cepheids. The expected role of GAIA in achieving better knowledge on Cepheids as well as methods of independent distance determinations involving Cepheids are also summarized.

1. Introduction

Classical Cepheids have been used as standard candles for establishing the cosmic distance scale for almost a century. Although many more distance indicators are known among the celestial objects, Cepheids have remained among the most popular ones because the precision that can be achieved when determining the distance of these supergiant radial pulsators with using the period-luminosity (P-L) relationship is considered to be superior to that offered by other distance indicators.

This paper summarizes the physical effects to be taken into account when attempting to reduce further the scatter in the P-L relation. General information on Cepheids is also given from an observer's viewpoint which supplements Bono (2003) paper describing mainly theoretical aspects.

2. General properties of Cepheids

The most characteristic feature in the pulsation of Cepheids is its regularity which results in existence of various relationships between the physical properties of these variables (P-L, period-amplitude, period-age relations, etc.).

Individual Cepheids, however, often show subtle deviations from regularity. The most typical among them is the secular or sudden change in the pulsation period. The origin of the secular period changes may be stellar evolution (relative accuracy of the value of the pulsation period is as good as 10^{-7}.) Sudden phase jumps in pulsation or slightly different alternating periods can be observed in Cepheids belonging to binary systems (Szabados 1992).

Evolution toward decreasing temperature causes increase in the pulsation period, while evolution along a track toward higher temperature in the HR diagram results in period decrease. Because of the width of the instability region, the P-L relation is only an approximation, the more adequate form is, in fact, the P-L-colour relationship. The width of the instability region depends on the luminosity for Cepheid pulsation: according to the recent terminology, Cepheids populate an instability *wedge* in the HR diagram (Fernie 1990a).

Evolutionary calculations (e.g. Schaller et al. 1992) also show that intermediate mass stars can enter the instability region three times during the post-main sequence evolution. The first crossing occurs when the star evolves into red giant from the main sequence. Subsequent evolution carries the star along a track that enters or crosses the instability wedge again. In this later evolutionary phase Cepheids are more luminous than during the first crossing, and the redward part of this loop corresponds to higher luminosity than the blueward.

Since more massive stars evolve faster, the evolutionary period change is more apparent for more massive, i.e. longer period Cepheids. It is not only the period that can vary, intriguing variations in the amplitude of the pulsation have also been observed. The secular decrease in the photometric and radial velocity amplitudes of Polaris (Kamper & Fernie 1998, and references therein) and the similar behaviour of Y Ophiuchi (Fernie 1990b) cannot be explained by the exit of these Cepheids from the instability region because neither Polaris, nor Y Oph is at the edge of the instability wedge.

Tiny changes in amplitudes or in the shape of the light curve can only be revealed if accurate observations are available and the time coverage is satisfactory. The present situation is unfavourable in this respect, even the brightest Cepheids are not observed frequently enough. The increasing number of robotic telescopes may considerably improve the situation.

The number and the brightness distribution of the known Cepheids shows that only a minor fraction of the Galactic Cepheids has been discovered up to now. Interstellar absorption in the Galactic plane works against revealing Cepheids in the optical band, except for the brightest representatives. Curiously enough, even the sample of bright Cepheids was not complete until the Hipparcos mission: the Cepheid type brightness variation of CK Cam ($\langle V \rangle = 7.^{\mathrm{m}}6$) and V411 Lac ($\langle V \rangle = 7.^{\mathrm{m}}9$) was revealed from the Hipparcos photometry (ESA 1997).

The histogram of brightness distribution of photoelectrically observed Galactic Cepheids (based on the latest edition of the General Catalogue of Variable Stars) is shown in Figure 1. The left panel shows the distribution of photoelectrically observed Cepheids. Among stars fainter than 11^{m}, there are almost 200 Cepheids which have never been measured photoelectrically. Information on these faint Cepheids has been derived from photographic observations, therefore even their light curves are not known accurately. The right panel shows the brightness distribution of the whole Galactic Cepheid sample. When plotting this histogram, the $\langle V \rangle - m_{\mathrm{pg}}$ colour index was assumed to be 1 mag.

GAIA photometry means a big step forward. On the one hand, the multicolour photometry is essential for deriving the colour and colour excess of the stars, as well as the magnitude correction necessary because of the interstellar absorption. On the other hand, discovery of a large number of new, mostly faint, Cepheids is expected in our Galaxy. Strangely enough, our knowledge on the Cepheid population of the Magellanic Clouds is more complete than on the

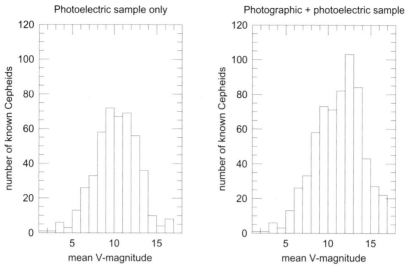

Figure 1. The number distribution of the Galactic Cepheids as a function of the mean apparent brightness. The histograms show the selection bias in discovering faint Cepheids. *Left*: photoelectrically observed Cepheids. *Right*: the same as in the left panel supplemented with the sample for which only photographic photometry is available (as to the colour index correction, see the text on the facing page.)

local Galactic one. The number of known classical Cepheids is about 800 in the Galaxy, while more than 2000 Pop. I Cepheids are known in the LMC and more than 1000 in the SMC. Estimating from the spatial density and taking into account the obscuration by interstellar matter, GAIA can observe about 6000 Galactic Cepheids, majority of them will be new variables (Eyer & Cuypers 2000).

An important subgroup of classical Cepheids is the sample of double-mode pulsators. Again, the number of known Galactic double-mode Cepheids (20) is much smaller than the number (over 100) of their counterparts in the Magellanic Clouds. It is worth mentioning that all double-mode Cepheids in the Magellanic Clouds have been discovered as the result of the EROS, MACHO, and OGLE microlensing projects (Beaulieu et al. 1997; Alcock et al. 1997; Soszynski et al. 2000, respectively), i.e. with the help of extensive precision photometry. GAIA photometry is promising in this respect, too: discovery of many new double-mode pulsators is expected among the faint Galactic Cepheids from this huge photometric dataset.

The luminosity of a Cepheid pulsating with a given period also depends on the mode of pulsation. Cepheids oscillating in the first overtone are more luminous than fundamental mode pulsators of the same period. In most cases the pulsation mode can be determined from Fourier decomposition of the light curves (Mantegazza & Poretti 1992), if well covered phase curves constructed from good quality photometric observations are available.

Since GAIA is primarily an astrometric project, the measurements will result in precise trigonometric parallax, i.e. distances. Then Cepheids of known distance can be used for calibrating the P-L relation. During completing this task, one has to take into account the pulsation mode and the crossing number for each Cepheid, both having an influence on luminosity.

While the mode of pulsation can be determined from the light curve, the crossing number can be derived, at least in principle, from the atmospheric chemical abundance and the sign of the secular period change of the individual Cepheids. However, further corrections may be necessary when determining luminosity of Cepheids. Dependence of luminosity on the abundance of heavy elements has been investigated both theoretically and empirically (Fiorentino et al. 2002; Udalski et al. 2001) with controversial results. By the time when GAIA data are available, this controversy will have been hopefully resolved.

It is essential that reddening in the given direction can be estimated from the intensity of the diffuse interstellar band at 8620 Å (Munari 1999).

An additional effect to be taken into account is the contribution of the companion(s) to the measured brightness. In this respect, each Cepheid has to be studied individually. Fortunately, astrometric and spectroscopic capabilities of GAIA will facilitate such study.

3. Cepheids in binary systems

Cepheids belonging to binary systems are key objects for the calibration of the P-L relation. On the one hand, negligence of the companion falsifies the apparent magnitude inferred from the observations as well as the trigonometric parallax (Szabados 1997), so lack of proper treatment of companion(s) has an adverse effect on the precision of the P-L relation if there are binaries among the calibrating Cepheids. And usually there are a lot, in view of the frequent occurrence of binaries (and multiples) among Cepheids (Szabados 1999). On the other hand, binarity allows to determine the luminosity of the Cepheid irrespective of the pulsation (e.g. Evans 1992b).

Binary stars are very frequent among Pop. I stars, and Cepheids are not exceptions in this respect. Companions have been searched for mainly at the brightest Cepheids. Majority of these secondaries are blue stars, so IUE was instrumental in revealing such companions (Evans 1992a). That survey was complete for Cepheids brighter than 8th magnitude in V. In lack of instrument dedicated for studying the UV spectral region, it is the variations in radial velocity that indicate binary nature of fainter Cepheids (Szabados 1996; Szabados & Pont 1998). Such observations are, however, time consuming because the orbital period for a pair involving a supergiant cannot be shorter than a year. Moreover, the number of individual data points has to be sufficiently large in order to separate the radial velocity variations into pulsational and orbital components. GAIA spectroscopy means a substantial contribution to the discovery and study of binary Cepheids. Even if the inclination of the orbit is not favourable for revealing binarity from radial velocity variations, the intensity ratios of selected metallic lines and lines of Paschen series may indicate existence of a companion, if the temperature difference between the components is large enough.

Figure 2 shows the selection bias in the census of binaries among classical Cepheids. The obvious selection effect is due to the fact that fainter stars are rarely chosen as targets for deep studies. Though the number of known binaries

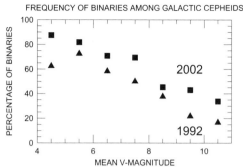

Figure 2. The selection bias in revealing companions to Cepheids. Frequency of binaries is relatively lower among the fainter variables. Triangles denote the situation valid in 1992, while the recently obtained percentage is marked with squares. Though many new binaries were found among the Galactic Cepheids in the last decade, the selection effect is still with us. Note the very high incidence of binaries among the brightest Cepheids.

has been slowly increasing (the progress achieved is seen in the figure), there is still a considerable deficiency in the number of binaries among Cepheids fainter than 8th magnitude. Based on the well studied bright variables, occurrence of binaries (including systems consisting of more than two components) among classical Cepheids seems to be as high as 60-80 %.

The milliarcsecond accuracy achieved by Hipparcos has been at the limit of detecting astrometric orbit of the nearest/widest binaries involving a Cepheid component (Szabados 1997). When interpreting the much more precise astrometric measurements to be obtained with GAIA, orbital motion of the Cepheids members in binaries has to be taken into account in order to avoid false parallax and proper motion values.

Among the companions to Cepheids no white dwarfs have been detected yet. Landsman et al. (1996) showed that any white dwarf remnant of a Cepheid companion should still be hotter than 15 000 K. If such a compact star is found to form a binary system with a Cepheid, the comparison of the white dwarf cooling age and the evolutionary age of the Cepheid would provide a reliable estimate of the mass of the white dwarf whose progenitor was a massive star.

Recently three Cepheids in eclipsing binaries were found among the variables in LMC (Alcock et al. 2002). GAIA photometry will facilitate the search for similar objects in the Milky Way galaxy. A Cepheid in an eclipsing system provides a direct method of distance determination, independent of trigonometric parallax or pulsation.

4. Cepheids in Galactic clusters and associations

Cepheids belonging to star clusters or associations are also very important in calibrating the P-L relation because their luminosity can be determined by main-

Table 1. Cepheids in Galactic clusters and associations

Cepheid	Member in	Type	Remark	Reference
T Ant	anon.	cluster		[1]
VY Car	Car OB1	assoc.	SB	[2]
GT Car	anon.	assoc.		[3]
SU Cas	Cas OB2	assoc.	SB	[4]
CEa Cas	NGC 7790	cluster	VB	[5]
CEb Cas	NGC 7790	cluster	VB	[5]
CF Cas	NGC 7790	cluster		[5]
DL Cas	NGC 129	cluster	SB	[6],[7]
V Cen	NGC 5662	cluster	coronal member	[8]
δ Cep	Cep OB6	assoc.		[9]
SU Cyg	anon.	cluster	SB	[10]
V1726 Cyg	anon.	cluster		[11]
T Mon	Mon OB2	assoc.	doubtful membership, SB	[12]
CV Mon	anon.	cluster		[13]
S Nor	NGC 6087	cluster	SB	[6]
TW Nor	Lyngå 6	cluster		[6]
QZ Nor	NGC 6067	cluster		[14]
V340 Nor	NGC 6067	cluster		[6]
UY Per	King 4	cluster		[15]
GY Sge	anon.	assoc.		[16]
U Sgr	M25	cluster	SB	[17]
WZ Sgr	anon.	cluster	SB	[3]
BB Sgr	Cr 394	cluster	doubtful membership, SB	[18]
KQ Sco	anon.	assoc.	doubtful membership	[19]
RU Sct	Trumpler 35	cluster	SB	[20]
EV Sct	NGC 6664	cluster	binary?	[6]
V367 Sct	NGC 6649	cluster	double-mode Cepheid	[6]
SZ Tau	NGC 1647	cluster	coronal member, SB	[21],[22]
RZ Vel	Vel OB1	assoc.		[19]
SW Vel	Vel OB5	assoc.		[19]
S Vul	Vul OB2	assoc.	doubtful membership, SB	[23],[24]
SV Vul	Vul OB1	assoc.	SB	[25]

References to the table: [1] Turner & Berdnikov (2002); [2] Turner (1977a); [3] Turner et al. (1993); [4] Turner & Evans (1984); [5] Sandage & Tammann (1969); [6] Mermilliod et al. (1987); [7] Gieren et al. (1994); [8] Turner (1982); [9] de Zeeuw et al. (1999); [10] Turner et al. (1998a); [11] Turner et al. (1994); [12] Turner (1976); [13] Turner et al. (1998b); [14] Moffett & Barnes (1986); [15] Turner (1977b); [16] Forbes (1982); [17] Sandage (1960); [18] Turner & Pedreros (1985); [19] Turner (1979); [20] Turner (1980b); [21] Efremov (1964); [22] Turner (1992); [23] Turner et al. (1986); [24] Turner (1980a); [25] Turner (1984) sequence fitting, a widely used method for obtaining distance of stellar aggregates. The list of Cepheids, members in Galactic clusters or associations can be found in Table 1. Most of them are *bona fide* members in the stellar aggregate, but in some cases the relation has to be confirmed. These doubts will be removed with GAIA's decisive membership data on each star in or near the direction of a cluster/association, from the precisely determined individual celestial position, parallax, proper motion and radial velocity values.

Cepheids, simultaneously members in star clusters/associations and binary systems provide an opportunity for cross-checking the three independent methods of distance determination that are applicable for these variables.

Cepheids situated in clusters of the Magellanic Clouds are also important for connecting the distance scales based on the cluster main-sequence fit and the Cepheid pulsation. Fortunately, many of the LMC clusters contain Cepheids, and in some of them, quite a few such pulsating variables are members: 21 Cepheids are in NGC 1866, 14 in NGC 2031, and 6 in NGC 2136 (Welch, Mateo, & Olszewski 1993).

Acknowledgments. Research grants OTKA T029013 and T034854 are gratefully acknowledged. This paper was partly prepared during the author's stay at Merate Observatory in the framework of the Italian-Hungarian TéT cooperation (project I-24/1999).

References

Alcock, C., Allsman, R.A., Alves, D., et al. 1997, ApJ 482, 89

Alcock, C., Allsman, R.A., Alves, D., et al. 2002, ApJ 573, 338

Beaulieu, J.P., Krockenberger, M., Sasselov, D.D., et al. 1997, A&A 321, L5

Bono, G. 2003, in GAIA Spectroscopy, Science and Technology, U.Munari ed., ASP Conf. Ser. 298, pag. 245

de Zeeuw, P.T., Hoogerwerf, R., de Bruijne, J.H.J., Brown, A.G.A., & Blaauw, A. 1999, AJ 117, 354

Efremov, Y.N. 1964, Perem. Zv. 15, 242

ESA 1997, The Hipparcos and Tycho Catalogues, ESA SP-1200

Evans, N.R. 1992a, ApJ 384, 220

Evans, N.R. 1992b, ApJ 389, 657

Eyer, L. & Cuypers, J. 2000, in The Impact of Large-Scale Surveys on Pulsating Star Research, Proc. IAU Coll. 176, L. Szabados & D.W. Kurtz ed.s, ASP Conf. Ser 203, pag. 71

Fernie, J.D. 1990a, ApJ 354, 295

Fernie, J.D. 1990b, PASP 102, 905

Fiorentino, G., Caputo, F., Marconi, M., & Musella, I. 2002, ApJ 576, 402

Forbes, D. 1982, AJ 87, 1022

Gieren, W.P., Welch, D.L., Mermilliod, J.-C., Matthews, J.M., & Hertling, G. 1994, AJ 107, 2093

Kamper, K.W. & Fernie, J.D. 1998, AJ 116, 936

Landsman, W., Simon, T., & Bergeron, P. 1996, PASP 108, 250

Mantegazza, L. & Poretti, E. 1992, A&A 261, 137
Mermilliod, J.-C., Mayor, M., & Burki, G. 1987, A&AS 70, 389
Moffett, T.J. & Barnes, T.G.III 1986, MNRAS 219, 45P
Munari, U. 1999, Baltic Astr., 8, 73
Sandage, A.R. 1960, ApJ 131, 610
Sandage, A.R. & Tammann, G.A. 1969, ApJ 157, 683
Schaller, G., Schaerer, D., Meynet, G., & Maeder, A. 1992, A&AS 96, 269
Soszynski, I., Udalski, A., Szymanski, M., Kubiak, M., Pietrzynski, G., Wozniak, P., & Zebrun, K. 2000, Acta Astron. 50, 451
Szabados, L. 1992, in Complementary Approaches to Double and Multiple Star Research, Proc. IAU Coll. 135, H.A. McAlister & W.I. Hartkopf ed.s, ASP Conf. Ser. 32, pag. 255
Szabados, L. 1996, A&A 311, 189
Szabados, L. 1997, in HIPPARCOS Venice'97, ESA SP-402, pag. 657
Szabados, L., Pont, F. 1998, A&AS 133, 51
Szabados, L. 1999, in Precise Stellar Radial Velocities, J.B. Hearnshaw & C.D. Scarfe ed.s, ASP Conf. Ser. 185, pag. 211
Turner, D.G. 1976, ApJ 210, 65
Turner, D.G. 1977a, AJ 82, 163
Turner, D.G. 1977b, PASP 89, 277
Turner, D.G. 1979, A&A 76, 350
Turner, D.G. 1980a, ApJ 235, 146
Turner, D.G. 1980b, ApJ 240, 137
Turner, D.G. 1982, PASP 94, 1003
Turner, D.G. 1984, JRASC 78, 229
Turner, D.G. 1992, AJ 104, 1865
Turner, D.G. & Berdnikov, L.N. 2002, AAS 200.0703
Turner, D.G. & Evans, N.R. 1984, ApJ 283, 254
Turner, D.G. & Pedreros, M. 1985, AJ 90, 1231
Turner, D.G., Ibrahimov, M.A., Mandushev, G.I., Berdnikov, L.N., & Horsford A.J. 1998a, JRASC 92, 145
Turner, D.G., Leonard, P.J.T., & Madore, B.F. 1986, JRASC 80, 166
Turner, D.G., Mandushev, G.I., & Forbes, D. 1994, AJ 107, 1796
Turner, D.G., Pedreros, M.H., & Walker, A.R. 1998b, AJ 115, 1958
Turner, D.G., van den Bergh, S., Younger, P.F., Danks, T.A., & Forbes D. 1993, ApJS 85, 119
Udalski, A., Wyrzykowski, L., Pietrzynski, G., Szewczyk, O., Szymanski, M., Kubiak, M., Soszynski, I., & Zebrun, K. 2001, Acta Astron., 51, 221
Welch, D.L., Mateo, M., & Olszewski, E.W. 1993, in New Perspectives on Stellar Pulsation and Pulsating Variable Stars, Proc. IAU Coll. 139, J.M. Nemec & J.M. Matthews ed.s, Cambridge Univ. Press, pag. 359

The Cepheid and RR Lyrae instability strip with GAIA

Giuseppe Bono

INAF - Rome Astronomical Observatory, Via Frascati 33, 00040 Monte Porzio Catone, Italy

Abstract. We present recent results concerning distance determinations based on the two most popular primary distance indicators, namely classical Cepheids and RR Lyrae. We discuss the problems affecting the Cepheid distance scale, and in particular the dependence of fundamental Period-Luminosity (PL) and Period-Luminosity-Color (PLC) relations on the metal content. The key advantages in using the K-band PL relation of RR Lyrae stars when compared with the M_V vs [Fe/H] are also presented. We outline the impact that GAIA's spectroscopic measurements will have not only on the distance scale but also to constrain the gradients of metals and α-elements (Thévenin 2003) across the Disk and the Halo as well as current theoretical predictions concerning Galactic models.

1. Introduction

The bulk of stars with spectral types ranging from late A to late G are pulsating variables located inside the so-called Cepheid instability strip. When moving from brighter to fainter objects inside this strip we find Classical Cepheids and RR Lyrae stars. These objects are the prototypes of young, intermediate-mass and old, low-mass standard candles. Owing to these intrinsic features and to the fact that they can be easily identified due to their large luminosity variation they are robust stellar tracers. Between these two groups of variables are located Type II Cepheids. Their periods are similar to classical Cepheids but they are old, low-mass stars in the Asymptotic-Giant-Branch phase (double-shell burning). This evolutionary phase is substantially shorter when compared to RR Lyrae and Cepheids.

Toward fainter magnitudes we find the δ Scuti stars (Breger 2002) and the Oscillating Blue Stragglers (Bono et al. 2003). These objects are intermediate-mass stars burning Hydrogen in the core or in a thick-shell, i.e. evolved off the main sequence. Obviously these variables outnumber the previous ones, but the luminosity amplitudes are smaller and the mode identification is still debated in the literature. Stars with spectral types ranging from late K to M and low surface gravities are once again pulsating variables located in the Mira instability strip. The physical mechanism driving the pulsation instability is the same, but the pulsation properties are substantially different. The most common variables located inside this strip are the semiregular variables and the Long-Period-Variables (Feast 2003).

Instruments on board of GAIA have been optimized to supply accurate trigonometric parallaxes and physical parameters (chemical composition, radial velocity, effective temperature, surface gravity, reddening, binarity) for stars with spectral types ranging from F to K. This means that photometric and spectroscopic data collected by GAIA will have a fundamental role to improve current knowledge on pulsating variables located inside the Cepheid and the Mira instability strips.

The microlensing experiments (EROS, OGLE, MACHO, PLANET), aimed at the detection of baryonic dark matter have already provided a large amount of photometric data in the direction of the galactic Bulge. These data substantially increased the number of variables for which are available accurate estimates of periods, mean magnitudes and colors. However, the observables that GAIA plan to measure are mandatory (Perryman 2003) to constrain the accuracy of theoretical predictions concerning evolutionary and pulsation properties of galactic stars, the formation and evolution of the Galaxy and its interaction with nearby dwarf galaxies as well as chemical evolution models.

In the following we discuss the role that GAIA spectroscopy will have on Cepheid and RR Lyrae distance scales as well as on the use of these objects as stellar tracers. Finally, we briefly outline the impact that GAIA will have on outer Disk and Halo stellar populations.

2. Classical Cepheids

The Cepheid distance scale is the crossroad for the calibration of secondary distance indicators, and in turn for estimating the Hubble constant H_0. The strengths and the weaknesses in using the PLC relation, that supplies individual Cepheid distances, or the PL relation, that supplies ensemble distances, have been widely discussed in the recent literature (Sandage et al. 1999; Tanvir 1999). Pros and cons in using optical, near-infrared (NIR), or Wesenheit magnitudes ($W = V - 2.45\,[V - I]$) reddening free magnitudes have also been lively debated during the last few years (Feast 1999; Freedman et al. 2001; Bono 2003). However, the critical issue concerning Cepheid distances is to assess on a firm basis whether the zero-point and the slope of PL and PLC relations do depend on the metal content. During the last few years the number of theoretical and empirical investigations focused on this problem are countless. Empirical findings seem to suggest that the PL relation in the optical bands presents a mild dependence on the metallicity. In particular, metal-rich Cepheids at fixed period appear *brighter* than metal-poor ones (Sasselov et al. 1997; Kennicutt et al. 1998; Sandage et al. 1999; Macri et al. 2001; Fouqué et al. 2003).

On the other hand, hydrodynamical envelope models that account for the coupling between pulsation and convection predict that metal-rich Cepheids, at fixed period, are *fainter* than metal-poor ones (Bono et al. 1999a,b; Alibert et al. 1999). This means that we are facing with a substantial discrepancy between theoretical predictions and empirical data. Current scenario is further *jazzed up* by multi-band analyses of Magellanic Cloud (MC) Cepheids that support the theoretical sign (Groenewegen & Oudmaijer 2000; Groenewegen 2000). At the same time, recent empirical investigations based on accurate individual distance and reddening determinations of galactic Cepheids appear to suggest that both the zero-point and the slope of the PL relation in the optical bands do depend on the metal content (Saha et al. 2003, in press; Tammann & Sandage 2003, in press). *Paucis verbis* the Cepheid PL and PLC relations in the optical bands are not universal. This means that to estimate the distance of external galaxies are necessary PL and PLC relations based on Cepheids that present the same mean metallicity. As a consequence, galactic Cepheids might be crucial to improve the intrinsic accuracy of distance determinations, since the metallicity distribution of these objects, in contrast with Magellanic ones, is quite similar to the mean metal abundance of external spiral galaxies where the two HST key projects identified Cepheids (Freedman et al. 2001; Saha et al. 2001).

However, it is worth mentioning that theoretical and empirical results do suggest that the PL and the PLC relation of First Overtone (FO) Cepheids marginally depends on metal content. In a recent investigation Bono et al. (2002a) found that predicted and empirical Wesenheit function as well as PL_K relations provide quite similar mean distances to the MCs. This finding seems quite promising, since these distance evaluations are marginally affected by systematic uncertainties. In fact, K-band and Wesenheit magnitudes are marginally affected by uncertainties on reddening corrections and presents a mild dependence on metallicity (Bono et al. 1999b). The width in temperature of FO Cepheids is roughly a factor of two narrower than for fundamental (FM) Cepheids. Therefore distances based on former variables are marginally affected by the intrinsic spread of the PL relation typical of the latter ones. Moreover, the period range covered by FOs is ≈ 1 dex shorter than for FM Cepheids, thus FO PL and PLC relations are hardly affected by changes in the slope when moving from long to short-period variables (Bauer et al. 1999; Bono, Caputo & Marconi 2001).

Why GAIA might play a crucial role to improve the Cepheid distance scale? To nail down the systematic uncertainties affecting the Cepheid distance scale accurate trigonometric parallax measurements for a sizable sample of galactic Cepheids are required . However, according to the above discussion accurate and homogeneous *measurements* of Cepheid metal abundances are mandatory to properly address the problem. The most recent spectroscopic investigation concerning the chemical composition of galactic Cepheids do rely on a sample of roughly 100 objects (Andrievsky et al. 2002b, and references therein). According to these authors the metallicity distribution across the galactic Disk might be split into three different zones:

i) the inner region, ranging from ≈ 4 to ≈ 7 kpc for which they found a metallicity gradient of $d[\text{Fe/H}]/dR_G \approx -0.13 \pm 0.03$ dex kpc^{-1};

ii) the central region, ranging from ≈ 7 to ≈ 10 kpc, for which the gradient is flatter and equal to $\approx -0.02 \pm 0.01$ dex kpc^{-1};

iii) a small portion of the outer Disk (toward the galactic anticenter), ranging from ≈ 10 to ≈ 12 kpc, with a gradient of $\approx -0.06 \pm 0.01$ dex kpc^{-1}.

Moreover and even more importantly, Andrievsky et al. (2002) found a discontinuity in the metallicity distribution located at roughly 10 kpc from the galactic center. These findings somehow supports the results obtained by Twarog et al. (1997) on the basis on photometric (Strömgren) indicators and by Caputo et al. (2001) on the basis of multiband (Johnson) Cepheid pulsation relations. However, no firm conclusion can be drawn concerning the metallicity gradient, since current photometric and spectroscopic data for Cepheids located in the outer Disk ($d \geq 12$ kpc) are scanty.

It goes without saying that homogeneous high resolution spectra for a complete sample of galactic Cepheids together with accurate distances might be the *Panacea* not only for the Cepheid distance scale but also to constrain dynamical models of the galactic Disk as well as its chemical evolution. Note that the occurrence of shallow metallicity gradients of iron and of iron-group elements might be caused, according to current predictions by gas infall from the halo, by gas viscosity in the Disk, or by a central bar structure (Portinari & Chiosi 2000; Andrievsky et al. 2000a). This scientific goal can be easily reached by GAIA, since a spectrograph with $R \approx 15\,000$ should supply accurate spectra ($S/N \geq 50$) Data plotted in Figure 1 show that short-period, galactic FO Cepheids present an absolute I magnitude ranging from -1.5 to -2.0. Therefore if we assume that the outskirt of the galactic Disk is roughly located between 20 and 25 kpc ($DM \approx 17.0$ mag) and a mean reddening that ranges from $E_{B-V}=0.5$ to 1.0, then the apparent magnitude of fainter Cepheids should range from $I \sim 16$ to $I \sim 17$ mag. This means that a substantial fraction of galactic Cepheids are brighter than the limiting magnitude ($V \sim 17.5$ mag) for which GAIA will supply accurate chemical compositions and radial velocities (Munari 2003, Munari et al. 2003).

Note that GAIA spectra will be collected at least over five consecutive years, therefore they can also be adopted to identify a substantial fraction of binary Cepheids (Szabados 2003). Finally, we mention that this project can be hardly accomplished with current generation of multi-fiber, wide field of view spectrographs such as FLAMES/GIRAFFE@VLT, since the spatial density of Cepheids is too low.

3. RR Lyrae variables

RR Lyrae stars are very useful objects, since they are robust, low-mass standard candles and reliable tracers of old stellar populations. They are ubiquitous across the galactic spheroid and thanks to their pulsation properties (peculiar shape of the light curves, narrow period range), they can also be easily identified in Local Group (LG) galaxies. Although, RR Lyrae stars present several advantages, distance estimates based on different calibrations (Baade-Wesselink method, HB models) of the M_V vs [Fe/H] relation taken at face value present a difference that is systematically larger than the empirical uncertainties (Walker 2000, 2003 in press; Cacciari 2003). This might indicates that current RR Lyrae distance determinations are still affected by systematics. Fortunately enough, empirical evidence dating back to Longmore et al. (1990) suggest that RR Lyrae stars do obey to a well-defined PL relation in the K-band (PL_K). The use of this relation might overcome some of the problems affecting the RR Lyrae distance scale, since the PL_K relation is marginally affected by evolutionary effects as well as by the spread in stellar mass inside the instability strip. This finding was further strengthened by a recent theoretical investigation (Bono et al. 2001) suggesting that RR Lyrae obey to a very tight PLZ_K relation connecting the period, the luminosity, the K-band absolute magnitude, and the metallicity. This approach seems very promising and should allow us to supply during the

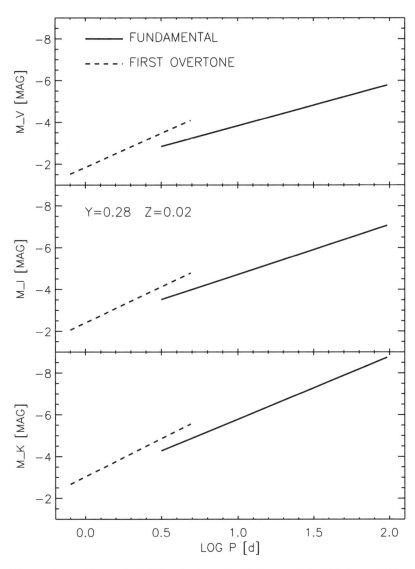

Figure 1. Theoretical PL relations in V (*top*), I (*middle*), and K (*bottom*) band at solar chemical composition (Bono et al. 1999b). Solid and dotted lines display FM and FO PL relations.

next few years an accurate calibration of the M_V vs [Fe/H] relation over the metallicity range covered by RR Lyrae ($-2.2 \leq$[Fe/H]≤ 0).

During the last few years RR Lyrae are becoming very popular, since the Sloan Digital Sky Survey (SDSS) detected an overdensity of candidate RR Lyrae and of A-type stars located approximately 50 kpc from the Galactic center. According to this empirical evidence Ivezic et al. (2000) and Yanny et al. (2000) suggested that such a clump is the northern tidal stream left over by the Sagittarius dwarf spheroidal (dSph). Independent observational (Ibata et al. 2001; Martinez-Delgado et al. 2001; Vivas et al. 2001) and theoretical (Helmi & White 1999; Helmi 2002) investigations support this hypothesis. The observations of such extra-tidal stellar remnants in dSph resembles the tidal debris recently detected in a large number of Galactic Globular Clusters (GGCs, Leon et al. 2000; Odenkirchen et al. 2002). On the other hand, dSph galaxies apparently host large amounts of Dark Matter (DM), and indeed the mass-to-light ratio in these systems range from $(M/L)_V \sim 5$ (Fornax) to ~ 100 (Ursa Minor), whereas in GGCs the M/L ratio is $\approx 1-2$. As a consequence, the study of the radial distribution of RR Lyrae can supply tight constraints on the tidal interaction that these interesting systems undergo with the Milky Way.

On the basis of V-band time series data that cover a large sky area (100 deg^2) Vivas et al. (2001) identified and measured the mean magnitude of 148 RR Lyrae stars and more than 50% of this sample belong to the clump identified by Ivezic et al. (2000). These data provided the first firm evidence that the galactic Halo does not show smooth contours in density. In fact, they also detected two smaller overdensities in the Halo one of which located at R ≈ 17 kpc seems related to the GGC Palomar 5, while the other is located at R ≈ 16 kpc.

It is worth stressing, that a substantial improvement in the intrinsic accuracy of the M_V vs [Fe/H] and of the PLZ_K relation does not allow us to use RR Lyrae in the galactic Halo to constrain the dynamical interaction of dwarf galaxies and GGCs with the Milk Way. The detections of peculiar radial distributions is hampered by the limited number of RR Lyrae stars for which are available accurate spectroscopic measurements of radial velocities and chemical compositions (Suntzeff et al. 1994; Layden et al. 1996; Dambis & Rastorguev 2001). Moreover, current sample of RR Lyrae in the galactic Halo might also be affected by a selection bias. Data plotted in Figure 2 show that RR Lyrae in the halo might be peculiar. The period distribution and the mean period of fundamental pulsators ($<P_{ab}>\approx 0.539$ d) mimic the behavior of Oosterhoff type I clusters (see figure 3 in Clement et al. 2001)[1]. However, the number of FO RR Lyrae in the Halo is quite small ($N_{FOs}/N_{RR} < 0.1$) and the period distribution of fundamental pulsators does show a gap at $P \approx 0.5$ that is not present among RR Lyrae in the galactic Bulge and in GGCs (Bono et al. 2003). At present, it is not clear whether these peculiarities are intrinsic or due to a selection bias.

---

[1]GGCs that host a good sample of RR Lyrae stars are classified Oosterhoff type I clusters if the mean fundamental period is roughly equal to $<P_{ab}>\approx 0.55$ and the ratio between first overtones and the total number of RR Lyrae, N_{FOs}/N_{RR} is roughly equal to 0.2. A GGC is classified as an Oosterhoff type II if $<P_{ab}>\approx 0.64$ and $N_{FOs}/N_{RR} \approx 0.5$.

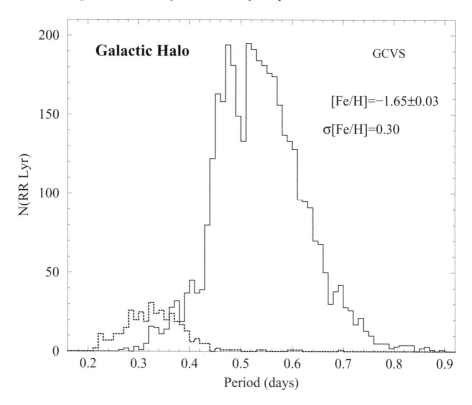

Figure 2. Period distribution of RR Lyrae stars in the galactic Halo according to the General Catalog of Variable Stars (GCVS). Solid and dashed line refer to fundamental and first overtone variables, respectively. The mean metallicity and the intrinsic spread in metallicity are labeled (Suntzeff et al. 1994).

Once again GAIA might play a crucial role to improve current empirical scenario. The unprecedented opportunity to supply accurate high resolution spectra down to a limit magnitude of $V \sim 17.5$ mag will allow us to trace the pulsation properties of RR Lyrae over a substantial portion of the Halo. The use of metallicity indicators based on the GAIA multi-band photometric system should also allow us to extend the spectroscopic analysis from the Bulge to the outermost regions of the halo ($DM \approx 19 - 20$).

Finally, it is worth mentioning that we still lack an empirical estimate of the dynamical mass of a Horizontal-Branch (HB) star, since no binary system has been detected that include one of these objects. The detection of a few of these systems would be of paramount relevance to constrain the input physics (equation of state, opacities, nuclear burning rates) of evolutionary and pulsation models, and in turn to properly address long-standing stellar astrophysical problems such as the second parameter problem and the HB morphology (Castellani 1999).

4. Final remarks

The compelling results obtained by the two HST key projects concerning the estimate of the Hubble constant contributed to the diffuse believe that problems affecting the calibration of both primary and secondary distance indicators have been settled. Recent findings concerning the dependence of the Cepheid PL relation on the metal abundance, as well as the nonlinearity of the RR Lyrae M_V vs [Fe/H] relation (Caputo et al. 2000) cast some doubts on this view. The mismatch between distance determinations based on different standard candles further strengthens this working hypothesis. Data listed in Table 1 clearly show that distance determinations to the Coma cluster based on different zero-points and secondary indicators range from 34.64 to 35.29 mag, while H_0 evaluations range from 60 to 84 km s^{-1} Mpc^{-1}. Note that the Coma cluster will play a fundamental role to improve the accuracy of the Hubble constant, since it is the nearest galaxy cluster not affected by local motions.

The above discussion brings forward the evidence that distance indicators may require new detailed empirical and theoretical investigations to nail down the deceptive errors affecting current distance determinations. In the near future different roots may shed new lights on this long-standing problem:

i) the use of the white-light interferometer, FGS3, on board of HST recently provided a new accurate estimate of the trigonometric parallax of δ Cephei (Benedict et al. 2002), and in turn a new calibration of the PL relation. This instrument during the next few years might supply accurate geometric distances for a handful of nearby Cepheids. This means that the zero-point *and the slope* of both the PL and the PLC relations can be improved;

ii) the new CCD camera (ACS) on board of HST should allow the detection of FO Cepheids in external galaxies where FM Cepheids have already been measured. This instrument could supply the unique opportunity to cross-check independent distance determinations based on the same group of variable stars;

iii) During the next few years ground-based survey telescopes aimed at detecting near Earth asteroids, such as *Pan-STARRS* (Kaiser 2002)[2] and *LSST*[3] will be equipped with detectors that cover a sky area ranging from one to several square degrees. Therefore a detailed sampling of stellar populations down to $V \sim 24 - 27$ mag might be accomplished in the near future. The same outcome applies for the wide field imagers that are already available on telescope of the 8m class such as SUPRIME@SUBARU or will become available in a few years such as LBC@LBT. The new multi-band time series data will allow a complete census of RR Lyrae and Cepheids belonging to the Galaxy as well as to LG galaxies.

In this possible scenario, GAIA gives the unprecedent opportunity to supply accurate trigonometric parallaxes, as well as accurate measurements of radial velocities, and chemical compositions for a large amount of galactic stars. During the next ten years ground-based telescopes of the 8m class equipped with multi-object spectrographs will supply accurate estimates of stellar parameters for

---

[2]For further information visit http://www.ifa.hawaii.edu/k̃aiser/pan-starrs/pressrelease/

[3]For further information visit http://www.lsst.org/lsst_home.html

Table 1. Compilation of distance determinations to the Coma cluster, according to different primary and secondary distance indicators.

Method[a]	Target(s)	ZP[b]	$(m-M)_0^c$	H_0^d	Ref.[e]
GCLF	NGC4874 IC4051	Virgo[f]	35.05 ± 0.12	69 ± 9	1
SBF_K	NGC4874	Ceph[g]	34.99 ± 0.21	71 ± 8	2
$SBF_{K'}$	NGC4889	Ceph[h]	34.64 ± 0.25	85 ± 10	3
SBF_I	NGC4881	Leo-I[i]	35.04 ± 0.31	71 ± 11	4
SBF_I	NGC4881	Ceph[j]	35.05 ± 0.53	73 ± 19	5
TF_H	20 galaxies	Ceph[h]	34.94 ± 0.13	73 ± 4	6
TF_I	...	Ceph[j]	34.66	$84 \pm 13/86 \pm 14$	5
FP_I	81 galaxies	Ceph[j]	34.67 ± 0.15	$83 \pm 6/86 \pm 6$	5
$D_n - \sigma$	81 galaxies	Ceph[j]	34.89 ± 0.16	$75 \pm 5/78 \pm 5$	7
$D_n - \sigma(K)$	24 galaxies	Leo-I[l]	34.90 ± 0.14	75 ± 6	8
SNIa	5	Virgo[m]	35.05 ± 0.49	70 ± 15	9
VM[n]	...	Virgo[n]	35.29 ± 0.11	60 ± 6	10

[a] Globular Cluster Luminosity Function (GCLF); Surface Brightness Fluctuation (SBF); Tully-Fisher (TF) relation; Fundamental Plane (FP); $D_n - \sigma$ or Faber-Jackson relation; Supernovae type Ia (SNIa). [b] Zero-point. [c] True distance modulus and relative error as given by authors. [d] Hubble constant (km s^{-1} Mpc^{-1}) and relative error as given by authors . [e] References: 1) Kavelaars et al. (2000); 2) Liu & Graham (2001); 3) Jensen et al. (1999); 4) Thomsen et al. (1997); 5) Freedman et al. (2001); 6) Watanabe et al. (2001); 7) Kelson et al. (2000); 8) Gregg (1997); 9) Capaccioli et al. (1990); Tammann et al. (1999). [f] Weighted-average true distance modulus based on Cepheids, TRGB, PNLF, and SBF, μ_0(Virgo)=30.99±0.03. They adopted a recession velocity of $V_r \approx 7100\pm200$ km s^{-1}. [g] Six nearby spiral galaxies for which are available HST Cepheid distances ($V_r \approx 7186 \pm 428$ km s^{-1} by Han & Mould 1992, hereinafter HM92). [h] Cepheid distances to M31 and Virgo Cluster (V_r by HM92). [i] Average SBF distance to NGC3379 in the Leo-I group based on Cepheids (V_r by HM92). [j] Revised Cepheid distances to Leo-I group, Virgo and Fornax clusters (Key Project, the adopted V_r values are 7143 and 7392 km s^{-1}). [k] Twelve nearby spiral galaxies for which are available HST Cepheid distances ($V_r \approx 7143$ km s^{-1}). [l] Unweighted-average true distance modulus based on Cepheids, TRGB, PNLF, and SBF, μ_0(Leo I)=30.17±0.01 ($V_r \approx 7200 \pm 300$ km s^{-1}). [m] Maximum-magnitudes vs rate-of-decline for Novae in M31 ($\mu_0 = 24.30 \pm 0.20$) and Virgo ($\mu_0 = 31.30 \pm 0.40$, $V_r \approx 7130 \pm 200$ km s^{-1}). [n] They adopted various (6) secondary methods (Jerjen & Tammann 1993). The ZP is based on the Cepheid distance to the Virgo cluster ($\mu_0 = 31.60 \pm 0.09$).

stellar populations in stellar systems such as globular clusters and nearby dwarf galaxies. However, the stellar density in the outer Disk as well as in the Halo is too low to be interesting targets for these instruments.

Finally, we mention that the selection of the GAIA photometric bands is crucial to improve the accuracy of stellar parameters we plan to supply. The estimates of stellar abundances strongly depend on the accuracy of effective temperature and surface gravity. Moreover and even more importantly, the calibration of new metallicity and reddening indicators are two outstanding legacies we are looking for from the GAIA mission.

Acknowledgments. I am indebted to my collaborators for helpful discussions and suggestions. This work was supported by MIUR/Cofin 2001 under the project: "Origin and Evolution of Stellar Populations in the Galactic Spheroid".

References

Alibert, Y., Baraffe, I., Hauschildt, P., & Allard, F. 1999, A&A 344, 551
Andrievsky, S.M. et al. 2002a, A&A 381, 32
Andrievsky, S.M., Kovtyukh, V.V., Luck, R.E., Lepine, J.R.D., Maciel, W.J., & Beletsky, Y.V. 2002b, A&A 392, 491
Bauer, F. et al. 1999, A&A 348, 175
Benedict, G. F. et al. 2002, AJ 124, 1695
Bono, G. 2003, in Hubble's Science Legacy: Future Optical-Ultraviolet Astronomy from Space, K.R. Sembach, J.C. Blades, G.D. Illingworth, & R.C. Kennicutt ed.s, ASP Conf. Ser., in press (astro-ph/0210068)
Bono, G., Caputo, F., Castellani, V., & Marconi, M. 1999b, ApJ 512, 711
Bono, G., Caputo, F., Castellani, V., Marconi, M., & Storm, J. 2001, MNRAS 326, 1183
Bono, G., Caputo, F., & Marconi, M. 2001, MNRAS 325, 1353
Bono, G., Caputo, F., Marconi, M., & Santolamazza, P. 2002, in Observational Aspects of Pulsating B- and A Stars, C. Sterken & D.W. Kurtz ed.s, ASP Conf. Ser. 256, pag. 249
Bono, G., Groenewegen, M. A. T., Marconi, M., Caputo, F. 2002a, ApJL 574, L33
Bono, G., Marconi, M., & Stellingwerf, R. F. 1999a, ApJS 122, 167
Bono, G., Petroni, S., & Marconi, M. 2003, in Interplay between Periodic, Cyclic and Stochastic Variability in Selected Areas of the HR Diagram, C. Sterken ed., ASP Conf. Ser., in press (astro-ph/0212183)
Breger, M. 2002, in Observational Aspects of Pulsating B- and A-Stars, C. Sterken & D.W. Kurtz ed.s, ASP Conf. Ser. 256, pag. 17
Capaccioli, M., Cappellaro, E., della Valle, M., D'Onofrio, M., Rosino, L., & Turatto, M. 1990, ApJ 350, 110
Caputo, F., Castellani, V., Marconi, M., & Ripepi, V. 2000, MNRAS 316, 819
Caputo, F., Marconi, M., Musella, I., & Pont, F. 2001, A&A 372, 544

Castellani, V. 1999, in Globular Clusters, C. Martinez Roger, I. Perez Fournon, & F. Sanchez ed.s, Cambridge Univ. Press, pag. 109

Clement, C.M. et al. 2001, AJ 122, 2587

Dambis, A.K., & Rastorguev, A.S. 2001, AstL 27, 108

Feast, M. 1999, PASP 111, 775

Feast, M. 2003, in GAIA Spectroscopy, Science and Technology, U.Munari ed., ASP Conf. Ser. 298, pag. 257

Fouqué, P., Storm, J., & Gieren, W. 2003, in Stellar Candles for the Extragalactic Distance Scale, D. Alloin & W. Gieren ed.s, Springer-Verlag, in press (astro-ph/0301291)

Freedman, W.L. et al. 2001, ApJ 553, 47

Gregg, M.D. 1997, New Astr 1, 363

Groenewegen, M.A.T. 2000, A&A 363, 901

Groenewegen, M.A.T., & Oudmaijer, R.D. 2000, A&A 356, 849

Han, M., & Mould, J.R. 1992, ApJ 396, 453

Helmi, A. 2002, Ap&SS 281, 351

Helmi, A., & White, S.D.M. 1999, MNRAS 307, 495

Ibata, R., Irwin, M., Lewis, G.F., & Stolte, A. 2001, ApJL 547, L133

Ivezic, Z. et al. 2000, AJ 120, 963

Jensen, J. B., Tonry, J. L., & Luppino, G. A. 1999, ApJ 510, 71

Jerjen, H., & Tammann, G. A. 1993, A&A 276, 1

Kaiser, N. 2002, AAS 201, 122.07

Kavelaars, J.J., Harris, W.E., Hanes, D.A., Hesser, J.E., & Pritchet, C.J. 2000, ApJ 533, 125

Kelson, D.D. et al. 2000, ApJ 529, 768

Kennicutt, R.C.Jr. et el. 1998, ApJ 498, 181

Layden, A.C., Hanson, R.B., Hawley, S.L., Klemola, A.R., & Hanley, C.J. 1996, AJ 112, 2110

Leon, S., Meylan, G., & Combes, F. 2000, A&A 359, 907

Liu, M.C., & Graham, J.R. 2001, ApJL 557, L31

Longmore, A.J., Dixon, R., Skillen, I., Jameson, R. F., & Fernley, J. A. 1990, MNRAS 247, 684

Macri, L.M. et al. 2001, ApJ 559, 243

Martinez-Delgado, D., Aparicio, A., Gomez-Flechoso, M. A., & Carrera, R. 2001, ApJL 549, L199

Munari, U. 2003, in GAIA Spectroscopy, Science and Technology, U.Munari ed., ASP Conf. Ser. 298, pag. 51

Munari, U., Zwitter, T., Katz, D. & Cropper, M. 2003, in GAIA Spectroscopy, Science and Technology, U.Munari ed., ASP Conf. Ser. 298, pag. 275

Odenkirchen, M., Grebel, E.K., Dehnen, W., Rix, H.-W., & Cudworth, K.M. 2002, AJ 124, 1497

Perryman, M.A.C. 2003, in GAIA Spectroscopy, Science and Technology, U. Munari ed., Asp Conf. Ser. 298, pag. 3

Portinari, L., & Chiosi, C. 2000, A&A 355, 929

Saha, A., Sandage, A., Tammann, G.A., Dolphin, A.E., Christensen, J., Panagia, N. & Macchetto, F.D. 2001, ApJ 562, 314

Sandage, A., Bell, R.A., & Tripicco, M.J. 1999, ApJ 522, 250

Sasselov, D.D., et al. 1997, A&A 324, 471

Suntzeff, N.B., Kraft, R.P., & Kinman, T.D. 1994, ApJS 93, 271

Szabados, L. 2003, in GAIA Spectroscopy, Science and Technology, U.Munari ed., ASP Conf. Ser. 298, pag. 237

Tammann, G.A., Sandage, A., & Reindl, B. 1999, in 19th Texas Symposium on Relativistic Astrophysics and Cosmology, Proceedings on CD-ROM, (astro-ph/9904360)

Tanvir, N. R. 1999, in Post-Hipparcos Cosmic Candles, A. Heck & F. Caputo ed.s, Kluwer, pag. 17

Thévenin, F., Bijaoui, A. & Katz, D. 2003, in GAIA Spectroscopy, Science and Technology, U.Munari ed., ASP Conf. Ser. 298, pag. 291

Thomsen, B., Baum, W.A., Hammergren, M., & Worthey, G. 1997, ApJL 483, L37

Twarog, B.A., Ashman, K.M., & Anthony-Twarog, B.J. 1997, AJ 114, 2556

Vivas, A. K., QUEST collaboration, 2001, ApJL 554, L33

Walker, A.R. 2000, in The Impact of Large-Scale Surveys on Pulsating Star Research, IAU Coll. 176, L. Szabados & D.W. Kurtz ed.s, ASP Conf. Ser. 203, pag. 165

Watanabe, M., Yasuda, N., Itoh, N., Ichikawa, T., & Yanagisawa, K. 2001, ApJ 555, 215

Yanny, B. 2000, ApJ 540, 825

Miras and other cool variables with GAIA

Michael Feast

Astronomy Department, University of Cape Town, Rondebosch, 7701, South Africa

Abstract. A general review is given of Mira and some other cool variables stars, concentrating on those aspects on which GAIA, and more particularly GAIA spectroscopy and radial velocities, will have a major impact.

1. Introduction

The observation of Miras and other late type variables with GAIA will bring rather special rewards, but it will also bring special problems in analysing and interpreting the data. In reviewing some of these promises and problems I hope it will become clear what a vital part in the success of the mission will be played by the spectroscopy: radial velocities and spectral data. I shall also mention ground based observations which will be need to make full use of GAIA data.

Miras are large amplitude, AGB variables with rather regular periods which range from about 100 days to 1000 days or more. At the longer periods they are often OH/IR sources. They can be broadly divided into oxygen-rich and carbon-rich objects (O- and C- Miras).

It has been estimated (Eyer & Cuypers 2000) that GAIA will discover and measure about 150 000 Mira variables. For a significant fraction of these, GAIA will obtain parallaxes, proper motions, radial velocities, spectral data, narrow and broad band photometry and periods. For many of the stars observations will cover the whole light cycle.

Miras are of central importance to a number of key issues in astronomy. Amongst these are the following: (*1*) they appear to define the end point of the AGB when mass-loss plays a major role in stellar evolution. This is a time in the life of a star which is not properly understood; (*2*) in old or intermediate age populations, the Miras are the brightest individual stars and hold out the possibility of calibrating the populations of nearby galaxies in which they can be isolated and studied; (*3*) they show a well defined period-luminosity relation making them important galactic and extragalactic distance indicators; (*4*) Mira periods are related to age and/or metallicity. They are therefore important traces of galactic structure and kinematics.

2. The O-Miras

Consider first the O-Miras. In the LMC these variables show a narrow period-luminosity relation in the infrared. At K (2.2 microns) the scatter about this relation is only 0.13mag (Feast et al. 1989). So far as we have been able to tell the slope of this relation is the same everywhere (i.e. in the SMC, in globular clusters and for Miras with individual distances from proper motion companions, Whitelock et al. 1994, Wood 1995, Feast & Whitelock 1999, Feast, Whitelock & Menzies 2002). Hipparcos allowed the zero-point of this relation to be calibrated directly from Mira parallaxes (Whitelock & Feast 2000a) and also indirectly through Miras in globular clusters with distances based on Hipparcos subdwarf parallaxes (Feast et al. 2002). GAIA will allow one to study this relation in great detail in the galactic field, provided the necessary ground-based infrared observations have been made (a programme along these lines has been started at the South African Astronomical Observatory by P.A. Whitelock, F. Marang and the writer). For the Miras with good coverage of the light curve by GAIA it will also be possible to discover whether one can replace the infrared photometry with some of the GAIA photometry. It may be that the GAIA photometry, together possibly with the spectra, will show evidence for a period-luminosity-colour relation of which there was a hint in the infrared photometry (Feast et al. 1989).

The galactic kinematics of O-Miras are particularly interesting and Hipparcos led to some some unexpected results. It has long been known that the galactic kinematics of O-Miras are a function of period (e.g. Feast 1963) and this allows the galactic kinematics of old and intermediate age stars to be studied as a function of age and/or metallicity in finer detail than is possible in other ways. Hipparcos improved this discussion in three ways: (i) the Hipparcos calibration of the PL(K) relation, mentioned above, allowed one to obtain the distances of many local Miras; (ii) the Hipparcos proper motions could then be used together with (ground-based) radial velocities to obtain space motions (Feast & Whitelock 2000a). Full space motions are essential if we are to study galactic kinematics free of assumptions; (iii) the Hipparcos photometry together with ground-based infrared photometry showed that the short period O-Miras divided into two groups with different mean colours at the same period (Whitelock, Marang & Feast 2000, Whitelock 2002). These two sequences were called the short-period (SP) -red and -blue groups. These two groups differ in their kinematic properties and it will be essential in any future work to distinguish between them. In the case of GAIA observations, this may perhaps be done spectroscopically since the SP-red stars have a later mean spectral type than the SP-blues at a given period. It may also perhaps be possible using the GAIA photometry. On the basis of infrared colours alone which show the division into two groups, though less clearly, and on the basis of kinematics one can show that Miras in globular clusters belong to the SP-blue group which, together with longer period Miras, can be called the main Mira sequence. These Miras in clusters, and by implication the Miras in the main Mira sequence generally, lie at the tip of the AGB and are the brightest objects, bolometrically, of their populations.

What of the SP-red Miras? They have different colours, spectral types and kinematics from the SP-blue Miras (see Whitelock et al 2000 and Feast & Whitelock 2000a for details). Their kinematics associate them with longer period Miras on the main Mira sequence. At present the best guess is that they are not yet at the end of their AGB lives but will evolve into Miras of longer period on the Mira main sequence. There is some evidence (Whitelock and Feast

2000a) that at a given period the SP-red Miras are brighter than the SP-blues. This would be consistent with the above discussion. Clearly we can anticipate that GAIA will clarify this whole picture using its combination of parallaxes, proper motions, radial velocities, spectral types and photometry. This will be a considerable contribution both to our understanding of AGB evolution and to the study of our own and other galaxies.

When the SP-red stars are omitted, the rest (i.e. the main Miras sequence, including the SP-blue variables), show a monotonic dependence of kinematics on period. In particular, the space motions show a smooth change in V_θ (the velocity in the direction of galactic rotation) from $133 \pm 19 \,\mathrm{km\,s^{-1}}$ at a mean period of 173 days to $223 \pm 4 \,\mathrm{km\,s^{-1}}$ at a period of 453 days (see Feast & Whitelock 2000a and Feast 2002). This is for stars which are mostly within about 1 kpc of the Sun. This dependence opens up the possibility of studying galactic kinematics of homogeneous populations of old and intermediate age stars on a much finer grid than is otherwise possible. Some guidance as to the nature of this grid is provided by Miras in globular clusters. These are confined to the more metal-rich clusters and to the shorter period Miras. In addition there is a rather clear period-metallicity relation for Miras in clusters (see Feast & Whitelock 2000b). To understand this relation we need to know whether the initial masses of the Miras in clusters (i.e. the turn-off masses) are a function of metallicity. This in turn requires us to know whether all the metal-rich globulars are the same age or whether age depends on metallicity. GAIA will make a major contribution to settling this issue by providing much improved estimates of globular cluster distances, either directly or through subdwarf parallaxes. However it seems worth noting that this will probably also depend on there being much more detailed ground-based spectroscopic work to determine chemical abundances in subwarfs both in the field and in clusters.

Apart from the globular cluster Miras there is rather little direct evidence on the initial masses of Miras. R Hya is a member of the Hyades moving group (Eggen 1985) and this suggests an initial mass of $\sim 2 M_\odot$. However this Mira is unusual, its period having varied from about 500 days to 385 days in the last 340 years (see e.g. Zijlstra, Bedding & Mattei 2002), probably due to the star being in a shell-flashing stage (see below). General arguments based on galactic kinematics and scale heights (e.g. Olivier, Whitelock & Marang 2001) suggest masses of $\sim 2 M_\odot$ for Miras with periods in the range ~ 400 to 800 days and $\geq 4 M_\odot$ for periods of ≥ 1000 days. In the Magellanic Clouds (Nishida et al. 2000) there are intermediate age clusters with turn-off masses of $\sim 1.5 M_\odot$ which contain C-Miras of period ~ 500 days. It may very well be that C- and O-Miras of the same period have the same initial masses but this has not yet been definitely established. We can expect GAIA to make an important contribution to the mass question. For instance the space motions from GAIA should allow the isolation of new moving groups some of which may contain Miras. We already know of a few Miras with common proper motion companions. Such objects are potentially important for the determination of limits to the initial mass of the Mira as well as its luminosity and, importantly, its chemical composition which is difficult to derive directly for a Mira in view of its cool and complex atmosphere. Incidentally, it is not clear that the Hipparcos catalogue has been properly searched yet for Miras with proper motion companions.

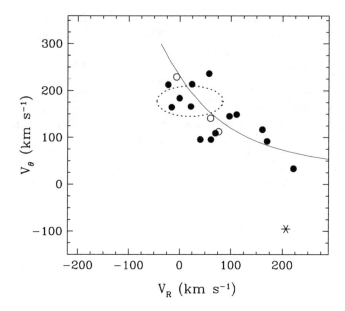

Figure 1. The correlation between V_θ and V_R for short period, blue-sequence, Miras. The solid line is from the simple model of Feast & Whitelock (2000a). The dotted oval shows the region occupied by the 'Hercules' stream (see Fux 2001 and Feast 2002). The asterisk is for S Car which is on a highly eccentric retrograde orbit.

The dependence of the mean motion of Miras in the direction of galactic rotation (V_θ) on period shows their importance for problems of galactic structure and dynamics. A quite unexpected result was found from the motion of Miras radially outwards from the galactic centre (V_R). Dividing the Miras into groups according to period, it was found that for Miras in the groups with mean periods of 228 days or more there is a small net outward motion of a few km s^{-1}. The significance of this is not at present very clear. However a rather startling result was found for the shortest period group (mean period 173 days) when the SP-red varables were omitted. These stars show a marked asymmetric drift ($V_\theta = 133 \pm 19$ km s^{-1}, compared with a circular velocity of 231 km s^{-1}, Feast & Whitelock 1997). Whilst this group is small (18 stars) it shows rather clearly a mean outward motion, V_R, of 75 ± 18 km s^{-1}. In addition the individual space motions show that there is a rather good correlation between V_θ and V_R in this group and all 11 stars which have an asymmetric drift greater than 65 km s^{-1} having positive values of V_R (see Figure 1 and Feast and Whitelock 2000a, Feast 2002).

Since many of these stars, all of which are within about 1 kpc of the Sun, are on highly eccentric orbits and will pass through, or close to, the Galactic Bulge, it seems likely that they are members of a bar-like distribution. It is known (Whitelock & Catchpole 1992) that the Miras in the Bulge itself belong to a central bar structure. A simple model would fit this suggestion for the local

Miras although the possibility that they are part of a stream due to an infalling satellite galaxy, cannot be ruled out (see Feast and Whitelock 2000a and Feast 2002 for details). It is evident that a detailed study of this phenomenon requires space motions for short period Miras over a wide region of the Galaxy. This will be possible with GAIA parallaxes, proper motions, radial velocities and periods. It remains to be seen whether auxilliary data will be require to separate out the SP-red Miras which have different kinematics.

Related to this problem is that of the Galactic Bulge itself where, as just mentioned, Miras show a bar-like distribution. The Bulge contains Miras of a wide range of period and GAIA should allow the kinematics and structure of the Bulge to be explored in detail as a function of period using positions and space motions.

3. The C-Miras and S-type Miras

An extensive discussion of the carbon-rich Miras is beyond the scope of this talk. However the parallaxes, proper motions, radial velocities, periods and spectra which GAIA will provide should answer a number of outstanding problems connected with these objects. The space motions of C-Miras are of importance in their own right and will show how closely these stars are related to the O-Miras of the same (or a different) period. Whilst it seems likely that C-Mira evolve from O-Miras this is not yet certainly established. If indeed this is their evolutionary path, it is still uncertain whether they change period in the process.

Space motions derived by GAIA will also be important for the understanding of the Miras with spectra of type S. These show strong zirconium oxide bands and lines of s-process elements are enhanced. These stars, which are oxygen-rich, are generally treated together with the O-Miras which have strong TiO bands. There is no clear dividing line between Miras with M type and those with S type spectra, the classes merging into one another (the MS stars). The extensive spectral and other data on O-Miras from GAIA may make it possible to divide them into groups which are more clearly defined and should lead to a much clearer understanding of this whole group of stars, their place in stellar evolution and their use in Galactic and extragalactic astronomy.

4. The low amplitude, semiregular variables

In this review a detailed discussion of the low-amplitude semiregular cool variables (the SR stars) has been omitted. However GAIA will almost certainly bring considerable order into this field which at present is in a rather unsatisfactory and confused state. In globular clusters, SR variables evolve with increasing period and luminosity (see e.g. Whitelock 1986, Feast 1989). They may change pulsation mode as they do so. In the LMC there are several sequences of semiregular, or low amplitude, variables in the period - luminosity plane, probably indicating several different pulsation modes (Wood 2000). The GAIA data will presumably reveal similar sequences in our own Galaxy (though infrared photometry of the relevant objects will probably be necessary to do this). With this done the kinematics, based on space motions, of the semiregulars divided according to particular period-luminosity sequence and to period, will help establish their evolutionary relationship. In view of the somewhat surprising results discussed above for the short period O-Miras, it seems very desirable that any kinematic analysis of these stars uses the full space and velocity co-ordinates. The spectra together with the photometry also opens up a largely unexplored area.

5. Deviations from the Mira Period-Luminosity relation

If our ideas based on the Magellanic Clouds, globular clusters and the Hipparcos results on local Miras, are correct, it can be anticipated that in any volume of space in the Galaxy most of the regular, large amplitude, cool variables measured by GAIA will be Miras on the PL relation (either O- or C- Miras). Together with infrared photometry GAIA will establish the slope of this relation and its zero-point. The parallaxes will also enable one to find Mira or Mira-like stars that lie off the PL relation. Some stars of this kind are expected and their luminosities, space motions and spectra will be of great interest.

The SP-red Miras which seem to lie above the main PL relation were mentioned in Section 2. But there are other types of Mira or Mira-like stars as well which are expected to lie off the PL relation. Towards the end of their AGB evolution, stars with initial masses in the range 4 to 6 solar masses can undergo Hot-Bottom-Burning (HBB). In this process the base of the H-rich convective envelope dips into the H-burning shell. The luminosity can then rise above that predicted by the classical relation between core-mass and luminosity. Carbon is burned to nitrogen and the beryllium transport mechanism results in an overabundance of lithium at the surface. Whitelock (2002, see also Whitelock & Feast 2000b and references there) has pointed out that most of the AGB variables in the Magellanic Clouds in which Smith et al. (1995) found lithium to be strong lie above the Miras PL relation. The frequency of occurrence of these stars and their kinematics are obviously of great interest since they gives clues to lifetimes and initial masses. Whitelock (2002) points out that the HBB phenomenon can explain a curious result found for LMC Miras. The evidence, which she summarizes, suggests that most LMC Miras with relatively thick dust shells and periods in the range 420 to 1300 days lie on an extension of the (bolometric) PL relation defined by shorter period Miras. These longer-period stars may be of sufficiently low mass that they never undergo HBB. However she points out that the few Miras with periods greater than 420 days and rather thin dust shells which seemed to indicate a break in the PL relation at that period in earlier work (Feast et al. 1989) may in fact be higher mass stars in a HBB phase. A full understanding of these phenomena is important not only for stellar evolution theory but also for the use of Miras as extragalactic distance indicators. For instance Whitelock (2002) suggests that the 641 day variable in IC1613 (Kurtev et al 2001) which lies well above the Mira PL may be in the HBB phase.

It has long been thought that the slow changes in period of some Miras (e.g. R Hya, see above) are due to the star undergoing helium shell flashing (thermal pulsing Wood & Zarro 1981). This is expected to be accompanied by changes in luminosity (e.g. Iben & Renzini 1983). Thus we may expect to find variables both above and below the PL relation due to this phenomenon. However the behaviour of R Hya is complex and the thermal pulse model has recently been challenged by Zijlstra et al. (2002) who suggest instead an envelope relaxation model.

Finally one should mention the possibility of finding dust-enshrouded OH/IR Miras below the PL relation. There is some evidence for such stars in the region of the Galactic Centre (e.g. Blommaert et al. 1998, Wood et al. 1998) but the bolometric luminosities of these stars are difficult to estimate accurately.

6. Red supergiant variables

Especially near the Galactic plane, GAIA will measure red supergiant variables. These objects show bolometric or infrared period-luminosity relations in, for instance, the Magellanic Clouds and M33 (Kinman, Mould & Wood 1987, Mould et al. 1990, Feast 1992). More recently, and of more direct relevance for GAIA, period-luminosity relations in the I-band have been established in Per OB1, LMC, M33 and M101 (Pierce, Jurcevic & Crabtree 2000, Jurcevic, Pierce & Jacoby 2000). In the I-band these red supergiant variables are about three magnitudes brighter than the classical Miras at a given period. The range in periods in Per OB1 suggests that the PL relation there is an evolutionary sequence in contrast to the Mira PL relation which is a mass/metallicity sequence, as discussed above. The PL relation established by Pierce et al. has a considerable r.m.s. scatter (0.42 mag). This may be partly observational. It remains to see whether this PL relation is influenced by inital mass, age, or metallicity . These stars should be detected by GAIA to large distances even in the galactic plane and should be excellent tracers of the distribution and kinematics of young objects. The radial velocity component will be vital to study the kinematics properly.

7. Some special challenges

There will be a number of special challenges to be met in the interpretation of GAIA observations of Mira variables. One challenge is of course that we are moving into a little explored area. For instance, there are few, if any Miras that have been studied systematically round their light cycles for spectroscopic and radial velocity variations in the spectral region of interest to GAIA. The complex structure of Mira atmospheres and its variation with phase is not yet properly understood. References to high resolutions studies and their interpretation, from the early work of Merrill to the present are conveniently summarized by Alvarez et al. (2001); see also the summary of Lebzelter and Hinkle (2002). In the optical and infrared regions doubling of absorption lines is seen. Typically the lines are split by $\sim 20\,\mathrm{km\,s^{-1}}$ in the optical region. At resolutions too low to show the splitting clearly, the variation in measured absorption line velocity round the cycle is probably $\sim 10\,\mathrm{km\,s^{-1}}$ or less, though there have been no studies in the optical region as extensive as those of Joy (1954) on Mira Ceti itself. There are variations of absorption line velocities with excitation potential, presumably due to the different depths of formation of the lines. The line doubling is thought to be due to shock waves in the atmosphere. These are also believed to excite the emission lines seen at some phases. In the past there has been some uncertainty about the definition of the actual radial velocity of a Mira in space. This is now generally taken to be the mean of the velocities of the two OH maser peaks which are found in some Miras. From a large body of data, generally involving only a few optical velocities for any given Mira, it is found that the mean absorption line velocity is too positive by $4\,\mathrm{km\,s^{-1}}$ with some dependence on period (Feast and Whitelock 2000a). This offset may be due to most of the optical velocities having been obtained in the brighter half of the light cycle. Perhaps of more relevance, so far as GAIA spectroscopy is concerned, is the fact that some Miras, at least, show emission in the CaII infrared lines near maximum light. At least in some cases these lines have inverse P-Cygni profiles with velocity separations of $20-30\,\mathrm{km\,s^{-1}}$ (Merrill 1934, 1960). Clearly, this must be taken into account in deriving mean velocities and space motions using these lines. The systematic study of the CaII lines round the light cycles of many Miras will be very revealing of the complex dynamics of Mira atmospheres.

A challenge, which affects particularly the astrometry, is the large angular size of the Miras. The diameters of Miras are a strong function of wavelength (e.g. Labeyrie et al. 1977). In a broad optical band, which is probably relevant to GAIA, the Mira R Leo and the Mira-like (SRa) star W Hya have mean angular diameters of 74 and 84 mas. These stars do not have circular symmetry and show evidence for asymmetric light distributions over the discs, possibly due to large star-spots (Lattanzi et al. 1997 and references there). The diameters of these stars are greater than the expected point spead function of GAIA. They are also nearly a factor of ten greater than their Hipparcos parallaxes (or the parallaxes derived from the $PL(K)$ relation of Feast et al. 2002). Thus the diameters of Miras are always larger than their parallaxes and astrometry may be affected by motions of the photocentre of the star due to changing shape or light distribution. Such effects might be incorrectly interpreted as due to motion in a binary. Whilst this is a serious concern, it should be noted that one of the fastest growing areas in astronomy at the present time is stellar interferometry. One may hope that by the time GAIA results are available, interferometry will have led to a good understanding of the shapes and surface structure of Miras and, in particular, the time scale on which these change, something on which we have little information at present.

8. Conclusions

The combination of GAIA astrometry, radial velocities, spectra, photometry and periods will have a profound and unique effect on our understanding of the nature and evolution of cool variable stars, particularly those on the AGB. Also the kinematics of these objects will allow us to study the structure and evolution of Our Galaxy in a way previously impossible. Hipparcos already led to surprises in this area and gave some foretaste of what GAIA will accomplish.

It would be rather valuable if ground-based observers gave some thought now to parallel programmes which would be completed by the time the results from GAIA become available. In the optical/infrared field some obvious programmes in this category are; interferometry, intensive infrared photometry and high resolution optical spectroscopy.

Acknowledgments. I would like to thank Patricia Whitelock for discussions and advice.

References

Alvarez, R. et al. 2001, A&A 379, 305

Blommaert, J.A.D.L., van der Veen, W.E.C.J., van Langevelde, H.J., Habing, H.J. & Sjouwerman, L.O. 1998, A&A 329, 991

Eggen, O.J. 1985, AJ 90, 333

Eyer, L., & Cuypers, J. 2000 in The Impact of Large-Scale Surveys on Pulsating Star Research, IAU Coll. 176, L. Szabados & D.W. Kurtz ed.s, ASP Conf. Ser. 203, pag. 71

Feast, M.W. 1963, MNRAS 125, 367

Feast, M.W. 1989, in The Use of Pulsating Stars in Fundamental Problems of Astronomy, IAU Coll. 111, E.G. Schmidt ed., Cambridge Univ. Press, pag. 205

Feast, M.W. 1992, Kon. Ned. Akad. Weten. Eerste Reeks, deel 36, 18

Feast, M.W. 2002, in Mass-losing Pulsating Stars and their Circumstellar Matter, Y. Nakada & M.Honma ed.s, Kluwer, in press (astro-ph/0207194)

Feast, M.W., Glass, I.S, Whitelock, P.A., & Catchpole, R.M. 1989, MNRAS 241, 375

Feast, M.W. & Whitelock, P.A. 1997, MNRAS 291, 683

Feast, M.W. & Whitelock, P.A. 1999, in Post-Hipparcos Standard Candles, A. Heck & F. Caputo ed., Kluwer, pag. 75

Feast, M.W. & Whitelock, P.A. 2000a, MNRAS 317, 460

Feast, M.W. & Whitelock, P.A. 2000b in The Evolution of the Milky Way, F.Matteucci & F. Giovannelli ed.s, Kluwer, pag. 229

Feast, M.W., Whitelock, P.A. & Menzies, J.W. 2002, MNRAS 329, L7

Fux, R. 2001, A&A 373, 511

Iben, I.Jr. & Renzini, A. 1983, ARA&A 21, 271

Joy, A.H. 1954, ApJS 1, 39

Jurcevic, J.S., Pierce, M.J. & Jacoby, G.H. 2000, MNRAS 313, 868

Kinman, T.D., Mould, J.R. & Wood, P.R. 1987, AJ 93, 833

Kurtev, R. et al. 2001, A&A 378, 449

Labeyrie, A. et al. 1977, ApJL 218, L75

Lattanzi, M.G., Munari, U., Whitelock, P.A. & Feast, M.W. 1997, ApJ 485, 328

Lebzelter, T. & Hinkle, K.H. 2002, in Radial and Nonradial Pulsations as Probes of Stellar Physics, C. Aerts, T.R. Bedding & J. Christensen-Dalsgaard ed.s, ASP Conf. Ser. 259, pag. 556

Merrill, P.W. 1934, ApJ 79, 183

Merrill, P.W. 1960, in Stellar Atmospheres, J.L. Greenstein ed., The Univ. of Chicago Press, pag. 509

Mould, J.R. et al. 1990, ApJ 349, 503

Nishida, S. et al. 2000, MNRAS 313, 136

Olivier, E.A., Whitelock, P.A. & Marang, F. 2001, MNRAS 326, 490

Pierce, M.J., Jurcevic, J.S. & Crabtree, D. 2000, MNRAS 313, 271

Smith, V.V., Plez, B., Lambert, D.L. & Lubowich, D.A. 1995, ApJ 441, 735
Whitelock, P.A. 1986, MNRAS 219, 525
Whitelock, P.A. 2002, in Mass-losing Pulsating Stars and their Circumstellar Matter, Y. Nakada & M. Honma ed.s, Kluwer, in press (astro-ph/0207168)
Whitelock, P.A. et al. 1994, MNRAS 267, 711
Whitelock, P.A. & Catchpole, R.M. 1992, in The Center, Bulge and Disk of the Milky Way, L. Blitz ed., Kluwer, pag. 103
Whitelock, P.A. & Feast, M.W. 2000a, MNRAS 319, 759
Whitelock, P.A. & Feast, M.W. 2000b, Mem. Soc. It. Ast. 71, 601
Whitelock, P.A., Marang, F. & Feast, M.W., 2000, MNRAS 319, 728
Wood, P.R. 1995, in Astrophysical Applications of Stellar Pulsation, R.S. Stobie & P.A. Whitelock ed.s, ASP Conf. Ser 83, pag. 127
Wood, P.R. 2000, Pub. Ast. Soc. Aust. 17, 18
Wood, P.R., Habing, H.J. & McGregor, P.J. 1998, A&A 336, 925
Wood, P.R. & Zarro, D.M. 1981, ApJ 247, 247
Zijlstra, A.A., Bedding, T.R. & Mattei, J.A., 2002, MNRAS 334, 498

Observing RV Tauri and SRd variables with GAIA

Glenn M. Wahlgren

Atomic Astrophysics, Lund Observatory
Box 43, SE-22100, Lund, Sweden

Abstract. The RV Tauri and SRd stars are luminous, pulsational variables of spectra types F-K. Due to their scarcity and location in the H-R diagram, the RV Tauri stars are considered to be in the post-AGB phase of stellar evolution, while the SRd class is considered to be less homogeneous in its membership. Our interpretation of these classes in terms of the evolutionary state of their members and any relationship between the classes is not well understood. The luminosity of their membership is of critical importance to both the understanding of the observations and theoretical modeling, yet remains an elusive quantity for galactic field stars. The GAIA mission represents the most comprehensive approach to data collection, involving astrometry, photometry, and spectroscopy, and it is anticipated that GAIA will have a tremendous impact upon our understanding of these types of stars.

1. Introduction

The GAIA mission will be particularly well suited to the study of variable stars and their contributions to building an understanding of the structure and evolution of our Galaxy. Luminous variable stars offer a means of determining distances, while information of their metallicity and orbital characteristics provide data on the temporal development of the Milky Way. In this light the Cepheids and Mira variables have been particularly useful, and the relevance to them of the GAIA mission has been addresses by Bono (2003), Feast (2003) and Szabados (2003). Often overlooked in this context are the less regular pulsators of the RV Tauri and semi-regular (SRd) classes. Their relatively high luminosity and distinct lightcurves allow them to be recognized to great distances. Unfortunately, this usefulness is offset by their relative scarcity and an elusiveness to be, as yet, defined in terms of a particular phase of stellar evolution or even stellar mass. The discussion which follows is presented in the spirit of how the capabilities of the GAIA satellite can be exploited to better understand the RV Tauri and SRd classes of variable stars. More thorough background information on the nature of these stars from different observational and theoretical perspectives has been presented by, for example, Fokin (2001), Pollard et al. (1996, 1997), Wahlgren (1993), and Wallerstein & Cox (1984).

1.1. Classification considerations

In various renditions of the H-R diagram the RV Tauri stars are typically found at high luminosity, nestled between the Cepheid instability strip and the Mira variables. In reality, their extent is far greater in luminosity. Their canonical characteristics, as listed in the General Catalogue of Variable Stars (GCVS, Kholopov 1985), are that their lightcurves have periods between 30 and 150 days and typically display alternating deep and shallow minima with amplitudes of up to 4 magnitudes. The mean light level can be constant (subclass RVa) or display a secondary period of large amplitude (2 mag) and long duration (hundreds of days) (subclass RVb). Spectroscopic subclasses have also been developed (Preston et al. 1963), based on low-dispersion spectroscopy. The SRd variables are also defined as luminous stars of spectral type F, G, or K. Their lightcurves appear to span a longer period range (30 – 1100 days) but also display amplitudes up to 4 magnitudes. In effect, the only criteria that have been used to distinguish between the two classes is that the lightcurves of the SRd variables may be more irregular and may lack the characteristic alternating minima of the RV Tauri stars. In addition, radial velocities of the RV Tauri stars, as a class, tend to be higher than those of the SRd stars (Rosino 1951), and the spectra of the SRd stars present more persistent (Rosino 1951) or brighter (Preston et al. 1963) hydrogen emission lines at lower spectral resolutions.

The RV Tauri and SRd classes are typically discussed together as a result of commonalities in their definitions. The great majority of RV Tauri stars were classified before the work of Preston et al. (1963) and did not benefit from any spectroscopic criteria. Stars of the SRd class were first listed separately in the second edition of the GCVS (Kukarkin et al. 1958) but their only discriminating feature among the semi-regulars was that they are of spectral types F, G, or K. As a result of the loose classification criteria both classes comprise a mixture of members with different stellar evolutionary stage or mass. The SRd class is recognized to be diverse, comprising high mass luminous stars and less luminous giants and weak-lined stars.

Many of the faintest variables have no spectroscopic information, along with poor lightcurve sampling, and require further data to confirm their classifications. Figure 1 presents the number of variables as a function of mean apparent photographic magnitude, using data taken from the GCVS (Kholopov 1985). The peaks of the distributions of the two classes are separated by 2.5 magnitudes. The few faintest RV Tauri stars are all found in the constellation Sagittarius and are no doubt heavily reddened. For only the brighter half of the RV Tauri distribution do stars have spectral information (photometric or spectroscopic). The GAIA radial velocity spectrometer (RVS) will be able to provide data for nearly all of the variables of Figure 1, which will allow for the discovery of errant members.

1.2. Evolutionary interpretation

As a result of their location in the H-R diagram and scarcity in number, the RV Tauri stars have been regarded as being in the post-AGB phase of stellar evolution. Support for this picture comes from stellar evolution calculations (cf. Gingold 1974) that show low mass (0.6 $M_\odot$) stars to evolve to high luminosity and sketch out blue loops in the luminosity-$T_{\rm eff}$ diagram during a period of

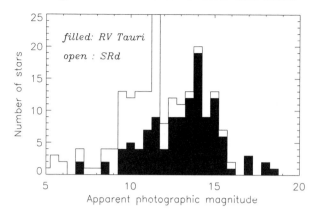

Figure 1. The number of variable stars are plotted as a function of the average value of the minimum and maximum apparent photographic magnitudes from the GCVS. For those relatively few stars with visual magnitude data a correction factor of 0.9 mag has been added to place them on the photographic magnitude scale.

helium shell flashes. The short duration of these loops, on the order of thousands of years, accounts for the few variables found in this region of the H-R diagram. This explanation is most reasonable for the RV Tauri stars found to reside in globular clusters as well as the field stars displaying galactic Halo characteristics. But whether this explanation can be extended to the class as a whole is doubtful in light of the apparent heterogeneity of the class members. A concerted effort needs to be undertaken for the catalogued variables to sift through the lightcurve data and acquire spectroscopic information in order to understand the different types of stars that comprise the RV Tauri and SRd classes.

Any evolutionary interpretation requires reliable estimates of luminosity. Due to their great distances, luminosities of the variable stars among the galactic field have been determined spectroscopically. Trigonometric parallaxes have been attempted for these stars by the Hipparcos astrometric satellite. Of the 122 RV Tauri (78 SRd) variables listed in the GCVS only 11 (22) are present in the Hipparcos catalogue and of these only 4 (12) were measured to have a parallax as large as its uncertainty, in essence providing no reliable distances.

Early period-luminosity (PL) relations for RV Tauri stars have been shown to be problematic (Wahlgren 1992, 1993). More recently, RV Tauri stars were detected in the Large Magellanic Cloud (Alcock et al. 1998) and a PL relation was developed. However, there is a strong selection effect built into this PL relation: the observations were limited in period and magnitude. The derived PL relation is based on stars that are comparable to the most luminous galactic RV Tauri stars of shorter period. Future observations of the LMC should be able to sample to fainter magnitudes and longer periods, possibly detecting additional RV Tauri/SRd like stars. Other deep surveys of regions within our galaxy (Derue et al. 2002) and in other galaxies (Antonello et al. 2002) are detecting variable stars with the characteristics of semi-regulars, and can be taken as an indication that many more of these variables are to be discovered in the Milky Way by GAIA.

In recent years a significant improvement in understanding, for at the least a subgroup of these variables, has resulted from extensive elemental abundance analyses. In two series of papers, samples of bright RV Tauri (cf. Giridhar et al. 2000a) and SRd (cf. Giridhar et al. 2000b) variables were observed with high spectral resolution. The elemental abundance patterns observed in the RV Tauri stars present convincing evidence that deficiences for most elements are not solely the result of galactic age, as was generally believed, but are also influenced by atmospheric or circumstellar chemistry. Abundance deficiencies were found to be correlated with an element's condensation temperature (T_{cond}) in a manner such that the elements of highest T_{cond} are the most deficient. These would be the first elements to condense onto dust grains in the circumstellar environment and produce deficiencies in abundance, as detected through atomic lines in the stellar photosphere. Neutral lines of iron-group elements, of which there are many in the GAIA spectroscopic region, are associated with T_{cond} that can lead to deriving moderate deficiencies of metallicity if this process is at work. Elements having low T_{cond}, such as carbon, sulfur and zinc, do not as readily condense out of the atmosphere and are therefore a better indicator of the initial stellar composition and population type. The more metal deficient, [Fe/H] $\leq$ –1.0, RV Tauri stars and the SRd stars in general do not display the T_{cond} dependent abundance deficiencies. The large range of luminosity displayed by the RV Tauri stars implies that they could have evolved directly from the red-giant branch or asymptotic giant branch (AGB). AGB stars that have undergone the third dredge-up phase might be expected to show enhancements of s-process elements, yet these were not detected, thereby implying that these RV Tauri stars do not fit the classic picture of evolution from thermal-pulse AGB stars. A cautionary remark for the extrapolation of this work to the class as a whole is that these results are based upon a small number of targets, selected for analysis according to the criteria that they display weak spectral lines, indicative of a metal-poor stellar population. To the extent that these stars represent the classes in general is still an open question since the majority of the variables are found at fainter magnitudes and both classes comprise members displaying unique characteristics.

2. GAIA's approach to the RV Tauri and SRd variables

The great attractiveness of the GAIA mission, in the context of these variable stars, is that multiple visits will be carried out to collect data via astrometry, photometry, and spectroscopy in near simultaneity. Ground-based observing programs typically involve only one of these domains, necessitating the assumption that observed characteristics are to a great extent repeated from one pulsational cycle to the next. This assumption can not be supported for the more irregular pulsators and those cooler RV Tauri stars for which TiO absorption bands appear at unpredictable strength at phases of low flux.

The approach taken to the analysis of GAIA observations will fall under one of two broad categories: (*i*) the study of previously catalogued variables, and (*ii*) the identification of new candidates. The first catagory will provide information which when considered with previous and perhaps concurrent observations will address topics specific to a star, such as its atmospheric dynamics and evolutionary phase, and the more class-oriented topics, including the creation of a PL relation and any discussion regarding the relationship between the two classes. The latter category will eventually improve the statistical foundations of areas such as galactic distribution and associated populations and the variables' place in the scheme of galactic chemical evolution.

The expected sampling frequency of photometric and spectroscopic data is anticipated to be approximately 20 visits per year, although not equally spaced in time, for a total of 100 photometric data points (each band) over the lifetime of the mission. These parameters are similar to those of the Hipparcos mission, from which we can draw parallels from the lightcurve data of the few RV Tauri and SRd stars observed.

- A variable mean level for the lightcurve can be identified (ex: U Mon)

- From photometric data collected over the lifetime of the mission multiple consecutive pulsational periods can be folded together to produce a lightcurve that displays the alternating pattern of the minima for variables having a constant mean lightlevel.

- Fourier analysis of the photometric data for SRd variables can be used to study the rate of change of the period, which is related to stellar structure and evolution (Percy & Kolin 2000).

The photometric lightcurves of these variables are not unique to their classes, especially at the GAIA sampling rate, and spectral information will be required to recognize legitimate RV Tauri and SRd stars from among the potentially many new candidates. The onboard photometric system will assist in this endeavor, especially for faint targets, but may not be sufficient depending upon how the system is ultimately defined. Detection of these variables would benefit from filterbands that sample the flux for possible hydrogen Balmer Hα line emission and at a wavelength close near 1 μm to identify circumstellar dust emission that is associated with some RV Tauri stars.

The spectral characteristics associated with the RV Tauri and SRd variables in the proposed GAIA wavelength interval, 8480–8740 Å, are: the presence and behaviour of emission lines from iron-group elements (Ti, Fe), distortions of the CaII triplet line profiles from those of non-variable giants of similar spectral type, and abundances for elements of different $T_{\rm cond}$. The presence and behaviour of the emission lines (periodic variability in strength and wavelength), along with distortions of the CaII lines by emission, are linked to the passage of shock fronts through the stellar atmosphere. Figure 2 presents sample spectra for part of the GAIA wavelength region, obtained with the Nordic Optical Telescope (NOT) SOFIN spectrometer at a resolving power of R = 23 000. This value of R is comparable to the upper limit being considered for spectroscopy from GAIA's radial velocity spectrometer. At resolving powers as low as R = 10 000 the distortions of the CaII profiles are lost and the emission lines shown in Figure 2 may rather be interpreted as high points of the continuum flux.

The influence of resolution on the ability to extract elemental abundances can be surmised from Figure 3. In this figure the observed spectrum of U Mon

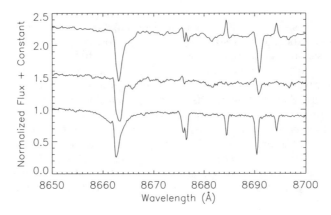

Figure 2. A segment of the GAIA spectral interval for variables R Sct (RVa, *top*), AC Her (RVa, *center*), and WY And (SRd, *bottom*). The spectra show distortions of the CaII 8662 Å line and emission from TiII and FeII.

is compared, for the resolving powers R = 23 000 and 10 000, with synthetic spectra computed for the cases of solar abundance levels and enhancements of the rare earths by two orders of magnitude. One notices that many of the observed features are not present in the synthetic spectra as a result of missing line opacity from both atomic and molecular lines. At the higher spectral resolution individual lines from non-iron group elements can be identified and analysed, while at the lower resolution the line blending will effectively limit the analysis to the most prominent spectral lines.

3. Summary - the impact of GAIA

The capabilities of the GAIA satellite in terms of astrometry, photometry and spectroscopy will contribute enormously to our understanding of the RV Tauri and SRd variable stars. Presented below is a charting of some of the more obvious applications of the expected GAIA mission. Anticipated, but not listed, are the spinoffs that inevitably result from large surveys of data and the benefits that will be realized from coordinated observations with ground or space observatories utilizing other regions of the electromagnetic spectrum. At the time of this writing the final configurations for the photometric system and spectral resolution have not been specified, as evidenced by several annotations.

```
Astrometry:

    parallaxes, with colors for de-reddening => luminosity, distance
              , with photometric periods      => PL, P-Amp relations
              , and appropriate red filter    => PL-C relation

    plus proper motion and radial velocity => space velocity, orbit
                                              and population type
```

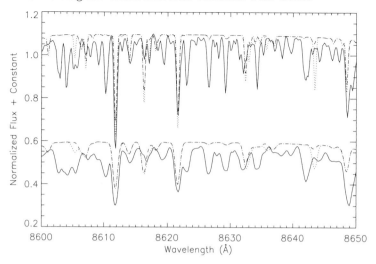

Figure 3. The observed spectrum of U Mon (RVB) at R = 23 000 (*upper*, solid) and R = 10 000 (*lower*, solid) are compared with synthetic spectra computed with solar abundances (dashed) and a 2 dex enhancement of the lanthanide elements (dotted).

```
Photometry:

  magnitudes  => lightcurve period, amplitude, and shape
                 (impacts detection and classification of new targets,
                 reclassification of previously catalogued variables)
              => indicator of CS reddening (with 1-micron filter)

  colors      => spectral class indicator for faint targets

Spectroscopy: spectral resolution dependent

  low R (10000)  => spectral type, variable subtype classification
                 => limited chemical abundance analyses due to
                    line blending (ex: Ca, Ti, Fe)

  high R (20000) => spectral type, variable subtype classification
                 => more extensive abundance analyses
                 => line profiles telling of atmospheric dynamics
                    via distortions of CaII and Fe-group emission
                 => radial velocity for pulsation and binarity
```

References

Alcock, C., et al. 1998, AJ 115, 1921

Antonello, E. et al. 2002, A&A 386, 860

Bono, G. 2003, in GAIA Spectroscopy, Science and Technology, U.Munari ed., ASP Conf. Ser. 298, pag. 245

Derue, F., et al. 2002, A&A 389, 149

Feast, M. 2003, in GAIA Spectroscopy, Science and Technology, U.Munari ed., ASP Conf. Ser. 298, pag. 257

Fokin, A. B. 2001, in Stellar Pulsation - Nonlinear Studies, M. Takeuti & D. D. Sasselov ed.s, Kluwer Academic Pub., pag. 103

Gingold, R.A. 1974, ApJ 193, 177

Giridhar, S., Lambert, D.L., & Gonzalez, G. 2000a, ApJ 531, 521

Giridhar, S., Lambert, D.L., & Gonzalez, G. 2000b, PASP 112, 1559

Kholopov, P.N. 1985, General Catalogue of Variable Stars, 4th ed., Moscow, Nauka Publishing House

Kukarkin, B.V., Parenago, P. P., Eframov, Yu., J., & Kholopov, P. N. 1958, General Catalogue of Variable Stars, 2nd ed., Moscow, Academy of Sciences Publishing House

Percy, J.R., & Kolin, D. L. 2000, J. AAVSO 28, 1

Pollard, K.R., Cottrell, P. L., Kilmartin, P.M., & Gilmore, A.C. 1996, MNRAS 279, 949

Pollard, K.R., Cottrell, P. L., Lawson, W. A., Lbrow, D.M., & Tobin, W. 1997, MNRAS 286, 1

Preston, G.W., Krzeminski, W., Smak, J., & Williams, J.A. 1963, ApJ 137, 401

Rosino, L. 1951, ApJ 113, 60

Szabados, L. 2003, in GAIA Spectroscopy, Science and Technology, U.Munari ed., ASP Conf. Ser. 298, pag. 237

Wahlgren, G.M. 1992, AJ 104, 1174

Wahlgren, G.M. 1993, in Luminous High-Latitude Stars, D. Sasselov ed., ASP Conf. Ser. 45, pag. 270

Wallerstein, G., & Cox, A.N. 1984, PASP 96, 677

The accuracy of GAIA radial velocities

Ulisse Munari

Astronomical Observatory of Padova – INAF, Asiago Station, 36012 Asiago (VI), Italy

Tomaž Zwitter

University of Ljubljana, Dept. of Physics, Jadranska 19, 1000 Ljubljana, Slovenia

David Katz

Observatoire de Paris, GEPI, 5 place Jules Janssen, 92195 Meudon, France

Mark Cropper

Mullard Space Science Laboratory, University College London, Holmbury St Mary, Dorking, Surrey RH5 6NT, United Kingdom

Abstract. The accuracy of GAIA radial velocities is explored as a function of spectral resolution (17 200, 8 600, 4 300 and 2 150) and target magnitude, on the base of observations conducted with ground based telescopes as well as via simulations on grids of synthetic spectra. The higher resolutions better performs all the way down at least to V=17.5 mag for mission-averaged radial velocities, both for MMS and Astrium designs. Only the 17 200 resolution however allows the radial velocities to match the better than 0.5 km s^{-1} accuracy of tangential motion for the 40 million best astrometric targets.

1. Introduction

The main reason to place a spectrograph on GAIA is to provide the 6th component of the phase-space, i.e. the radial velocity. For some thousands nearby and high-velocity stars, radial velocities will also be required to account for perspective acceleration.

To properly support kinematic findings from tangential motions, radial velocities should obviously parallel the precision of the latter.

Tangential motions of about 40 million objects will be determined by GAIA with an error less than 0.5 km s^{-1} (cf. ESA-SCI2000-4). This require a high precision of radial velocities at the bright end of the magnitude range of interest, i.e. a high enough spectral resolution. It could be argued that high resolution spectra could be obtained from the ground for these 40 million objects (for ex. by the RAVE project currently under planning). Even if such a task will ever be carried out successfully, however it will be able to observe each star only a

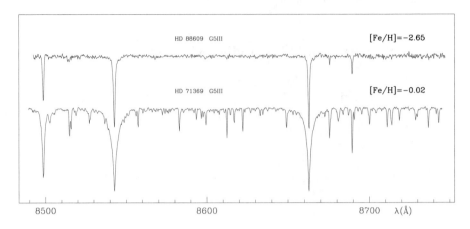

Figure 1. Example of Disk and Halo G5 giants.

very few times, at most. But obtaining a precision of 0.5 km s^{-1} for *barycenter* radial velocities requires a very careful handling of binarity, atmospheric activity, pulsations, spots on rotating surfaces, and many other assorted disturbances. This means a lot of spectra well distributed over a long period of time (it is worth notice that the orbital velocity of two solar type stars in a 5 yr period binary is 11 km s^{-1}). Only a multi-epoch survey like the one performed by GAIA will have the time coverage and the number of spectra to properly account for these and others disturbing effects. How delicate and time consuming is to obtain *barycenter* radial velocities accurate to 0.5 km s^{-1} or better, is well illustrated by the IAU standard radial velocity stars. In spite of the efforts with high resolution spectrographs over the last decades, the uncertainty on their velocities is still 0.3 km s^{-1} on the average, as detailed in Table 1 (from data of the IAU Commission 30 as summarized in *The Astronomical Almanac* for 2003).

On the other end, going the faintest possible is great for galactic population studies. For such investigations, a precision of 10 km s^{-1} at $V=17.5$ mag is considered enough for both the Bulge and the Halo, as emerged in this conference and earlier at various meetings of the GAIA RVS working group (RVS-CoCo-002, -003 and -004 available from *LiveLink*).

Being accurate at the bright end and going as faint as science requires sounds like conflicting requirements. Conflicting are also the types of corresponding targets to observe. The stars making up the astrometric 0.5 km s^{-1} club will be dominated by nearby metal rich stars, while the Halo targets used in tracing the galactic formation history will mainly score metal poor giants. The first type posses wide-wings CaII lines and hundreds of lines in between carrying a great deal of radial velocity information. The second type show instead a useless flat continuum and just three narrow CaII lines (see Figure 1 for examples).

One could think that lowering the spectral dispersion to the aim of increasing the S/N could turn into better radial velocities at the faint magnitudes. Such a reasoning may be valid in photometry, where all photons counts the same way. However, in radial velocity business, the radial velocity information is carried *only* by the lines and not by a flat continuum, whatever high its S/N

Table 1. Distribution in radial velocity accuracy of IAU radial velocity standards.

accuracy (km s^{-1})	<0.1	0.1	0.2	0.3	0.4	0.5	0.6
N. of IAU stand.	4	15	13	12	6	15	6

might be. Thus, increasing the accuracy of radial velocities requires increasing the spectral resolution (as Figure 2 well illustrates). Obviously, photons in the continuum are required to see lines on it, and for a fixed exposure-time, slit-less and spinning satellite like GAIA this means that a higher resolution stops to pay dividends at some certain faint magnitude. To determine how faint is the balance magnitude is the aim of this paper.

On basic principles, it can be anticipated that the balance magnitude will be different for Halo, Bulge and Disk stars given the different density and types of absorption lines in their spectra (cf. Figure 1). However, this has not been explored in proper details by existing investigations which concentrated on solar abundance stars, leaving it for future work. The effect of metallicity and gravity on the accuracy of radial velocities as function of the spectral resolution needs to be carefully considered because, should large differences arise, the relative strength of the GAIA science case for the Halo and for the Disk would then guide the choice.

Observations and simulations supplement each other to provide a comprehensive understanding of the feasibility and potential of GAIA radial velocities: real observations deal better with systematics and cosmic scatter of targets at the brightest end of the magnitude range, while simulations are mandatory at the lowest S/N ratios and for the exploration of individual factors isolated from the rest. However, a great deal of work still needs to be done in several fields, including: (*a*) accuracy of the wavelength calibration, with simulations so far assuming error-free calibration; (*b*) accounting in the simulations for cosmic scatter, binarity, chromospheric activity, pulsation, etc., particularly at the faint end of the magnitude range; (*c*) extending the simulations to the metallicities typical for Halo objects; (*d*) run the simulations also for the bi-dimensional whole-field-of-view solution case; (*e*) from observed spectra, testing of the mission-end accuracy achievable on stars with active and rotating photospheres (i.e. non periodic phenomena affecting only part of the lines); (*f*) from observed spectra, testing the accuracy of barycenter velocities for various type of pulsating stars and binary stars (i.e. periodic phenomena affecting all lines).

The content of this paper is entirely based on the classical star-by-star extraction, calibration and cross-correlation approach. However, regions of the sky with the highest stellar densities could produce such a large overlapping of the spectra to spoil the star-by-star approach. In this case it could be necessary to cross-correlate at the same time the whole, bi-dimensional focal plane, using astrometry and photometry to build up the reference bi-dimensional frame. First

Table 2. $V - I_C$ index of representative GAIA targets.

lum.class	F5	G0	G5	K0	K5	M2
I	0.40	0.63	0.74	1.10	1.39	2.38
III	0.44	0.66	0.89	0.93	1.59	1.91
V	0.51	0.69	0.72	0.83	1.33	2.05

exploratory attempts in this direction are going to be undertaken (Zwitter and Munari, in preparation).

2. S/N of GAIA spectra

The S/N per pixel of the extracted mono-dimensional spectrum of a star of magnitude I_C for a single passage over the GAIA spectrograph focal plane can be expressed as:

$$\frac{S}{N} = \frac{519 \cdot 10^{-0.4 I_C} \zeta A D t}{\sqrt{519 \cdot 10^{-0.4 I_C} \zeta A D t + n A \zeta \Phi t (Z + B) + n t K + n N R^2}} \quad (1)$$

where 519 is the number of ph. cm^2 s^{-1} Å^{-1} from a I_C=0.0 mag star (the Cousins I_C being the band closest to the central wavelength of the GAIA range), ζ is the overall telescope + spectrograph + CCD efficiency, A is the mirror collecting area (in cm^2), D is the spectral dispersion (in Å pix^{-1}), n is the spectrum width perpendicular to dispersion (in pixel per single CCD, including the drift induced by precession of satellite spin axis if not accounted for by a de-rotating mechanism), Φ is the area on the sky of one pixel (in arcsec^2), Z is the zodiacal light (in ph. s^{-1} arcsec^{-2}. Its surface brightness given in the ESA-2000-SCI-4 is I_C=21.8 mag arcsec^{-2}, assuming Sun's color for the zodiacal light. For values as function of the ecliptic coordinates see Zwitter 2002), B is the continuum background light caused by overlapping stars (in ph. s^{-1} arcsec^{-2}), K is the dark current (in e^- s^{-1} pix^{-1}), N is the number of CCDs crossed by the spectrum during a passage over the focal plane, R is the read-out noise (in e^- pix^{-1}) of a single CCD, and t is the crossing time of the focal plane (in sec). Values for two instrumental configurations are:

	ζ	A	n	Φ	N	K	R	t
MMS original	0.35	5250	3	1.0	2	∼0	3	30.2
Astrium 6CCDs	0.35	2500	3	1.5	6	∼0	3	16.5

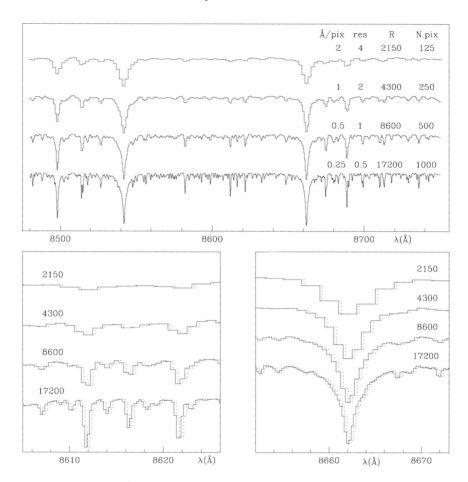

Figure 2. Resolution effects on cross-correlation. The top panel shows spectra the same K0 III star at four different resolution. The lower two panels represent zoomed parts of the top one. The added dotted lines show a 10 km s^{-1} shift between object and template.

3. Accuracy from real observation of single stars

The most comprehensive observational works to date on the accuracy of GAIA radial radial velocities for *single* stars is that by Munari et al. (2001a, hereafter MAT01). Munari et al. (2001b) is instead the first paper of a series devoted to binary stars (a second paper in the series is in preparation by Zwitter at al.).

MAT01 selected 24 IAU standard radial velocity stars with spectral types uniformly distributed from F7 to M2, rotational velocities typical for field stars at these spectral types ($V_{\rm rot} < 10$ km s^{-1}) and metallicities close to solar (<[Fe/H]>=−0.06).

Table 3. Accuracy of GAIA radial velocities (ϵ, in km s^{-1}) derived from real spectra secured at S/N=12, 33, 110 and resolving powers 17 200, 8 600, 4 300, 2 150 (cf. Munari et al. 2001a). The conversion magnitudes (computed with the help of Eq.1) refer to an average GAIA target (a K0 star), suffering from a modest reddening (E_{B-V}=0.1), and the Astrium-$6CCDs$ configuration. They are given for single passage (^{e}V) as well as mission averaged (^{m}V).

Å/pix	resol.	S/N=12			S/N=33			S/N=110		
		^{e}V	^{m}V	ϵ	^{e}V	^{m}V	ϵ	^{e}V	^{m}V	ϵ
0.25	17200	12.5	15.4	1.1	10.8	14.2	0.6	8.4	12.6	0.4
0.5	8600	13.2	16.2	3.1	11.6	15.0	1.6	9.2	13.4	1.1
1	4300	14.0	16.9	11	12.3	15.7	5.6	9.9	14.1	4.5
2	2150	14.8	17.6	40	13.1	16.4	18	10.6	14.9	14

MAT01 explored four dispersions for GAIA spectra: 0.25, 0.5, 1.0 and 2.0 Å pix^{-1}. The spectrographs of the Asiago observatory were set up to deliver exactly a FWHM=2.0 pixels for the instrumental PSF. Therefore, the explored resolutions were 0.5, 1, 2 and 4 Å equivalent to resolving powers 17 200, 8 600, 4 300 and 2 150. A total of 782 spectra were collected at the telescopes, exposed to have S/N=12, 33 or 110 for the final extracted mono-dimensional spectra. The S/N=12 was chosen as the lowest limit at which the noise is still dominated by the photon statistic (and thus adding up n similar spectra each exposed for a time t produces an added spectrum equivalent to a single exposure lasting for a time nt), and the S/N=110 to represent the best GAIA observations. Going below S/N=12 with ground-based observations would not be much useful for two basic reasons: (a) the instrumental noise pattern (CCD cosmetic, dark current, read-out noise, spectrum width) would quite possibly be different from those of the GAIA spectrograph, and at that such low photon fluxes, the instrumental noise starts to affect the end product, (b) the background in conventional ground-based spectrographs (both scattered lights as well as the solar spectrum of background sky during the bright moon phases during which slit-spectrographs are usually mounted at the telescopes) is high, variable and dispersed (GAIA background will mainly be undispersed zodiacal light, with dispersed overlaping stars growing in importance with increasing surface stellar density).

We already remarked on the unimpressive accuracy of the radial velocities of IAU standards: the 24 objects selected by MAT01 score an average ±0.23 km s^{-1} error. This means that, in general, the cross-correlation among any two of them would carry an uncertainty of 0.33 km s^{-1}, which is far too large for this type of investigation. MAT01 therefore devised an observing strategy and a network of cross-correlation runs that did not depend on the actual knowledge

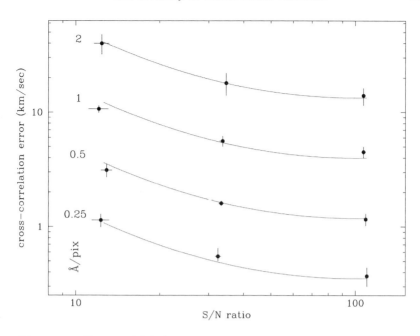

Figure 3. Errors of GAIA radial velocities as a function of dispersion and S/N as estimated by Munari et al. (2001a) on the base of real spectra. The curves are drawn from Eq.(2) the dots come from Table 3.

of the radial velocities of the template stars, requiring only their constancy over a few hour time (which should be the case for IAU standard stars).

During first pass solutions, GAIA will select the most appropriate cross-correlation template for each object star using the color information from the contemporaneous multi-band photometry. This choice will grow better with passing epochs given the improved photometry and increasing S/N of the co-added epoch spectra in the ground-based GAIA database. Once the mission will be over, all spectra will be re-measured according to the best characterized template. However, radial velocity data obtained during the mission will suffer from a spectral mis-match between template and object that will reduce with time. To simulate the effect of mis-match, MAT01 cross-correlated each object will all the others in the explored F7 to M2 range (all these spectral types are anyway dominated by the same CaII and FeI lines). The final MAT01 results are summarized in Table 3 and displayed in Figure 3.

The observational data in Table 3 are accurately fitted by the simple law:

$$\lg \epsilon = 0.6 \times (\lg \frac{S}{N})^2 - 2.4 \times \lg \frac{S}{N} + 1.75 \times \lg D + 3 \quad (2)$$

where ϵ is the radial velocity error (in km s^{-1}), S/N is the signal-to-noise ratio (per pixel on the stellar continuum for the extracted mono-dimensional spectrum) and D is the spectral dispersion (in Å pix^{-1}). This equation (cf. Figure 3 where it is used to plot the fitting curves) clearly indicates that at least in the

$S/N \geq 10$ regime, the factor most influencing the accuracy of radial velocities is the dispersion. The up-turn or flattening of the relations at higher S/N is confirmed also by simulations based on grids of synthetic spectra (see next Section), also if less marked. The flattening is much more evident with the real data for two main reasons: (a) the mismatch is wider compared to synthetic simulations (becoming the dominant source of error at the higher S/N), and (b) real spectra have finite uncertainties in wavelength calibration (as will be the case with GAIA spectra), while synthetic simulations run so far have none. This is another source of error that a higher S/N cannot compensate for.

Concluding this section, it is clear from real spectra that, even in the case of minimal spectral mismatch and errors in wavelength calibration, a dispersion of at least 0.5 Å pix^{-1} (resolution R$\geq$10 000) is required to reach an accuracy of 0.5 km s^{-1} in the mission-averaged radial velocities of bright stars.

4. Accuracy from simulations based on synthetic spectra

Katz (2000) and Katz et al. (2002) have performed the first investigations of the accuracy of GAIA radial velocities via simulations on synthetic spectra. Zwitter (2002) has expanded the work on simulations with results in agreement with those of Katz and collaborators. Zwitter considered three types of stars: an A8 V, a K1 V and a G6 I. Maximum mis-matches allowed between template and object were $\triangle T_{\text{eff}}$=250 K, $\triangle \log g$=0.5, $\triangle$[Fe/H]=0.5, $\triangle V_{\text{rot}}$=10 km s^{-1}. The explored mis-match is typical for mission-end conditions of relatively bright *single* and *inactive* targets, and it is less than that adopted for the previously discussed investigation based on real spectra. All synthetic spectra are taken perfectly wavelength calibrated, with no cosmic spread in individual chemical abundances, [α/Fe] or micro-turbulent velocity. No perturbing effects due to binarity, pulsation, intrinsic variability or active photosphere were introduced. The overlapping by background stars was not taken into account, but the effect of zodiacal light was considered in detail.

Zwitter (2002) study confirms that all the way down to V=17.5 mag the 0.25 Å pix^{-1} dispersion (17 200 resolving power) provides the most accurate radial velocity results (cf. Figure 4). His original study was tailored to the MMS design for the GAIA spectrograph. Later he has carried out similar investigations for the Astrium designs that were presented unpublished to meetings of the GAIA RVS working group. The results for the Astrium designs are pretty similar to those for the MMS configuration, with the only notable difference of a shift toward brighter magnitudes by $\triangle$mag$\sim$0.6 for Astrium-6CCDs configuration. Table 4 reports the mission averaged results for two spectrograph designs (combining Zwitter's results for K1 V and G6 I stars and zodiacal light with a surface brightness of I_C=21.8 mag arcsec^{-2}).

5. Conclusions and directions for future work

A great deal of work has been put so far into the investigation of the accuracy of GAIA radial velocities. Independent results obtained with telescopes and with simulations are reassuring similar and support the idea that GAIA will impact

Table 4. Accuracy of GAIA radial velocities (in km s^{-1}) derived from simulations on synthetic spectra of various S/N and resolving power (cf. Zwitter 2002). The conversion between magnitudes and S/N refers to an average GAIA target (a K0 star) suffering from a modest reddening (E_{B-V}=0.1), a zodiacal light of I_C=21.8 mag arcsec^{-2} and no overlapping by background stars.

	V mag											
MMS original	14.0		15.0		16.0		17.0		18.0		19.0	
Astrium-6CCDs	13.5		14.5		15.4		16.4		17.4		18.4	
Å/pix	S/N	err.	S/N	err.	S/N	err.	S/N	err.	S/N	err.	S/N	err.
0.25	60	0.16	27	0.26	12	0.54	4.8	1.29	1.9	3.20	0.8	12.7
0.5	102	0.21	48	0.34	23	0.68	9.3	1.45	3.8	3.42	1.5	10.9
1	169	0.49	85	0.62	42	0.97	18	1.89	7.5	4.10	3.0	10.2
2	264	0.60	144	0.81	76	1.34	35	2.57	15	5.36	6.0	12.7

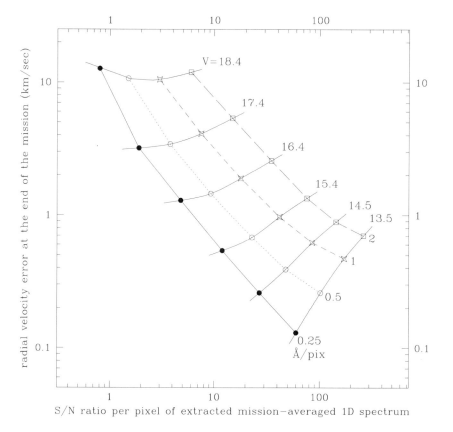

Figure 4. Plot of Table 4 data for the Astrium-*6CCDs* configuration.

Table 5. Mission-end accuracy of GAIA radial velocities (in km s^{-1}) for solar abundance, single and inactive G-K stars from combined results of real star observations and simulations with synthetic spectra for the Astrium-*6CCDs* configurations.

		\multicolumn{5}{c}{V}				
res.	Å pix^{-1}	14	15	16	17	18
17200	0.25	0.4	0.7	1.6	3.7	8
8600	0.5	1.1	1.3	2.0	3.3	7
4300	1	3.7	3.8	4.0	5.0	10

in an unimaginable way the radial velocity business. Table 5 is a preliminary attempt to combine into a homogeneous and unique frame the results of observations and simulations, giving preference - for the reasons above explained - to observations at the higher S/N and to simulations at the lowest.

Even though the grand picture is now clear (in particular about the need of adequate spectral resolution, see also discussion in Desidera and Munari 2003), a lot of details are still hidden, requiring further work (both with telescopes and simulations) before fine tuning of the spectrograph design and operation mode is achieved. At present the most relevant topics that need to be addressed appears to be: • re-run similar simulations for stars with Halo metallicities, which sport no absorption lines expect core-saturated and narrow CaII triplet lines; • increase the realism of simulations by considering various degree of binarity, variability, pulsation, active surfaces, emission lines and cosmic spread in atmospheric parameters; • explore in greater detail the case of the 40 million stars with tangential motions with an accuracy greater than 0.5 km s^{-1}; • investigate the gain that could be provided by a simultaneous cross-correlation of the whole, two-dimensional field of view as opposed to the case of the star-by-star extraction, calibration and cross-correlation approach; • to explore the concept of co-adding before measurment of extremely faint spectra suffering from spectrum overlap with nearby stars (continously changing for the spin axis precession).

References

Desidera, S. & Munari, U. 2003, in GAIA Spectroscopy, Science and Technology, U.Munari ed., ASP Conf. Ser. 298, pag. 85

Katz, D. 2000, PhD thesis, University of Paris VII

Katz, D., Viala, Y., Gomez, A., & Morin, D. 2002, in GAIA: A European Project, O.Bienaymé and C.Turon ed.s, EAS Pub. Series 2, pag. 63

Munari, U., Agnolin, P. & Tomasella, L. 2001a, Baltic Astron. 10, 613

Munari, U., Tomov, T., Zwitter, T., Milone, E.F., Kallrath, J., Marrese, P.M., Boschi, F., Prša, A., Tomasella, L. & Moro, D. 2001b, A&A 378, 477

Zwitter, T. 2002, A&A 386, 748

Stellar rotation from GAIA spectra

Andreja Gomboc

University of Ljubljana, Department of Physics, Jadranska 19, 1000 Ljubljana, Slovenia

Abstract. Stellar rotation influences our understanding of stellar structure and evolution, binary systems, clusters etc. and therefore the benefits of a large and highly accurate database on stellar rotation, obtained by GAIA, will be manifold. To study the prospects of GAIA measurement of projected rotational velocities $V_{rot} \sin i$, we use synthetic stellar spectra to simulate the determination of $V_{rot} \sin i$ at different resolutions (R=5 000–20 000) and S/N (10–300). Results on the accuracy of $V_{rot} \sin i$, presented here, show that GAIA will be capable to measure also low rotational velocities (~ 10 km s^{-1}), provided that the resolution is higher than 10 000.

1. Introduction

At present there are approximately 20 000 stars with measured $V_{rot} \sin i$ (Glebocki & Stawikowski 2000). Sorting them by stellar type and rotational velocity (Munari & Katz 2002, Soderblom 2001) shows that early type stars (O, B, A, early F) have high rotational velocities (50–400 km s^{-1}) and late type stars (late F, G, K, M) are slow rotators (in majority with $V_{rot} \sin i$ <50 km s^{-1}), which is mainly attributed to the fact that late type stars have convective envelopes, while early type stars do not. In view of measuring stellar rotational velocities with GAIA, it is necessary to estimate the accuracy of measured $V_{rot} \sin i$, which helps us in answering the question, whether slow rotators can be discerned from non-rotators with GAIA spectra.

2. Scientific questions related to stellar rotation

Data obtained by GAIA with millions and millions of stellar spectra and measured rotational velocities will certainly crucially contribute to the current knowledge on stellar rotation and its influence on various aspects of stellar physics. To shortly name a few (cf. Maeder and Meynet 2000, Soderblom 2001, Patten and Simon 1996):

- the effect of stellar rotation on stellar structure and evolution: rotation of a star influences its structure, luminosity, position on the HR diagram, life time, etc. Rotation can via induced mixing lead to He and N enrichment, chemically peculiar stars, it can cause turbulence, influence stellar winds and spots;

- the impact of differential rotation on stellar structure, surface phenomenology, models for generating magnetic fields etc;
- stellar rotation as the indicator of age (for low mass ZAMS stars) through the magnetic breaking and the indicator of photospheric activity;
- in binary systems: stellar rotation is important in understanding a number of questions, like the synchronization between rotational and orbital periods and the effectivness of tidal energy dissipation;
- rotational velocity as an indicator of angular momentum of solar type stars and how it is affected by the presence of massive distant planets;
- stellar rotation as an indicator of age, total mass and binding energy in open clusters and as a clue to test various theories of fragmentation of the parent cloud;
- the orientation of rotational axis - are they really randomly oriented and therefore not connected with the rotation of the Galaxy - which is the question that only wide GAIA statistics can answer.

3. Estimating GAIA accuracy on $V_{\rm rot} \sin i$

The important issue of GAIA contribution to stellar rotation physics is the accuracy of obtained $V_{\rm rot} \sin i$. Here are presented results of simulating the determination of rotational velocities on synthetic stellar spectra. We used Kurucz synthetic stellar spectra database in the GAIA wavelength region.

To start the simulation we choose the spectrum with known $T_{\rm eff}$, $[Z/Z_\odot]$, $\log g$, radial velocity and original $V_{\rm rot} \sin i$, artificially add Poison distributed noise and afterwards fit it with noise-free spectra with various $V_{\rm rot} \sin i$. As the best fitting spectrum we take the spectrum with minimum χ^2 and repeat the test N times. The accuracy is determined as the standard deviation error between the original and recovered $V_{\rm rot} \sin i$:

$$\sigma^2 = \frac{1}{N} \sum_{i=1}^{N} \left[V_i(orig) - V_i(rec) \right]^2$$

We used synthetic spectra of four star types: late G type giant ($T_{\rm eff}$=4750, $[Z/Z_\odot]$=−0.5, $\log g$=1.0), K dwarf ($T_{\rm eff}$=4750, $[Z/Z_\odot]$=−0.5, $\log g$=4.5), G type giant ($T_{\rm eff}$=5500, $[Z/Z_\odot]$=−0.5, $\log g$=2.0) and early F type star ($T_{\rm eff}$=7250, $[Z/Z_\odot]$=0.0, $\log g$=4.5), all with $V_{\rm rot} \sin i$=10 km s^{-1} at different resolutions: R=5 000, 8 615, 17 230 and 20 000. Noise added corresponds to signal to noise ratio in the range of S/N=10–300 and the number of trials is N=5 000.

As the fitting spectra we first use the spectra with stellar parameters ($T_{\rm eff}$ $[Z/Z_\odot]$, $\log g$, $V_{\rm rad}$) exactly the same as in the original spectrum, so that they differ only in $V_{\rm rot} \sin i$. The obtained accuracy in such simulations is shown in Figure 1: as expected the accuracy is much better for high resolution and improves with better S/N. It should be stressed though, that this is the most ideal case, since other stellar parameters are precisely known.

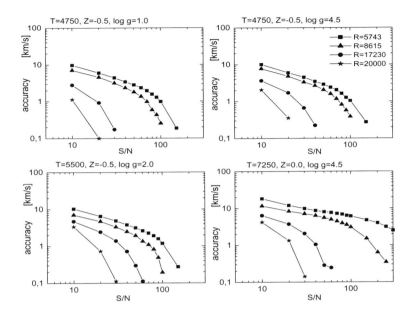

Figure 1. Estimated accuracy of $V_{\rm rot}\sin i$ for four types of stars at resolutions R=5 743, 8 615, 17 230, 20 000 as a function of S/N. Other stellar parameters ($T_{\rm eff}$ [Z/Z$_\odot$], $\log g$, V_{rad}) are presumed to be exactly known, i.e. the same in the original and fitting spectra.

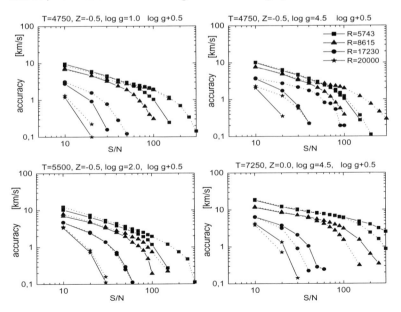

Figure 2. The same as Figure 1. Dotted lines show the accuracy obtained if $\log g$ is offset from its true value by 0.5.

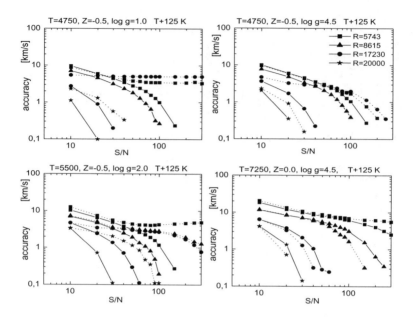

Figure 3. The same as Figure 1. Dotted lines show the accuracy obtained if T_{eff} is offset from its true value by 125 K.

Due to possible uncertainties in these, the accuracy of rotational velocity will generally become worse. To estimate the influence of those uncertainties on the accuracy of obtained rotational velocity, we performed the same simulations by fitting the original spectrum with spectral templates that had one of the parameters (T_{eff} [$Z/Z_\odot$], $\log g$ or V_{rad}) offset from its true value.

The least crucial factor turns out to be the uncertainty in $\log g$. The dotted curves in Figure 2 show the accuracy of rotational velocity if fitted by spectra having $\log g$ offset for 0.5 from its original value.

More crucial is an accurate temperature. An error of 250 K in some cases keeps the rotational velocity error at 5-10 km s^{-1}, not improving with S/N, while an error of 125 K (Figure 3) is small enough that at least at high resolutions the accuracy is better than 1 km s^{-1} at high S/N.

One of crucial factors is an accurate metallicity. An error of 0.5 dex in metallicity leads to inaccurate rotational velocities (error >10 km s^{-1}), an error of 0.25 dex is about a factor of 2 better and 0.1 dex is quite good for high resolutions (Figure 4). Note that the accuracy of the rotational velocity does not seem to suffer so much if the metallicity is underestimated than if it is overestimated by the same amount.

Crucial is also the accuracy of the obtained radial velocity. As shown in Figure 5, the error in radial velocity of 10 km s^{-1} leads to an error in rotational velocity of more than 10 km s^{-1}. If the radial velocity error is 5 km s^{-1} the error of rotation velocity is about 2 times smaller.

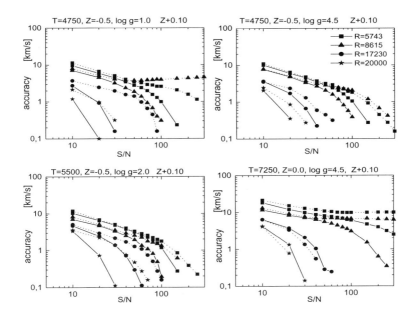

Figure 4. The same as Fig. 1. Dotted lines show the accuracy obtained if $[Z/Z_\odot]$ is offset from its true value by 0.1.

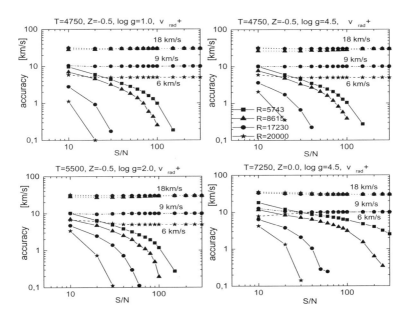

Figure 5. The same as Figure 1. Dashed and dotted lines show the accuracy obtained if the radial velocity is offset from its true value by 18, 9 and 6 km s^{-1} (top to bottom).

4. Conclusions

Table 1 gives the summary of stellar rotation accuracy obtained by simulations based on synthetic stellar spectra for different resolution and signal to noise ratio in the continuum. Columns 3-7 report the rotational velocity errors if the template used to recover the spectrum is ideal or mismatched for a given amount in temperature, metallicity, surface gravity or radial velocity.

Table 1. Estimated accuracy of $V_{\rm rot}\sin i$ (in km s^{-1}). ΔRV in km s^{-1}.

R	S/N	ideal	$\Delta\log g=0.5$	$\Delta T_{\rm eff}=125$ K	$\Delta[Z/Z_\odot]=0.1$	ΔRV=5
5 000	10	10-20	10-20	10-20	10-20	
	100	2-7	2-8	2-10	2-15	
10 000	10	7-12	7-12	7-12	8-13	
	100	0.2-3	1-3	2-5	2-8	
20 000	10	1-4	1-4	2.5-4	1-4	5-10
	100	<0.1	<0.1	<0.1	<0.1	5-10

The results we obtained can be used only as a rough indicator of GAIA capabilities. We used synthetic stellar spectra without allowing for modeling uncertainties, chemical composition peculiarities etc. The results of simulations show that effects of combined errors in two of the template parameters (e.g. $T_{\rm eff}$ and $[Z/Z_\odot]$, or $[Z/Z_\odot]$ and $\log g$) on the accuracy of rotational velocity are not always easily predictable. One should also be aware that the crowding of stellar spectra in GAIA focal plane may smear out some line profile details, which may be crucial in determining the rotational broadening of spectral lines. Nevertheless, we believe that measuring even small rotational velocities with GAIA is not out of reach, provided that the resolution is better than 10 000, or preferably better than 15 000.

References

Glebocki, R. & Stawikowski, A. 2000, AcA 50, 509
Maeder, A. & Meynet, G. 2000, ARA&A 38, 143
Munari, U. & Katz, D. 2002, RVS-CoCo-002 (Livelink)
Patten, B. M. & Simon, T. 1996, ApJS 106, 489
Soderblom, D. R. 2001 in Encyclopedia of Astronomy and Astrophysics, A. Baker, D. Emerson, N. Hankins & J. Matthias ed.s, Inst. of Phys. Pub., pag. 3137

Chemical abundances from GAIA spectra

Frédéric Thévenin

O.C.A., BP 4229, F-06304 Nice Cedex 4, France

Albert Bijaoui

O.C.A., BP 4229, F-06304 Nice Cedex 4, France

David Katz

Observatoire de Paris-Meudon, Plaace Jansen, F-92195 Meudon Cedex, France

Abstract. To derive stellar parameters and accurate abundances from GAIA spectra is a difficult challenge which depends on the adopted spectral resolution. We explored its influence on the restored atmospheric parameters of late-type stars in order to help to constraint the spectrograph resolution for the satellite.

1. Introduction

Intended for radial velocities, the design of the GAIA spectrograph presents a complex problem when it comes to defining the optimum resolution for both RV and other astrophysical aims. On one hand, we would like to maximize the resolution in order to allow the extraction of the largest amount of astrophysically interesting quantities. On the other hand, high resolution affect limiting magnitude and increase spectral overlapping.

One of the major goals of GAIA is to determine the chemical evolution of the Galaxy. This requires the determination of both iron and α (or s and r) element abundances. If GAIA does not resolve the iron lines, which are more difficult, then our spectral analysis will determine a global chemical composition mostly dominated by calcium. This will allow for an accurate determination of radial velocity and Ca abundances, but will not allow for an accurate study of the chemical evolution of the Galaxy. Iron absorption lines exist in the GAIA spectral range. However, for Halo stars, which are metal poor, the intensity of these lines becomes weak. This sets the minimum resolution that we need for the instrument.

To constrain exciting chemical abundance models, the errors on the abundances have to be lower than 0.1 dex. This forces us to adopt a maximum error of 80 K on the effective temperature and a maximum error of 0.2 dex on the log of the surface gravity. Combined with radial and tangential velocities, these accurate abundances would make significant improvements in our knowledge on the history of the Galaxy.

2. How to determine the required resolution

There are two methods for determining the required stellar atmospheric parameters (T_{eff}, $\log g$, and $[Z/Z_\odot]$) and the detailed abundances, each of which requires a different resolution. In the first, we can determine all parameters directly from the spectra. The alternative is to determine T_{eff} and $\log g$ from photometric and astrometric observations and to use the spectra exclusively for the detailed chemical abundances. The difficulty with the first is that it requires enough high resolution for the instrument, while the difficulty in the second is that it requires detailed knowledge of the stellar absorption. Thus, we suggest that we use both techniques in order to insure that we have the required data for as many stars as possible.

It is beyond the scope of this contribution to discuss the quality of synthetic spectra used for the stellar analysis but we remind that many studies have shown the importance to compute them with non-LTE codes (see Asplund et al. 2000, Thévenin & Idiart 1999, Idiart & Thévenin 2000). Also improved atomic and molecular data have to be done (see Walhgren et al. 2003, Plez 2003).

The GAIA spectral region is formed with weak and strong lines of neutral and ionized metals and also of the Paschen hydrogen lines. This spectral richness gives a diversity of the response of line absorptions with the variation of T_{eff}, $\log g$ and $[Z/Z_\odot]$ thanks to Boltzmann and Saha laws. These responses of line intensities with the atmospheric parameters are well detailed in table IV of Cayrel & Jugaku (1963). So, a priori, the GAIA spectral range can be used for the detailed analysis of late-type stars without help of photometry and astrometry. We note from this cited table IV that the variation of line intensities are strongly non linear with the three atmospheric parameters. In the following we shall concentrate only on stars cooler than $T_{\text{eff}}=7500$ K for which the continuous opacity is dominated by the H$^-$ ion.

Therefore, we have to explore the correlation between the lines in GAIA spectra and the three atmospheric parameters of the stars in order to construct a three dimensional classification. In two dimensions, such exercises are well-known as *stellar classification* based on spectral line ratio; depth of two lines having the same variation with surface gravity g and having different excitation potential give a line ratio which depends only with the temperature. Same exercise can be done with lines having the same excitation potential in order to eliminate the effect of the temperature but having different relation with the surface gravity (see table IV of Cayrel & Jugaku); therefore the line ratio becomes a pure gravity indicator. These two-dimensional classifications are of course limited because they depend strongly on the abundance of each element used to construct the line ratios and therefore cannot be used for quantitative determination of the three atmospheric parameters. In consequence, we developed a technique using the spectrum as a whole to recover quantitative characteristics of the atmospheres of observed stars in three dimensions.

We used a grid of 163 spectra computed with the Kurucz LTE code and covering the HR diagram from 4500 K $< T_{\text{eff}} < 7500$ K and $1.0 < \log g < 5.0$, for abundances varying from solar to metal-poor as -3.0 dex with irregular steps of 500 K, 0.5 and 1.0 by mean for the three parameters.

In order to establish that the recovered information (T_{eff}, $\log g$, $[Z/Z_\odot]$) is present in GAIA spectra we computed eigenvalues of the variance-covariance matrix of the spectra. Ten significant eigenvalues has been found showing that the information on the three varying parameters exists in the set of spectra and that the non-linearity is strongly present because more than three eigenvalues were found. This confirms our feeling on the richness of the spectra based on the table IV of Cayrel & Jugaku.

What is the best? Is there a linear combination of the pixels that can give a projection of the spectra on three axes (T_{eff}, $\log g$, $[Z/Z_\odot]$) which is not too strongly depending on the non-linearity of the stellar atmosphere physics? We define the following notation: for each spectra S_i, the three atmospheric parameters are T_i, G_i, M_i, corresponding to $\Theta(k=1, 2$ or $3)_i$ and we want to find a linear combination of the spectra to recover immediately by projection the three parameters of the observed star: T*, G*, M*. If α_i is the value to apply to spectra to get the parameter Θ we can write: $B(k) = \Sigma_{i=1,I} \alpha_i S_i(k)$ and the estimator is $\underline{\Theta}_i = \Sigma_{k=1,K} B(k) S_i(k) = \Sigma_j c_{ij} \alpha_j$

As we see, we recover the variance-covariance matrix in this mathematical expression. The estimator has to maximize $\Sigma \underline{\Theta}_i \Theta_i / \Sigma \underline{\Theta}_i^2$ We plug the grid itself (with spectra having S/N = 1000) in order to have a look at how the technique restored the grid (in terms of T_i, G_i, M_i) and with which errors.

The results are not so bad with sigma errors of what we can expect with classification criteria. Decreasing the resolution of the spectra increase drastically the error on restored parameters simply because of the non-linearity of the physics of the stellar atmospheres. The only one parameter which is reproduced with small errors, even with a resolution lower than 5 000 is the temperature; surface gravities and metallicities are recovered with less precision. Below R=15 000 such technique is not good.

This confirms the impossibility to find a linear combination of the observables, e.g. the pixels, to extract very accurately the three fundamental parameters.

In order to perform better T*, G*, M* of the observed stars, let's go back to a well-known approach which consists in comparing the grid of models with the observed spectrum using a technique of the minimum of distances (Thévenin & Foy 1983 , Cayrel et al. 1991, Katz et al. 1998). The path of the grid is important because of the non-linearity. We know that such techniques are very powerful but the exercise we explore here is until which low resolution it is acceptable to work for our goal on the chemical evolution of the Galaxy. For this we decreased the resolution of spectra by powers of 2. Distances was ponderated in order to increase the diagnostic of the minimum distances as used by the technique of the Objective Analysis, i.e. with weighting factor depending on the distance of the observed vector and the points of the grid: $\exp^{-d_i/a}$, where d_i is the distance to the atmospheric model i with a parameter a which depends on the S/N. More details will be given in a forthcoming paper.

The exercise have been done with a test-model having S/N=1,000 with decreasing resolution from 500 000 to 5 000. Result are good enough until the resolution 15 000. For lower resolution the errors on the restored parameters increase drastically excepted for T_{eff}. From this exercise we conclude that R=12 000 seems to be a minimum for the resolution of RVS at the present state of the work. Now if we use a spectrum with a S/N=50, the restored parameters are poor below R=15 000[1] For all tests, a smaller path of the grid will probably decrease these errors (depending on the non-linearity of the physic of the atmospheres). The conclusion of this exercise is that the choice of the resolution is crucial for chemical abundances of high quality.

3. The alternative using results from photometry and astrometry

If we use photometry and astrometry in order to derive T_{eff} and $\log g$, measuring spectral abundances becomes a problem of measuring the equivalent width of line elements. This is a well-known problem of stellar atmospheres and the error on the measure of the equivalent width is in relation with the S/N and the resolution by the following formula: $\delta W = \sqrt{2n}\eta\delta x$, where n is the sampling of the spectrum, $\eta=1/(S/N)$ and δx is the pixel size in Å.

Example: for a star with $V = 13.0$, S/N=50 at 15 000 of resolution, using a $n=3$ sampling we obtain an error of $\delta W = 0.015$ Å. This shows that the minimum W we can measure on a spectra for a weak line is ≈ 0.30 Å, which corresponds to an error of $\Delta log N = 0.18$ on the corresponding abundance.

4. Conclusions

In conclusion, the resolution of the GAIA spectrograph is a crucial point for any detailed abundance analyses. If we want to get more information than the three fundamental parameters the spectral resolution has to be above $\approx$ 12 000. Techniques of the minimum of distances are the right way to extract the requested informations from spectra and other techniques like neuronal network (Bailer-Jones 2003) can be of great help.

References

Asplund, M., Nordlund, Å., Trampedach, R. & Stein, R.F. 2000, A&A 359, 743
Bailer-Jones, C. 2003, in GAIA Spectroscopy, Science and Technology, U. Munari ed., ASP Conf. Ser. 298, pag. 199
Cayrel, R. & Jugaku, J. 1963, An.Ap. 26, 495
Cayrel, R., Perrin, M.N., Barbuy, B. & Buser, R. 1991 A&A 247, 108
Idiart, T. & Thévenin, F. 2000, ApJ 541, 207
Katz, D., Soubiran, C., Cayrel, R., Adda, M. & Cautain, R. 1998, A&A 338, 151
Plez, B. 2003, in GAIA Spectroscopy, Science and Technology, U.Munari ed., ASP Conf. Series 298, pag. 189
Thévenin, F. & Foy, R. 1983, A&A 122, 261
Thévenin, F. & Idiart, T. 1999, ApJ 521, 753
Walhgren, G.M. & Johansson, S. 2003, in GAIA Spectroscopy, Science and Technology, U. Munari ed., ASP Conf. Ser. 298, pag. 481

---

[1]Figures corresponding to these exercises will be sent on request (thevenin@obs-nice.fr)

Wolf-Rayet stars as seen by GAIA

Andrzej Niedzielski

Torun Centre of Astronomy, N. Copernicus University, Torun, Poland

Abstract. GAIA will determine distances to all known WR stars in Galaxy. Due to its astrometric precision GAIA will also detect components, if present, in most of possible WR binary systems. This mission will therefore be crucial in our understanding of physics of WR stars. In this contribution we review basic properties of WR stars, summarize the results of Hipparcos mission, and predict the result of spectroscopic observations in the GAIA spectral range.

1. Short characteristic of WR stars

The Wolf-Rayet (WR) stars are defined by the shape of their spectra, which are dominated by strong and wide emission lines that represent a wide range of ionization and excitation. Classification scheme of WR-star spectra was originally introduced by Beals (1938). According to the strongest emission lines present in their optical spectra the WR stars are divided into two major spectral types: nitrogen (WN) and carbon (WC). WN stars show mainly emission lines of helium and nitrogen, while he spectra of WC stars are dominated by carbon and helium emission lines. The very rare WO stars are an extreme case of WC stars with highest ionization states and strong emission lines of oxygen dominating their optical spectra. The spectral types of WR stars are further subdivided into subtypes based on line ratios in the optical spectral range: WN 1-11, WC 4-11 and WO 1-5. The ionization degree decreases with increasing subtype. A more recent 3-dimensional classification scheme of WR stars has been proposed by Smith et al. (1996).

The distribution of WR stars in the Galaxy, very similar to that of O-stars, as well as their coincidence with HII regions and duplicity with OB stars proves that WR stars are young, massive objects. This conclusion is also supported by finding WR stars in regions of recent star formation in other galaxies. However, some stars with very similar spectral features exist also among Population II objects, namely central stars of planetary nebulae (CSPN). These stars, contrary to the more common Population I counterparts, are usually designated as [WR].

Detailed studies show that masses of WR stars range from 5 to 25 $M_\odot$. The WN stars are more massive with average mass of 22 $M_\odot$, while average mass of WC stars is 12 $M_\odot$. The observed emission line spectra are created in dense, fast expanding winds accelerated by the hot and luminous stars (T=40–100 kK, $\log(L/L_\odot)$=5/6) up to V_∞=3000 km s^{-1} and over. Resulting mass-loss is of the order of 10^{-5} $M_\odot$ yr^{-1}. The WN stars are the brightest, reaching $M_V = -7$.

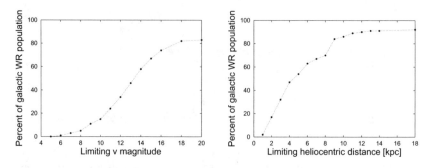

Figure 1. The population of known galactic WR star is largely accessible to GAIA instruments.

With so peculiar spectral characteristics and high luminosities WR stars are quite easily identified by spectroscopic, or narrow-band photometric observations in our and other galaxies even at large distances.

Presently we know 234 Population I WR stars in Galaxy (van der Hucht 2001a, 2001b): 134 WN stars, 87 WC stars, 10 WN/WC stars and 3 WO ones. We do not know was is the total WR population in Galaxy but we estimate it to be a few thousands stars. The population of WR stars in LMC counts 135 members (Breysacher et al. 1999, Massey et al. 2000): 109 WN stars, 24 WC ones, 1 WN/WC star and 1 WO star. Also the population of WR stars in SMC is known. van der Hucht (2001b) lists 11 stars, most of them (10) of WN class. The WR populations in LMC and SMC are assumed to be complete.

WR stars were detected also in some more distant Local Group galaxies (NGC 6822: 4 WN stars, IC 10: 28 WR stars, IC 1613: 1 WR star, M 31: 49 WR stars and M 33: 141 WR stars according to van der Hucht 2001b).

2. Hipparcos observations of Wolf-Rayet stars

Hipparcos observed 67 WR stars down to $V=12$ mag (~30% of known WR stars in Galaxy). The results of these observations, less spectacular then expected, are summarized in Moffat et al. (1998) and Marchenko et al (1998). The parallax to only one WR star was measured with reasonable accuracy: γ^2 Vel=HD 68273 was found to lay at $d=258$ pc distance ($\pi=3.88\pm0.53$ arcsec, Perryman et al. 1997), while the distance estimated from spectrophotometric parallax would be $d=450$ pc. Hipparcos found also two new visual binaries with separations below 1 arcsec: WR 31 with separation between components of 0.635 arcsec and WR 66 with separation of 0.396 arcsec. Important quantitative finding is also the fact that positions of WR stars were determined by Hipparcos with 20% errors larger than overall. In the detailed kinematical study of Moffat et al. (1998), 8 new runaway WR stars were reported. Another interesting finding was also that about 60% of studied WR stars show photometric variability (Marchenko et al. 1998).

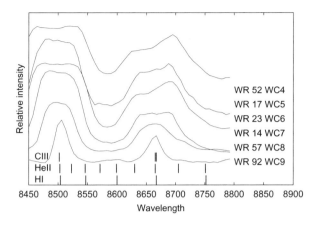

Figure 2. WC stars as observed in GAIA spectral window (intensity not to scale). Spectra from Vreux et al. (1989).

3. Wolf-Rayet stars in GAIA spectral window

The UV, optical and NIR spectra of WR stars are overcrowded by many emission lines of different intensity. The narrow GAIA spectral window appears, at a first glance, somewhat away from the richest line regions. However, this is not the case. As can be seen in Figure 2 and in Niedzielski and Nugis (2003) even the 250 Å GAIA spectral window contains many interesting lines. The near infrared spectral region of WR stars has been studied only fragmentally and with insufficient spectral resolution for complete identification studies. Only a few WR stars have been observed in the 8 000–10 000 Å range (Swings & Jose 1950, Kuhi 1966, Vreux et al. 1983, Vreux et al. 1989, Vreux et al. 1990). Swings & Jose (1950), Edlen (1956) and Vreux et al. (1983) presented identification of strongest emission lines in spectra of WR stars in wavelength range 6500-8800 Å.

3.1. Nitrogen sequence stars

The GAIA spectroscopic range include the hydrogen Paschen lines from P13 to P16. These lines are blended with He I lines of the same principal quantum numbers and with He II lines of the $n - 6$ series with $n = 24, 26, 28, 30$ and 32. Some of the odd n lines of He II $n - 6$ series in this are unblended and the difference of even n and odd n members of He II $n - 6$ series delivers information on H/He ratios in WN stars. Nugis & Niedzielski (1995) concluded that for WN stars the best estimates of the hydrogen-to-helium ratios can be obtained from the neighbouring HeII $n - 6$ series lines ($n \geq 12$). Many nitrogen lines possibly present in the GAIA spectral window are listed in Swings & Jose (1950). More detailed discussion on WN stars in GAIA spectral window is presented by Niedzielski & Nugis (2003).

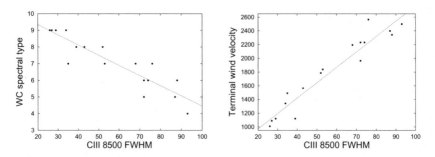

Figure 3. FWHM (in Å) of CIII line at 8500 Å correlates very well with WC spectral subtype (*left*) and with terminal wind velocity (in km s^{-1}) determined from UV P-Cygni profiles (*right*).

3.2. Carbon sequence WR stars

The spectra of WC stars are dominated by strong carbon lines: CIII 8500.32 and a blend of CIII lines 8663.65 and 8665.22 (Edlen 1956). Edlen (1956) lists also possible hydrogen-like CIII line at 8410 and CII lines at 8696.7 and 8682.2 Å. More carbon and oxygen lines are listed in Swings & Jose (1950), among them CII 8590.5 and 8594.8 and OIII 8461.9 positively identified in spectrum of HD 184738, a [WR] star.

The variation of width and shape of CIII lines with WC spectral type is presented in Figure 2. Starting from narrow, well defined Gaussian profile in late type WC stars both these strong lines widen towards earlier spectral types and show flat-topped profiles, characteristic for extended optically thin envelopes. The variation of shape of CIII blend at ∼8665 with spectral type proves contribution from higher ionization component, possibly CIV 8699.5 (or CIV 8707 blend) in early type WC stars. Contribution from CII lines at 8682.6, 8696.7 to the red wing of the blend in late type stars (especially in WC 8) is clear as well. Since both discussed CIII features are present and strong enough in all spectral types of WC stars their width change can be used to determine spectral types of WC stars and, for example, for first estimates of terminal wind velocity v_∞ from GAIA spectroscopy. Empirical relations between WC spectral type (from van der Hucht 2001a) vs. FWHM(CIII 8500) is presented in the left panel of Figure 3 (correlation $r=0.90$). Linear fit to these data allows one to estimate WC spectral type from relation:

$$Sp^{WC} = -0.061 \times \text{FWHM}_{\text{CIII}8500}[\text{Å}] + 10.56. \qquad (1)$$

Linear fit to empirical relation presented in Figure 3 right panel (correlation r=0.96) between V_∞ (from Niedzielski & Skórzyński 2002) and FWHM(CIII 8500) gives estimated terminal wind velocity as:

$$V_\infty^{WC}[\text{km s}^{-1}] = 22.45 \times \text{FWHM}_{\text{CIII}8500}[\text{Å}] + 521.55. \qquad (2)$$

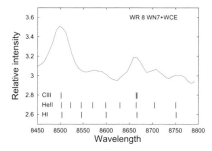 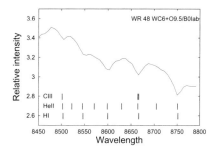

Figure 4. Both binary and composed WN/WC stars can be studied with GAIA spectroscopy. Spectra from Vreux et al. (1989).

3.3. Binary and exotic WR stars

Since in the GAIA spectral window interesting lines of both WN and WC stars are present together with Pashen series lines of hydrogen also binaries with OB companions and composed spectrum WN/WC objects can be studied from GAIA spectroscopy. In the case of the former stars the Pashen lines make radial velocities measurements of the OB companion possible while typical WR features allow one to follow changes in RV of the other component. The latter case, of composed spectrum WN/WC stars is interesting as well, since carbon and helium lines of the WC star are superimposed on hydrogen (and possibly nitrogen) spectrum of the WN component. In Figure 4 an example of composed spectrum object WR 8 (WN7+WCE) is presented together with the binary system WR 48 (WC6+O9.5/B0Iab).

4. Summary

GAIA will observe all known WR stars in Galaxy and will detect many new since our sample is complete only to $d = 3\text{-}5$ kpc. GAIA will obtain distances to all known and to many new WR stars: to 1% for ~ 100 WR stars ($\sim 50\%$ of known population), to 10% for ~ 100 WR know and other ~ 100 new.

GAIA astrometry will resolve many binaries and will determine masses for WR components.

GAIA will obtain 100 medium band photometric measurements for all WR stars what will allow for variability studies. GAIA medium band photometry will pre-select peculiar stars and will find many new WR.

In addition GAIA will observe all WR stars through its spectral window in a very interesting but poorly studied far-red region. GAIA spectra will allow to determine ~ 100 radial velocity measurements for most of known galactic WR stars therefore many new spectroscopic binaries are expected. Due to location of GAIA spectra window, radial velocity studies of all types of WR stars will be possible. In the case of WC stars spectral types and terminal wind velocities will be obtained immediately. H/He ratio determinations will be possible for single and binary WN stars. However, detailed spectral classification of WN stars will require additional medium band photometry.

Acknowledgments. We thank J.M. Vreux for making available the unpublished NIR spectra of WR stars.

References

Beals, C.S. 1938, Trans. IAU 6, 248
Breysacher, J., Azzopardi, M., & Testor, G. 1999, A&AS 137, 117
Edlen, B. 1956, Vistas in Astr. 2, 1456
Kuhi, L.V. 1966, ApJ 145, 715
Marchenko, S.V. et al. 1998, A&A 331, 1022
Massey, P., Waterhouse, E. & DeGioia-Eastwood, K. 2000, AJ 119, 2214
Moffat, A.F.J. et al. 1998, A&A 331, 949
Nugis, T. & Niedzielski, A. 1995, A&A 300, 237
Niedzielski, A. & Skórzyński, W. 2002, Acta Astron. 52, 81
Niedzielski, A. & Nugis, T. 2003, in GAIA Spectroscopy, Science and Technology, U. Munari ed., ASP Conf. Ser. 298, pag. 439
Perryman, M.A.C. et al. 1997, A&A 323, L49
Smith, L.F., Shara, M.M. & Moffat, A.F.J. 1996, MNRAS 281, 163
Swing, P. & Jose, P.D. 1950, ApJ 111, 513
van der Hucht, K.A. 2001a, New Astr. Rev. 45, 135
van der Hucht, K.A. 2001b, in The Influence of Binaries on Stellar Population Studies, D. Vanbeveren ed., Kluwer, pag. 141
Vreux, J.M., Dennefeld, M. & Andrillat Y. 1983, A&AS 54, 437
Vreux, J.M., Dennefeld, M. & Andrillat, Y., & Rochowicz K. 1989, A&AS 81, 353
Vreux, J.M., Andrillat, Y. & Biemont, E. 1990, A&A 238, 207

session 6
DOUBLE AND BINARY STARS
chair: M. Cropper

P. Niarchos, W. Van Hamme, R. Wilson and E. Milone.

Clock-wise from top-left: S. Söderhjelm, F. Thévenin, M. Steinmetz, S. Ansari, L. Magrini and A. Henden

Fundamental stellar parameters from eclipsing binaries

Eugene F. Milone

RAO, The University of Calgary, Physics & Astronomy Dept., 2500 Univ. Dr., NW, Calgary, AB, T2N 1N4, Canada

Abstract. Reviews by Wood (1963), Popper (1980) and Andersen (1991) of eclipsing binary data indicate that between only three and four dozen eclipsing binary systems at any given epoch have had sufficiently 'well-determined absolute dimensions' to challenge contemporary stellar models and evolution. With many more observations, and improved analysis techniques, a new compilation might expand this list by a modest factor. However, the potential of data from GAIA expand such a list by at least an order of magnitude has been demonstrated in a paper by Munari et al. (2001), which used GAIA-like ground-based radial velocity data and Hipparcos & Tycho photometry alone to determine parameters to better than a few percent for three target systems. Since GAIA is designed to have improved astrometric and multi-passband photometry and radial velocity spectroscopy capabilities, the combined data will enable us to 'close the loop' by providing sufficiently precise parallaxes to check distances derived from the analyses of the observable curves. These can lead to iterative improvements in the models, distances, in the fundamental parameters, and in such refinements as empirical limb-darkening, gravity-brightening, and albedoes. In addition to aiding the determination of the fundamental parameters of stars GAIA's instrumentation will illuminate our understanding of interacting binaries and their evolution.

1. Introduction: why binaries are important

Bradstreet's (1993) popular software program, *Binary Maker 2*, tells it plainly: 'Binaries for the Masses'. Direct measurement of mass requires acceleration – and thus other mass. To observe acceleration we usually measure velocities. If the signature features of both components are measurable with high precision, (making the system a 'double-lined spectroscopic binary', or SB2 system), the period of the system and the orbital elements can be established from the radial velocity curves, the ratio of the amplitudes of which will also reveal the mass ratio. Because the plane of the orbit is generally inclined to the plane of the sky by some angle i, the radial velocity curves yield only projected velocities, $V_{1,2} \sin i$ in our line of sight. The integrated curves then provide only the projected semi-major axis, $a \sin i$; Kepler's 3rd law provides only the projected sum of the masses. Finally, with the mass ratio, the RV curves yield only projected masses: $M_{1,2} \sin^3 i$. This is as far as we can go with RVs alone, but the degeneracy on the semi-major axis and the masses is lifted if the system is a visual or an eclipsing binary. The most precise values of the fundamental data are provided by these types of stars, notwithstanding important developments in interferometry, with the important exception of the Sun. In particular, we can determine the inclination, i, with other orbital elements, thus providing confirmation of the latter obtained from radial velocities, and yielding the masses unambiguously.

If in an eclipsing system, each component can probe the light distribution of the other's surface, resulting in precise and accurate radii and the relative radiative properties of the stars; moreover, limb-darkening, gravity-brightening, and macular effects of all kinds are open to investigation. The limb-darkening is a probe of the opacity and the temperature gradient in a star's atmosphere; for interactive systems, mutual geometric and irradiation effects (the classical 'oblateness effect' due to tidal distortion; the 'reflection effect' due to mutual irradiation; 'gravity-brightening' due to the higher surface brightness at quadrature compared to conjunction aspects); and, when the Roche geometry is measurable with precision, the photometric mass ratio.

2. GAIA and fundamental data from binaries

Eclipsing binaries (hereafter EBs) account for $\sim 1/3$ of all variable stars, according to the 4th edition of the General Catalogue of Variable Stars, although the categories are not mutually exclusive. To see how GAIA's EB data can transform the field and advance astrophysics, we review the sources and compilations of fundamental data to the present.

2.1. Improvements in kind and quality of data

Prior to 1911, there were ~ 10 EBs with computed orbits. The development of methods to analyze the light curves by H. N. Russell provided the means; and the first great effort to improve our knowledge of the fundamental properties of stars through analysis of many systems came through Russell's graduate student, Harlow Shapley (1915), who over 2 $1/2$ years obtained 10 000 observations with Princeton's polarizing photometer, performed analyses of 190 orbits, and produced a catalogue of 90 systems with orbital elements. Of these, only 31 with were accorded 'first grade' status orbits (on the basis of having the best visual and photographic data), 24 were judged to have second grade and 17 third grade orbits. The Russell model and analyses were used for all this work.

The next fundamental improvement came in the compilation of Kopal & Shapley (1956): orbital elements for 34 detached, 34 semi-detached, and 15 contact systems, in the categories that had been developed by Kopal a few years earlier. The Russell model was used except for the 'contact systems,' which we would today refer to as 'over-contact systems,' where Roche model geometry was employed. In this catalogue, intermediate limb-darkenings were used for the first time.

As noted, deriving the masses and sizes to high precision requires both spectroscopy and photometry. The attempts to sort out the data and solutions that gave the best results were undertaken at basically two institutions in the 20th century.

The *Catalogue of the Orbital Elements of Spectroscopic Binary Stars* of the Dominion Astrophysical Observatory was produced in eight editions, beginning in 1908. The 6th catalogue (Batten 1967) has 737 systems, only 36 of a quality (i.e., having 'definitive orbits'). For the 7th (Batten et al. 1978), 8th (Batten et al. 1989), and the on-line 9th (2002, http://sb9.astro.ulb.ac.be), no longer produced exclusively by DAO, the numbers of systems (and a quality) are: 978 (45), 1469 (55), and 1707 (221), respectively. In the 6th and 7th catalogues, the grading is done mainly on the basis of the precision and reliability of the elements; a usually indicates that $M\sin^3 i$ or f(M) are given to at least 3 significant figures. In the 8th catalogue, only two significant figures are given for $M\sin^3 i$ values. For b (in the 9th ed., 4) quality, the numbers are: 303 for the 8th and 368 for the 9th.

The 'Finding List' of the University of Pennsylvania similarly appeared in many editions. The 4th edition (Koch, Sobieski, & Wood 1963) and the 5th (Wood et al. 1980) contained all eclipsing systems with well-determined light elements, noting those in need of more observational attention. The 4th and 5th editions contained 1266 and 3546 eclipsing binary entries, respectively.

The Graded Photometric Catalogue (Koch, Plavec & Wood 1970) produced also at the Univ. of Pennsylvania, contained 219 systems, of which only 2 were graded (observational) quality A. This catalogue used observational quality alone, not solution method or analysis details, as the basis for assessment but complicated and partially eclipsing systems got less weight. The criteria for a weighting of 5 were: $m.e. \leq 0.006$; ≥ 1 obs. per long. degree; multiwavelength photoelectric light curves; and for 4 they were: $0.007 \leq m.e. \leq 0.011$, $0.8 - 0.9$ obs./deg, and 1-passband photoelectric light curve. All visual and photographic estimates were rejected.

Even though the spectroscopic binaries catalogue emphasized the quality of the solutions, and the eclipsing binaries graded catalog the quality of the data, it is still fair to say, with Koch et al. (1970) that for graded catalogues "observational precision and reliability are the *sine qua non* for accurate elements."

The extraction of orbital elements from the light and radial velocity curves is only the first step, however. In combination, the fundamental parameters emerge from the analyses as a self-consistent set. Progress in this area is certainly present, but the number of systems with parameters sufficiently accurate and precise to challenge stellar structure and evolution has been remarkably low and relatively constant over time. This suggests that theory has kept pace with observation.

Aitken (1935) compiled a list of 321 spectroscopic binaries of which 22 are eclipsing; of these, only 9 are listed with unprojected masses.

Wood's (1963) review of eclipsing binary data listed only 42 binaries with 'well-determined absolute dimensions' (but no mass or radius to better than one decimal place) and only 11 binaries with reliably computed limb-darkening.

Poppers (1980) review demonstrated subsequent advances in observational material by listing 36 eclipsing systems for which masses and radii could be considered precise to two decimal places. He noted the relative paucity of late-type systems in the compilation.

Andersen's (1991) summary gives ~ 45 carefully selected binaries (with M and R determinations to $\sim 2\%$). Of these, only 6 were unchanged from Popper's paper, 18 had improved elements, and 21 were new. The number of what we may call 'well-determined systems' was therefore larger than in Popper's list, but not by a order of magnitude. Thus, there is ample opportunity for GAIA to produce improved numbers of well-determined elements.

2.2. Requirements for fundamental parameters

We expect that the photometry has 1–2% precision, is carefully reduced and well-transformed; and that the spectroscopy is free of systematic effects and of sufficient S/N. With these conditions, with no intrinsic variability, and with an adequate model and modeling code, the geometric elements, and the radii & masses of the stars can be determined to below the 1% level. This assumes that tidal and irradiation effects, the classical 'oblateness' and 'reflection' effects of bygone days, and the limb-darkening are all adequately modeled, even in nominally 'detached' systems. The next problem is the temperature scale to provide well-calibrated temperatures, and luminosities.

If absolute visual magnitudes can be determined from parallaxes and good quality photometry, the bolometric corrections set the level of precision of the (bolometric) luminosity. From the luminosity and the surface areas, which come from the light curve analysis, the relative visual surface brightness is derived. But in order to tie the intrinsic brightness and color of each component to the observed light of the system, the latter must be corrected for interstellar extinction. Fortunately, the reddening correction for nearby systems is small, hence a color index-surface brightness correlation may be accurately determinable to finer precision than in the past. For bolometric corrections, well-calibrated infrared and ultraviolet data are the only true solution. The large numbers of stars in closely set bins of intrinsic color index that GAIA should produce may permit determinations for many systems, from the select number of cases for which the non-visual data are or become available.

For interactive stars, and for magnetically active stars, the result may be problematic. Ultimately, such activity places limits on the precision of the the stellar parameters for those systems to perhaps 1% or more in some cases.

The basic extraction procedure can be flow-charted as follows.

Precision photometry, spectroscopy, astrometry
$\Downarrow$
R, M, T, L to ~ 1% precision
+
Accurate distance, reddening, metallicity
$\Downarrow$
CMDs, PCRs, age
$\Downarrow$
Tests of models, isochrones, evolutionary tracks
+
Explanation(s) for intrinsic scatter on CMD due to other effects
+
Statistics in mass-transfer cases; over-contacts

A binary component-populated Hertzsprung-Russell diagram (Figure 1a on the left, and 1b on the right, taken from Andersen (2002), shows the best-observed systems, where the fundamental parameters are determined to 2% or

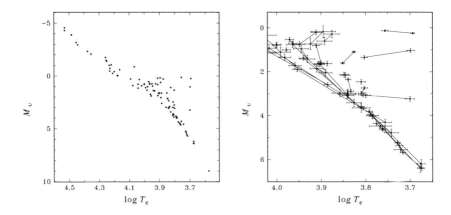

Figure 1. Hertzsprung-Russell diagram of the components of the best-observed binary star systems, where the fundamental parameters are determined to $\sim 2\%$ or better. *Left*, the entire set; *right*, a portion of the lower part with binary components joined by lines. Note the paucity of cool, low luminosity stars (from Andersen 2002 and included here with permission).

better. Less than 100 separate stars are shown. Thus, in a sense, although meeting a much higher standard of precision in the data and accuracy in the results, in numbers of 'precise' results, we are essentially at the stage where Shapley left the field in 1915, and Kopal in 1956.

Figure 1b, an enlarged portion of Figure 1a, indicates with lines the components of each binary. The two systems on the upper right are identified as giants, the rest are nominally main sequence stars, but the scatter here indicates that many of the stars, even though nominally dwarfs, have evolved with respect to the zero-age main sequence. This is important because mass-luminosity relations drawn from plotting all these systems together have a scatter due to evolution and composition effects, as Andersen (2002) notes.

In the absence of mass loss, g changes only with radius for a given mass. Plots of $\log g$ vs. mass for the components of a binary provide one set of tests for theoretical models because models with the same metal abundance must fall on a line with the same slope as the line joining the components of different mass. This test may indicate, for example, that some effect such as convective overshoot may be required to fit the data adequately. Another test, involving $\log g$ vs. $\log T_{\text{eff}}$, is shown in Figure 2, also taken from Andersen (2002), and discussed in detail by Torres et al. (1997). It demonstrates for three well-studied field systems, the effects of metal abundance.

Although different assumptions about helium content and details of the convective overshooting lead to divergent predictions, one need not agree with all aspects of the solutions or with the computed model tracks to note the effect of differing [Fe/H], for a given treatment. We note here that GAIA's discrimination in [Fe/H] is expected to be 0.15 dex.

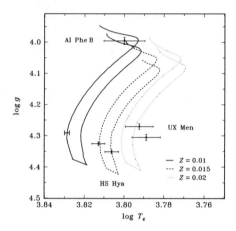

Figure 2. A $\log g$ vs. $\log T_{\text{eff}}$ plot of tracks and components of three of the best determined binary stars systems: AI Phe, HS Hya, and UX Men. All three systems have primary (more massive) stars that are nearly identical. The secondary stars are also quite similar. The two halves of each track were computed for masses differing by 3%, and the horizontal error bar on the hottest track (that for AI Phe A, which has evolved out of the diagram) represents an observational uncertainty of 0.6% in mass (from Andersen 2002, with permission).

Now we come to two critical questions: (i) can GAIA permit us to derive fundamental data as well or better than previous work? and (ii) can it add substantially to the list of very well-determined systems from ground-based data?

2.3. GAIA light- and radial velocity simulations

Munari et al (2001) fitted the Hipparcos and Tycho photometry and RV data from echelle spectroscopy obtained at Asiago to match the wavelength range and resolution proposed for the GAIA RV spectrometer of three detached binaries: V505 Per, OO Peg, and V570 Per. Figure 3 shows the fittings for the first two systems. Although the photometric precision is lower and the number of passbands fewer than that proposed for the GAIA photometer (15 at present writing), the analyses yielded uncertainties of $\sim 1\%$ for the mass ratio, q, better than $\sim 2\%$ for the masses, $\sim 0.5\%$ for a, 1-4% for the radii and 1-2% for the temperatures for the two systems shown in Figure 3. The poorly conditioned data of V570 Per, with only a few points in the Hipparcos passband in each of the minima, and essentially no useful data in the Tycho passbands, yielded percentages in the same quantities of: 1.3%, $\sim 2.4\%$, 0.7%, 10–25%, and 2-3%, respectively. The V570 Per data mark a kind of worst case, but even there, the preliminary results could be used in the modeling of a more extensive and precise data set. Therefore, the answer to the first question of the preceding section appears to be 'yes!'.

Extrapolating from Hipparcos data, the number of eclipsing systems expected to be observed by GAIA is, conservatively, $\sim 400\,000$; of these $\sim 25\%$

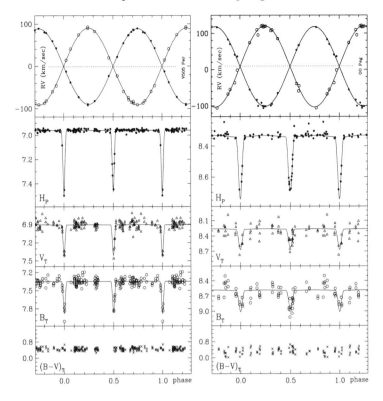

Figure 3. Fittings of the detached system light and radial velocity curves of OO Peg and V505 Per, the data of which mimic the expected phase coverage and radial velocity precision anticipated from GAIA (from Munari et al. 2001).

will be SB2s. If only 1% of them can provide M and R to 1 to 2% precision, we will still have 1000 systems capable of yielding fundamental data for tests of stellar structure and evolution. Thus the answer to the second question also appears to be 'yes!'

2.4. Closing the Loop

GAIA parallaxes can provide distances independent of reddening. However, EB analysis may determine radii to 1% or occasionally better. So, if GAIA spectroscopy can provide an accurate temperature for at least one component, and if GAIA photometry is good enough for EB analysis to yield precise values of $L_{\lambda_{1,2}}$, ΔT and $R_{1,2}$, then better reddening corrections and $CI - \sigma$ relations may be determinable from sufficient numbers of nearby systems, particularly those with good BCs. This will lead in turn to improved distances of more remote systems for which parallaxes will be not as precise and for which reddening will have to be taken into account. This scheme requires a detailed $CI - \sigma$ relation based on the nearby stars. Of course, the achievement of this goal may turn out to be complicated. Although Wesselink (1969) found no dependence of surface brightness on luminosity in the best data set of his day, the numerous high precision results expected from GAIA may show otherwise, but the sheer numbers of systems should allow the determination of any dependences.

3. GAIA and the boon to binary star studies

GAIA is expected to detect millions of binaries! Once detected, new binaries can be followed up by ground-based observations, and analyses can make full use of GAIA's expository photometry, spectroscopy, and astrometry.

Quite apart from providing fundamental data, GAIA will greatly increase what we know of the statistics of systems in which the stars are interacting to various degrees and in various ways, as for example, binaries in which the 'Algol paradox' is seen (where the apparently more massive component is less evolved than its less massive companion), and those in which both objects are well-evolved (as, for example, the 'cool Algols', only a handful of which are currently known; Popper 1992).

Generally, for evolved systems with evidence of mass exchange or systems in close enough proximity for one component to have influenced the evolution of the other, radiative properties must be modeled at least as well as geometric properties. The resulting masses and radii can indicate where the stars are with respect to the RLOF radius of each component. Precise photometry will permit sources of light curve asymmetries and transient effects to be identified and underlying temperatures and luminosities to be determined. For these interacting, late-type systems, spot and envelope effects must be modeled, especially of systems showing strong activity cycles, as in RS CVn-like systems. Improved statistics of these systems and improved parameters, especially of partially eclipsing systems, may help to demonstrate the presence of *companion-reinforced accretion process* (Eggleton 1992) in these systems.

Overcontact systems are numerous. By far and large, they are members of the old disk population, enjoy good visibility because they are brighter than single stars of this population, and are seen to eclipse over a large range of inclination. The period-color relations computed for different ages and metals by Rubenstein (2001) indicate that in all cases the systems become redder with age and with increasing metals. Sharpening of these relations requires comparisons to precise multi-wavelength photometry of large numbers of well-measured systems. The luminosity functions of over-contact systems in Baade's Window, investigated by Rucinski (1997) with OGLE data, can be extended further; good reddening corrections and the improved distances from GAIA should permit the further honing of these objects as standard candles.

For young binary star systems, pre-main sequence models can be tested with the numbers of such systems that GAIA may be able to find. Palla (2001) nicely summarizes the usefulness of the handful of systems thus far known, none of them containing components with less than solar mass.

Basri (2000) notes that while the dearth of brown dwarfs in systems of large luminosity ratios appears to be real, brown dwarfs are being found in increasing numbers both in infrared surveys and in low-luminosity binaries. From this he concludes that instead of a "brown dwarf desert" we have a brown dwarf "desert island." We can expect GAIA to provide some eclipsing systems which will provide precise radii for these objects.

Finally, just as the binary condition provides an explosion of information about both stars over that attainable for a single star, the information gain of stars in an ensemble can be even more explosive. The brilliant work of Perryman et al. (1998) on the Hyades can be expanded manyfold, over a range of hundreds of parsecs, and in this work binaries will again play a major role. The upshot will be a great treasure of fundamental properties of stars covering a range of chemical composition, and the kinematic and composition homogeneity of these ensembles can be tested in detail for each cluster.

4. Summary

GAIA's combined astrometric, multi-passband photometry, and spectroscopic capabilities, if fully realized, will enable us to:

- 'close the loop' by providing sufficiently precise parallaxes to check distances derived from the analyses of the EBs and SB2s observable curves;
- determine reddening & improve $CI - \sigma$ relations;
- permit iterative improvements in the models, distances, and in the fundamental parameters;
- in such refinements as empirical limb-darkening, gravity-brightening, and albedoes;
- and improve statistics of types of binaries and their properties.

The discussion of GAIA's contributions could extend beyond these points to include the challenges to models of abundant high-quality solutions, especially of low-mass systems. The difficulties of fitting low-mass stars, for instance, is compounded by the necessary treatment of spots and other active region phenomena. It remains moot that the treatment of these phenomena does not affect the solutions in any fundamental way. Modeling packages of the future need to explore options to test different configurations of spots, aided by ground-based infrared photometry and polarizing spectroscopic observations of specific systems to arrive at well-optimized solutions. Finally, solutions of EB & SB2 in clusters will have even more value than at present, because the precise determinations of temperature and chemical composition as well as the radii and masses will greatly aid the improvement of theoretical models.

Acknowledgments. It is a pleasure to acknowledge the help of Paola Marrese, Ulisse Munari, Michael Williams, and Tomaž Zwitter in the preparation of this paper. Dimitri Pourbaix provided binary statistics for the 9th Spectroscopic Binary catalogue. GAIA related modeling was supported by the Natural Sciences and Engineering Research Council of Canada. Travel was supported by the University of Calgary Research Grants Committee and by a grant to U. Munari, for both of which the author is thankful.

References

Aitken, R. 1935, The Binary Stars, McGraw-Hill; reprint 1964, Dover

Andersen, J. 1991, A&A Rev. 3, 91

Andersen, J. 2002, in Observed HR Diagrams and Stellar Evolution: the Interplay Between Observational Constraints and Theory, T. Lejeune & J. Fernandez ed.s, ASP Conf. Ser. 274

Basri, G. 2000, ARA&A 38, 485

Batten, A.H. 1967, Sixth Catalogue of the Orbital Elements of Spectroscopic Binary Systems, Pub. DAO 8, No. 8

Batten, A.H., Fletcher, J.M. & Mann, P.J. 1978, Seventh Catalogue of the Orbital Elements of Spectroscopic Binary Systems, Pub. DAO 15, No. 5

Batten, A.H., Fletcher, J.M. & MacCarthy, D.G. 1989, Eight Catalogue of the Orbital Elements of Spectroscopic Binary Systems, Pub. DAO

Bradstreet, D.H. 1993, Binary Maker 2.0 Light Curve Synthesis Program, Norristown, Contact Software

Eggleton, P. 1992, in Evolutionary Processes in Interacting Binary Stars, Y. Kondo, R.F. Sistero & R.S. Polidan ed.s, IAU Symposium 151, pag. 167

Koch, R.H., Sobieski, S. & Wood, F.B. 1963, A Finding List for Observers of Eclipsing Variables, 4th ed., Univ. of Penn. Press

Koch, R.H., Plavec, M. & Wood, F.B. 1970, A Catalogue of Graded Photometric Studies of Close Binaries, Univ. of Penn. Printing Office

Kopal, Z. & Shapley, M.B. 1956, Catalogue of the Elements of Eclipsing Binary Systems, Jodrell Bank Annals 1, fasc. 4; reprinting, undated: John Sherratt & Son

Munari, U., Tomov, T., Zwitter, T., Milone, E.F., Kallrath, J., Marrese, P.M., Boschi, F., Prša, A., Tomasella, L. & Moro, D. 2001, A&A 378, 477

Palla, F. 2001, in The Formation of Binary Stars, H. Zinnecker & R.D. Mathieu ed.s, IAU Symp. 200, pag. 472

Perryman, M.A.C., Brown, A.G.A., Lebreton, Y., Gomez, A., Turon, C., de Strobel, G. Cayrel, Mermilliod, J.-C., Robichon, N., Kovalevsky, J. & Crifo, F. 1998, A&A 331, 81

Popper, D.M. 1980, in ARA&A 18, 115

Popper, D. M. 1992, in Evolutionary Processes in Interacting Binary Stars, Y. Kondo, R.F. Sistero & R.S. Polidan ed.s, IAU Symp. 151, pag. 395

Rubenstein, E.P. 2001, AJ 121, 3219

Rucinski, S.M. 1997, AJ 113, 407

Shapley, H. 1915, A Study of the Orbits of Eclipsing Binaries, Contr. Princeton Univ. Obs. No. 3

Torres, G., Stefanik, R.P., Andersen, J., Nordstrom, B., Latham, D.W. & Clausen, J.V. 1997, AJ 114, 2764

Wesselink, A.J. 1969, MNRAS 144, 297

Wood, F.B. 1963, in Basic Astronomical Data, Univ. of Chicago Press, pag. 370

Wood, F.B., Oliver, J.P., Florkowski, D.R. & Koch, R.H. 1980, A Finding List for Observers of Interacting Binary Stars, 5th ed., Univ. of Penn. Printing Office

Toward optimal processing of large eclipsing binary data sets

Robert E. Wilson

Astronomy Department, Univ. of Florida, Gainesville, FL 32611, USA

Stuart B. Wyithe

Harvard College Obs., 60 Garden st, Cambridge, MA 02138, USA

Abstract. We examine two areas of importance for optimal processing of large quantities of eclipsing binary data from GAIA and related projects. Area 1 concerns efficient, effective, and automatic processing, while Area 2 concerns choices of binaries for follow-up observations with large optics. Strategies in both areas will influence the ultimate usefulness of the data for confrontations with single and multiple star evolution theory and for distances. In Area 1 we outline an Expert Systems approach to decision making in regard to adjusted parameters, starting values, constraints, stopping criteria, adequacy of the model, and incorporation of miscellaneous information in a variety of forms. Area 2 concerns relative usefulness of binaries according to morphological type, kind of eclipse, and several other considerations. Our main points are that semi-detached and overcontact binaries should be preferred to detached binaries, and that complete eclipses are far preferable to partial eclipses for accurate masses, luminosities, radii, and distances.

1. Introduction

Eclipsing binaries (EB) are distance indicators, via comparison of apparent and absolute magnitudes, but are not *standard* candles in the ordinary astronomical sense, as their luminosities are not taken from a calibration of recognized standard objects. Implicit within the EB distance method (hereafter EBD) is individual measurability of *MLR* (mass, luminosity, radius) for suitably observed and well conditioned binaries. EB distance indicators have advantages over standard candles in that the troublesome calibration step is eliminated and object to object differences are of no consequence. The EBD concept dates from Russell (1948) or perhaps earlier, and has spurred several recent distance estimation projects as a consequence of reviews by Paczynski (1997, 2000) that re-awakened interest in the idea. An EB luminosity (as well as mass and radius) is measurable due to a confluence of three kinds of data:

(i) a light curve allows estimation of relative system dimensions, local relative surface brightness, luminosity ratio, orbital geometry, and aspect. That is, it provides a picture to correct geometrical scale, and a *gray scale*. In a slightly restricted sense it actually provides a color picture, since we can convert local emission properties to color if necessary. Naturally some light curves do this much better than others, as discussed below and in the literature. For a review of EB light curve analysis, see Wilson (2000);

(ii) double lined spectra establish of the absolute scale. With Doppler measures in kilometers per second, integration over time provides the orbital scale in kilometers. Combining (i) and (ii), we have absolute star dimensions. The mass ratio can be estimated from light curves in certain highly favorable cases, so only single lined spectra may be needed (Sect. 4.3). Such photometric mass ratios, q_{ptm}, can be very accurate for binaries that show complete eclipses, and are confidently known to be semi-detached or overcontact. Radial velocity mass information and a q_{ptm} are mutually supportive for an EB with double lined spectra;

(iii) spectra tell radiative properties and, with the help of a stellar atmosphere model, absolute emission. Further information drawn from light curves specifies how emission varies with location due to gravity brightening and reflection.

Combination of (i), (ii) and (iii) for favorable examples provides everything needed to compute radiative emission in physical units and without use of standard objects of any kind. EBD relies only on geometry and physics and accordingly is nearly as primary and direct as a trigonometric distance, so EB and trigonometric distances can serve as mutual checks. EB distances are aperture-limited, whereas trigonometric distances are baseline-limited. EBD is useful to much greater distances than the trigonometric method, given realistically accessible parallax baselines in the immediately foreseeable future.

2. Main EB optimization areas

After GAIA has generated large numbers of EB light and velocity curves and spectra, further optimal processing will be in two basic areas, both very important for maximization of overall value. Area 1 is that of efficient, effective, and automatic extraction of astrophysical parameters. In this context, *efficient* refers to optimization of computer usage, *effective* to optimization of accuracy, and *automatic* to avoidance of personal intervention. Also very important is Area 2, selection of optimal EBs for mass, luminosity, radius and distance (*MLRD*) estimation in follow-up observations with large telescopes. *MLR* data are important for confrontations with single and multiple star evolution theory. Distances are important for setting the internal scale of our Galaxy, for locating specific structures within our Galaxy, for finding distances of the Magellanic Clouds and perhaps other galaxies in the Local Group (and thereby contributing to determination of the cosmic distance scale), and for serving as checks on trigonometric distances. Although success in Area 1 will allow essentially *all* recorded EBs to be processed, follow-up with large aperture telescopes for data of highest achievable *MLRD* accuracy will be practical only for a modest subset of EBs. Modelers of EB observables are well aware of disguised problems that can undermine apparently straightforward analyses even for highly accurate data, so selection of follow-up candidates should be based on simulations, so as to identify binaries with intrinsically strong light/velocity solutions that are resistant to such ills. We now discuss Areas 1 and 2.

3. Area 1: dealing with large data sets

For the individually observed binaries treated in typical journal papers, there are strong arguments not to automate:

(*a*) many superficially normal close binaries have unexpected features, and humans are much better than machines at dealing with the unexpected;

(*b*) several dissimilar kinds of data must be brought to bear on typical problems, and the kinds differ from one binary to the next. Experienced persons can make the required strategy and value judgments needed for optimal use of varied kinds of information;

(*c*) system selection – we often do not know what we are dealing with until the end (and maybe not even then!). Rules for selection ultimately derive from experience;

(*d*) which parameters are to be free? Can we change as we proceed? On-the-spot judgments are needed.

Strategies for all of these sub-problems are hard to program. Their automatic treatment to full satisfaction is a distant goal. Even the ordinary automation of feeding the output of an iterative solution step back into the input – without personal inspection – is risky for EBs in view of the points mentioned just above. However for enormous data sets we have to automate or face information overload. Best would be to construct a computer program that does everything an ideal experienced person does. Not only would such a program be intricate and have to provide for input in a wide variety of forms, but its author would have to be thoroughly experienced with EBs or take the advice of experienced persons. Still the program's logic might be controversial, as disagreements among experts are common. One can reasonably expect, however, that some decades of applications may produce EB solution programs that match and even exceed performances of expert persons. We will then have entered and benefited from the established field of Expert Systems.

3.1. Expert Systems: a machine acting as an experienced person

There is a large literature on Expert Systems and associated software, whereby highly optimized programs apply user-supplied rules in processing data, although astronomical applications to date are limited.

So what does an experienced EB person do? The person can

(*1*)... judge whether to follow up on a given EB (perhaps make further observations with larger optics or other instruments, or go beyond a routine analysis). Like the person, expert software could flag systems that deserve special attention. An expert person or program can examine

    a. scatter,
    b. evidence for orbital eccentricity,
    c. supporting information, such as spectra, cluster membership, and companions,
    d. form of eclipse (complete vs. partial),
    e. eclipse depths,
    f. morphological type,
    g. proximity effects (tides and irradiation),
    h. abnormalities;

(2)... decide on logical constraints (usually based on morphological type);
(3)... decide on parameters to be adjusted and set values for those that will not;
(4)... decide, at each iteration, what parameters should continue to be followed, and possibly do multiple subset solutions;
(5)... decide when to stop iterations;
(6)... evaluate quality of results, based on
   a. intrinsic variance of data and solution variance,
   b. convergence of iterations,
   c. standard errors of parameters, and
   d. flatness of residual vs. phase (or time) graph.

Considerations in all of the above are:
   – what parameters are highly correlated?
   – what model failures might be exposed by the overall collection of data?
   – recognition of trapping into local minima,
   – rules for several kinds of weighting.

Programming all of the above is not easy, considering that:
   – experts are hard to find, and even experts may have misconceptions,
   – various kinds of information are in distinctive forms,
   – ordinarily subjective judgments must be programmed to be objective.

In overview, the idea is to have a machine act as an expert person, but enormously faster and without mistakes.

4. Area 2: Optimal EB types for MLR and distances

Some EBs are far more useful than others for $MLRD$ – the diamonds to be distinguished from ordinary rocks. Binary star morphology arose mainly from ideas of Kuiper (1941) and Kopal (1954, 1955, 1959) and consists of three types: *detached* binaries (DB), *semi-detached* (SD), and *overcontact* (OC), with all three types well represented in binary star catalogs. There is some evidence for a fourth type of *double contact* (Wilson 1979), although double contact binaries are rare and not very suitable for accurate $MLRD$ work. Significant differences between the Kuiper and Kopal definitions continue in present use of these names, as discussed by Rucinski (1997) and by Wilson (2001). Briefly the issue concerns whether contact and detachment refer to star-on-star or star-on-lobe. Here we use the star-on-lobe meaning, so a contact component is in accurate contact with its limiting lobe but not necessarily with the other star, and systems that exceed their limiting lobes will be called overcontact rather than contact. A limiting lobe is the largest *closed* equipotential around a given star, so there is loss of matter to the companion (lobe overfow) if one star exceeds its lobe even slightly. Most of the recent $MLRD$ literature has stressed the value of well-detached binaries, but we can ask if justification for that preference is self-evident. The thesis here will be: not only is the justification *not* self-evident, but there are important advantages to SD and OC binaries that have largely gone un-recognized, with certain SDs and OCs accordingly being the desirable diamonds of $MLRD$ research.

4.1. Perceived advantages of well-detached binaries

Tides and Reflection. One can surmise some unstated reasons behind the tacit preference for well-detached binaries in published *MLRD* work. One reason could be that tides and reflection (proximity effects) are small in the well-detached case and do not degrade light and velocity solutions. However tides and reflection have been solved problems for over 30 years (e.g. Lucy 1968; Wilson & Devinney 1971; Mochnacki & Doughty 1972), and they actually somewhat strengthen results by providing extra information. They certainly do not weaken results. Reflection could possibly have some minor adverse consequences for radial velocities, as one cannot expect spectra of irradiated surfaces to be identical to non-irradiated surfaces with the same T_{eff}. However, insofar as the problem can be handled in terms of local T_{eff} and $\log g$, it has been examined and the consequences for velocity measures turn out to be small (Van Hamme & Wilson 1994). The WD light/velocity program (Wilson & Devinney 1971; Wilson 1979, 1990) computes radial velocity proximity effects routinely at user option. Therefore proximity effects are not a significant problem for light or velocity solutions, and in reality are mildly helpful.

Light Curve Steadiness. DBs are considered to have relatively steady light curves, but many SDs and OCs also have steady light curves. We expect to have a million GAIA binaries and can focus on the steady ones. The most important consideration for overall light curve scatter usually is ordinary observational noise. Figure 1 shows an OC light curve *with excellent coherence over many cycles*, and it is not an unusual case – many similar and even better ones could be shown. Many OCs are very stable.

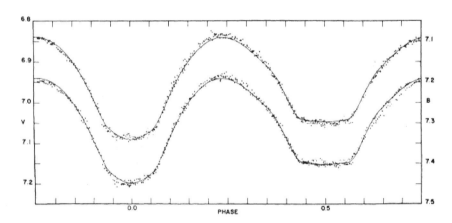

Figure 1. Paczynski's (1964) light curves of the OC binary AW UMa, illustrating excellent coherence over 328 cycles. Note that the full width of the scatter band is less than $0^{m}_{.}02$. Figure reprinted from the Astrophysical Journal (from Wilson & Devinney 1973, ApJ 182, 539).

Ease of Absolute Flux Prediction. Estimation of T_{eff} and $\log g$ from spectra is a very important part of EB luminosity and distance measurement that may be perceived as less of an impediment to accuracy for DBs than for SDs and

OCs, where we have *surface distributions* of those quantities rather than one T_{eff} and one $\log g$ for each star. However that is another solved problem, as there are computer programs (e.g. Linnell & Hubeny 1994) that reproduce spectra of tidally distended and irradiated close binaries. So finding T_{eff} and $\log g$ for SDs and OCs may require some extra work, but entails little loss of accuracy. The key point is that light curve computation basically begins from a reference point on each star (usually the $+z$ pole) for which we have radiative intensity as a function of direction and wavelength, with other surface points linked to the reference point via effects such as gravity brightening and reflection. Computation then ends with integrated flux for a distant observer. That is, we need to be explicit in defining the meaning of parameters T_{eff} and $\log g$ – they apply at the reference point. A practical problem is to find the scaling factor that converts relative fluxes thus derived to absolute fluxes. Rigorous computation of that factor ensures generation of absolute (cgs) SD and OC light curves that are essentially as accurate as those of DBs. Proper scaling relations now exist (Wyithe & Wilson 2003, in prep.) to convert bandpass-integrated output to absolute flux seen at a distance.

4.2. SD and OC binaries: accuracy considerations

Well known in data fitting problems is the importance of keeping the number of estimated parameters reasonably small. Every assignment of a definite value from external evidence or confidently known theory solidifies results on remaining (i.e. adjusted) parameters. Also very important in strengthening solutions is adoption of functional relationships based on well established theory, such as the relation between lobe size and mass ratio for SDs. See Wilson (1979) for discussions of such applied relationships as constraints on solutions. Having noted that SDs and OCs are economical of parameters in several ways, we next examine that issue in several contexts.

Eccentricity. A great merit of SD and OC binaries is that their orbits are circular in essentially all examples, so that we have *a priori* knowledge that the orbital eccentricity, e, is zero. Perusal of many light curves of true Algol type EBs (the classical SDs) finds secondary eclipses always separated half a cycle from primary eclipses, as expected for circular orbits. While half-cycle separations happen also for eccentric orbits viewed down their major axes, half-cycle separations for Algols *as a class* cannot be so explained. Some binaries that have been *called* Algols violate the $e = 0$ rule, but almost certainly it is because they are not SD and therefore not true Algols. If we see clear evidence for an eccentric orbit, it is telling us that the binary is detached, and probably rather well detached. Here we save two otherwise free parameters, e and ω, thanks to the strong tidal circularization that goes with lobe filling and lobe excess. These are indeed important parameters for several reasons. For example, non-zero e upsets the straightforward connection between eclipse depth and surface brightness that circular orbit solutions lean upon. A connection remains for eccentric orbits but it is greatly complicated, and thereby effectively undermined, by eccentric orbit geometry and its inevitable uncertainties. Argument of periastron ω introduces both its own uncertainties and an annoying practical problem for *automatic* solutions, in that there is no preferred initial estimate. For most light curve parameters, safe starting values can be guessed from simple rules, with relative radii ($r = R/a$) usually between 0.1 and 0.5, inclinations between 70° and 90°, and eccentricities near zero for potentially useful EBs. However ω's range unbridled over 2π radians and, although rough estimation algorithms can be written, they require inputs such as eclipse width and eclipse phase that are difficult to measure by automatic means without having done a full light curve solution. Of course there is no need for initial estimates if we have a full solution.

Non-synchronism. Here we may save rotation parameters for one or both stars thanks to tidal synchronization. Likelihood of synchronous rotation increases sharply with approach of a star's surface to its Roche lobe (e.g. Stothers 1973), so the contact component of an SD and both components of an OC basically have known photospheric rotation states. There could be small side issues, such as minor departures from uniform rotation, but with little effect on light/velocity solutions. However nature has not been entirely kind in keeping SDs simple, as the detached components of some SDs have experienced spin-up by transfer of matter from their lobe filling stars, a process that converts orbital angular momentum to spin angular momentum. Rotation statistics for 26 and 36 Algols respectively are in Wilson (1988) and Van Hamme & Wilson (1990). Identification of *very fast* rotation via spectral line broadening should be straightforward, but effects of modest levels of asynchronous rotation on *MLRD* accuracy need to be quantitatively assessed via simulations.

Star to star streams, hot impact areas, and circumstellar disks. Lobe overflow in a typical SD results in a stream of gas from the contact star that may strike the detached star directly, or may loop around to self-intersect and subsequently fall onto the detached star, or may form a quasi-steady circumstellar disk hydrodynamically. Such phenomena are absent from sensibly detached binaries because there is no stream, and absent from OCs because there is no place to put either a stream or disk, so they are basically SD phenomena and are diagnostic of the SD condition. Many Algol type binaries clearly have some combination of these features, as shown by emission lines with characteristic Doppler behavior, and one may conjecture that they are present in all Algols at some level, although not always readily detectable. Light curve solutions of some SDs may be significantly degraded by such complications, but spectra can tell which SDs have disks and streams prominent enough to undermine results.[1] Disk/stream emission lines are weak in most Algols, often being detected only in total eclipses of the bright component, so the flux in bright lines may be negligible in broad band and medium band photometry. In overview, streams and disks are a serious problem only for a minority of Algols and other SDs, except for cataclysmic variables which are unlikely candidates for accurate *MLRD* work. Streams and disks do not exist in OCs.

Observing efficiency – how many points in eclipse? A glance at the many EB light curves produced as by-products of gravitational lensing surveys (e.g. Grison et al. 1995, Alcock et al. 1997, Udalski et al. 1998), or perusal of EB papers, or visualization of the relevant geometry, shows more points in eclipse with SDs and especially with OCs than with DBs. Many of the DB light curves in Udalski et al. have stiletto-like eclipses that contain only a very small fraction

---

[1]Strong stellar winds can produce accretion disks and complicated emission line spectra but mainly for rare binaries, too exotic for accurate *MLRD* applications.

of the data points. Coupling this point with the fact that by far most light curve information is in the eclipses, we realize that SDs and OCs make relatively efficient use of observing time compared to DBs, which is not a small point.

Interchanged radii (aliasing). Among light curve solution problems is a fairly well known one that we call *aliasing*, whereby approximately correct radii are found but they are interchanged. Simulations with synthetic light curves (Wyithe & Wilson 2001, 2002) show that aliasing is less of a problem with SDs and OCs than with DBs.

4.3. Overall considerations – often neglected, seldom stressed

Complete vs. partial eclipses. Intuition suggests that *complete eclipses must lead to far more reliable light curve solutions than do partial eclipses*. The point is well known, although well known is not the same as widely appreciated. Simulations of solution statistics for large numbers of synthetic binaries (Wyithe & Wilson 2001, 2002) confirm intuition, as does experience with individual binaries. The essential argument can be cast in terms of the number of independent functional relations that come from light curves of completely eclipsing and partially eclipsing binaries, *vis-à-vis* the number of geometrical free parameters, and is given in Wyithe & Wilson (2003, in prep.), or more briefly in Wilson (1978). Basically the number of empirical relations matches the number of parameters for complete eclipses, while for partial eclipses there effectively are fewer relations than parameters.

Value of photometric mass ratios. The existence of q_{ptm}'s for SDs and OCs can be explained quite simply. In the SD case, the dimensionless mean radius of the lobe filling star, R/a, allows measurement of lobe size, R_{lobe}/a, which is a well known function of mass ratio. The radii in an OC, taken together, measure how the OC 'dumbbell' figure divides and also measure overcontact level. Of course, 'how the dumbbell divides' depends on mass ratio, so again we have a q_{ptm}. OCs are better than SDs for q_{ptm} because *two* radii contribute information while only one radius contributes for SDs. The q_{ptm} concept essentially does not apply to DBs, where neither radius contributes. Since radii are far more reliably measured for complete than for partial eclipses and both kinds of q_{ptm} derive from radii, by far the most accurate q_{ptm}'s are from completely eclipsing systems. *Variation due to tidal deformation (ellipsoidal variation) gives almost no q_{ptm} information*, as shown by simulations and contrary to a common and occasionally published misconception. An easy way to see this point is to notice that some binaries with very little ellipsoidal variation have rather strongly determined q_{ptm}'s, while light curves of others with substantial ellipsoidal amplitudes yield no mass ratio information.

Proximity effects – do they help or hurt? Exploitation of well-understood hydrostatic theory does allow somewhat strengthened solutions for SDs, OCs, and even to a small extent for mildly detached binaries. That is, we know how to compute static tides and can utilize that knowledge to bolster solutions. One should not over-emphasize this point, as by far most EB light curve information is in eclipses, but the old view that tides (and reflection) are nuisances to be circumvented fails to recognize their usefulness as solution handles at a modest level.

Third light. A spatially unresolved light source in or near an EB reduces light curve amplitude (in magnitudes) with no effect on form, and disregard of third light (l_3) can lead to significantly wrong results. One can solve for l_3, but the hold on it may be weak, with uncertainties introduced into other parameters through correlations. The uncertainties can be major for large l_3. As a rule, partially eclipsing solutions are already weak and cannot afford a significant l_3 problem. Can l_3 be estimated from a spectral energy distribution (SED)? In principle, yes, but in practice an adequate SED usually is not at hand and its analysis is another, perhaps difficult, fitting problem. Third light is a potential problem for any EB, regardless of type, and one must remember that triple, quadruple, etc. systems are not rare. Algol, the prototype of EBs, has a third star considerably more luminous than the photometric secondary star. A practical course of action is to be vigilant for evidence of l_3 and discard binaries for *MLRD* work where it is suspected.

Simultaneous solutions. GAIA's multi-band photometry should be analyzed as a coherent set for each EB, as some parameters are logically identical in all bands. Separate analyses would introduce multiple values for i, q, etc. and needlessly increase the number of free parameters and weaken overall results. Similarly, simultaneous light/velocity solutions save on parameters and improve results. Necessary weighting schemes are discussed in Wilson (1979). *Automatic* weighting will require programming that might best be incorporated within Expert Systems logic.

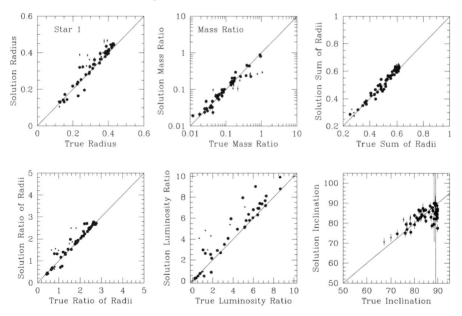

Figure 2. Parameters derived from simulated EB light curves of SD binaries compared with known values. Large dots are for complete eclipses and small dots for partial eclipses. The standard deviation in the maxima of each synthetic binary is $0\overset{m}{.}02$. Error bars are 1σ.

4.4. Solution simulations

Extensive simulations that can be used to compare the usefulness of DB and SD light curve solutions are discussed and illustrated by Wyithe and Wilson (2001, 2002), and analogous OC simulations will be in Wyithe and Wilson (2003, in prep.). However those are for data of survey accuracy and we also need simulations for follow-up accuracy, which are shown for SD solutions in Fig. 2. Note that complete eclipse results follow considerably more tightly along the 45° lines than do partial eclipse results. Space does now allow discussion, but the conclusions of Wyithe and Wilson (2002) essentially apply.

References

Alcock, C., et al. 1997, AJ 114, 326

Grison, P., et al. 1995, A&AS 109, 447

Kopal, Z. 1954, Jodrell Bank Ann. I, 37

Kopal, Z. 1955, Ann. d'Ap. 18, 379

Kopal, Z. 1959, Close Binary Systems, John Wiley & Sons

Kuiper, G.P. 1941, ApJ 93, 133

Linnell, A.P., & Hubeny, I. 1994, ApJ 434, 738

Lucy, L.B. 1968, ApJ 153, 877

Mochnacki, S.W., & Doughty, N.A. 1972, MNRAS 156, 51

Paczynski, B. 1964, AJ 69, 124

Paczynski, B. 1997, in The Extragalactic Distance Scale, M. Livio ed., Cambridge U. Press, pag. 273

Paczynski, B. 2000, AAS 196, 1001

Rucinski, S.M. 1997, IBVS 4460

Russell, H.N. 1948, Harvard Obs. Monograph No. 7, 249

Stothers, R. 1973, PASP 85, 363

Udalski, A., Soszynski, I., Szymanski, M., Kubiak, M., Pietrzynski, G., Wozniak, P., & Zebrun, K. 1998, Acta Astron. 48, 563

Van Hamme, W. & Wilson, R.E. 1990, AJ 100, 1981

Van Hamme, W. & Wilson, R.E. 1994, Mem.Soc.Astron.Ital. 65, 89

Wilson, R.E. 1978, ApJ 224, 885

Wilson, R.E. 1979, ApJ 234, 1054

Wilson, R.E. 1988, in Critical Observations vs. Physical Models for Close Binary Systems, K. C. Leung ed., Gordon & Breach, Montreux, pag. 455

Wilson, R.E. 1990, ApJ 356, 613

Wilson, R.E. 2000, Eclipsing Binary Stars, in MacMillan Encyclopedia of Astronomy and Astrophysics, Bristol, Institute of Physics, pag. 706

Wilson, R.E. 2001, IBVS 5076

Wilson, R.E., & Devinney, E.J. 1971, ApJ 166, 605

Wyithe, J.S.B., & Wilson, R.E. 2001, ApJ 559, 260

Wyithe, J.S.B., & Wilson, R.E. 2002, ApJ 571, 293

Stellar atmospheres in eclipsing binary models

Walter Van Hamme

Department of Physics, Florida International University, Miami, FL 33199, USA

Robert E. Wilson

Astronomy Department, University of Florida, Gainesville, FL 32611, USA

Abstract. Stellar atmosphere emergent intensity is needed to model binary star observables, preferably by a compact, fast, and accurate method. In our strategy, only small numerical files are needed, overall light curve computation speed is affected only slightly, and results reproduce those of full stellar atmospheres with good fidelity. The strategy is based not on effective wavelength but on integrated bandpass response, as approximated by series of Legendre polynomials. The only interpolation is in $\log g$. Figures show the quality of Legendre representation of atmosphere intensities. Importance of speed, portability, and modularity are discussed. A unique feature is smooth transition between atmosphere and blackbody treatment so as to avoid radiative discontinuities in very high and very low temperature regions. One can thereby apply atmosphere computations to stars that have some surface parts hotter or cooler than existing atmosphere models allow.

1. Introduction

Essential for modeling eclipsing binary (EB) observables is the emergent intensity in the direction of the observer at various surface points. In the case of light curves, local intensities are needed to obtain differential fluxes and, after surface integration, the total light from each star. Local differential flux is a weight factor for radial velocity proximity effects (e.g. Wilson & Sofia 1976). Intensities and fluxes depend on local effective temperature (T_{eff}), surface gravity ($\log g$), composition and wavelength (or in terms of actual measurements, bandpass). Several groups have improved the stellar atmosphere part of binary star light curve computation. Milone, Stagg & Kurucz (1992) introduced a version of the WD program (Wilson & Devinney 1971; Wilson 1979, 1990) with a routine based on Kurucz (1979) stellar atmospheres. They computed ratios of model atmosphere to blackbody fluxes in a T_{eff}, $\log g$, wavelength grid. After fluxes are integrated over bandpass response curves (see Kallrath et al. 1998 for available bands) for one of several discrete surface gravities, the procedure interpolates in temperature to obtain the atmosphere to blackbody flux ratio. Linnell and co-workers (Linnell 1984; Linnell et al. 1998) generate a synthetic spectrum and sample its continuum across each band. Fluxes and intensities are integrated over filter transmission curves to obtain bandpass quantities. The Linnell et al. (1998) approach requires significant computation time, especially for solutions of large sets of light curves (say as an analog of the 1459 solutions by Wyithe & Wilson 2001, 2002), as a full spectrum based on many local atmosphere intensities is calculated at each phase. Orosz & Hauschildt (2000) developed a light curve synthesis program that makes direct use of the Hauschildt et al. (1999b) atmosphere intensities. Their program contains a grid of bandpass-integrated intensities ($UBVRIJHK$) at various temperatures, $\log g$'s and zenith angles (θ). Linear interpolation in T_{eff}, $\log g$ and $\mu = \cos\theta$ yields intensity. Our work also is based on bandpass integrations. An effective wavelength approach may be adequate for spectral regions that have reasonably simple behavior but, as commented in many papers and recently emphasized by Linnell et al. (1998), it is not adequate for the Johnson U band, which is strongly influenced by the Balmer jump and the confluence of high number Balmer lines in B, A, and F-type stars. In general, bandpass-integrated light curves are an improvement on 'effective wavelength' light curves for almost any band.

GAIA promises to deliver light and velocity information on one million EBs, so efficient extraction of astrophysical information will require fast programs. We developed a compact and portable scheme to implement pre-computed atmosphere model intensities. It has now been put into the WD binary star program without increasing computation time significantly.

2. Method details

We used normal emergent intensities in CD-ROMs 16 and 17 described in Kurucz (1993), for microturbulent velocity 2 km s^{-1}. Intensities are given at 1221 wavelengths from 9 to 160 000 nm and 11 $\log g$'s from 0.0 to 5.0 (cgs). Temperature ranges depend on $\log g$ and abundance [M/H], with the largest range from 3500 K to 50 000 K. Temperature limits and the 19 abundances are in Kurucz (1993). Bands currently used and references for bandpass response are in Wilson & Van Hamme (2003).[1]

For each atmosphere model we integrated intensities over each band, weighted by response function, and similarly integrated blackbody intensities. Irrespective of abundance or surface gravity, Legendre polynomials represent intensities very well as functions of T_{eff} over modest-sized subintervals. Experiments showed the optimum number of subintervals to be 4, with beginning and end points dependent on bandpass and model. Accordingly four T_{eff} subintervals were bounded by lower (T_{l}) and upper (T_{h}) effective temperatures. We then scaled T_{eff} according to $\phi_{\text{T}} = (T_{\text{eff}} - T_{\text{l}})/(T_{\text{h}} - T_{\text{l}})$ and made Least Squares Legendre fits of degree m with ϕ_{T} as independent variable. Selection of m was based on the number of points in a subinterval. Values of m equal to one third the number of points, up to a maximum of $m = 9$, worked well. The number of coefficients to be estimated was $m + 1$, so the maximum number of Legendre coefficients was 10. Use of the end point of each subinterval as the starting point

---

[1] Monograph available at FTP site ftp.astro.ufl.edu, directory pub/wilson/lcdc2003

of the next greatly reduced discontinuities at boundary points. Mean deviations between actual and computed intensities rarely exceeded a few percent, with typical deviations of just a few times 0.001 in $\log_{10}$ of intensity. Figure 1 shows model intensities and Legendre fits for one band, Figure 2 shows the root mean square residual for each model and band.

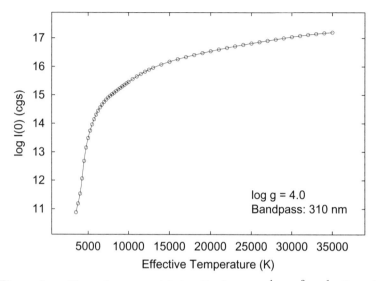

Figure 1. Normal emergent intensity in $\mathrm{erg\,s^{-1}\,cm^{-3}\,sr^{-1}}$ of a solar composition atmosphere for the band and $\log g$ indicated. Circles are stellar atmosphere intensities and lines represent the Legendre fits in separate intervals (3500, 4250 K), (4250 K, 7500 K), (7500 K, 7750 K) and (7750 K, 35 000 K).

A file was made with a block for each of the 19 compositions, each block listing the temperature limits and Legendre coefficients for every band, $\log g$, and temperature sub-interval. With 11 $\log g$'s, 25 bands, 19 compositions, 4 temperature subintervals, and 10 Legendre coefficients with 2 temperature limits per subinterval, the data file contains 250,800 numbers. A similar Legendre blackbody file spans 500 to 500 000 K and is very small, as the only dimensions are bandpass and temperature sub-interval. The files can be incorporated into other binary star programs to calculate model and blackbody intensities.

3. Atmosphere to blackbody transitions

In many close binaries, especially those with tidally distorted components, at least one of the stars has part of its surface outside the range of available atmosphere models. Figure 3 shows how T_{eff} and $\log g$ change along the central meridian of the primary star in TU Muscae, an O8 overcontact binary. Between about 55° and 75° co-latitude, T_{eff} and $\log g$ are outside the range of Kurucz atmosphere models. To abandon atmosphere models for the entire star would be a severe adverse consequence of the range limitation, yet a simple blackbody

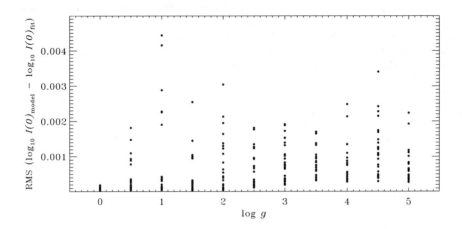

Figure 2. Root mean square residuals of model and fitted intensities. For each $\log g$, there are data points for each of the 25 bands.

patch would impose an artificial discontinuity that could introduce very bad effects in computed light curves. One expects similar problems in stars with magnetic spots, in Algols, where the contact star's temperature can be very low over parts of the surface, and in late-type W UMa overcontact binaries with low gravity connecting necks.

A novel feature of the procedure is that, to avoid any discontinuity, we establish an atmosphere to blackbody transition region in T_{eff} and in $\log g$. We call the transition operation *ramping*. If the T_{eff}, $\log g$ combination is outside the range of atmosphere applicability, the program smoothly connects atmosphere model intensities to bandpass blackbody intensities over built-in ranges in $\log g$ and T_{eff} whose limits can easily be changed. There are two alternative kinds of ramps to ensure near-linearity, with the transition gap a function of $1/T_{\text{eff}}$ or $\log_{10} T_{\text{eff}}$, depending on the location of the band's effective wavelength (for the mid-point of the transition interval) with respect to the blackbody peak. Below the peak we ramp in $1/T_{\text{eff}}$ and above we ramp in $\log_{10} T_{\text{eff}}$. There are transition regions also in $\log g$ for cases where $\log g$ drops below 0.0 or goes above 5.0. Two ramp parameters set the low and high $\log g$ intervals and another two set the low and high T_{eff} intervals. Some cases can have transitions in *both* $\log g$ and T_{eff} (double ramping). At the lower T_{eff} limit (3500 K in most cases), there is a fixed temperature transition interval. At the upper T_{eff} limit, which depends on $\log g$, the transition interval scales with the temperature upper limit. Ramping is illustrated in Figure 4.

4. Overall assessment and planned improvements

Our Legendre-generated intensities reproduce stellar atmosphere intensities with errors typically smaller than astrophysical uncertainties in the original atmospheres. The only interpolation is in $\log g$. Light curve program run time is not

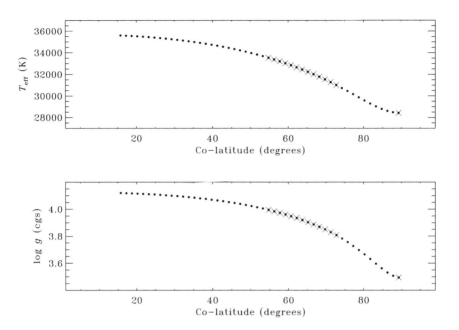

Figure 3. T_{eff} and $\log g$ along the central meridian of the primary component of TU Mus. For $\log g = 3.5$, atmosphere models up to $T_{\text{eff}} = 31\,000$ K are available; for $\log g = 4.0$, we have model atmospheres up to $T_{\text{eff}} = 39\,000$ K. An atmosphere to blackbody transition region (crossed points) is needed in the co-latitude band between 55° and 75°. Note that this band is not near the high temperature polar region because of the specifics of the upper temperature limit – $\log g$ dependence.

noticeably affected, an important advantage when the program will be applied to massive amounts of observational data from instruments like GAIA.

We intend to add bandpasses, for example those of the GAIA photometric systems, as well as lower temperature atmosphere models. The atmosphere models of Hauschildt and collaborators (Hauschildt et al. 1999a,b) are suitable candidates.

References

Hauschildt, P. H., Allard, F. & Baron, E. 1999, ApJ 512, 377
Hauschildt, P. H., Allard, F., Ferguson, J., Baron, E. & Alexander, D. R. 1999, ApJ 525, 871
Kallrath, J., Milone, E. F., Terrell, D., & Young, A. T. 1998, ApJ 508, 308
Kurucz, R. L. 1979, ApJS 40, 1

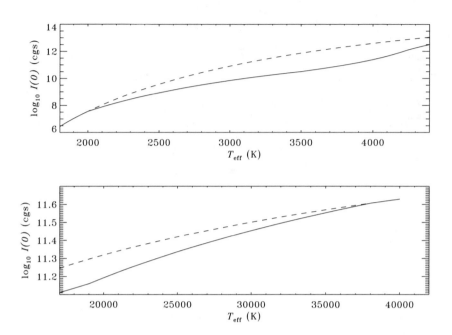

Figure 4. *Top:* 310 nm band atmosphere to blackbody transition for a $\log g = 2.5$, solar composition model. The continuous line is the relation actually used, the dashed curve is blackbody intensity. Full atmosphere treatment is above 3500 K, full blackbody treatment is below 2000 K. *Bottom:* N band transition from 19 000 K to 39 000 K.

Kurucz, R. L. 1993, in Light Curve Modeling of Eclipsing Binary Stars, E.F. Milone ed., Springer-Verlag, pag. 93
Linnell, A. P. 1984, ApJS 54, 17
Linnell, A. P., Etzel, P. B., Hubeny, I. & Olson, E. C. 1998, ApJ 494, 773
Milone, E. F., Stagg, C. R. & Kurucz, R. L. 1992, ApJS 79, 123
Orosz, J. A. & Hauschildt, P. H. 2000, A&A 364, 265
Wilson, R. E. 1979, ApJ 234, 1054
Wilson, R. E. 1990, ApJ 356, 613
Wilson, R. E. & Devinney, E. J. 1971, ApJ 166, 605
Wilson, R. E. & Sofia, S. 1976, ApJ 203, 182
Wyithe, J. S. B. & Wilson, R. E. 2001, ApJ 559, 260
Wyithe, J. S. B. & Wilson, R. E. 2002, ApJ 571, 293

Observational tests of the GAIA expected harvest on eclipsing binaries

Tomaž Zwitter

University of Ljubljana, Department of Physics, Jadranska 19, 1000 Ljubljana, Slovenia

Abstract. GAIA observations of eclipsing binary stars will have a large impact on stellar astrophysics. Accurate parameters, including absolute masses and sizes will be derived for $\sim 10^4$ systems, orders of magnitude more than what has ever been done from the ground. Observations of 18 real systems in the GAIA-like mode as well as with devoted ground-based campaigns are used to assess binary recognition techniques, orbital period determination, accuracy of derived fundamental parameters and the need to automate the whole reduction and interpretation process.

1. Introduction

GAIA observations of eclipsing binary stars will be of utmost importance to advances in stellar astrophysics. For no other class of objects one could determine fundamental stellar parameters, i.e. absolute mass, size and surface temperature distribution with a comparable accuracy. Solutions of wide detached binaries can be used to accurately position them on the absolute H-R diagram. Identical age of both components places useful constraints on the theoretical isochrones for the given metallicity and rotational velocity which will also be derived from GAIA observations. Components in short period systems are closer and mutually disturbed, so their evolution is different from that of single stars. But accurate surface temperatures and sizes derived from binary solutions fix their luminosity and are useful to gauge their distance even for objects that are too far for the astrometric capabilities of the satellite (see Wyithe & Wilson 2002).

Availability of on-board spectroscopy is vital to the study of eclipsing binaries. Semi-major axis and stellar masses could not be determined in any other way, and additional information on metallicity and rotational velocity helps in physical interpretation. One might argue that this information could be obtained by ground-based follow-up observations. In our experience this is not feasible. In Asiago we launched an intensive campaign to spectroscopically observe eclipsing binaries discovered by Hipparcos (Munari et al. 2001 [hereafter M2001], Zwitter et al. 2003, in press, [hereafter Z2003]). After three years we barely finished the spectroscopic coverage of the first 18 systems. Hipparcos discovered nearly 1000 systems, GAIA will see hundred thousands. These objects are distributed over the whole sky, so fiber optic spectroscopy cannot reduce the required observing time significantly.

The strength of the GAIA mission is in the numbers. GAIA will observe $\sim 4 \times 10^5$ eclipsing binaries brighter than $V = 15$, $\sim 10^5$ of these will be double-lined systems (M2001). Even if the stellar parameters will be determined at 1% accuracy only for 1% of them this is still 25-times more than what has been obtained from all ground-based observations in the past (cf. Andersen 1991). Moreover most of the GAIA binaries will be of G-K spectral type (cf. Zwitter & Henden 2003) where there exists only a small number of systems with accurate solutions.

Various aspects of eclipsing binaries are discussed by Milone (2003), Wilson & Wyithe (2003), Van Hamme & Wilson (2003) and other contributions to this conference. Here we focus on our experience obtained from real stars that were observed in the GAIA-like mode. We start with discussion of how an object is recognized to be a binary and a determination of its orbital period. Next we discuss the accuracy of derivation of its fundamental parameters and the possibility to detect intrinsic variability of stars in binaries. We close with some general remarks on the types of binaries that will be discovered. We stress that huge numbers of objects call for completely automated reduction and possibly even interpretation techniques.

2. Orbital period from multi-epoch observations

A large fraction of binary stars with orbital periods over a month that are closer than 1 kpc will be discovered astrometrically. Systems with periods of up to 10 years will be recognized due to their non-linear proper motion and those with periods of over a century will be resolved (ESA SP-2000-4). Systems with orbital periods of less than a month will be mainly discovered by their photometric and spectroscopic variability.

GAIA is unique because it will re-observe the same region of the sky many times over. The number of transits for the spectroscopic focal plane will be around 100 with extremes a factor 2 higher or lower. The transits are not distributed evenly in time (see Figure 1). This should pose no problems in analysis of binary stars if the satellite rotation and precession periods are kept incommensurable. The sampling permits a good phase coverage of all orbital periods that are shorter than the mission lifetime. Also the duration of individual focal plane passages is just 100 seconds, so orbital motion smearing is negligible.

Photometric variability does not need to be a consequence of the binarity of the source: pulsations, rotating and time-dependent stellar spots, as well as different types of semi-regular variables will be common among the G and K stars that will be the most frequent type of objects observed. The best way to recognize that the detected photometric variability is indeed due to binarity is by establishing its repeatability and light curve shape; so the orbital period needs to be determined. The same is true for spectroscopic observations. The exceptions are of course double-lined binaries where a quarter-phase spectrum with well separated lines immediately suggests a binary nature of the source.

Potentials of photometry and spectroscopy for determination of orbital period are different, with spectroscopy being always preferable. As an example let us examine a detached system, GK Dra (Figure 2). It was discovered by the Hipparcos satellite. But a search for orbital period from the 124 Hipparcos ob-

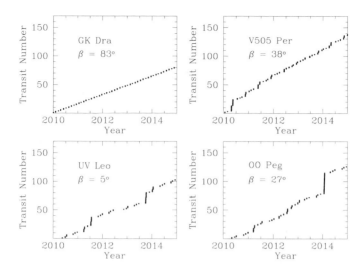

Figure 1. Accumulation of transits over the spectroscopic focal plane during the 5-year mission lifetime for four examples of binary stars. Dynamics of observations depends on the ecliptic latitude of the target (β). GK Dra is located close to ecliptic pole, so the transits are almost periodic. Coverage of stars at other ecliptic latitudes is more patchy.

servations proved unsuccessful. There are several possible periods with the most likely solution of 16.96-days (Figure 2, top). This value, which is also quoted in the Hipparcos catalogue, is immediately shown to be wrong by only 35 spectroscopic observations in the GAIA spectral window obtained by our GAIA-like ground-based observing campaign (Figure 2, middle; Z2003). The advantage of spectroscopic monitoring is in the fact that radial velocities are constantly changing with orbital phase. So every point contributes to the orbital period search. In the case of photometry the light curve out of eclipses is flat, so determination of orbital period is based only on a couple of points within the eclipses. An extensive photometry can of course resolve this problem, but note how a periodogram from over 1300 photometric observations (Figure 2, bottom) is still more ambiguous than the one obtained from only 35 spectroscopic observations in the GAIA spectral window. The strength of spectroscopy for the orbital period determination can also be seen from the light curves in Figure 3.

Photometric information obtained by the GAIA satellite will be far superior to that of Hipparcos. GAIA will reach much fainter magnitudes and observe in many broad and narrow photometric bands. But the number of epoch observations will be similar to that of Hipparcos. The coverage of GK Dra to be obtained by GAIA is given in Figure 4. Note that only a single observation in broad band filters falls within eclipses. The eclipse coverage in intermediate passband filters is better. Broad- and narrow-band photometry can be used to measure orbital inclination as well as relative sizes and absolute temperatures of both stars. But the orbital period itself will be much easier to determine

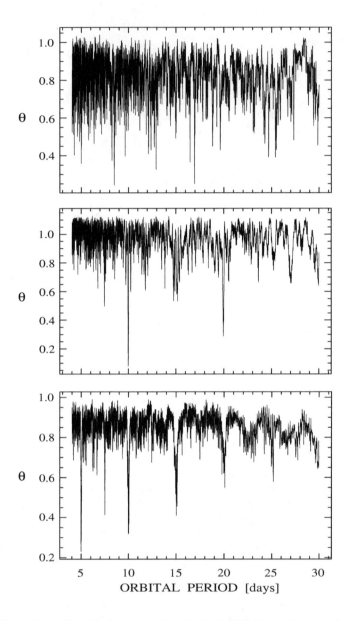

Figure 2. Search for orbital period of GK Dra using a phase dispersion minimization method (Stellingwerf 1978) on three data sets. *Top:* 124 Hipparcos observations in the H_P-band. *Middle:* 35 spectroscopic measurements of radial velocity of the primary star. *Bottom:* 1323 ground-based V band photometric observations (Dallaporta et al. 2002a). Hipparcos data favour the 16.96 day period, while spectroscopy and dedicated photometry identify the correct value of 9.97 days.

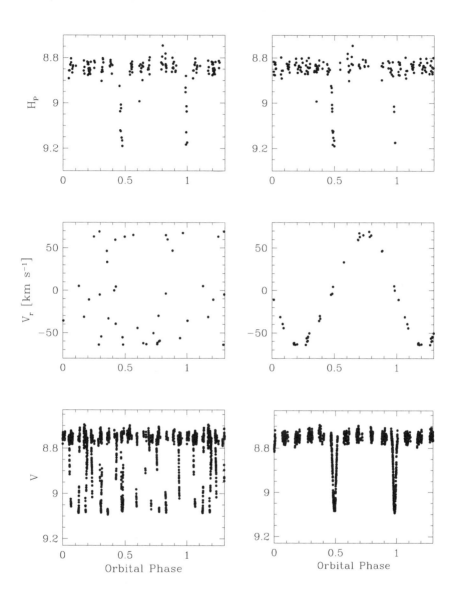

Figure 3. Orbital phase plots of GK Dra for orbital period of 16.96 days (*left* panels) and 9.9742 days (*right* panels). *Top:* Hipparcos observations in the H_P-band. *Middle:* spectroscopic measurements of radial velocity of the primary star. *Bottom:* ground-based V band photometric observations.

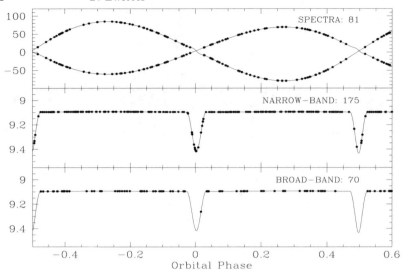

Figure 4. Phase coverage of spectroscopic, narrow- and broad-band photometric observations for a typical eclipsing double-lined spectroscopic binary. Data points mark individual passages of GK Dra over spectroscopic, narrow- and broad-band filter focal planes between 1 Jan 2010 and 31 Dec 2014. Note a nice phase distribution of spectroscopic observations and a modest photometric coverage of eclipses.

from spectroscopy. A total of 81 spectra are well distributed over orbital cycle, so the orbital period, semi-major axis, and both masses will be unambiguously measured from spectroscopic radial velocity measurements.

3. Accuracy of fundamental parameters

As mentioned earlier we are observing 18 Hipparcos binaries in the GAIA-like mode. This means we are trying determine their orbital solution and fundamental parameters using only Hipparcos (H_P, B_T, V_T) photometry and ground based spectroscopy in the GAIA spectral range. Spectroscopic data are extracted from a single Echelle order observation with the 1.8-m telescope of the Asiago observatory. It turns out that such an approach is realistic as the accuracy of the solution is limited by a rather small (~ 100) number of Hipparcos photometric measurements resulting in a poor coverage of eclipses. This will be also the case with GAIA. The results we obtain should present a lower limit to the expected GAIA accuracy, as we are using only rather noisy Hipparcos photometry, while GAIA photometry will have excellent precision and a much larger number of photometric bands. Figure 5 presents the light curve shapes of the overcontact binary V781 Tau obtained in different narrow and broad passbands. Note that the photometric accuracy for objects brighter than $V \sim 18$ will be better than 0.01 mag, so even rather subtle differences in the light curve shape and magnitude level for this binary with $T_1 - T_2 = 170$ K will be easily discernible.

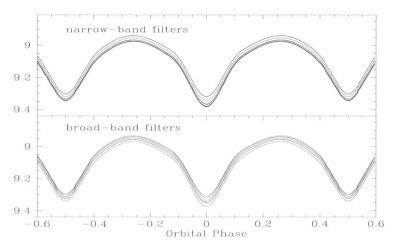

Figure 5. Simulated light curves of an overcontact binary V781 Tau for the narrow- (*top*) and broad-band (*bottom*) GAIA filters proposed by Munari (1999). Each curve will be sampled with ~ 100 points with errorbars not exceeding 0.01 mag at $V=18$.

In Table 1 we quote accuracies of fundamental parameters for 6 systems published so far (M2001, Z2003). We note that relative errors in most parameters are 2% or lower. The exceptions are individual radii which are not well determined due to a scarce photometric coverage of eclipses. Solutions of UV Leo and GK Dra also have large uncertainties. This is due to their intrinsic variability (see below). V781 Tau is a binary which fills its Roche lobe up to the L2 point. Temperatures and sizes of such binaries can be accurately determined. So the distance can also be calculated with a remarkable accuracy.

4. Intrinsic variability

Many stars of G and K spectral types are intrinsic variables. This is true also for binary members. With this goal in mind the observations in that GAIA-like mode that use only Hipparcos photometry with ~ 100 observations of each star were supplemented by devoted ground-based photometric campaigns (Dallaporta et al. 2000, 2002a,b Mikuz et al. 2002, Frigo et al. 2002).

UV Leo is a detached system with surface spots which cause a variation of system brightness by ~ 0.04 mag (Mikuz et al. 2002). The system also showed a sudden change in the orbital period in Feb. 1981, possibly due to a passage of a low-mass third body. Devoted photometry of GK Dra (Figure 3, bottom right) shows unusually large scatter. It turns out that the differences between the binary solution and observations are not due to noise but point to an intrinsic variability of δ-Sct type (Figure 6).

A limited number of photometric and spectroscopic observations obtained by GAIA will make it difficult to study intrinsic variability of the binary components. Still a large number of photometric bands will easily point to temper-

Table 1. Accuracy of fundamental parameters obtained from observations in the GAIA-like mode. Quoted errors are formal mean standard errors to the solution.

object type sp. type	V570 Per detached F5	OO Peg detached A2	V505 Per detached F5	V781 Tau overcontact G0	UV Leo detached G0	GK Dra detached G0
a	0.7%	0.5%	0.5%	0.2%	1.2%	0.5%
$mass_1$	2.3%	1.7%	1.5%	1.8%	7%	3.5%
$mass_2$	2.5%	1.8%	1.6%	1.8%	6%	3.3%
T_1	150 K	150 K	40 K	50 K	100 K	100 K
T_2	180 K	180 K	60 K	30 K	100 K	100 K
R_1	10%	4%	1.4%	0.4%	2%	1.5%
R_2	25%	4%	3%	0.3%	2%	1.7%
distance	6%	6%	7%	1.5%	20%	10%

ature changes that do not repeat with orbital cycle and are so due to intrinsic variability of the binary components. Interesting cases could be picked for detailed follow-up observations. These include new interacting binaries (Cropper & Marsh 2003).

5. Non-eclipsing and non-spectroscopic binaries

So far we discussed eclipsing double-lined spectroscopic binaries. In majority of cases we will be less fortunate. The systems could be too faint to obtain any useful spectroscopy, non-eclipsing or single-lined. We discuss these in turn.

For systems fainter than $V = 15$ spectroscopic radial velocities will be difficult to measure even in double-lined cases. These faint objects will far outnumber the bright spectroscopic binaries. Binarity will have to be established from eclipses or a reflection effect, both measured with a large number of photometric passbands but in a limited number of epochs. Some problems with the determination of orbital period of such systems have been discussed in sect. 2. Here we only note that the binarity of these sources could be in general easily established due to a large photometric accuracy. In many cases the orbital ephemeris could also be derived, therefore permitting ground-based spectroscopic follow-up observations at quarter phases establishing absolute masses and dimensions of the binary components. The easiest to recognize will be systems close to contact, where much of the system information could be recovered from photometry alone.

Eclipses will be rather uncommon, especially in the wide detached cases. But for a double-lined non-eclipsing spectroscopic binary the mass ratio can still be easily calculated. And in not too wide binaries the reflection effect can be measured from accurate photometry. This constrains the temperatures and inclination of the system. The inferred spectral classification can be finally checked against the properties of the spectra obtained close to quadratures where the spectral lines are well separated.

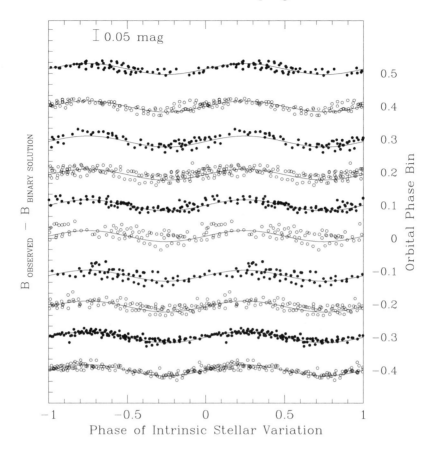

Figure 6. The difference between the observed B magnitudes of GK Dra (Dallaporta et al. 2002a) and the ones generated from the binary system solution, folded on an intrinsic stellar variation period of ~ 170 minutes. Differences pertaining to different orbital phase bins ($P_{orb} = 9.97$ days) are marked by different symbols and vertically offset for clarity. Note that a sinusoidal variation with a peak-to-peak amplitude of ~ 0.05 mag is present throughout the orbital cycle and is maintaining its phase. Intrinsic variability of binary components will be common among GAIA binaries.

Most binaries with the mass ratio below 0.3 will be single-lined, permitting to derive only a spectroscopic mass function.

6. Some remarks on reduction and interpretation procedures

GAIA will discover huge numbers of spectroscopic and eclipsing binaries. The numbers are orders of magnitude larger than everything collected in the last century from the ground. In many cases the observations obtained by GAIA will be good enough to determine system parameters at 1-2% accuracy level. They will have an immense impact on theories of stellar structure and evolution.

Such a large data set requires an automation of all stages of reduction and interpretation. No-one could recognize photometric eclipses or winging radial velocity curves in hundred-thousands or even millions of systems by eye. But even interpretation and classification has to be completely automatic with only the most unusual cases to be marked for human inspection. Wyithe & Wilson (2001, 2002) successfully classified some photometric eclipsing binaries from the OGLE database with semi-automatic procedures. Prsa (2003) obtained some encouraging results for double-lined eclipsing binaries. Clearly development of reliable classification and analysis procedures is one of the major tasks facing the scientific community before the launch of GAIA.

Acknowledgments. Generous allocation of observing time by Astronomical Observatory of Padova and financial support from the Ministry of Education, Science and Sport of Slovenia are acknowledged.

References

Andersen, J. 1991, A&A Rev. 3, 91

Cropper, M. & Marsh, T. 2003, in GAIA Spectroscopy, Science and Technology, U.Munari ed., ASP Conf. Ser. 298, pag. 407

Dallaporta, S., Tomov, T., Zwitter, T. & Munari, U. 2000, IBVS 4990

Dallaporta, S., Tomov, T., Zwitter, T. & Munari, U. 2002a, IBVS 5312

Dallaporta, S., Tomov, T., Zwitter, T. & Munari, U. 2002b, IBVS 5319

Frigo, A., Piccoli, P., Setti, A., Tomasoni, S., Tomov, T., Munari, U., Marrese, P.M. & Zwitter T. 2002, IBVS 5359

Mikuz, H., Dintinjana, B., Prša, A., Munari, U. & Zwitter, T. 2002, IBVS 5338

Milone, E.F. 2003, in GAIA Spectroscopy, Science and Technology, U.Munari ed., ASP Conf. Ser. 298, pag. 303

Munari, U. 1999, Balt. Astron. 8, 73

Munari, U., Tomov, T., Zwitter, T., Milone, E.F., Kallrath, J., Marrese, P.M., Boschi, F., Prša, A., Tomasella, L. & Moro, D. 2001, A&A 378, 477

Prša, A. 2003, in GAIA Spectroscopy, Science and Technology, U.Munari ed., ASP Conf. Ser. 298, pag. 457

Stellingwerf, R.F. 1978, ApJ 224, 953

Van Hamme, W. & Wilson, R.E. 2003, in GAIA Spectroscopy, Science and Technology, U.Munari ed., ASP Conf. Ser. 298, pag. 323

Wyithe, J.S.B. & Wilson, R.E. 2001, ApJ 559, 260

Wyithe, J.S.B. & Wilson, R.E. 2002, ApJ 571, 293

Wilson, R.E. & Wyithe, S.B. 2003, in GAIA Spectroscopy, Science and Technology, U.Munari ed., ASP Conf. Ser. 298, pag. 313

Zwitter, T. & Henden, A.A. 2003, in GAIA Spectroscopy, Science and Technology, U.Munari ed., ASP Conf. Ser. 298, pag. 489

GAIA Spectroscopy, Science and Technology
ASP Conference Series, Vol. 298, 2003
U. Munari ed.

Open questions in binary star statistics and the contribution of GAIA to solve them

Jean-Louis Halbwachs

Observatoire Astronomique de Strasbourg, 11 rue de l'université, F-67000 Strasbourg, France

Frédéric Arenou

Observatoire de Paris–Meudon, bat 11, 5 place J. Janssen, F 92195 Meudon Cedex, France

Anne Eggenberger, Michel Mayor & Stéphane Udry

Geneva Observatory, 51 chemin des Maillettes, CH-1290 Sauverny, Switzerland

Abstract. The statistical properties of binary stars are still poorly known, since it is difficult to prepare samples of double stars with homogeneous and corrigible selection effects. However, it seems that the distributions of mass ratios, periods and eccentricities are rather complex and that binaries are generated by at least two different formation processes. The GAIA radial velocity survey and the GAIA astrometry will dramatically change the deal, by providing enormous samples of spectroscopic binaries and of long-period binaries covering almost the whole range of mass ratios. For this purpose, the best resolution of the spectrograph should be R= 20 000 rather than R= 5000.

1. Introduction

Since the distributions of parameters such as mass ratios, orbital periods (or semi-major axes), and eccentricites are the touchstone of the formation models of binary stars, they were hardly searched by several generations of astronomers. The major difficulty in deriving them is the correction of the selection effects altering the samples of binaries. This is the reason of the success of the study by Duquennoy & Mayor (1991, DM91 hereafter), which is still the leading reference today: their 'complete' (i.e. unbiased) sample contains only about 2 dozens of spectroscopic binary orbits, 16 visual binary orbits and 30 common proper motion (CPM) binaries, but they were able to properly take the selection effects into account. Therefore, the limitation of their results comes from the statistical uncertainties related to small counts. Their main results were (i) a nearly constant or slightly increasing distribution of mass ratios for binaries with periods shorter than about 10 years (Mazeh et al. 1992), (ii) a distribution of secondary masses similar to the initial mass function for long-period binaries, and (iii) a unique log-normal distribution of periods.

We are now revising the study of DM91, by improving the selection of the unbiased sample thanks to the parallaxes measured by the Hipparcos satellite, and by adding K-type stars and cluster stars. The analysis of spectroscopic binaries (SB) is completed (Halbwachs et al. 2002), and, with about twice more SB than in DM91, it is now obvious that the reality is more complicated than they thought: a large proportion of binaries with nearly equal-mass components (twins) was found with periods shorter than 50 days, and, even when the period is larger than 50 days, twins have orbits less eccentric than binaries with mass ratios less than 0.8. The distribution of periods is also probably not unique, since the log-normal distribution doesn't fit our new sample. Therefore, binaries are certainly produced by several different processes, and a lot of work must still be done in order to derive the range of efficiency of each of them.

Since the features above emerge by simply doubling the sample, what will still appear when very large samples will be available? GAIA will measure several times the radial velocities of millions of stars over a timespan of about 5 years. If the accuracy of the individual measurements is sufficient, a very large sample of SB with known, and then corrigible selection effects is expected. The properties of this sample are investigated in the next section.

The binaries with periods larger than the duration of the mission will also be observed as astrometric or as 'visual binaries', when their components will be separated. Since the treatment of close binaries will be rather complicated, it is hard to already evaluate how many of them will be found for any period. On the contrary, the case of binaries with components entirely separated is easy to treat, and this is done in Section 3. Surprisingly, the detection of wide companions is today far to be complete, even among the nearest stars. Although we failed to find any new companion of nearby stars in the Tycho-2 catalogue (Høg et al. 2000), which is limited to stars brighter than 12 mag, it is obvious that a lot of companions fainter than about 16 are still to be discovered. Lépine, Shara & Rich (2002) identified some of them recently, but their survey is limited to proper motions beyond $0''.10$ yr^{-1} and separations closer than $1''.5$. Therefore, when GAIA will be launched, even the companions of the nearby stars will probably not have been all detected.

2. The GAIA spectroscopic binaries

2.1. The model

Our aim is to extrapolate the SB counts in our nearby solar-type survey to the SB which will be discovered with GAIA. For that purpose, the following hypotheses are used:

- the density profile perpendicular to the galactic plane obeys the law

$$\rho(z) = \rho_0 / \cosh^2(z/(2h)) \qquad (1)$$

where ρ_0 is the stellar density of F7-G-K main sequence stars on the galactic plane; according to the star counts from Hipparcos, $\rho_0 = 11.10^{-3}$ $*/\text{pc}^3$. It comes from the velocity dispersion derived from Hipparcos (Holmberg &

Flynn 2000) that the scale height h is 220 pc for solar-type stars. The distribution of the spectral types and absolute magnitudes among the F7-G-K main sequence stars was taken from the Hipparcos stars with parallaxes larger than 37 mas, since their selection is complete (Halbwachs et al. 2002);

- the proportion of binaries with periods shorter than 5 years is 12 %. The distributions of mass ratios and of periods are taken from Halbwachs et al. (2002). Two distributions of mass ratios are used, one for the binaries with periods shorter than 50 days, and one for the others;

- from the Coravel survey in open clusters, we assume that the SB will be detected and that their orbits will be derived when the semi-amplitudes of radial velocities fit the condition :

$$K_1\sqrt{1-e^2} > \begin{cases} 4\,\epsilon_{\rm RV} & \text{if } q < 0.9 \\ 23\,\epsilon_{\rm RV} & \text{else} \end{cases} \quad (2)$$

where $\epsilon_{\rm RV}$ is the error of the individual radial velocity measurements and q the mass ratio. $\epsilon_{\rm RV}$ depends on the G magnitude of the stars, and also on the resolution of the spectrograph, as indicated in a table provided by D. Katz for solar-type stars. We assume that the interstellar extinction only depends on the projection of the distance of the star on the galactic plane, and that it is 1 mag kpc^{-1}.

Our model is a bit straightforward, but it should be sufficient to derive reasonable approximations, and to help in the choice of the most efficient spectrograph resolution. This question is not easy : for bright stars, the measurements performed with R= 5000 are less accurate than those obtained with R= 20 000, but the errors increase then less when G becomes fainter; so the measurements of the faintest stars are more accurate with the former resolution than with the latter.

2.2. Properties of the GAIA spectroscopic binaries

The model above was used to derive the numbers of SB orbits obtained with GAIA, as well as their distributions of mass ratios and of periods. The results are summarized in Figure 1. It appears that the largest sample of SB would be obtained with R= 20 000, and this resolution should also provide the best detection rate for small mass ratios and for long periods. As a consequence, the numbers of SB with brown dwarf secondaries, would be 200, 500 and 1200 when R= 5000, 10 000 and 20 000 respectively.

Since the resolution R= 5000 provides the smallest RV errors for faint magnitudes, it could be considered as the most suitable for investigating the variations of binary rate in the galactic plane. However, this argument for a low resolution hardly stands up, since the distribution function of detected binaries is larger for R= 5000 than for R= 20 000 only for distances larger than about 1400 pc. The number of detectable SB beyond this limit is very small, anyway : 300 when R= 5000, instead of 160 when R= 20 000! Therefore, the advantage is quite negligible.

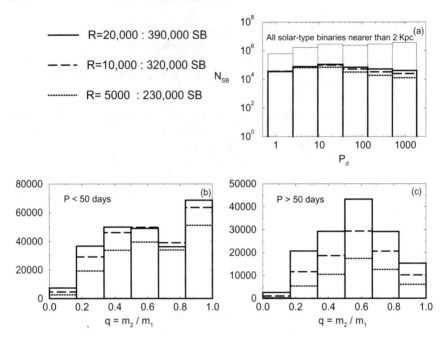

Figure 1. The spectroscopic binaries observed with the GAIA radial velocity survey, assuming different spectrograph resolutions.

In conclusion, although the number of SB could be smaller in reality since we neglected the stars which would be lost due to overlapping of the spectra, R= 20 000 seems to be the most promising resolution for statistical studies of short period binaries.

3. The common proper motion and parallax binaries

3.1. The wide binary model

Now we consider the other side of the period distribution, which is the domain of the wide separations. The model presented in the previous section is adapted to the separation of components in GAIA astrometry, by adding the following hypothesis :

- the proportion of solar-type stars which are primary components of a binary system is 65%; this is the minimum frequency according to DM91. The distribution of periods is the log-normal distribution of DM91, but the semi-major axes are restricted below 0.2 pc. The secondary masses are taken from the initial mass function of Kroupa (2001);

- the secondary components are detected if brighter than $G=20$ mag, and when the apparent separation is larger than the limit :

$$s_{\min} = \max(1'', 500'' \times 10^{-G_1/5}) \qquad (3)$$

This limit looks a bit pessimistic, since, in reality, many pairs will be resolved although their separations are below; however, we want to be sure that both components will get broad band photometric measurements, and that the mass ratio of the system will be derived.

This model was used to generate synthetic binaries and to select those which could be separated with GAIA. Since we want to evaluate the ability of GAIA to provide a sample adequate for the derivation of the distribution of mass ratios, we limit the synthetic binaries to the primary magnitude $G_1 < 16$ mag. In order to avoid optical pairs, the distance of the binaries is also limited to 400 pc. Therefore, the condition of common proper motions and parallaxes will remove almost all non-physical binaries, as we will see hereafter.

3.2. The optical pair model

The model used to generate wide binaries is adapted to the optical pairs : the stellar density is extrapolated assuming the IMF, and we added the distribution of tangential velocities of stars, and the acceptation criteria of wide binaries. For simplicity, it is assumed that the distribution of velocities is isotropic, with the dispersion for each axis :

$$\sigma_{V_\alpha} = \sigma_{V_\delta} = \min(19,\ 40 \times (B-V) + 5) \text{ km s}^{-1} \quad (4)$$

The acceptation condition on the distances, D, is at the 3σ-level :

$$|D_1 - D_2| < 3 \times \sqrt{\sigma_{D_1}^2 + \sigma_{D_2}^2} + 0.2 \text{ pc} \quad (5)$$

since we adopted 0.2 pc as the maximum separation between the components. For the tangential velocities, $\vec{v}$, the criterium is

$$|\vec{V}_1 - \vec{V}_2| < 4.74 D \sqrt{-2(\sigma_{\mu_1}^2 + \sigma_{\mu_2}^2)\ln 0.001} + V_{\text{parabolic}} \quad (6)$$

where D is in pc, the proper motion errors $\sigma_{\mu_{1,2}}$ are in "/yr, and the velocities are in km s^{-1}. The $\ln 0.001$ term means that the probability to exclude a physical binary with this criterium is only 0.1 %. $V_{\text{parabolic}}$ is the maximum difference between the velocities of components of a system bound by gravitation.

3.3. Properties of the GAIA wide binaries

We obtained a sample of 250 000 binaries with CPMπ components, $G_1 < 16$ mag and $D < 400$ pc. The amount of optical pairs is only 110. The distributions of periods and mass ratios of these systems are shown in Figure 2. The sample is adapted to the study of binaries with periods longer than 1000 years, and any secondary mass above the hydrogen ignition limit is accessible.

4. Conclusion

GAIA will provide very large samples of double stars covering the whole range of periods. The radial velocity measurements should permit the derivation of the statistical properties of short-period binaries with an accuracy that could

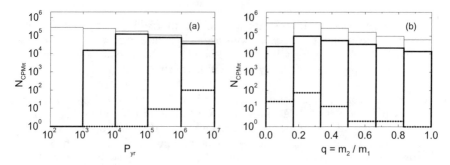

Figure 2. The wide binaries with solar-type primary components brighter than $G = 16$ and nearer than 400 pc. The histograms in thin line refer to all binaries, the 250 000 binaries separated by GAIA are represented in thick line, and the dotted line represents the 114 optical pairs with compatible proper motions and parallaxes.

not be achieved from on-ground survey, due to the huge amount of measurements which would be necessary. However, a large resolution should be chosen in order to explore the range of the small mass ratios (around the brown dwarf limit or beyond). It will then be possible to evaluate the contributions of the various processes which have been proposed for close binary formation and orbital evolution (cf. Bate et al. 2002). The study of wide binaries will also be feasible using a very large sample. Therefore, reliable statistical material will be available to debate questions so different as the frequency of wide companions around close twins (suspected to be larger than around single stars or other close binaries, Tokovinin 2000), the amount of brown dwarfs or of white dwarfs hidden in binary systems, among field stars but also among cluster stars (see Adams et al. 2002), or the disruption of wide pairs due to interactions with molecular clouds.

Acknowledgments. It a pleasure to thank Arnaud Siebert for valuable suggestions.

References

Adams, T.A., Davies, M.B., Jameson, R.F. & Scally, A. 2002, MNRAS 333, 547
Bate, M.R., Bonnell, I.A. & Bromm, V. 2002, MNRAS , 336, 705
Duquennoy, A. & Mayor, M. 1991, A&A 248, 485
Halbwachs, J.L., Mayor, M., Udry, S. & Arenou, F. 2002, A&A 397, 159
Høg, E., Fabricius, C., Makarov, V.V., Urban, S., Corbin, T., Wycoff, G., Bastian, U., Schwekendiek, P. & Wicenec, A. 2000, A&A 355, 27
Holmberg, J. & Flynn, C. 2000, MNRAS 313, 209
Kroupa, P. 2001, MNRAS 322, 231
Lépine, S., Shara, M.S. & Rich, R.M. 2002, AJ 123, 3434
Mazeh, T., Goldberg, D., Duquennoy, A. & Mayor, M. 1992, ApJ 401, 265
Tokovinin A. 2000, A&A 360, 997

GAIA and the spectroscopic binaries: what to expect in terms of orbit determination?

Dimitri Pourbaix

Institute of Astronomy and Astrophysics, Université Libre de Bruxelles CP 226, Bld du Triomphe, B-1050 Bruxelles

Sylvie Jancart

Institute of Astronomy and Astrophysics, Université Libre de Bruxelles CP 226, Bld du Triomphe, B-1050 Bruxelles

Abstract. Thanks to its Radial Velocity Spectrometer, GAIA will potentially discover lots of new spectroscopic binaries. The effects of the scanning law and the precision of the radial velocities on the orbit determination are investigated. Special care that would improve the situation of short period binaries is also suggested.

1. Introduction

Whereas the astrometric precision of GAIA will still supersede what will be achievable with ground-based observations within the forthcoming decade, the precision of the GAIA radial velocities (RV) is low with respect to today data. As long as the aim of these GAIA RV is to investigate the dynamics of our galaxy, such a limited precision is compensated by the number of RV measurements and stars. The situation is rather different if one aims at studying spectroscopic binaries for which one looks for orbital parameters.

We present the effects of the scanning law and the RV precision on the spectroscopic orbit determination problem using synthetic data of real binary systems.

2. Scanning law

The GAIA nominal scanning law (NSL) was described by Lindegren (2001) who also wrote a FORTRAN code that gives the observation epochs and attitudes of the satellite for any location on the sky. We here focus on the output of Lindegren's code for the RVS field.

The number of passages of a region of the sky in front of RVS as a function of the ecliptic coordinates is mapped in Figure 1. The first thing worth noting is the large variation of that number, resulting in a variation of the final RV precision for any star, regardless of its multiplicity. For a given instantaneous precision, the final RV of a star located $51°$ above or below the ecliptic will be about twice as precise as for a similar star in the ecliptic plane. In the case of spectroscopic binaries, systemic velocities and amplitudes of the RV curves (K) will suffer the same variations, thus also affecting the mass function estimates.

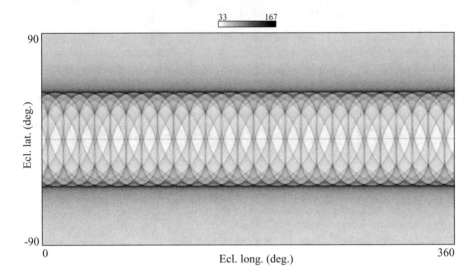

Figure 1. Number of RV measurements according to the GAIA nominal scanning law described by Lindegren (2001).

The longest time gap, i.e. the longest time interval without any RV measurement, changes also. Depending on the location on the sky, that gap ranges from 39 to 195 days. Whereas the map of the number of observations exhibits a strong regularity, the map of the gap exhibits a sinusoidal behavior with an amplitude of 51° and a period of 360° on top of the otherwise regular pattern.

Given the NSL, what percentage of spectroscopic binary orbits can one expect to recover?

3. Simulations

In order to answer that question, 1281 spectroscopic binary orbits were retrieved from the 9th Catalogue of Spectroscopic Binary Orbits[1]. They all fulfill the $K \geq 10$ km s^{-1} and $P \leq 1400$ days (where P is the orbital period) conditions. The distribution of some of the orbital parameters is given in Figure 2 (left panel).

Owing to the repetitive pattern in the NSL (as long as the number of observations is concerned), one can limit the analysis to the $[0, 15°] \times [0, 60°]$ grid. For computing time reasons, a 1° step is also adopted. At each node, radial velocities are generated for all binaries, assuming a Gaussian noise of 1, 5, or 10 km s^{-1} and orbits are subsequently fitted.

---

[1] http://sb9.astro.ulb.ac.be

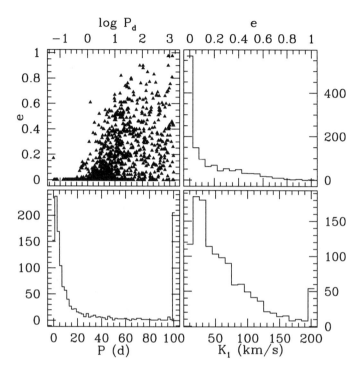

Figure 2. Distribution of the eccentricity, orbital period and amplitude of the 1281 selected systems.

The algorithms by Scargle (1982) and Stellingwerf (1978) were both used to obtain a periodogram and thus derive a first guess of the period (see sect. 4 for further discussion). The percentage of recovered periods is plotted in Figure 3. The period is assumed to be recovered if it is off by less than 10% with respect to the genuine period used to generate the data.

Although the number of observations changes by a factor of four over the whole grid, the pattern of the scanning law (right most panel of Figure 3) is barely visible at a noise level of 1 km s^{-1} and vanishes completely at lower precisions. Since obtaining the period is a *sine qua non* step in order to derive the whole set of orbital parameters, it is fair to conclude that the noise is the main limiting factor, much more constraining that the NSL.

What about the other parameters? Once P is accurately guessed, K is recovered in more than 93% of the cases, regardless of the noise and the number of observations.

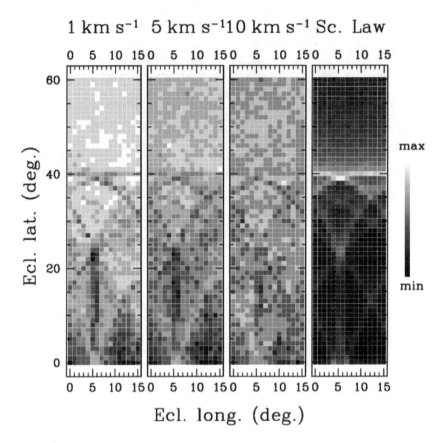

Figure 3. Map of the percentage of recovered periods. At the noise level of 1, 5, and 10 km s^{-1}, min and max are respectively 77–84, 63–70 and 41–46.

4. The problem of short periods

According to Eyer & Bartholdi (1999), the Nyquist frequency for the GAIA radial velocity spectrometer should be at least 1000 d^{-1}. There should therefore be no problem in investigating binaries with periods down to a few minutes, say one hour.

As illustrated in Figure 4, very short period binaries will be troublesome. Indeed, the great regularity of the NSL is likely to cause lots of aliases for which there will be no way to decide whether one identifies the true period of the binary or one of its aliases. Unless some additional informations about the (likely) period, e.g. from the position in the HR diagram, are made available, one should better limit the periodogram to period larger than 1 day, even if it means missing a lot of orbits.

This actually raises the question of what we do expect from GAIA: a census of the binaries and an orbital solution for some of them or an orbital solution for all of them even if some are plain wrong (from an astrophysical standpoint). We support the former policy.

In the sample used in Sect. 3, 152 binaries have periods shorter than 1 day, i.e. 12%. They are thus part of the 34% of orbits that could not be recovered. Nevertheless, the noise is still the main cause of failures.

5. GAIA on SB9

So far, the simulations aimed at assessing the effect of the scanning law. How would GAIA behave on the 1672 systems listed in SB9, assuming a noise of 5 km s^{-1}? In this second phase of simulations, the systems have their real position on the sky and no orbital solution with $P < 1^d$ is looked for. This latter criterion prevents from reaching the right period in 9% of the cases.

The period is well recovered for 46% of the systems, in which cases 98% of the amplitudes are accurately estimated too. For 38% of the sample, the variation of the RV was too small (i.e. the standard deviation of the RV does not exceed three times the adopted noise) and the orbit determination was aborted. Other cases where the procedure was also aborted include multiple significant peaks in the periodogram (10%) or no peak high enough to be considered as significant.

6. Conclusions

Even though the scanning law exhibits rather large variations over the sky in, e.g. the number of observations and the longest time gap between consecutive observations, there is almost no noticeable effect in the determination of the

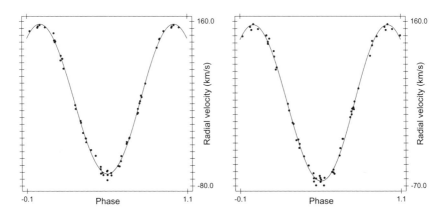

Figure 4. *Left:* synthetic data (noise of 5 km s^{-1}) and fitted orbit. *Right*: another sample of synthetic data generated with the same orbit but the derived period is $0^d.073$ instead of the expected $0^d.17$.

period of spectroscopic binaries. The precision of the radial velocities is the main limiting factor in the period search.

For instance, simulations show that more than 50% of the spectroscopic binaries do not need/require a binary model, either because the radial velocity variations are consistent with pure noise or because several periods are equally likely as far as data fitting is concerned.

With the theoretical NSL, short period binaries (period below 1 day) are all likely to be missed even at large signal to noise ratio. This is caused by the strong regularity of the NSL. Short period spectroscopic binaries would benefit from some irregularity in the observation distribution either in GAIA itself (slow drift in the scanning law) or thanks to some complementary ground-based observations.

Acknowledgments. This research was supported in part by an ESA/PRODEX Research Grant.

References

Eyer, L. & Bartholdi, P. 1999, A&AS 135, 1
Lindegren, L. 2001, SAG-LL-35 (Livelink)
Scargle, J. D. 1982, ApJ 263, 835
Stellingwerf, R. F. 1978, ApJ 224, 953

Distance-effects in the census of binaries with GAIA: principles and qualitative results

Staffan Söderhjelm

Lund Observatory, Box 43, SE-22100 Lund, Sweden

Abstract. The present contribution describes some simulations of the observation of binaries with GAIA. The results do not differ fundamentally from earlier estimates, e.g. in GAIA's Concepts and Technology Study Report (ESA-2000-SCI-4), but they are based on more realistic assumptions and much more extensive numerical work.

1. Introduction

There are four main steps in these simulations. Firstly, the binaries are taken from a 'population-synthesis type' galaxy model, where 12 Gyr of evolution of a galactic disk with binaries is simulated. In the second step, the astrometric observations of these binaries are simulated in pixel-detail, while for the RVS instrument only some magnitude- and color-dependent mean errors in the final radial velocities are assumed. In the third step, these simulated observations are used as inputs to several different solution models, as appropriate for binaries of successively shorter period. Finally, the results of these solutions are compared with the known binary parameters, to give some measures of the solution 'success-rates'.

2. The Galaxy model

A key insight for the observation of binaries with GAIA is that a large fraction of the astrometric orbits will be for systems with a degenerate component. Instead of adding this fraction more or less ad hoc, a Galactic disc model with stellar evolution is used in order to produce these binaries from basic initial mass-distributions.

So far, the model consists only of a thin disk and a thick disk, since it was originally created to provide a reasonable fit to the Hipparcos and Tycho data. In the present runs, a constant Star Formation Rate (0-12 Gyr) and a time-independent power-law IMF (with -2.3 slope for masses above 0.8 $M_\odot$, and -0.3 down to 0.025 $M_\odot$) is used, with a wide and almost age-independent metallicity-distribution. One fourth of the 'objects' produced by this IMF are accepted as single stars, while 75 % are made into binaries with a q-distribution function that is mainly gaussian [N(0.23,0.42)] as given by Duquennoy & Mayor (1991), but with a narrow peak close to 1.0 (cf Söderhjelm 2000; Halbwachs et al. 2002). The distribution of the semi-major axes of the binaries is again close to that given by Duquennoy & Mayor [N(1.5,1.5) in $\log a(\mathrm{AU})$], with cutoffs above

6 and below -3]. For the eccentricities, an $f(e) \sim e$ is used at large periods, with rapid circularization below some 10 days. Although most of these parameters are 'reasonable', no serious attempt has been made to fit them to existing observational data. Crude comparisons with Hipparcos/Tycho have verified approximately the numbers of bright, resolved binaries, but there are complex selection effects in the Hipparcos Input Catalog that need more thorough study.

For periods above some 5 years, each component in the binary evolves independently, while at shorter periods, there are epochs where one or the other star exceeds its Roche-lobe, and mass-transfer has to be taken into account. This has been done only very schematically so far, but there are rapid codes (cf. Hurley, Tout & Pols 2002) that may be introduced at a later stage. To specify the observable mass, temperature and luminosity of each component in a binary at the present epoch, we need stellar evolutionary tracks for 12 Gyr for all masses and all metallicities. To avoid interpolation problems, the analytical approximations by Hurley, Pols & Tout (2000) are used, but with two remaining caveats. Firstly, so many calls for stellar data are made in the simulations that a precomputed 'grid'-version was still found more practical. Secondly, the Hurley et al. models go only down to a bona fide stellar mass-limit around 0.08 $M_\odot$. For the low-mass part, there is thus an extra grid with data due to Baraffe et al. (1998), and consequent continuity problems. Then there is the standard problem going from 'theoretical' ($M_{\rm bol}, T_{\rm eff}$) to 'observable' (M_V, $V-I$) coordinates, and again only 'illustrative' transformations are used, based mainly on the data in Bessell, Castelli & Plez (1998).

3. GAIA observation simulations

The galaxy model gives in each direction a list of binaries, which are then 'observed' by a model GAIA. For the present study, only systems within 400 pc from the galactic plane were used, but with the data collected in 'rings' with successively larger distances from the Sun.

Each object is scanned a number of times in different position-angles, and at each such epoch, a realistic series of astrometric observations is simulated. Only a 'constant' (2-dimensional) PSF is used, but with its size scaled by a color-dependent 'effective wavelength'. The actual positions of the binary components are calculated from the orbital elements, and the true light-distribution on the CCD are given by convolution with the (color-scaled) PSFs. Simulated observations (with Poisson and readout noise) are derived, normally 16 samples per CCD on 10 CCDs per scan, each sample an average over 12 across-scan pixels. This 16×12 window (0.7×1.6 arcsec on the sky) is centered on the geometrical centre of the binary, but in order to cover also wider pairs, 6x10 windows centered on each component are used when required.

Partly because the details of the RVS (mainly the resolution) is still undecided, but mainly because the radial velocity mean errors are so dependent on the detailed spectra, only a very crude modelization has been used. The mean error (for the primary component) for a specific star is given as a product of three terms. A main one, dependent on the magnitude, a spectral-type one, dependent on the intrinsic color, and a 'dilution' one, to take into account the light from the secondary. A typical 'good' (G-K star) RV mean error is assumed to be about 0.5 km s^{-1} below 10th magnitude, 5 km s^{-1} at G=14.0 and 20 km s^{-1} at G=15.5, but the overly schematic color/spectrum dependence is probably the weakest point.

4. Illustrative solutions

The simulated observations are now put through a series of different solution models. In reality, these models have to be much more complex, in order to e.g. combine data of different kinds, iteratively refine color-dependent calibrations and with little supervision search large regions of the parameter space for orbital binaries. The present versions are standard iterative least squares solutions (with numerical partial derivatives) for various parameter sets. The 1-dimensional (x) astrometric pixel-counts at each scan are fitted in a global model which gives also the y-coordinates of the components, enabling an accurate correction for the parts of the PSF falling outside the observation-window. The iterations are always started close to the known correct results, and problems with local minima and poor convergence should thus be much less severe than in any real reductions. These solutions do illustrate the basic principles, however, and may provide some upper limits to the real solution-efficiencies. They are listed here in order of decreasing binary period.

Resolved doubles. For the widest pairs, with 0.01–2 arcsec separation, a standard 12-parameter model (5 astrometric parameters plus a magnitude for each component) is used. Even for the closest pairs, the individual colors of the components are assumed to be approximately known (e.g. from photocentre shifts with color), giving somewhat optimistic resolution limits.

Curved photocentre motions. For unresolved binaries with periods around 10–30 years, even a small photocentre semi-major axis can give a measurable curvature in the position observations. It is easy to solve for additional quadratic and cubic terms in the proper motions, giving 9 instead of 5 astrometric parameters in a 'single-star' solution.

Orbit solutions without RV. For periods shorter than 10 years, one may try to determine the full orbit for the photocentre. There are 5 astrometric parameters plus a magnitude for the center of gravity, plus 7 standard orbital elements. The full knowledge of good starting parameters simplifies the problem very much, and the number of good orbits is certainly overestimated.

Orbit solutions with RV. Using also the RV observations, an attempt is made to solve for the full astrometric orbit as above, plus the radial velocity amplitude and the system velocity, 15 parameters in all.

RV orbits. At even shorter periods, the photocentre orbit is inperceptibly small, but a pure 5-element spectroscopic orbit may still be derived in addition to the 'single star' (=systemic) astrometric solution.

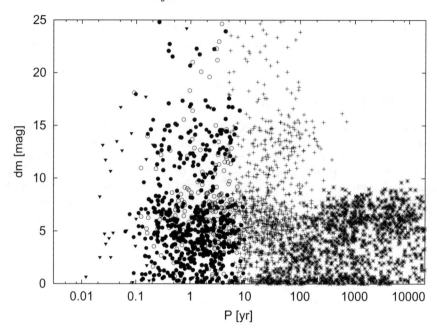

Figure 1. 'Acceptable' solutions as obtained with different solution models for bright Pop I stars 400-1000 pc from the Sun. The asterisks are resolved pairs, the crosses pairs with non-linear photocentre motion. The open circels are photocentre orbits, the solid circles photocentre orbits with also a RV amplitude. Finally, the triangles signify pure radial velocity orbits with imperceptible astrometric deviations. Each symbol would correspond to some 1000 systems if scaled to the full GAIA mission.

5. Solution results

As an example, Fig 1 shows the successful solutions of different kinds for a sample of bright nearby stars ($G < 14.5$, $d = 400 - 1000$ pc). The resolved pairs can not have Δm larger than about 10, but the photocentre orbits are often obtained for systems with invisible white or brown dwarf components.

For each kind of solution, one may define a 'success-rate', the number of acceptable solutions divided with the attempted ones. The 'absolute' value of such a rate depends a lot on the detailed input- and solution-parameters, but their changes with distance and magnitude should be of some relevance. For systems close to the Sun, orbit sizes and/or separations are large, and the success-rates are generally high. Figure 2a shows such rates as derived from the data in Figure 1, and in Figures 2b, 3a and 3b, the data derived from successively larger distances. Naturally, the resolved systems have larger and larger periods (at the same angular separation) as the distances increase, which tends to give in the end a 'gap' around 100 year period where few binaries can be detected. For the orbital pairs, the astrometric signature shrinks with increasing distance, and in Figure 3 the short-period edge of the orbits without RV is clearly shifted

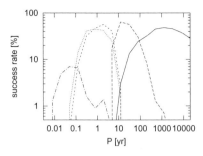
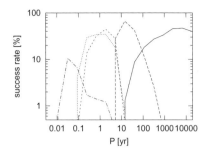

Figure 2. Logarithmic plot showing the fraction of successful solutions of different kinds. The full line is for resolved systems, the long-dashed for curved proper motion systems, the short-dashed and dotted for astrometric orbits without and with RV, and the dot-dashed line finally for pure RV orbits. To the *left* are the results for $G<14.5$ systems at $d=400$-1000 pc, to the *right* at $d=1000$-2500 pc.

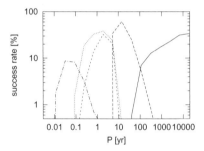
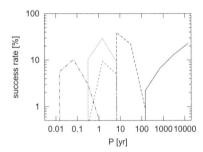

Figure 3. Success-rate curves as in Figure 2, but now for systems at $d=2.5$–6.3 kpc (*left*) and $d=6.3$–16 kpc (*right*).

to the right, while the solutions with RV are less affected (as expected, the pure RV orbits are solved equally well at all distances).

In Figures 4 and 5, the same 'ring' distances are selected, but now for a sample of faint ($G = 17.5 - 20.5$) stars. The main difference is that no RV data are available, and thus only three categories of solutions remain. The 'gap' between the resolved and the curved proper motion systems is very apparent in Figure 5, as well as the practical absence of astrometric orbits (because of the combination of small size and poor accuracy).

6. Conclusions

The simulations presented illustrate a few well-known points bearing on the census of binaries with GAIA. This census will be very complete in regions within a kpc from the sun, where binaries at all periods from less than a day (eclipsing) to millions of years (CPM-pairs) will be detected. The completeness diminishes with distance and with faintness, and there will be period-intervals

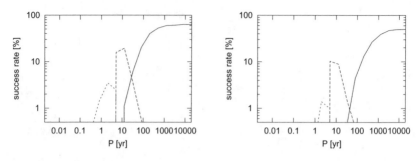

Figure 4. Sucess-rate curves as in Figure 2, but for $G = 17.5 - 20.5$ stars at d=400–1000 pc (*left*) and d=1000–2500 pc (*right*).

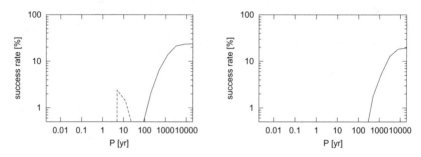

Figure 5. Sucess-rate curves as in Figure 2, but for $G = 17.5 - 20.5$ stars at d=2.5–6.3 kpc (*left*) and d=6.3–16 kpc (*right*).

with poor or no coverage. One such example is the 50-200 year range of periods beyond say 5 kpc, or the < 1 year range for faint stars without radial velocity observations. All statistical binary star studies will have to study the selection effects carefully, and further simulations along the present lines can be used to gain valuable insights. There is work to be done however both in the Galaxy modelling (especially the addition of short-period binary evolution to include the eclipsing binaries) and in a more realistic modelling of the RVS observations.

References

Baraffe, I., Chabrier, G., Allard, F., Hauschildt, P.H. 1998, A&A 337, 403
Bessell, M.S., Castelli, F. & Plez, B. 1998, A&A 333, 231
Halbwachs, J.L., Mayor, M., Udry, S. & Arenou, F. 2002, A&A 397, 159
Duquennoy, A. & Mayor, M. 1991, A&A 248, 485
Hurley, J.R., Pols, O.R. & Tout, C.A. 2000, MNRAS 315, 543
Hurley, J.R., Tout, C.A. & Pols, O.R. 2002, MNRAS 329, 897
Söderhjelm, S. 2000, Astron. Nachr. 321, 165

Binary stars with GAIA and the mass-luminosity relation

Oleg Malkov

Institute of Astronomy of the Russian Academy of Sciences, 48 Pyatnitskaya St., 119017 Moscow, Russia

Dana Kovaleva

Institute of Astronomy of the Russian Academy of Sciences, 48 Pyatnitskaya St., 119017 Moscow, Russia

Abstract. Comparison of observational radii of main sequence components of eclipsing binaries with ones of isolated stars of the same spectral type shows noticeable difference. For A to late-F stars observable radii of eclipsing binaries exceed those of single stars, and the situation is reverse for earlier type stars. It leads, in particular, to a difference in luminosity. Possible effects are proposed to explain this feature. We conclude that the existing empirical mass-luminosity relation based for masses higher than about 1.5 $M_\odot$ almost exclusively on eclipsing binary data can hardly be applied for construction of the initial mass function for single stars of intermediate and high masses unless GAIA will provide accurate data for a large enough number of wide binaries.

1. Introduction

The well-known fact is that dynamic stellar masses to construct calibrating relations (as the mass-luminosity relation, MLR) and, further, to restore fundamental distributions (one of the most important: the initial mass function, IMF) from observational ones, are provided by binaries of certain types. Main of these types are double-lined eclipsing binaries (EB, close pairs), and visual binaries (wide pairs, components are evolving like single stars). Let us stress two points: (*a*) here and further we will refer only to main sequence stars, and (*b*) the word *isolated*, from evolutionary point of view, can be applied both to single stars and to components of wide pairs.

Present-day empirical knowledge of the MLR for low-mass stars is mainly based on data on visual binaries, due to low number of higher quality data on EB components and resolved spectroscopic binaries in this mass region (cf. Malkov, Piskunov & Shpil'Kina 1997). The empirical MLR for higher masses based on high accuracy data on main sequence (MS) components of double-lined EB stars seems to be much better determined (Figure 1a).

However, let us imagine that by some cause we cannot use the EB data for construction of the MLR. In this case, what information can be obtained only from the binaries of other types for this purpose? The number of points in Figure 1b representing ground-based observations for stars with masses larger

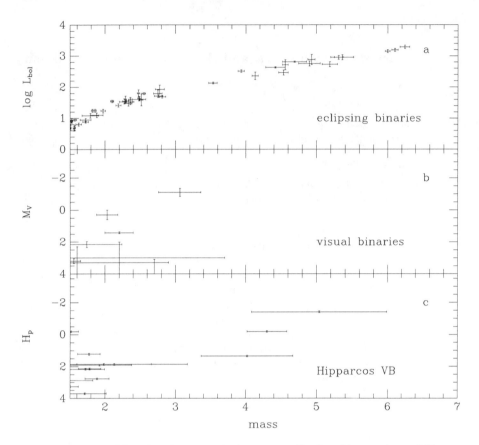

Figure 1. Mass-luminosity relation. *a*: eclipsing binary data, MS only; *b*: visual and resolved spectroscopic binaries, ground based data; *c*: visual binary Hipparcos data by Martin & Mignard (1998), and Söderhjelm (1999).

than 1 $M_\odot$ as well as the accuracy of these data are clearly unsufficient to work up the shape of the MLR in this mass region. And even the data from Hipparcos have not improved the situation significantly (Figure 1c). Particularly, the data for wide binary components does not allow to make any reasonable conclusions about the shape of the MLR for masses larger than about 1.5 $M_\odot$. Thus, present-day empirical knowledge of the MLR for masses larger than 1.5 $M_\odot$ is completely based on data about components of close binaries.

2. Why the eclipsing binary set is not an ideal base for the MLR

We have demonstrated that the contribution of EB stars data in the mass-luminosity relation for masses >1.5 $M_\odot$ is crucial. However, this relation is usually applied to 'normal', i.e. isolated stars, and used to restore fundamental

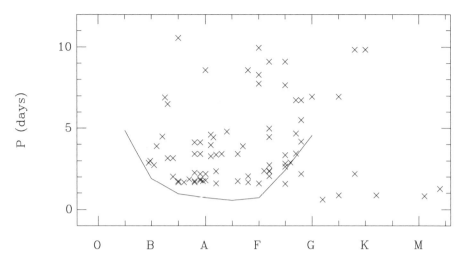

Figure 2. Rotation period of isolated stars (*line*, McNally 1965) and orbital period of eclipsing binaries (*points*, Malkov 1993)

distributions for them. Now the key question is: are we sure that there is no difference between observational parameters of EB components from those of isolated stars? Below we discuss rotational effects as one of possible reasons for such doubts.

Rotation: isolated stars and eclipsing binaries. Isolated early-type stars are rapid rotators. On the other hand, almost all pairs with $P < 15^d$ are synchronized, so components of close binaries rotate slower. Figure 2 illustrates this statement. The line approximates mean dependence of rotation period on spectral type for single stars from McNally (1965), points represent orbital periods of EB stars from the catalogue of astrophysical parameters of binary systems of Malkov (1993). One may see that for spectral classes F and earlier rotational periods of close pair components are systematically larger than those of isolated stars (if we suppose all these close pairs to be synchronized).

Observational effects of rotation. Rotation changes stellar evolution and global parameters of a star. Due to non-spherical shape of rotating stars observed effective temperature, radius and magnitude depend on the line of sight. Rotation axis of isolated stars are randomly oriented; EB components are, by definition, mainly observed toward the equator (cf. Abt 2001). All this may result in difference between observational parameters of close pair components and those of isolated (single and wide pair components) stars.

3. EB components in comparison with isolated stars

Further we list some examples of observational evidences of difference between characteristics of set of isolated stars and those of set of EB components.

Radii. The comparison between radii of close pairs components and radii of single stars may serve as one of such evidences. In Figure 3 radii of 3500 MS single

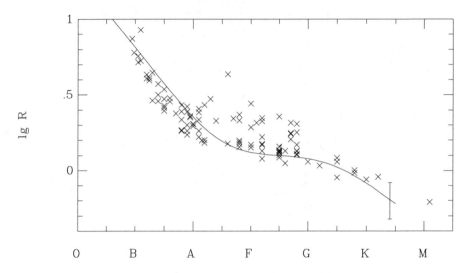

Figure 3. Radii of MS single stars (*line*, Pasinetti-Fracassini et al. 2001) and those of EB components (*points*, Malkov 1993). Typical scatter of single stars radii is indicated by the error bar.

stars from the 3rd ed. of CADARS (Catalogue of Stellar Diameters; (Pasinetti-Fracassini et al. 2001) are approximated by a spline. The points indicate radii of 114 MS detached EB components from Malkov (1993). One may see that the radii of single stars are systematically larger for B stars, and lower for A-F spectral types.

Bolometric correction scales. Most of published bolometric correction (BC) scales based on photometric data, including recent ones, are in rather good agreement. A noteworthy difference (cf. Flower 1996) is the scale of Habets & Heintze (1981, hereafter H&H). The H&H scale of BC is based on superposition of data of components from wide and close binaries: the authors have erroneously considered these stars as having similar nature and similar observational features. The resulting difference between photometric BC and H&H scales is most significant for spectral types A to late F.

Effective temperatures. Ribas et al. (1998) have compared effective temperatures of detached eclipsing binaries computed from Hipparcos trigonometric parallaxes and from photometric determinations. The sample contains all well-studied detached double-lined eclipsing binaries with accurate Hipparcos parallaxes. They cover the spectral range from early B to early K. A small but systematic trend for stars with spectra later than A0 is observed between the two temperature determinations.

Radiative fluxes. The comparison of radiative fluxes in the V band of the components of eclipsing binaries with Hipparcos parallaxes and of single star set was made by Popper (1998). His results lead to the conclusion that for EB components with spectra later than F4 the radiative fluxes are systematically smaller, whereas they are larger than those of single stars for components earlier than A0.

4. Results of comparison of EB components with isolated stars

As a result of the tests just described on radius, BC scale, effective temperature and flux, the spectral range can be divided into three parts: G-K-M, A-F and B types. For late spectral types (G-K-M) no definite conclusion can be drawn in view of the paucity of observationla data, and only the other two spectral subdivisions will be considered.

A-F types. For this spectral range, according to the results of comparison of the radii (see Figure 3), EB components have larger radii than isolated stars of the same spectral type. The BC-scale comparison shows that EB components have higher bolometric luminosities than isolated stars. This can be due to their larger radii or/and higher temperatures. Comparison of T_{eff} obtained for EB components with standard photometric calibrations with those obtained using Hipparcos parallaxes leads to the conclusion that they have higher temperatures and/or larger radii than isolated stars. Radiative flux test indicates that radii of EB components are larger than radii of isolated stars for spectra F4 and later. Thus, all the tests show similar results: EB components of A-F type have *larger* radii than corresponding isolated stars. This result is in a qualitatevely agreement with considerations discussed in Sect. 2.

B types. For this spectral range direct comparison of radii of EB components with those of isolated stars indicates that EB components have smaller radii or/and higher temperatures. Neither BC-scales test nor comparison of T_{eff} for this spectral range lead to definite conclusion due to small sample and low accuracy. However, comparison of radiative fluxes indicates that the radii of EB components of B type are smaller than those of isolated stars. These results seem to disagree with rotation effects hypothesis as well as with the results obtained for A-F spectra. The possible explanation for this 'B star paradox' can be, for instance, as follows. Relative radii ($r = R/a$) in B systems are, in average, larger than in pairs of other spectral types; small increase of radius R moves the system to 'semi-detached' and thus excludes it from the statistics of detached MS systems. On the other hand, systems with larger a (sufficient to keep a 'detached' status) have lower probability of discovery.

5. Current MLR and construction of the initial mass function

Thus, we found that our empirical knowledge of the mass-luminosity relation for stars with masses larger than 1.5 $M_\odot$ (Figure 1a) based on data on close binary components (cf. Kovaleva 2001) evidently cannot be applied to restore parameters and distributions for isolated stars. On the other hand, present-day knowledge of the same MLR based on data for isolated stars – components of wide binaries, that would be correct to use for mentioned purposes – is very poor. Let us remind, in addition, that not even the general behaviour of the MLR itself but its precise shape is crucial for the purpose of restoring the initial mass function from observational luminosity function:

$$\frac{dN}{d\log(\text{mass})} = \frac{dN}{dM_V} * \frac{dM_V}{d\log(\text{mass})}$$

Evidently, the available data can give no hint for behaviour of the MLR' for masses >1.5 $M_\odot$. New observations of high accuracy are necessary to make conclusions on this most important subject.

6. Challenges for GAIA astrometry and photometry of visual binaries

What we need to receive from GAIA to fix the situation with derivation of the empirical MLR? The requirements are as follows:
- minimum number of observed systems: 10 000 (to collect a few hundreds systems with masses >1.5 $M_\odot$);
- mass uncertainties: better than 3% (the typical present-day accuracy achieved for EB components);
- accurate photometry to obtain visual luminosities.

7. Conclusions

For B0V-G0V stars observational parameters of components of eclipsing binaries systematically differ from those of isolated stars. Accurate observational data should be collected for visual binaries with mass >1.5 $M_\odot$ (and/or observational data for eclipsing binaries should be corrected for all rotation and selection effects, that does not seem to be a realizable task) to determine dynamic masses and luminosities of components and construct the MLR. Then the initial mass function should be revised.

References

Abt, H.A. 2001, AJ 122, 2008
Flower, P.J. 1996, ApJ 469, 355
Habets, G.M.H.J. & Heintze, J.R.W. 1981, A&AS 46, 193
Kovaleva, D.A., 2001, AZh 78, 1104
Malkov, O.Yu. 1993, Bull. Inf. CDS 42, 27
Malkov, O.Yu., Piskunov, A.E. & Shpil'Kina, D.A. 1997, A&A 320, 79
Martin, C. & Mignard, F. 1998, A&A 330, 585
McNally, D. 1965, The Observatory 85, 166
Pasinetti-Fracassini, L.E., Pastori, L., Covino, S. & Pozzi, A. 2001, A&A 367, 521
Popper, D.M. & Guinan, E.F. 1998, PASP 110, 572
Ribas, I., Gimenez, A., Torra, J., Jordi, C. & Oblak, E. 1998, A&A 330, 600
Söderhjelm, S. 1999, A&A 341, 121

session 7

GROUND-BASED AND SPACE SURVEYS, CONCLUDING REMARKS

chair: T. Zwitter

F. Boschi listening to M.A.C.Perryman during a coffee-break

Clock-wise from top-left: F. de Felice, F. Arenou, Y. Viala, G. Bertelli, C. Bailer-Jones and A. Spagna

Current status and future prospects of ground-based and space photometry

Arne A. Henden

USRA/USNO, Flagstaff Station, P. O. Box 1149, Flagstaff, AZ 86002 USA

Abstract. This is a review of the current status of ground based photometry. In particular, the paper will concentrate on all-sky, multiple filter, high-quality catalogs. The availability of space-based catalogs will be discussed, as well as the impact of all-sky catalogs on future space missions such as GAIA.

1. Introduction

Astronomers have been producing accurate photometry since the beginning of the last century. Early photocells were used to measure the brightness of naked-eye stars. After World War II, the photomultiplier tube (PMT) revolutionized the field, permitting the measurement of millions of objects even with small telescopes. Over the past two decades, the field of photometry has undergone yet another revolution with the use of the CCD detector. The CCD gives a multiplexing advantage over the PMT, and has a far wider spectral response and increased sensitivity. Now even a 20 cm telescope can observe objects fainter than visible on the Palomar Sky Survey photographic plates.

There is certainly an enormous amount of published photometry, ranging from microlensing monitoring to determination of the solar constant. Photometry can be thought of as extremely coarse spectroscopy, with the intent of accurately measuring the time-tagged flux from some object in specific passbands. Like spectroscopy, accurate photometry requires the use of standard reference objects whose magnitude has been accurately calibrated. These reference objects may be in the field of view of the program object, or may be elsewhere in the sky.

However, not every telescope is at a photometric site, nor can it be used at all times under photometric conditions. To make most efficient use of telescope time requires an all-sky catalog with sufficient density that any given field will have multiple reference stars with accurate photometry within it.

Since GAIA will produce an all-sky photometric catalog, it is of interest to discuss the need for such catalogs, the current status of both ground-based and space-based all-sky catalogs, and the proposed future programs that might influence all-sky photometry prior to the GAIA era.

Because many surveys are not discussed in refereed literature, Internet URLs are given instead of references to guide the reader to more detailed information.

2. Uses of all-sky catalogs

Traditionally, photometry has been performed by using a small set of standard stars, calculating extinction and transformation coefficients, and then using these coefficients to determine the magnitudes and colors of selected stars in the field of some program object. After the comparison stars have been calibrated, all further photometry is performed differentially with respect to those stars, thereby eliminating the requirement of photometric skies or at least of having to observe standard stars on every night.

This approach has been successful in the past because of the small number of program objects. A handful of trained astronomers observing at excellent sites could calibrate the required fields. However, needs are changing. Many more objects are being monitored now than in the past, so calibrating all of the fields is almost approaching an all-sky survey in itself. Using a large telescope such as Keck or VLT to calibrate fields, taking valuable time to observe standard stars, is a waste of aperture. Targets of opportunity often require rapid calibration of their comparison stars.

Therefore, the main purpose of an all-sky photometric catalog is to provide a means of performing differential photometry with respect to local standards, without the overhead of having to observe separate standard fields. Such differential photometry is then possible under non-photometric conditions and makes maximal use of telescope time. Some common examples are given below.

GRB afterglow research. Using satellites such as HETE-2, Integral and Swift, gamma-ray bursts are being localized to within a few arcmin, sometimes only minutes after the start of a burst. When reporting the discovery of an accompanying optical afterglow, observers often use a commonly available catalog, such as USNO-A, and pick a nearby star to use as the main comparison star. They then report their photometry with respect to that star. If no afterglow is found, they report an upper limit with respect to an ensemble of USNO-A stars. *Such practices should be discouraged.* USNO-A, USNO-B, GSC2, etc. are astrometric catalogs with very poor photometry. In every case, any reported photometry will have to be modified by the reader once the comparison stars are properly calibrated. USNO-A does not cover every bandpass used by typical observers, so some sort of crude transformation must be used to be able to publish a multiwavelength report. Finally, any comparison star chosen from USNO-A may or may not be a good color match to the afterglow candidate. If the star is red, it may be variable. It may be too faint or so bright that some observers cannot use it for their calibration, forcing the use of another, equally poorly measured, comparison star. Final calibration may take several days or weeks, depending on the weather at a particular site, or the phase of the moon or availability of a specific camera. Several sites may independently calibrate the field, a redundant activity that is a waste of telescope time. An all-sky catalog solves all of these problems.

Supernovae and novae and other transient objects. These objects are similar to GRB afterglows in that immediate calibration is important so that researchers can choose appropriate comparison stars and be able to report final photometry, rather than making a preliminary report that has to be modified later. As transient objects fade, the observer may need to choose other comparison stars in order to stay within the dynamic range of the detector. The transient object may change color with time, so a comparison star of different color may be important. Finally, you need to choose a non-variable comparison star. For GRB afterglows, most of the photometry is performed in the first few days, whereas for a SNe, monitoring may occur over a year or more. Comparison star variability may be more apparent with such a longer time span.

Differential photometry of moving objects. Few asteroids have periods short enough that the entire light curve can be covered during a single night. Some objects exhibit chaotic light curves, where monitoring over many cycles is important. At the same time, these objects move in the sky, so the photometric reference frame changes. Calibration could be performed along the entire strip at a later date, but this is a large impact on telescope time, instead of being able to study another asteroid or continue the study of the chaotic one. If calibration is performed on a piecemeal basis as the object moves across the sky, systematic errors can appear in the calibration.

An all-sky, multiwavelength catalog has far more uses than the few described above. Some data-mining activities are:

- Searching for variable objects. If the catalog encompasses multiple epochs, usually necessary for precision photometry, then it can be searched for objects with larger than normal photometric scatter. If multiple wavelengths are involved, then there are additional tests, such as the Welch-Stetson index, that can be applied to further constrain possible variability. Variable objects can be considered interesting by themselves, or may be objects to avoid when selecting comparison stars for other program objects.

- Historical archive of data. For example, the precursor of the peculiar nova V838 Mon appears to have been a typical F-star based on the multiple wavelength data available through 2MASS, DENIS, UCAC and POSS-II. No variability is apparent in the SAI photographic archive. Using historical archives, the amplitude of the outburst can be determined, as well as helping to decide when the object has returned to the quiescent state. The higher the accuracy of the archival data, and the more passbands present, the better understanding we have of these objects. For SNe, even a database of images might be important so that accurate image subtraction techniques can be used to remove the underlying galaxy from the SNe photometry. Waiting for the SNe to fade enough to provide a baseline galaxy image might take years.

- Finding peculiar objects, such as very blue or very red stars. Many photographic surveys for unusually colored objects have been made, such as the Palomar-Green survey for blue objects. Such blue objects may be quiescent counterparts for cataclysmic variables or extragalactic sources like quasars. Very red objects are likely to be Mira or semiregular variables, or may indicate highly reddened fields containing embedded young stars. These are all important areas of research, and having a photometric catalog to search to find new candidates is a huge time savings.

- Determinination of reddening in specific, small fields. There are several published extinction maps of our Galaxy. These maps are fine for a general idea of how the extinction varies with position. However, in a small CCD field of view, such coarse grids of extinction may be inadequate to describe the conditions along a particular line of sight. An all-sky catalog can provide knowledge of the colors of objects in the field, as well as such things as reddening-free color indices.

- Calibration of instrumentation. Widefield cameras are very difficult to flatfield properly. If you use a flatfield screen, you can have significant gradients across the field. If you calibrate using master sky flats, systematic gradients can appear depending on ground light sources or the inclusion/exclusion of the Milky Way. A check on the calibration is to image a field with accurate photometry and see if the wide-field photometry has systematic gradients. These gradients can then be removed after-the-fact, or the flatfielding process can be improved. When using a telescope with an unfamilar camera, such as at a national observatory, transformation coefficients need to be calculated in order to transform your raw filtered observations into the standard system. Often the national facility does not provide such coefficients, but instead requires the individual observer to obtain them. This may be fine if the observer has a long observing run and can take time from the main project to observe standard stars. For shorter runs, or even target-of-opportunity exposures, a simpler approach is to use objects with wide color range within the field of view and calculate the coefficients on the science exposure. If the photometric catalog provides high-accuracy colors over the entire sky, this local approach is possible.

- Support of astrometric analysis. To transform relative parallaxes to absolute parallaxes requires knowledge of the reference frame. Photometric parallaxes of the reference stars are often used. Other problems, such as color dependence on astrometric solutions, can be solved with an all-sky photometric catalog.

- Color space analysis, such as the determination of color loci of a class of objects. This is a means of selecting objects to study, such as white dwarf plus M-star pairs or RR Lyrae variables. Often such object classes occupy specific places in multiwavelength space and can be selected for or against. Finding members of such classes from photometric catalog datamining can be used by itself in a statistical sense, or can be used as an input catalog to a program on another telescope.

3. Requirements for an all-sky photometric catalog

Based on the comments above, it is fairly obvious that all-sky photometric catalogs are not only useful, but essential for modern astronomy. Creating such a catalog takes an enormous amount of time, not only telescope time but processing and reduction time. Since any given site can see only a fraction of the celestial sphere, to cover the entire sky requires at least two sites. With typical CCDs, either a shallow survey is performed, a survey in one color is made, or else many years are devoted to the project.

Assuming such a photometric catalog were to be designed today, what features should it have? To be useful, an all-sky photometric catalog must contain the following information:

- It must be homogeneous with minimal systematic errors, and must cover both northern and southern skies.

- It must have multiple passbands, preferably in the photometric system of the user. In fact, spectrophotometry would make the ideal catalog, since the user can then convolve the desired system bandpass with the spectral points to derive a magnitude in any wide-band filter system.

- The photometry must be accurate (0.01 mag or better on average).

- Both bright and faint stars need to be measured. Wide field followups might need 9th magnitude stars, for example, while 8 m images might saturate at 22nd magnitude.

- It must have dense coverage, so that many stars are included in each CCD field of view. This is a typical byproduct of going faint.

- The catalog should be complete with few missing objects, and resolve the galactic plane.

- Some sort of star/galaxy classifier is necessary to remove galaxies from consideration as photometric standards.

- Multiple epochs need to be contained in the catalog. This not only removes nightly systematics, but provides a means of identifying variable objects within the catalog.

- The catalog must be made available to the general community, either with a web-based interface or on archival media. These catalogs are large and might be good candidates for inclusion in the Virtual Observatory concept.

These requirements are non-trivial, and are the reason why no such catalog currently exists.

4. Available multiwavelength surveys

The astrometric catalogs such as USNO-B have photometric extractions from the photographic plates, but the quality of the data is quite low; typically 0.2 – 0.3 mag at best. *Astrometric catalogs should only be used for obtaining a rough idea of the brightness of objects.*

The Tycho2 catalog (http://www.astro.ku.dk/~erik/Tycho-2) gives B_T and V_T magnitudes for most of its stars. The photometric quality is reasonably good to about 10th magnitude, degrading quickly thereafter. The Bt and Vt passbands can be transformed into Johnson B_J and V_J with reasonable accuracy for normal stars. The Tycho2 catalog is therefore the only available 'deep' all-sky photometric catalog, and its usefullness depends on the simularity between the

B_T/V_T passbands and the desired passband. In addition, even at 10th magnitude, there are seldom Tycho2 stars in a typical CCD field of view, and if one or two do appear, a multiple exposure transfer from those bright objects down to the faintness of the program object must be performed.

Therefore, most observers currently use one of the photometric standard star sets, depending on the filter system used. There are typically a few hundred of such standards, usually placed near the celestial equator. These standards need to be observed; zeropoint, extinction and transformation coefficients calculated; and the standard system transferred into the field of interest. This requires several photometric nights with many observations of standards in order to obtain reasonable accuracy, a fairly major investment of telescope time. This is similar to the early use of astrometric standards along with transit telescopes to derive local astrometric standards in a region of the sky.

Other than Tycho2, the only other all-sky catalogs are of even brighter stars, such as the Bright Star Catalog or the WBVR catalog. These are of marginal use by most programs, though they can be used to calibrate wide-field, bright surveys.

5. GAIA's impact

GAIA, after its mission is accomplished, will create a catalog of high-accuracy photometry down to a limiting magnitude of 18, or fainter. Since it will have multiple filters (often multiple narrow-band filters as well), will not have the atmosphere to contend with, and will cover the entire sky with a single, stable instrument, the photometric catalog from GAIA will be unprecedented. However, there are several considerations to keep in mind:

- Launch is still a decade away, and the output products from the mission will be several years after that.
- Science in the intervening years still needs to be done.
- For several proposed space missions and ground-based projects prior to GAIA, knowledge of the sky for use as input catalogs is critical.
- What happens if (heaven forbid) GAIA fails to perform to specification? Several space missions have lost instruments, been boosted into improper orbits, spun out of control or even were destroyed. Promise of a future catalog is no reason to stop current survey activities.

6. Existing photometric databases

All of the existing databases, except for Tycho2, are compendiums of sequences scattered through the literature. As mentioned earlier, there are thousands of such sequences, generally surrounding objects of interest such as variable stars, open and globular clusters, or galaxies. Most of these sequences are in one of the two primary wide-band photometric systems: Johnson-Cousins *UBVRI*, or Sloan $u'g'r'i'z'$. However, since a variety of photometric systems have been used, some fields have only been calibrated in an alternative system such as the Strömgren *ubvy* narrow-band filters, or some fields have been calibrated in more than one system. In this section, only large databases containing the two primary wide-band systems will be mentioned.

GCPD (http://obswww.unige.ch/gcpd/gcpd.html). The Mermilliod General Catalog of Photometric Data contains about 10^5 stars culled from the literature, mostly with *UBV* photometry. The quality is uneven, and large gaps in spatial coverage and magnitude depth are present.

LONEOS (ftp://ftp.lowell.edu/pub/bas/starcats/loneos.phot). The LONEOS photometric catalog, like the GCPD, is culled from the literature. Brian Skiff has been selective in what data is included in the LONEOS catalog, so the quality is consistent. It contains primarily *BVR* photometry, with a median magnitude of about 14. The spatial sampling is relatively uniform, but sequences are spaced every few degrees.

Hipparcos (http://astro.estec.esa.nl/Hipparcos/catalog.html). While the Hipparcos mission itself is only complete to 8th magnitude, the catalog does include standard Johnson-Cousins magnitudes and colors for most surveyed stars.

GSPC2 (http://www-gsss.stsci.edu/gspc/gspc2.htm). The Guide Star Photometric Catalog 2 is an extension of GSPC1, going deeper and in BVR colors. The sequences are spaced every 6 degrees, with a few extra sequences inserted to be able to calibrate the POSS-II.

Henden (ftp://ftp.nofs.navy.mil/pub/outgoing/aah/sequence). The Henden sequences consist of about 1000 fields, chosen around known variable stars, ranging from about $-30°$ declination to the North Pole. Most sequences are *BV*, though a large number are *BVRI* and a few are *UBVRI*. Typical depth is about 19th magnitude.

7. Near-term surveys

Survey work is indigenous to astronomical research. Most CCD surveys are targeted, looking at some small region of the sky (HDF N/S), for some specific class of objects (MACHO, OGLE), or are related to GRB afterglow research (ROTSE, LOTIS, RAPTOR), supernovae (LOTOSS), or near-earth objects (LINEAR, LONEOS, NEAT). Most current, ongoing surveys are not designed for high-quality photometry. Instead, they are trying to go as faint and as rapidly as possible. These unfiltered surveys will not be mentioned further, even though they are performing their primary task admirably.

7.1. Single-filter surveys

CAMC (http://www.ast.cam.ac.uk/~dwe/SRF/camc.html). The Carlsberg Merdian Telescope, like FASTT and Bordeaux, is a converted transit telescope that uses a CCD at the focal plane. These transit telescopes were typically 15-20 cm aperture refractors. Using them with CCDs generally means limiting the bandwidth with some filter; usually *V*-like or *R*-like is chosen. With a single bandpass, data cannot be transformed onto any standard system. Instead, the ensemble photometry is zeropointed as best as possible to a standard filter, with the color difference between the selected filter/telescope/CCD system and the standard system being ignored. With several of these single-filter systems producing catalogs in different passbands, it is possible to combine the catalogs and transform the results onto a common standard system. The Carlsberg Merdian Telescope is undertaking a survey of the sky from −3 to +30 declination in the SLOAN r' passband, with a magnitude range between 9th and 17th. Currently they have released the $-3°$ to $+3°$ zone, including $6.3 \cdot 10^6$ stars.

UCAC (http://ad.usno.navy.mil/ucac/). The USNO CCD Astrometric Catalog is an all-sky astrometric survey that uses a 20 cm refractor plus 4k×4k CCD with a narrow filter (bandpass 579–642nm). It goes down to $R=16$ with better than 20 mas average accuracy for the positions. The photometry is poorly calibrated, though it is excellent within small regions. Most of the southern sky was released in 2001 (UCAC1), with the remaining southern sky to be released shortly. The survey is continuing in the northern hemisphere and a full catalog should be released within the next two years.

7.2. Multiple-filter surveys

There are several on-going multiwavelength surveys. Most of these use wide-field cameras and so have poor resolution, especially near the galactic plane. The exceptions are the two near-IR surveys.

DENIS (http://www-denis.iap.fr/denis.html). The DEep Near-Infrared Survey is an $I_C J K_s$ survey. Started in 1996, using the 1m ESO telescope at La Silla, Chile, data collection was completed in 2001. The survey covers the southern sky up to the celestial equator. This survey has reasonable internal photometric errors of about 0.03 mag.

2MASS (http://www.ipac.caltech.edu/2mass/). The Two-Micron All-Sky Survey was recently released to the general public. This survey covers the entire sky at JHK, and shows that calibration of the near-IR sky is actually better understood than for the better-known visible sky! Using 2MASS or DENIS along with optical colors yields good separation of star types in color loci.

ASAS (http://www.astrouw.edu.pl/∼gp/asas/asas.html). The All Sky Automated Survey (third edition) uses two 7 cm telephoto lenses plus 2k×2k CCDs at Las Campanas, along with a one degree field 25 cm telescope. ASAS will cover the southern sky to 14th magnitude in the V and I_C passbands. Variable stars in one quarter of the southern sky have been released, and a catalog of all stars is expected.

LOTIS (http://compton.as.arizona.edu/LOTIS/new.html). The Livermore Optical Transient Imaging System was originally developed as a fast-response unfiltered telescope for GRB afterglow research. It has recently been upgraded with $BVRI$ filters. There are four 10 cm telephoto lenses plus 2k×2k CCDs on a common mount. It will cover the northern sky to 14th magnitude, though the large pixels mean that the catalog will be confusion-limited near the galactic plane.

TASS (http://www.tass-survey.org). The Amateur Sky Survey consists of a set of telescope systems situated around the world. Each system contains two 10 cm telescopes plus 2k×2k CCDs; one has a V and one has an I_C filter. They will calibrate the entire northern sky from 7th to about 16th magnitude. A preliminary catalog will be released in 2003. This data will support the UCAC.

APT (http://newt.phys.unsw.edu.au/∼mcdba/apt.html). The Automated Patrol Telescope is a refurbishment of a 50 cm Baker-Nunn camera, sited at Siding Spring Observatory. A new 6k×6k CCD camera is being built, and will provide *BVRI* photometry to 17th magnitude for most of the southern sky. Like TASS, this data will also be used to support the UCAC.

7.3. Deeper multi-filter partial sky surveys

As the sky is surveyed to fainter limits, usually larger telescopes along with smaller fields of view are used. This generally means the entire sky visible from some site is not surveyed due to time constraints. Two excellent surveys are underway.

QUEST (http://www.astro.yale.edu/bailyn/quest.html). Using the CIDA 1.0 m Venezuelan Schmidt, the QUasar Equatorial Survey Team surveys the zone $-6°$ to $+6°$ around the celestial equator. The CCD drift-scan camera uses *BVR* filters and goes to roughly 19th magnitude. Objective prism spectra will also be available for many of the objects.

SDSS (http://www.sdss.org). The Sloan Digital Sky Survey uses a 2.5 m telescope at Apache Point Observatory (NM). This survey will give $u'g'r'i'z'$ photometry of sources between 14th and 22nd magnitude over at least 6500 square degrees of the northern sky away from the galactic plane. An initial data release of 3500 square degrees is expected to occur in January 2003.

8. Possible near-future surveys

The previous surveys all represent telescope systems that are already in place and actively surveying the sky. Some delays might be expected due to completion of the observing or of the data reduction, but final results are anticipated shortly.

This section discusses either proposed surveys or telescopes that could be used for surveys provided enough interest was shown. Any surveys from these systems would not be released for several years.

VST (http://twg.na.astro.it/vst/vst_homepage_twg.html). The Very Large Telescope Survey Telescope is a 2.6 m telescope under construction at the ESO-VLT site on Paranal (Chile), completion date April 2004. It will use a mosaic CCD camera to cover a one square degree field to support the VLT activities. An all-sky survey is possible.

VISTA (http://www.vista.ac.uk). The Visible and Infrared Survey Telescope for Astronomy is a 4 m telescope under construction at the ESO-VLT site that will become operational about 2006. The current funding supports an IR mosaic camera (*zJHK*), with an optical camera possible in the future. An all-sky survey is anticipated, but completion of the survey may take a decade.

DUCAC (http://www.nofs.navy.mil). The USNO 1.3 m telescope plus 6k×8k CCD mosaic camera covers 1.5 square degrees. This system may be used to produce a *BVI* photometric catalog to 20th magnitude over the northern sky (a Deep UCAC), requiring about three years to complete.

PanSTARRS (http://poi.ifa.hawaii.edu/~poi). The Panoramic Survey Telescope and Rapid Response System is a cluster of four 2 m-class telescopes with large CCD mosaics, either on Mauna Kea or Haleakala, with first light about 2006. The primary program is a search for near-earth objects, where each telescope covers a different region of the sky to obtain maximal sky coverage every night. However, it is possible to have all four telescopes point to the same region of sky and use different filters.

LSST (http://www.lsst.org/lssto/index.html). The Large Synoptic Survey Telescope was proposed in the Decadal Survey. This 8 m-class telescope will cover the entire northern sky every 3-4 nights to 24th magnitude. This may be accomplished in multiple passbands. First light is predicted around 2010.

9. Possible interim space missions

The best possible solution to the all-sky, multiwavelength survey problem is a space mission. One telescope/camera system would then be able to survey the entire sky, eliminating many of the systematic errors. With no atmosphere to contend with, blending of objects is lessened and better photometry is possible in the Galactic Plane. Extended bandpasses can be used: UV or some near-IR regions that are not possible from the ground could be added, thereby giving more astrophysical information about the objects cataloged.

DIVA (http://www.ari.uni-heidelberg.de/diva/). The Double Interferometer for Visual Astrometry is a German space mission designed to provide accurate parallaxes, proper motion and photometry/spectroscopy for some $4 \cdot 10^7$ objects to about 15th magnitude. Launch date is still uncertain.

FAME (http://www.usno.navy.mil/FAME/). This is a similar mission to DIVA, but with improved astrometric accuracy. It will provide multiwavelength photometry of $4 \cdot 10^7$ sources to about 15th magnitude. It is currently unfunded.

GALEX (http://www.srl.caltech.edu/galex/). The GALaxy Evolution eXplorer is a mission expected to be launched shortly. It will survey the sky in two UV wavelength bands: 135-180nm and 180-300nm with 4 arcsec resolution, down to roughly 21st magnitude. Combined with 2MASS/DENIS and some optical survey, full coverage of object spectral energy distributions from the UV to the near-IR is possible.

10. Summary

Tycho2 is the deepest all-sky photometric catalog currently available. It is only complete to 10th magnitude, and with two nonstandard filters, it should be considered a minimal photometric catalog. Several projects are underway to provide multicolor photometry to perhaps 14th magnitude over the next two years. Fainter limiting magnitudes with possibly more or different filters may be available in the 4-year time span. Space-based missions are most likely to give homogeneous datasets without systematics, but results from even DIVA or FAME are not to be expected for another decade. GAIA will supersede all of these efforts, but with results more than a decade away.

Faint and peculiar objects in GAIA: results from GSC-II

Daniela Carollo, Alessandro Spagna, Mario G. Lattanzi, Richard L. Smart

INAF - Osservatorio Astronomico di Torino, I-10025, Torino, Italy

Simon T. Hodgkin

Cambridge Astronomical Survey Unit, Institute of Astronomy, Cambridge, CB3 0HA, UK

Brian J. McLean

Space Telescope Science Institute, Baltimore, MD 21218, USA

Abstract. The 1 billion objects of GSC-II make up a formidable data set for the hunt of peculiar and rare targets such as late type stars, white dwarfs, carbon dwarfs, asteroids, variable stars, etc. Here we present a survey to search for ancient cool white dwarfs, which led to the discovery of several stars with peculiar spectral distributions and extreme physical properties. Finally, we discuss the impact of the GAIA mission with respect to these peculiar and faint white dwarfs.

1. The second Guide Star Catalogue

The Second Guide Star Catalogue (GSC-II) project (Lasker et al. 1995, McLean et al. 2000) is a collaborative effort between Space Telescope Science Institute (STScI) and Osservatorio Astronomico di Torino (OATo) with the support of the European Space Agency (ESA) – Astrophysics Division, the European Southern Observatory (ESO) and GEMINI. The aim of this project is the construction of an all-sky catalogue containing classifications, colors, magnitudes, positions and proper motions of ~ 1 billion objects down to the magnitude limit of the plates ($B_J \sim 22.5$). At the moment, GSC-II is one of the largest stellar catalogue, the only comparable one being USNO-B (Monet et al. 2002).

GSC-II is based on about 7 000 photographic Schmidt plates (POSS and AAO) with a large field of view ($6.4° \times 6.4°$). All plates were digitized at STScI utilizing modified PDS-type scanning machines with 25 μm square pixels (1.7 ″/pix) for the first epoch plates, and 15 μm pixels (1.0 ″/pix) for the second epoch plates. Each digital copy of the plates was analyzed by means of a standard software pipeline which performs object detection and computes parameters, features and classification for each identified object. Position and magnitude for each object was found from astrometric and photometric calibrations which utilized the Tycho2 (Høg et al. 2000) and the GSPC-2 (Bucciarelli et al. 2001) as reference catalogs.

The first public release of GSC-II (GSC2.2) was delivered in June 2001 and contains 445 851 237 objects down to $B_J < 19.5$ and $R_F < 18.5$, providing positions with an average accuracy of 0.2 arcsec, photographic photometry B_J and R_F with 0.15-0.2 mag accuracy and classification (stellar/extended objects) accurate to 90%.

2. Search for nearby Halo White Dwarfs

Cool white dwarf (WD) stars are the remnants of stars which were born when the Milky Way was very young. A WD cools and fades in a well defined manner, thus the WD luminosity function is imprinted with the star formation history of the Galaxy back to its very beginning. In particular, detecting the end of the WD sequence will provide a direct measure of the age of the Galaxy and its fundamental components, i.e. the Disk, Thick Disk and the Halo (Fontaine, Brassard & Bergeron 2001). The usefulness of WDs as stellar chronometers has been stimulated by the recent progress in the cooling models based on non-grey atmospheres and refinements of the internal physics (e.g. Hansen 1999, Chabrier et al. 2000) for both DA and non-DA WDs with $T_{\text{eff}} < 4\,000$ K, which are the temperatures expected for ancient Halo WDs. In this range, cooling WDs with hydrogen atmosphere start to become fainter but bluer because of the strong H_2 opacity due to the collision induced absorption (CIA) towards longer wavelengths, whereas helium atmosphere WDs continue to redden.

In addition, it has been suggested that Pop II WDs could contribute significantly to the baryonic fraction of the dark Halo. In fact they are obvious candidates for the MACHOs revealed by the LMC microlensing surveys (Alcock 2000) which seem to indicate that $\sim 20\%$ of the dark matter is tied up in objects with ~ 0.5 $M_\odot$. The most extensive survey to date (Oppenheimer et al. 2001) provides a lower limit on the space density of $\rho \sim 10^{-4}$ pc^{-3}, that is, 5 times larger than expected from the canonical stellar Halo, and $\sim 1\%$ of the expected local dark Halo density. These results are still a matter of debate. In fact Reid, Sahu and Hawley (2001) claimed that the kinematics of the Oppenheimer sample is consistent with the high-velocity tail of the Thick Disk. Moreover these stars have a spread in age that is more consistent with a Thick Disk population (Hansen 2001).

The aim of our survey is to search for Halo WDs using plate material from the GSC-II in the northern hemisphere and improve the measurements of Halo WDs space density. Also, we will confront the WD models with our sample of cool ancient objects in order to improve the cooling tracks of WDs with $T_{\text{eff}} < 4\,000$ K. Although WDs should be a typical component of the Halo, such objects are very difficult to observe because they are extremely faint. In fact, theoretical cooling tracks by Chabrier et al. (2000) predict an absolute magnitude of $M_V = 16.2$ and 17.3 for a 0.6 $M_\odot$ WD of 10 and 13 Gyr respectively (excluding the nuclear burning phases). Objects with these magnitudes are observable only within a few tens of parsecs with the GSC-II material which contains objects down to the plate limits of 22.5, 20.8 and 19.5 mag for the blue B_J, red R_F and infrared I_N plates respectively.

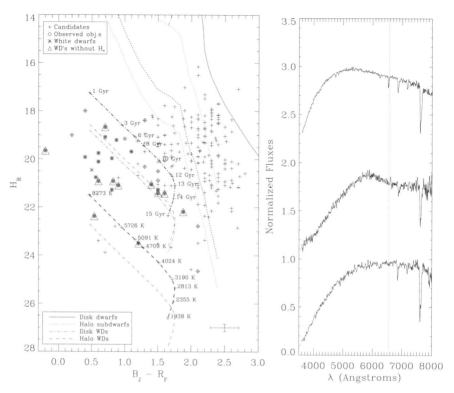

Figure 1. *Left*: the RPM diagram for an area of 800 square degrees. *Right*: spectra of new cool WDs. Dotted line indicates Hα position.

2.1. Plate material, processing and selection criteria

Our survey covers an area of ∼ 1300 square degrees which corresponds to 40 regions in the sky, mostly located toward the North Galactic Pole (NGP). In order to detect high proper motion objects, we processed POSS-II plates (blue, red, infrared) with epoch difference $\Delta t \sim$ 2-10 yr by means of the standard GSC-II pipeline. Also, we performed object matching and derived proper motions using the procedure described in Spagna et al. (1996), then faint ($R_F > 16$ mag) and fast moving ($0.3 < \mu < 2.5$ ″/yr) stars were identified. Each target was checked by a visual inspection of POSS-I and POSS-II plates in order to reject the false detections (e.g. mismatches and binaries) and to confirm its proper motion. Another very useful parameter for the selection of the targets is the reduced proper motion (RPM), $H = m + 5\log\mu - 5$. The RPM diagram, H_R vs. $(B_J - R_F)$, was adopted to identify faint objects with high proper motion and to separate Disk ad Halo WDs from late type dwarfs and subdwarfs. Figure 1 (left panel) shows the RPM diagram for a set of regions. Here, the thick solid and dotted lines show the locus of the Disk dwarfs and the Halo subdwarfs based on the 10 Gyr isochrones down to 0.08 M$_\odot$ from Baraffe et al. (1997, 1998) with [Fe/H]=0 and −1.5, respectively. Dashed and dot-dashed lines shows

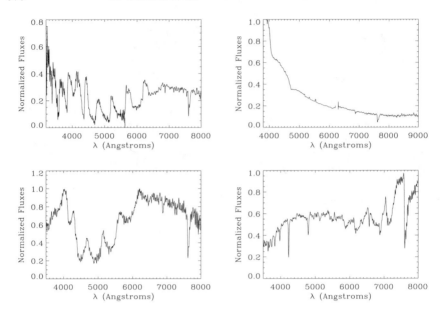

Figure 2. A sample of peculiar objects. *Top left*: a peculiar DQ WD with strong C_2 Deslandres-d'Anzabuja and Swan bands. *Top right*: a very hot magnetic WD candidate. *Bottom left*: a magnetic DQ WD. *Bottom right*: a binary system (WD+dM)

the cooling tracks of 0.6 $M_\odot$ WDs with hydrogen atmosphere from Chabrier et al (2000). We adopted mean tangential velocity (towards the NGP) of $V_T = 38$ km s^{-1} (Disk) and 270 km s^{-1} (Halo). Thin lines indicates the 2σ kinematics thresholds. Finally, spectral analysis is required for a confirmation of the nature of the selected candidate.

2.2. Spectroscopic follow-up and preliminary results

Low resolution spectroscopy is suitable to recognize the spectral type and the main chemical composition of the stars. Spectroscopic observations were carried out with the 3.5 m TNG (La Palma), the 4.2 m WHT (La Palma), and the 3.5 m at Apache Point Observatory (USA). Most of the targets were observed in the first semester 2002 at TNG using the low resolution spectrograph DOLORES (Device Optimized for Low Resolution) with the LR-B Grism1 which gave a nominal dispersion of 2.8 Å pix^{-1} and useful wavelength coverage from 3000 to 8800 Å. We performed spectroscopic follow-up for candidates from 800 square degrees (1/50 of the sky) which corresponds to $\sim 60\%$ of our total area. The number of Halo WD candidates after the selection criteria was 47 and we obtained 32 spectra plus JHK infrared photometry for 12 stars during 3 nights. The results are remarkable: of the 32 observed targets, 23 are WDs and 12 have no Hα line. We also found 4 M dwarfs, 2 subdwarfs, a binary system (dM+WD) and 3 interesting peculiar objects. The left panel of Figure 1 shows the RPM diagram for these 800 square degrees. The observed objects and those classified as white dwarfs are marked with different symbols. The right panel of Figure 1 shows a few confirmed cool WDs in our sample, including a 'coolish' DA (top spectrum) with a weak Hα line and two cool WDs, while Figure 2 shows the peculiar objects.

3. Peculiar objects and classification problems

An unexpected result of this survey is the discovery of a significant fraction of objects with a very complex nature. Some examples are presented in Figure 2, where the top left panel shows a peculiar DQ WD, with extremely strong C_2 absorption bands, while the bottom left shows a magnetic carbon rich WD. On the top right is a probable very hot magnetic WD and the bottom right an unresolved binary system WD+dM. We point out that all these cases could not be classified properly till their spectra became available. Even when spectra are available, the classification can be tricky for objects with extreme physical properties and no previous observations or good theoretical models. This was the case of the peculiar carbon rich WD named GSC2U J131147.2+292348 (Fig. 1, top left). The object is fast moving ($\mu \simeq 0.48$ arcsec yr^{-1}), and faint ($V \simeq 18.7$). A check on the SIMBAD database revealed that the star was not in the NLTT catalogue (Luyten 1979) but, quite surprisingly, was listed as a quasar candidate (object OMHR 58793) by Moreau & Reboul (1995), who measured an UV excess but did not detect any proper motion, perhaps because of a cross-matching error. The real nature of this object was realized after a thorough analysis of its spectrum with the support of infrared photometry (Carollo et al. 2002), even though the lack of adequate models in the literature was a serious problem.

4. The impact of the GAIA mission

The main impact of GAIA with respect to the faint and nearby objects, such as the large variety of WD types, will be the determination of accurate distances by means of the *trigonometric parallaxes* for *all* the objects detectable in the solar neighborhood down to $V \approx 20$ mag. Distances will directly provide the absolute magnitudes which will permit a robust, even if preliminary, identification of these objects as WDs. Thus, the fact that the information from the GAIA spectrograph and the medium band photometric system will not be available for the faintest objects is not as dramatic a problem as in the case of current ground based surveys.

We expect that GAIA will carry out a complete *census* of WDs, including the faintest ancient and cool WDs previously discussed. To this regard, GAIA will provide a complete and unbiased sample (i.e. not kinematically selected). Moreover, accurate tangential velocities will be derived from proper motions also in the cases where radial velocities are not available, and will help to separate the Halo and Disk WDs. Of course, a large fraction of non-DA WDs including a certain percentage of peculiar objects will be also detected by GAIA. Broad band photometry should help to identify these cases by means of their *anomalous* colors also for the dimmest objects without any further spectro-photometric data. Clearly, this is a challenging and non trivial issue for the classification task of the GAIA data reduction. Moreover, among the many science cases that GAIA data will address, these objects point out the logical necessity to complement the astrometric and spectro-photometric observations of peculiar or unclassified objects with a spectroscopic follow-up with large ground based telescopes. Fortunately, it will probably not be difficult to obtain observing time with 4-8 meters telescopes at the epoch when the first GAIA results will be delivered ($\sim$ 2015?).

Finally, we mention the fact that objects with such a peculiar spectral distribution as those shown in Figure 2 could be affected by residual systematic errors due to a chromaticity effect (Gai et al. 1998, Lindegreen 1998) not properly corrected by the standard astrometric calibrations. However, this does not seem a critical problem for nearby objects (i.e. having large values of parallax and proper motion). Also, the possibility that the multiple observations of these high proper motion objects are not correctly matched in crowded regions can probably be avoided by means of a robust matching algorithm.

References

Alcock, C. 2000, Science 287, 74
Baraffe, I. et al. 1997, A&A 327, 1054
Baraffe, I. et al. 1998, A&A 337, 403
Bucciarelli, B. et al. 2001, A&A 368, 335
Carollo, D. et al. 2002, A&A 393, L45
Chabrier, G. et al. 2000, ApJ 543, 216
Fontaine, G., Brassard, P. & Bergeron, P. 2001, PASP 113, 409
Gai M., Carollo D. & Lattanzi M.G. 1998, SAG-ML-007 (Livelink)
Hansen, B.M.S. 1999, ApJ 520, 680
Hansen, B.M.S. 2001, ApJL 558, L39
Høg, E. et al. 2000, A&A 355, 27
Lasker B.M. et al. 1995, in Future Possibilities for Astrometry in Space, M.A.C. Perryman, F. van Leeuwen & T.-D. Guyenne ed.s, ESA SP-379, pag. 137
Lindegreen, L. 1998, SAG-LL-024 (Livelink)
Luyten, W.J. 1979, A catalogue of stars with proper motions exceeding 0"5 annually, 2nd ed., Minneapolis, Univ. of Minnesota
McLean B.J. et al. 2000, in Astronomical Data Analysis Software and Systems IX, N. Manset, C. Veillet & D. Crabtree ed.s, ASP Conf. Ser. 216, pag. 145
Monet, D.G. et al. 2002, AJ , in press, (astro-ph/0210694)
Moreau, O. & Reboul H. 1995, A&AS 111, 169
Oppenheimer, B.R. et al. 2001, Science 292, 698
Reid, N.I., Sahu, K.C, & Hawley, S.L. 2001, ApJ 559, 942
Spagna, A. et al. 1996, A&A 311, 758

RAVE: the RAdial Velocity Experiment

Matthias Steinmetz[1]

Astrophysikalisches Institut Potsdam, An der Sternwarte 16, 14482 Potsdam, Germany

Abstract. RAVE[2] (RAdial Velocity Experiment) is an ambitious program to conduct an all-sky survey (complete to $V=16$ mag) to measure the radial velocities, metallicities and abundance ratios of 50 million stars using the 1.2 m UK Schmidt Telescope of the Anglo-Australian Observatory (AAO), together with a northern counterpart, over the period 2006 – 2010. The survey will represent a giant leap forward in our understanding of our own Milky Way galaxy, providing a vast stellar kinematic database three orders of magnitude larger than any other survey proposed for this coming decade. RAVE will offer the first truly representative inventory of stellar radial velocities for all major components of the Galaxy. The survey is made possible by recent technical innovations in multi-fiber spectroscopy, specifically the development of the 'Echidna' concept at the AAO for positioning fibers using piezo-electric ball/spines. A 1 m-class Schmidt telescope equipped with an Echidna fiber-optic positioner and suitable spectrograph would be able to obtain spectra for over 20 000 stars per clear night. Although the main survey cannot begin until 2006, a key component of the RAVE survey is a pilot program of 10^5 stars which may be carried out using the existing 6dF facility in unscheduled bright time over the period 2003–2005.

1. Introduction

In the first decade of the 21st century, it is being increasingly recognized that many of the clues to the fundamental problem of galaxy formation in the early Universe lie locked up in the motions and chemical composition of stars in our Milky Way galaxy (for a review see e.g. Freeman & Bland-Hawthorn 2002). Consequently, significant effort has been placed into planning the next generation of large-scale astrometric surveys like GAIA. Stellar spectroscopy plays a crucial role in these studies, not only providing radial velocities as a key component of the 6-dimensional phase space of stellar positions and velocities, but also providing much-needed information on the chemical composition of individual stars. Taken together, information on space motion and composition can be used to unravel the formation process of the Galaxy.

---

[1] for the RAVE Science Working Group

[2] http://www.aip.de/RAVE

However, the GAIA mission, which will provide astrometry, radial velocities and chemical abundance for up to 1 billion stars, is unlikely to be completed before the end of the next decade. Among the existing surveys, the Hipparcos and the Tycho-2 catalogs have compiled proper motions for 118 000 and for 2.5 million stars, respectively, but radial velocities have been completed for only a few ten thousands. The new release of the HST guide star catalog, GSC2.3, will include proper motions, and the astrometric satellite DIVA plans to compile proper motions for up to 40 million stars, but no radial velocities are available for these targets. To our knowledge no systematic survey is planned so far that includes radial velocities and that is capable of filling this gap in size and time between existing surveys and the GAIA mission.

With the successful demonstration of ultra-wide-field (40 $\deg^2$) multi-object spectroscopy (MOS) on the UK Schmidt telescope of the AAO, a major opportunity beckons to generate the first large-scale all-sky spectroscopic survey of Galactic stars with a radial velocity precision better than $2\,\mathrm{km\,s^{-1}}$ (see Figure 1).

In this paper we map out the case for RAVE (RAdial Velocity Experiment), an ambitious plan to measure radial velocities and chemical compositions for up to 50 million stars by 2010 using novel instrumentation techniques on the UK Schmidt telescope and on a northern counterpart.

2. The RAVE survey

RAVE is a large international collaboration involving a still growing list of astrophysicists from Australia, Canada, France, Germany, Italy, Japan, Slovenia, the Netherlands, the UK and the USA. The RAVE survey is split into two components: a pilot survey and a main survey.

- The pilot survey is a preliminary spectral survey, using the existing 6dF system at the UK Schmidt telescope of the AAO to observe about 100 000 stars in $\approx$ 180 days of unscheduled bright-time during the years 2003–2005. The 6dF spectrograph will target on the CaII triplet region (8480-8740 Å) favored by the GAIA instrument definition team. Spectra are taken at a resolution of $R = 4000$. Test observations indicate that an accuracy of $2\,\mathrm{km\,s^{-1}}$ can be achieved (see Figure 1).

  The target list would include a large fraction of the 118 000 Hipparcos stars that are accessible from the southern hemisphere as well as some of the 2 539 913 stars of the Tycho-2 catalog. The survey will focus on stars in the color range $0.4 < B - V < 0.8$. For these stars, useful photometric parallaxes can be derived if the trigonometric parallaxes are not available. The 6dF spectra will also provide useful estimates of the [Fe/H] value.

- The main survey will utilize a new Echidna-style multi fiber spectrograph (see Figure 2) at the UK-Schmidt telescope. It consists of a 2250-spine fiber array covering the full field of the Schmidt Telescope (40 $\deg^2$). The spines are hexagonally-packed on $\approx$ 7 mm centers. Each spine can be deformed by a piezo element resulting in a 15-arcmin patrol area. The key advantage of this new MOS design is its short reconfiguration time of

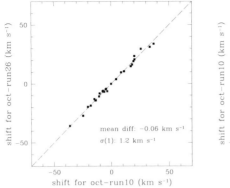 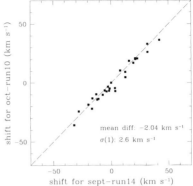

Figure 1. Test observation of the radial velocity of stars performed in bright time by K. Freeman on 6dF ($R = 4\,000$) in the second half of 2001. The figure correlates the radial velocity measured with the same fiber at two zenith distances (*left*), and at two different epochs one month apart (*right*).

≈ 5 min. Echidna will feed an efficient spectrograph using a high-efficiency Volume Phase Holographic (VPH) grating and will use a single 2k×4k red-optimized detector. With 250 Å coverage required for the CaII triplet region, a dispersion of 0.375 Å pix^{-1} will yield $R \approx 10\,000$ spectroscopy with 2-pixel sampling.

Again the survey will target the CaII triplet region. At $R = 10\,000$ we expect radial velocities at $1\,\mathrm{km\,s^{-1}}$ accuracy. Iron abundances [Fe/H] could be determined to 0.1 dex accuracy and useful estimates of the differential [α/Fe] abundance ratios for about half the stars ($V < 15$ mag).

The main study is expected to be performed throughout the period 2006-2010. Owing to the short reconfiguration time of the new Echidna MOS, about 22 000 stars can be observed per night in 30 min exposures, resulting in S/N=30 spectra at $V = 15$ mag. Throughout the 5 year campaign, a total of 25 million stars can thus be targeted. Subject to an equivalent instrument on the northern hemisphere, RAVE will yield a total of 50 million radial velocities, metallicites and, for a subset of $\lesssim 50\%$ of the sample, also abundance ratios.

The RAVE survey will provide a vast stellar kinematic database three orders of magnitude larger than any other survey proposed for this decade. The main data product will be a magnitude-limited survey of 26 million Thin Disk main sequence stars, 9 million Thick Disk stars, 2 million Bulge stars, 1 million Halo stars, and a further 12 million giant stars including some out to 60 kpc from the Sun. RAVE will offer the first truly representative inventory of stellar radial velocities for all major components of the Galaxy.

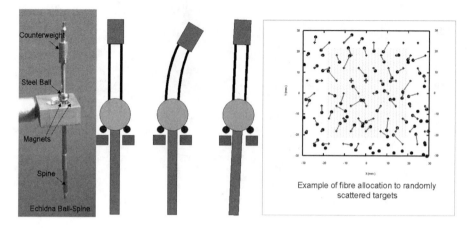

Figure 2. The Echidna concept. *Left*: single ball-spine of the Echidna MOS. *Center*: illustration of the work principle of an individual echidna spine. *Right*: example of fiber allocation to randomly scattered targets.

3. The RAVE science case

Within the cold dark matter (CDM) paradigm, the Galaxy built up through a process of accretion over billions of years from the outer Halo. Sophisticated computer simulations of structure growth within a CDM universe have now begun to shed light on how this process may have taken place (see e.g. Steinmetz & Navarro 2002). These advanced computer models do not only provide information about the structure and kinematics of the major stellar components of a galaxy but also on their chemical signatures and their stellar age distribution. In the context of these simulations, RAVE will revolutionize our understanding of the formation and evolution of all major components of the Galaxy: the Disk, the Bulge and the Halo.

3.1. Halo sub-structure

The details of galaxy formation are not well understood. In particular, CDM simulations actually predict far more infalling satellites than are currently observed (Moore et al. 1999, Klypin et al. 1999). The orbital timescales of stars in the outer parts of galaxies are several billion years and it is here we would expect to find surviving remnants of accretion. The disrupting Sagittarius dwarf spheroidal galaxy was discovered by Ibata et al. (1994) from a multi-fiber radial velocity survey. Five years later, Helmi et al. (1999) discovered a stellar stream within 1 kpc of the Sun after combining a radial velocity catalog with the Hipparcos database.

Both of these studies demonstrate the power of accurate radial velocities and proper motions in identifying cold stellar streams. We can expect RAVE to reveal evidence of many tens of similar streams both in the Halo, in the outer Bulge and within the Thick Disk. When combined with the next generation of surveys such as DIVA, this tally may extend to many hundreds of infalling systems.

3.2. Chemical signatures

A key aspect of RAVE will be the availability of chemical signatures like [α/Fe] and [Fe/H], in addition to accurate kinematics. The α elements arise from massive stars and the bulk of their mass is released in Type II supernova explosions. The Fe-peak elements are produced primarily by Type Ia supernovae which begin to dominate after a billion years or more. Unique signatures from abundance ratio pairs like ([α/Fe], [Fe/H]) may help to identify a common site of formation among widely separated stars. The use of chemical signatures can be extended to other components of the Galaxy, in particular, the Halo and the outer Bulge and the Thick Disk.

3.3. Bulges

The formation of stellar bulges, a major element of galaxy classification schemes, is not well understood. The Galactic Bulge stars are almost as metal rich as the Thin Disk but as old as the Thick Disk and parts of the Halo. Our current picture is that large bulges are formed from a rapid collapse of a spherical cloud, and that the small bulges are either formed from accretion or from the action of the central bar after the disk formed. In an alternative model, as favored by the CDM model of structure formation, bulges are the remnants of early gas-rich mergers between some of the first building blocks of a galaxy. A key constraint is the [α/Fe] abundance ratio, which has been determined for only a few dozen Bulge stars to date. A short star formation epoch either during the collapse of the Bulge or, as favored in the CDM model, in the progenitors from which a bulge is assembled is expected to lead to enhanced [α/Fe] for most of the stars; an extended star formation period during the Bulge assembly would imply [α/Fe] = 0.

3.4. Thick Disk

The stars in the Thick Disk are at least as old as those in the globular cluster 47 Tuc and it is widely believed to be a 'snap frozen' relic of the early disk shortly after the onset of disk dissipation. In this picture, an infalling satellite vertically heated the early disk to a scale height of 1 kpc. Another possibility is that the Thick Disk is made up of tidal debris from infalling satellites.

From the combined chemical and kinematic signatures, the RAVE survey should cleanly distinguish between competing models for the Thick Disk and thus end a many-year old debate on the origin of the Thick Disk. A major unknown in disk formation is whether the extent of the stellar disk is laid down during the major epoch of dissipation, or whether it grows with time. The RAVE survey will clearly establish whether the radial extent of the Thick Disk is comparable to or less than the Thin Disk. The chemical information will also be very important. The existence of an abundance gradient in the Thick Disk, as we observe in the Thin Disk, would argue against an infall origin; unique chemical signatures in the Thick Disk would argue for an infall origin.

3.5. Thin Disk

The largest fraction of the RAVE targets will be Thin Disk dwarfs and giants. Little is known about the dynamical state of the Thin Disk beyond 2 kpc of the

Sun. The existence of the inner bar and outer stellar warp is firmly established but many areas of astrophysics would benefit from their influence being understood in far more detail. Some external galaxies have optical disks, which appear to be lop-sided with respect to the dark Halo. Whether this is the case for the Galaxy is not known. The intrinsic brightness of the giants allows these important tracers to be observed throughout the entire optical extent of the Galaxy. The giants probe the large-scale dynamical state of the Galaxy, in particular the influence of the inner bar, the outer warp and the degree of eccentricity and lop-sidedness of the optical disk.

The vast number of dwarf stars in the RAVE survey will reveal the dynamical state of the Thin Disk and neighboring spiral arms within a few kiloparsecs of the Sun's position. This is crucial information if we are to construct an accurate model of the gravitational potential of the disk, and its distribution function. Recently, it has become clear that even the old stellar populations appear to show sub-structure. The RAVE survey will provide fundamental information on how different stellar populations deviate from dynamical equilibrium, and therefore constrain the formation history of the disk and its different components (e.g. spiral arms, stellar associations, etc).

3.6. Summary and conclusions

RAVE is an international project designed to survey radial velocities, metallicities and abundance ratios for the brightest 50 million stars in the Galaxy down to a completeness limit of $V=16$ mag. This is sufficiently deep to allow for kinematic and chemical studies of all major stellar components of the Galaxy. RAVE has a number of science goals addressing a wide range of priority areas in galactic structure and dynamics.

The RAVE survey is expected to be performed throughout the period 2003-2010. It also provides an opportunity to pre-empt some of the spectral work in the GAIA mission, providing results up to one decade earlier than that planned for the GAIA final release and probably still well ahead of the launch of GAIA (2012). Thus RAVE serves as an ideal real-data training set for the final design of the GAIA data reduction pipeline and may even influence some of the final design decisions.

References

Freeman, K. & Bland-Hawthorn, J. 2002, ARA&A 40, 487.
Helmi, A., White, S.D.M., de Zeeuw, P.T. & Zhao, H. 1999, Nature 402, 53
Ibata, R.A., Gilmore, G. & Irwin, M.J. 1994, Nature 370, 194
Klypin, A., Kravtsov, A.V., Valenzuela, O. & Prada, F. 1999, ApJ 522, 82.
Moore, B., Ghigna, S., Governato, F., Lake, G., Quinn, T., Stadel, J. & Tozzi, P. 1999, ApJL 524, L19
Steinmetz, M. & Navarro, J.F. 2002, New Astron. 7, 155.

The ING telescopes in the GAIA era

Romano L.M. Corradi, Danny J. Lennon & René G.M. Rutten

Isaac Newton Group of Telescopes, Apartado de Correos 321, E-38700 Santa Cruz de la Palma, Spain

Abstract. A brief introduction to the Isaac Newton Group of Telescopes at the Roque de los Muchachos Observatory is presented. Wide-field imaging and multi-object spectroscopy, together with adaptive optics imaging/spectroscopy, are the main areas of instrumentation that will be supported and developed in the future. The objective of this contribution is to invite the astronomical community to think about the possible role of our telescopes, like the 2.5m INT and the 4.2m WHT, for preparatory or complementary surveys in support to the GAIA mission.

1. The Observatory

The Observatorio del Roque de los Muchachos is located on the island of La Palma (Canary Islands), at a height of about 2400 m above sea level. The observatory hosts a large number of astronomical facilities used by scientists from many European countries. Night-time and solar telescopes are in operation, but also high-energy cosmic ray experiments are in use. This makes the observatory one of the main centres for ground based astronomy in the world, and certainly the main centre on European soil. This success is largely inspired by the excellent atmospheric conditions of the site. The observatory site is operated by the Instituto de Astrofisica de Canarias and an international scientific committee oversees its management. The facilities at the observatory, however, are operated and controlled independently by their owners.

For sun-lovers it may not come as a surprise that the weather conditions on the Canary Islands are good. But there are other reasons that make La Palma one of the best sites in the world for astronomical research. The Island of La Palma enjoys in fact a fairly unique natural environment. The steep, high mountainous island, stable trade winds and low inversion layer are responsible for the often cloudless skies at the mountain top. In particular the observatory is known to enjoy very good seeing conditions both for night time and day time observing; the median seeing is 0.69 arcsec. Furthermore, strict laws control light pollution on La Palma in a very effective way.

2. The Isaac Newton Group

The Isaac Newton Group of Telescopes (ING) was set up in the early 1980's as the flagship Northern-hemisphere optical observatory for the United Kingdom and The Netherlands through a bilateral partnership between the national research councils, now the UK Particle Physics and Astronomy Research Council, and the Netherlands Organisation for Scientific Research. Recently the partnership was extended to include the Instituto de Astrofisica de Canarias, who formally joined the ING in 2002.

The ING comprises three optical telescopes, the 4.2 m William Herschel Telescope (WHT), the 2.5 m Isaac Newton Telescope (INT) and the 1 m Jacobus Kapteyn Telescope (JKT), located on the Roque de los Muchachos Observatory on La Palma. The WHT has now been in operation for nearly 15 years, indicative of the relative youth of the observatory. The ING telescopes have played a central role in the success of astronomical research in the United Kingdom, The Netherlands, as well as in the host country, Spain. Scientific productivity of the telescopes, in particular that of the WHT, has been to the present date extremely high (both in quantity and quality) to international standards (see Benn and Sánchez 2001).

Currently, the ING is undergoing a dynamical restructuring plan in response to the impact on its budget following the UK's recent membership of ESO. As a consequence, the 1 m JKT telescope, in spite of its good image quality (it is presently equipped with a modern 2k×2K CCD providing a field of view of 10×10 arcmin^2 with a pixel of $0''.33$ as projected on the sky), will be closed in the middle of 2003 unless new partners willing to contribute to the running costs of the telescope appear. The other two larger telescopes will be maintained, but rapid developments in technologies and the construction of several 8 m class telescopes as well as the launch of dedicated satellites like GAIA imply that ING's telescopes must adapt to a new scientific role for the future. The future central role of the ING is seen to focus on two strands: (i) exploitation of adaptive optics, in particular at visible and near IR wavelengths, and (ii) multi-object imaging and spectroscopy over a wide field. Both these areas provide important science capabilities competitive with, and complementary to those at larger telescopes. With these areas of focus the ING optimally exploits the excellent properties of the La Palma observatory site and the quality of the telescope and its infrastructure, whilst offering a long-term development path that provides important instrumentation capability to the astronomical community.

2.1. Wide field imaging and spectroscopy at the ING

The 2.5m INT telescope is equipped with one of the few optical wide-field imagers presently available in the Northern hemisphere. Its Wide Field Camera works at the prime focus of the telescope and consists of a mosaic of four 2k×4k EEV CCDs, providing a field of view of 34×34 arcmin^2 with a sampling of $0''.33$ per pixel. Standard broad-band filters as well as a good number of narrowband emission-line filters are available. The performance of the instrument can be illustrated by the following figure: the exposure time needed to get a S/N ratio of 100 in the V band for a star of magnitude 20 (a 'typical' faint-end magnitude of the GAIA mission) is 130 sec, meaning that in principle the whole Northern hemisphere can be covered with 900 hours of observations, i.e. some 80 clear nights.

Figure 1. Aerial view of the Observatorio del Roque de los Muchachos. Telescopes are located next to the border of large crater of the national park "la caldera de Taburiente". In the background, the islands of Tenerife (left) and La Gomera (right) are also visible.

The 4.2m WHT telescope is also equipped with a Prime focus imager, consisting of two 2k×4K EEV CCDs, which corresponds to a field of view of 16×16 arcmin^2 with a pixel size of $0''\!.24$.

Follow-up spectroscopy of images taken with these wide-field cameras can be done at the ING with Autofib2/Wyffos, the multi-fibre robotic spectrograph at the WHT prime focus. Autofib2 can presently observe simultaneously up to 150 science targets over a field of 1 degree diameter (with an unvignetted field of 40 arcmin). The fibres feed the Wyffos spectrograph, which allows a range of resolutions up to $R \sim 10\,000$ thanks to the large choice of gratings available.

It is easy to imagine how these observing capabilities can be of help to complement or support some of the GAIA science goals. In this respect, astronomers from the European and worldwide community are welcome to apply for observing time at the ING telescopes.

2.2. Adaptive optics at the 4.2m WHT

Adaptive optics (AO) instrumentation is now an integral part of a growing number of telescopes around the world, including the WHT with the recent delivery of NAOMI. First results have shown the potential of NAOMI in delivering diffraction limited image quality in the near-IR. The scientific gains from image quality that defeats atmospheric turbulence are huge. Clearly observations from space are hard to beat from the ground, even with adaptive optics, except where novel instrumentation and a large telescope collecting area provide the cutting edge. This is where the WHT will be competitive. The WHT, placed on a site with excellent seeing conditions, is well positioned to builds its scientific use on AO techniques. For this reason plans are being developed for the deployment of a laser guide star system to promote the widest possible exploitation of excellent image quality through AO at the WHT.

3. Future developments at the observatory

The Roque de los Muchachos Observatory is Europe's premier astronomical observing site. Ongoing developments of new facilities will ensure that the observatory remains at the forefront of astronomical enterprise, and further strengthen European collaboration. In particular two facilities currently under construction that will have an important impact in the time when Gaia will be operational are the Spanish-led 10-m Gran Telescopio de Canarias (GTC) and the 17 m Cherenkov Telescope, MAGIC.

Looking ahead, plans to develop an extremely large telescope may well come to fruition on La Palma. Such large endeavour would most certainly involve many European countries and firmly plant the La Palma observatory as Europe's main Northern Hemisphere observing site. Further information on the ING can be found in our Web pages at http://www.ing.iac.es

References

Benn, C. R. & Sánchez, S. 2001, PASP 113, 385

Conference summary

Michael A.C. Perryman

Astrophysics Missions Division, ESA-ESTEC, 2200AG Noordwijk, The Netherlands

During the early 1990s, and based on the knowledge from the Hipparcos project, it became clear to European scientists that astrometric measurements from space, based on projected technology, would soon allow the measurement of the positions, parallaxes and annual proper motions of millions of stars at the tens of microarcsec level. This offered the prospects of measuring significant trigonometric distances out to about 10 kpc. Over a period of about 10 years, this has evolved, through the careful work of numerous scientists and industrial groups, into the prospects for the GAIA mission that we know today: the possibility of measurement of all stars to about 20 mag, with a positional accuracy of 3–4 microarcsec for those brighter than 12 mag, 10 microarcsec at 15 mag, and about 0.2 milliarcsec at 20 mag. At the present time it is not obvious how to design a space mission (based on realistic engineering principles and costs) that would perform significantly better than this, and for so many objects; at the same time, designing a system for this target performance at 15 mag provides the same information for all other stars of the same magnitude (and a corresponding degradation in accuracy due to photon noise for fainter objects, down to the detection limit of about 20 mag). In other words, the present GAIA design gives a certain performance for hundreds of millions of objects. The enormous scientific case for GAIA has been woven around these accuracy capabilities.

I trace my own interest in the prospects for on-board radial velocity measurements for GAIA to a proposal made by Roger Griffin to the UK SERC in June 1987: to construct two dedicated 1.5 m ground-based telescopes (in the northern and southern hemispheres), equipped with 'Coravel-type' radial-velocity spectrometers, and to measure the radial velocities of Hipparcos programme stars in time for their inclusion in the published Hipparcos Catalogue expected in about 1995. The proposal was not approved, nor was it approved when re-submitted in revised form to the SERC in 1990. Attempts, at about this time, to gain financial support for such a programme through ESO, and through ESA (on the grounds that it would be of great value in complementing the Hipparcos results) were also unsuccessful. Key programmes *were* subsequently submitted to and approved by ESO to gather such radial velocities systematically for Hipparcos stars (ESO Messenger No. 56, June 1989, pag. 12), but I am uncertain as to the wide availability of these and other complementary measurements made for the Hipparcos stars. Their systematic acquisition and uniform publication was never achieved.

With hindsight, it might seem surprising that such a programme should have generated so little scientific interest or support. For those involved in trying to stimulate these efforts — I recall that amongst others, Catherine Turon, as leader of the Hipparcos Input Catalogue Consortium, and Adriaan Blaauw, as chairman of the Programme Selection Committee for Hipparcos (see, e.g., ESA SP–234, 1985, pag. 267) were staunch advocates — it was disappointing that the Hipparcos Catalogue had to be published without the third component of the stellar space motions, the more so since we were conscious that their acquisition from ground would have cost a small fraction of the overall cost of the other two components from space! Let me quote directly Adriaan Blaauw's prescient remarks in his summary of the 1988 Sitges meeting on the Scientific Aspects of the Hipparcos Input Catalogue Preparation (pag. 500): '... there are, of course, things left one would like to see done. One of them, perhaps the most urgent, is in the field of radial velocities ... there is no really dedicated Hipparcos radial velocity programme ... May we be sure that, when the Hipparcos proper motions become available, we will not be faced with a deplorable lack of data on the third component of stellar motions?'

With the prospects of the GAIA astrometric results now a reality, I believe that there is much more awareness of the urgency and importance of acquiring complementary data (radial velocities and photometry) for the GAIA programme stars, and ensuring that they are obtained and published in a uniform and timely manner. Perhaps the following words in an early paper reporting the developments of the Sloan Digital Sky Survey still summarise the situation: 'Optical astronomy has long lagged behind other astronomical disciplines of more recent origin in the production of survey data of well-known and characterized completeness and in digital form, both for imaging and spectroscopy. This has had an adverse impact in many areas, since it is still true that for the most part an object is not 'understood' until it has been identified in the optical and its nature as revealed by optical observations is understood' (Gunn & Knapp 1993, ASP Conf. Series 43, pag. 267).

With GAIA advancing in prospects during the mid-1990s, and conscious of the importance and difficulties of acquiring these complementary data, we looked into the general approach of acquiring GAIA complementary data, including the radial velocities, from the ground (Perryman, 1994, in 'Astronomical and Astrophysical Objectives of Sub-Milliarcsecond Optical Astrometry', 211); for the radial velocities the relative advantages from ground and space (Favata & Perryman 1995, ESA SP–379, pag. 153); while Uli Bastian focussed on the ground-based option, in a concept which looks very much like the RAVE project discussed at this meeting (1995, ESA SP–379, pag. 165).

During the early phases of the design of GAIA, the decision was made to include a dedicated radial velocity instrument on board the satellite; this dedicated telescope was subsequently designed to include the medium-band photometric instrument as well. The objectives were to ensure that these data *would* be acquired within the context of the GAIA project, optimised accordingly, acquired at the same measurement epochs and, with the benefits of space, freed from the atmospheric and systematic zonal errors plaguing such measurements from ground. The collective optimisation of this telescope and instrument combination was the product of many discussions on the instrument's scientific goals and technological optimisation.

Following the acceptance of GAIA in the ESA scientific programme in October 2000, a management and working group structure for the scientific aspects of the mission was established, including those preparatory activities related to the radial velocity spectrometer. The scientific structure in place for GAIA devolves the first level of responsibility for study and optimisation of the various scientific tasks to a (largely autonomous) set of working groups. In addition to the radial velocity working group, the activities of which are the main theme of this conference, there are other groups associated with the instrument design and optimisation (photometry, on-board object detection), with the consideration and treatment of specific astronomical objects (variable stars, double and multiple stars, solar system objects, etc.), and with the central processing of the data itself (simulation, data analysis, etc.). Of course there is overlap in some of these tasks, and the collective optimisation, priorities, instrumental feasibility, and so on, are the collective tasks of the GAIA Science Team, in place to advise ESA on all aspects of the scientific conduct of this enormous mission.

We were indeed fortunate that David Katz and Ulisse Munari stepped forward to lead the radial velocity working group, setting up a scientific working group which now includes about 30 core members and some 30 associate members (incidentally David started his leadership of the group while a research fellow at ESA). The group quickly focussed on a subset of measurement and instrument possibilities. Based on earlier measurements made from the ground, Ulisse Munari made (to my knowledge) the first specific proposal to centre the GAIA radial velocity measurements around the near-IR triplet of Ca II at 860 nm (GAIA technical note UM–PWG–003, 1997; see also MemSAIt, 1992, 63, pag. 195), while a similar suggestion appears to have been made independently by Rudolf Le Poole (1997, personal communication).

A detailed list of tasks (and schedules) for the working group were established during 2001. Over the past year or so, four meetings of the working group have taken place, a number of technical reports have been written and circulated, and the group is now focusing its effort on arriving at a detailed technical and scientific baseline for the instrument. The working group feeds its conclusions and recommendations to ESA through the GAIA Science Team.

With the start of the industrial development phase during mid-2002 another organisational development has recently taken place. A group of scientific institutes, already involved in the radial velocity working group effort, have responded to the industrial bid to develop and optimise the GAIA payload. This work, led by Mark Cropper at MSSL, and involving individuals at MSSL (UK), University of Leicester, Observatoire de Paris, Padova & Asiago Astronomical Observatories, and University of Ljubljana, should ensure a significant advance in the technical design of the radial velocity instrument, as well as very efficient feedback into the instrument design from the working group effort.

The conference aims were described in the conference announcement as follows: *'They include a review of GAIA spectroscopy and its science goals, the spectrograph design and involved technology, synergy with GAIA photometry and astrometry, legacy to ground-based spectroscopy, review of tools for spectral data analysis and treatment, as well as radial and rotational velocities, astrometric radial velocities, model atmospheres, chemical abundance analysis, spectral peculiarities, signature of mass loss and interstellar medium, stellar pulsation, binarity and eclipsing stars, galactic kinematics, structure and evolution, and GAIA spectroscopic contribution to our understanding of stars'.*

The conference took place in the Residenza del Sole, in the beautiful valley of Gressoney Saint Jean, and the location provided excellent conference facilities, comfortable accommodation, and fine dining and relaxation facilities in a convivial setting, establishing an excellent ambiance for the meeting and the discussions which developed around it.

Introductory presentations set the scene for the GAIA mission as a whole (Perryman, Pace), with presentations providing some indications of the complexities of the astrometric measurements and analyses (Mignard, de Felice), and of the approach being developed for the overall data analysis (Ansari). The current status of the radial velocity spectrometer, from the perspectives of scientific motivation, spectral features of the spectral region, and accuracy (Munari), design/performance and simulations (Katz), and technical solutions (and problems!) (Cropper) were supported by presentations of the related problems of crowding (Zwitter), telemetry flow (Viala) and supporting data bases of synthetic (Zwitter) and observational (Sordo) spectral grids. The interest and richness of the proposed spectral region was illustrated forcefully by a number of speakers: for chemical abundance analyses (Thévenin), and for stellar rotation (Gomboc). Some hardware options for wavelength calibration were also presented (Desidera, Pernechele).

The importance of the radial velocity measurements for studies of the structure and dynamics of our Galaxy were nicely illustrated in various presentations (Gilmore, Bienaymé, Bertelli, Spagna) and the applications of pulsating and variable stars (Cepheids, RR Lyrae, and Mira variables) to the structure and distance scale (Szabados, Bono, Feast). Applications to specific object classes such as planetary nebulae (Magrini), Wolf-Rayet stars (Niedzielski), and RV Tau stars (Wahlgren) were also discussed.

The great interest in eclipsing binaries observed with GAIA was underlined in presentations by Milone, Wilson, and van Hamme. The huge scientific interest, and complexity, of binary star studies, was also addressed (Arenou, Halbwachs, Kovaleva, Pourbaix, Söderhjelm, Zwitter).

Efforts to optimize the astrometric, radial velocity and photometric measurements of GAIA involve a huge and systematic simulation effort, aspects of which were outlined by Babusiaux. Some of the work of the Photometry Working Group was reported by Jordi and by Vansevičius. Bailer-Jones summarised the GAIA project's ambitious plans for object classification and parameterization. One of the essential inputs to many of these various tasks are synthetic spectra, and we were fortunate to have three presentations by people deeply involved in modeling of stellar atmospheres (Nesvacil, Hauschildt, Plez). With the expected contribution of the GAIA results in the field of stellar models, it was good to have the participation and make close contact between these groups and the GAIA project.

Towards the end of the conference, we heard presentations on the photometric data that may be expected from ground-based programmes over the next years (Henden), on the possible role of the ING telescopes in the GAIA era (Corradi), and on ambitious plans for a ground-based radial velocity programme (RAVE) essentially supporting the DIVA and GAIA astrometry (Steinmetz). An informative summary of all the poster contributions was made during the meeting by Mark Cropper.

In trying to assess where the design of the radial velocity spectrometer stands at the time of this conference, it is worth stressing that there is not really a large amount of 'parameter space' available for optimisation: the collecting area, field of view, and detector efficiency and noise characteristics cannot be changed substantially: as a result, radial velocity accuracies of around 10–20 km s^{-1} at limiting magnitudes of around $V = 16 - 17$ mag appear feasible; significantly improved accuracies at the same magnitude limit, or a significantly fainter magnitude limit do not. Nevertheless there is still scope to tune the scientific goals a little, at least on superficial inspection: higher resolution ($R \sim 20\,000$) may yield improved spectral diagnostics for brighter stars (and a smaller number of stars to a brighter magnitude limit for radial velocities); while lower resolution ($R \sim 5\,000$) may yield less diagnostic information on brighter stars, but more stars to a fainter limit with radial velocity information. If the trade-off were that simple, it is important then to focus on the acquisition of data that can only be gathered from space, emphasizing that such data will only be available in the period 2015–20, before which time many ground-based developments will have occurred. It might be safe to reason that high-resolution spectroscopy on the brighter stars (to $V \sim 14$) may best be left to ground-based observations, and place GAIA's emphasis on acquiring the largest number of radial velocities for the faintest stars possible; and concentrating on the fact that the GAIA measurements will be homogeneous, contemporaneous with the astrometry and photometry, and at multiple epochs such that binary discrimination can be assisted. We must avoid launching an instrument which will have been superseded or made obsolete by ground-based measurements, and avoid focusing the scientific goals on topics that are better addressed from ground anyway. Some of GAIA's power lies firmly in the domain of detecting and classifying objects or evolutionary phases which are better followed up from the ground.

Such a line of argument does not necessarily imply that the lowest spectral resolution in the range currently under consideration (viz $R \sim 5\,000$) must be preferred; further detailed simulations on the problem of crowding, sky background, and extracting the radial velocity data must be undertaken. Given the simulation and design effort underway, we might hope that the eventual choice of the instrument parameters (such as resolution and binning) might emerge without too much contention over the next few months.

With a number of questions in the air, let us look first at the successes of the radial velocity group over its first year of existence. Notable is the fact that an organisational structure and a first assessment of tasks have been agreed. Major progress in studies and simulations of all aspects of the instrument have been made, and it is now rather clear how all of these elements are to be connected. The direct support of major scientific laboratories to the industrially-led design effort is a welcomed developed. Another major success, not discussed at the meeting, is that the wavelength range has been accepted by all; the scientific richness of the proposed region has been beautifully demonstrated, and no convincing alternatives have been suggested. The fact that GAIA, and its radial velocity initiative, have presumably inspired the contemplation of projects such as RAVE, may also be taken as welcome developments.

All of this should not be taken for granted, since the coordination of such an effort is non-trivial. Here some words of Steven Beckwith which may inspire (and perhaps console) our colleagues during these challenging times *'Most scientific advances today are cooperative efforts. They require tremendous organisation. That organisation employs the talents of some people who have the vision of what is possible, and others who are willing to believe they could achieve the vision through cooperative effort. The great advances most often are made by people who have the vision to carry out new exploration, the talents to understand what has to be done, and the temperament to lead a group of unsuspecting and often irrational and unpleasant people to carry out a task.'* (Beckwith, 1993, ASP Conf. Series 43, pag. 303).

Priority tasks facing the working group include fixing the spectrograph resolution (which impacts on many areas of the scientific performance and technical implementation, such as the on-board data handling, on-board compression schemes, and telemetry rate); achieving an accurate perception of what data will be acquired from ground over the next 20 years and what may be left safely for GAIA. A number of other detailed optimisations may be possible: extracting the region around the Ca II lines for the fainter stars to reduce the telemetry rate; including a rotation mechanism (and accepting the associated failure risks) to improve the signal-to-noise; and investigating the use of L3CCD technology (perhaps in photon-counting mode) to decrease the detector read noise. Some areas still open for more detailed study are the issue of superposition of spectra to form a mission-averaged value (do they add as expected, are specific hardware constraints or tests needed); detailed development of an error budget; and an overall calibration plan, including the problem of the radial velocity zero point calibration. A more global task for GAIA is achieving a proper connection between the spectroscopy and the photometry: what diagnostics are best obtained from one or left to the other and, further into the future, how will the classification and physical characterisation of each object make use of the various data coming from these different sources.

Let me finish by thanking the Scientific Organising Committee, so ably led and coordinated by Ulisse Munari, which succeeded in putting together an inspiring and timely meeting on this exciting subject. All participants were extremely grateful to the Local Organising Committee chaired by Federico Boschi and also to the Local Support Group, to Manager Davide Porro and his team at the Residenza del Sole, and to Anna Mello Rella of Serenissima Viaggi for taking care so enthusiastically of all the participants travel, accommodation, and excursion events.

At the end of the week, we were left with the impression of one part of the GAIA mission in excellent scientific and technical hands, with much work to be done, but with a clear direction and a strong spirit of enthusiastic collaboration evident. Such a meeting, with all of its scientific promise, makes us look forward to the launch of GAIA with even more eager anticipation.

I've scattered a few quotations throughout this summary, so let me finish with another: *'Although accurate measurement of proper motions is exacting and sometimes tedious work, it cannot be emphasized too strongly that it is of central importance to the development of a picture of the structure, kinematics, and dynamics of our Galaxy.'* (Mihalas & Binney, Galactic Astronomy). And to this we will include the task of measuring the radial velocities of millions of stars which lies ahead, and promises so much.

# POSTERS

The poster area

Clock-wise from top-left: J. Halbwachs, D. Pourbaix, A. Prša, T. Kaempf, P. Willemsen and C. Jordi

The significance of wavelength coincidence statistics in abundance determination within the context of GAIA

Salim G. Ansari

European Space Agency / ESTEC, The Netherlands

Abstract. In an attempt to qualify the presence of chemical elements, prior to carrying out a thorough chemical analysis, the Wavelength Coincidence Statistics (WCS) method was introduced on a purely statistical basis of the presence of spectral lines. In order to introduce some physical notion of the strength of these lines, the stellar $\log gf$, the temperature and line strengths are used as an extension to the method. This would allow sorting these lines on the basis of their strengths and then running a comparison check on whether an element may truly have a presence, or whether it is spurious. Within the context of GAIA, the WCS and its extension would be worthwhile considering. It's automation in conjunction with spectral line identification would at least allow the detection of stellar spectra with chemical anomalies for a subsequent abundance analysis with data of higher quality. This method will however only work if a significant number of unblended lines can be identified.

1. Introduction

Wavelength Coincidence Statistics (WCS) is based on a Monte Carlo technique of creating a set of *nonsense* wavelengths within a fixed tolerance range around measured laboratory wavelengths (Hartoog et al 1973) The number may vary between 100 to 10 000 sets of wavelengths, which are in turn used to statistically identify and assess chemical elements in a spectrum. The method has been successfully used on high resolution spectra in the optical region to identify the most interesting features in chemically peculiar stars (c.f Cowley and Aikman 1980). An extension to WCS was devised by Ansari (1987) where a physical parameter was introduced to provide a notion of the strength of a spectral line depending on its atomic characteristics and stellar temperature. The Intensity Parameter $<IP>$ may therefore be calculated:

$$<IP> = \log gf - \theta\chi \qquad (1)$$

where $\log gf$ denotes the line's oscillator strength, $\theta = 5040/T_{eff}$ and χ is the lower excitation potential in eV. By *sorting* chemical lines according to their $<IP>$ found in stars under various temperatures and examining the significance level of specific chemical elements, it is possible to quickly identify stars of interest for a more detailed chemical analysis. The method can be very relevant, especially when a very large number of stars are being investigated. In this

Table 1. We summarise the number of "strong" (out of a total of 3200) spectral lines per element based on their $<IP>$ values, which was roughly $<IP> -3$ at $T_{\text{eff}}=20\,000$. This gives some indication as to which elements one can expect in this spectral region for any given temperature. Note: we exclude the Paschen lines.

Element	No. of lines	Element	No. of lines
N 1	2	Cr II	3
Mg II	3	Cs I	26
Si I	12	Mn I	19
S I	15	Mn II	5
Ca I	6	Fe I	33
Ca II	3	Fe II	11
Sc I	8	Co I	2
Sc II	2	Ni I	6
Ti I	44	Ni II	1
V I	3	Sr II	3
Cr 1	17	Zr I	4

paper, we investigate only a set of atomic lines. We neither attempt to identify molecular lines in cool stars, nor identify atmospheric lines.

2. Relevance to GAIA

In an attempt to classify objects, based on their chemical composition, the WCS can be a powerful tool to allow further investigation of specific chemical species. The large number of spectra expected to be observed by GAIA will require an automated procedure to sift through a large number of data. Apart from an automated line identification routine, it is important to consider an algorithm that will run through these measured spectral lines and to extract, out of the large sample, some stellar objects of interest.

3. Analysis

WCS was applied to a set of representative spectra. We used a grid of models provided by T. Zwitter (private communication), which are based on grid models calculated by Munari & Castelli (2000) and Castelli & Munari (2001), selecting a random number of effective temperatures from cooler to hotter stars, around $\log g=4$ and solar abundance. The aim of this test was to see how effective WCS would be in identifying the most dominant chemical elements and to investigate its usefulness in quickly identifying the most typical features. We used spectra at a resolution of 17 200 (0.25 Å pix^{-1}) for the purpose. The GAIA wavelength window is roughly between 8450 and 8750 Å. The Vienna Atomic Line Database

Table 2. For each spectrum used, we list the detected chemical elements, the number of lines used and the significance achieved. At each temperature, we specifically extracted from VALD only the lines at that given effective temperature. The higher the significance, the more confident can we be in the presence of a chemical species. We set the tolerance to detect the spectral lines at ±200 mÅ. We used Zwitter's synthetic spectra at $\log g$=4.0, $V_{rot} = 0$ and $[Z/Z_\odot]$=0, except for $T_{\rm eff}$=4000 K, where we used the model with $[Z/Z_\odot]$=0.5. The number of usable iron-peak element lines diminishes rapidly at higher temperatures.

Element	No. of Lines	No. of Hits	Significance
$T_{\rm eff}$=4000 K			
Mg I	9	4	3.645
Si I	7	3	3.215
Ca I	6	2	1.679
Ca II	3	2	2.929
Ti I	31	17	7.780
V I	3	1	1.401
Co I	2	1	1.930
Cr I	12	5	3.492
Mn I	6	3	2.874
Fe I	58	33	11.413
$T_{\rm eff}$=7275 K			
N I	10	6	5.735
Mg I	5	3	4.018
Si I	18	9	8.986
S I	15	5	2.900
Ca II	3	3	7.496
Fe I	33	23	12.133
Fe II	6	2	2.630
$T_{\rm eff}$=9750 K			
N I	9	6	8.400
S I	8	3	4.854
Ca II	3	2	6.244
Fe I	2	2	5.155
Fe II	6	1	1.603

(VALD; Kupka et al.1999) lists some 3200 atomic lines in this region (see Table 1 for a summary of selected number of lines, based on the calculated $<IP>$ for a wide range of effective temperatures). The region is characterised by the Ca II triplet, which is dominant at all temperatures. In order to test the reliability of WCS at this resolution and at the same time to see how well the method could be use to detect present spectral lines and chemical species, we extracted spectral lines with the same parameters (temperature, $\log g$ and abundance) for each synthetic spectrum we used. We then created a set of measured spectral lines for each spectrum, which we then ran WCS on. The summary of results are given in Table 2. It is evident that the most predominant lines were identified very easily with the method, thereby ensuring that, if we had a very large sample of stars, it should be fairly straightforward to use this method effectively. This could be particularly significant while searching for stars with overabundant chemical elements (such as Silicon, Manganese or Magnesium.) However, due to the low resolution and possible low signal-to-noise ratios in the real observations, some of the weaker lines, which may become more difficult to detect, may lead to spurious WCS results. Therefore care must be taken in interpreting data and more observations in other wavelength regions would be necessary to carry out a deeper investigation.

4. Conclusions

The spectra that will be gathered by GAIA are expected to dramatically increase the sample of some of the stars with a peculiar chemical composition. By doing so, we hope to first establish a much larger set of sources that will in turn allow us to better classify these objects. Furthermore we hope that with a larger sample our model atmospheres can greatly improve leading to a better understanding of the source of such peculiarities. Based on spectra acquired by GAIA, a preliminary spectral line identification can firstly be carried out to establish the presence of certain chemical elements. WCS, among other possible algorithms, should be considered as an 'automated' process to identify the presence of certain chemical elements in order to help sub-classify large numbers of stars depending on their chemical composition. Furthermore, it should be mentioned that GAIA spectra are not suited for deeper chemical analysis, in the absence of a large number of spectral lines for any element and the chosen spectral resolution. In order to achieve a curve-of-growth analysis, it is important to have a wide range of line intensities within a chemical species, achievable only if high resolution spectra are available. It is expected that with a more complete list of spectral lines that would also include molecules, a deeper investigation can be carried out in the intricacies of the chosen spectral region that will also define the typical spectral characteristics expected over a wide range of spectral classes and other sources.

References

Ansari, S.G. 1987, A&A 181, 328
Castelli, F. & Munari, U. 2001, A&A 366, 1003
Cowley, C.R. & Aikman, G.C.L. 1980, ApJ 242, 684
Hartoog, M.R., Cowley, C.R. & Cowley, A.P. 1973, ApJ 182, 847
Kupka F., Piskunov N.E., Ryabchikova T.A. et al. 1999, A&AS 138, 119
Munari, U. & Castelli, F. 2000, A&AS 141, 141

NLTE line-blanketed CaII calculations for evaluation of GAIA spectroscopic performances

Innocenza Busà, Isabella Pagano

INAF - Catania Astrophysical Observatory, Catania, Italy

Marcello Rodonò

Department of Physics and Astronomy, University of Catania, Italy

Maria Teresa Gomez, Vincenzo Andretta & Luciano Terranegra

INAF - Capodimonte Astronomical Observatory, Naples, Italy

Abstract. NLTE line-blanketed calculations of the profiles of the Ca II near-IR triplet (8498, 8542 and 8662 Å) are performed for a grid of 27 photospheric models with $T_{\rm eff}$ in the range 4200-6200 K, $\log g$ between 4.0-5.0, metallicities 0.0, –1.0 and –2.0, and for four different resolving powers R = 5 000, 10 000, 15 000, and 20 000. The goal of the work is to test the effect of departures from LTE and the sensitivity of the NLTE profiles to changes in stellar parameters in the range of dispersion considered for the GAIA spectrometer.

1. Introduction

The Ca II triplet applications span from stellar chromospheres to active galaxies, from starbust regions to globular cluster metallicity analysis. Ca II triplet lines are used for the spectral classification of stars and for the identification of supergiants in the Galaxy. Moreover, these lines have a wide application as chromospheric activity indicators. For this reason, the GAIA mission and in particular its spectroscopic (8480–8740 Å region) performances are of great interest for the community. In particular, the choice of a medium (R=5 000) or high (R=20 000) resolving power, which is yet in discussion, will establish which kind of analysis will be possible with GAIA spectroscopic data. In this poster we will deal with NLTE line-blanketed simulations of the Ca II triplet lines at four R values from 5 000 to 20 000 showing which effects and which approaches become important at different resolutions.

2. Line profiles and EW sensitivity to stellar parameters

Ca II triplet line profiles have been calculated for a grid of 108 photospheric models with $T_{\rm eff}$=4200, 5200, 6200 K, $\log g$=4.0, 4.5, 5.0, [Fe/H]=0.0,–1.0, –2.0 (NextGen, Hauschildt et al. 1999) and resolution 5 000, 10 000, 15 000 and 20 000 using version 2.2 of the code Multi (Carlsson 1986) and the treatment of line blanketing as described in Busà et al. (2001).

For all three lines, we find that the sensitivity of the profiles to metallicity is quite strong and is higher for the hottest models. The dependence on T_{eff} is higher for lower metallicity while the dependence on gravity is very weak becoming significant only for low temperatures and metallicity.

Different values of spectral resolution give profiles that, as expected, differ mainly in the line cores.

The dependence of the Calcium triplet 'central depression' (CD = depth of the line from the continuum level to the core) on stellar parameters, when no convolution for instrumental profile is done, is very weak for both the three lines: the calculated CD values span a rather narrow range, from 0.80 up to 0.98 which means a maximum percentage variation of 22%. On the other hand, when we consider the convolution for instrumental profiles we find that the adopted resolving power dominates the observed CD and its dependence on stellar parameters. In order to look for a correlation of CD as function of (T_{eff}, [Fe/H], $\log g$, R), we considered a set of 216 CD values obtained considering 8 resolving power values spanning the range 5 000 to 25 000. The multilinear fitting relation found is:

$$CD_{\text{conv}} = A \cdot \log R + a \cdot T_{\text{eff}} + b \cdot \log g + c \cdot [\text{Fe/H}] + \quad (1)$$
$$+ d \cdot [\text{Fe/H}] \cdot T_{\text{eff}} + e \cdot [\text{Fe/H}] \cdot \log g + f$$

where:

- $A = 4.82 \cdot 10^{-1}$, $a = -1.29 \cdot 10^{-5}$, $b = -3.65 \cdot 10^{-2}$, $c = -4.91 \cdot 10^{-2}$, $d = 2.71 \cdot 10^{-5}$, $e = -5.64 \cdot 10^{-3}$ and $f = -1.07$ for 8662 Å line (correlation coefficient $r = 0.98$);

- $A = 4.52 \cdot 10^{-1}$, $a = -1.26 \cdot 10^{-5}$, $b = -3.80 \cdot 10^{-2}$, $c = -5.07 \cdot 10^{-2}$, $d = 2.65 \cdot 10^{-5}$, $e = -6.09 \cdot 10^{-3}$ and $f = -0.910$ for 8542 Å line ($r = 0.98$);

- $A = 5.30 \cdot 10^{-1}$, $a = -1.41 \cdot 10^{-5}$, $b = -3.17 \cdot 10^{-2}$, $c = -4.27 \cdot 10^{-2}$, $d = 2.58 \cdot 10^{-5}$, $e = -1.33 \cdot 10^{-3}$ and $f = -1.37$ for 8498 Å line ($r = 0.99$).

The strength of the A coefficient, with respect to the other coefficients, put in evidence that the observed CD is dominated by the adopted R value.

Furthermore, we find that the measured equivalent width EW (which does not depend on R) of the above profiles is well correlated with the stellar parameters. A multilinear fitting relation is in fact obtained with correlation coefficient of 0.99 for both the 8662, 8498 and 8542 Å lines:

$$EW(NLTE) = a \cdot T_{\text{eff}} + b \cdot \log g + c \cdot [\text{Fe/H}] + \quad (2)$$
$$+ d \cdot [\text{Fe/H}] \cdot T_{\text{eff}} + e \cdot [\text{Fe/H}] \cdot \log g + f$$

where:

- $a = -2.75 \cdot 10^{-4}$, $b = 6.6 \cdot 10^{-2}$, $c = 3.46 \cdot 10^{-1}$, $d = 8.90 \cdot 10^{-5}$, $e = 1.88 \cdot 10^{-2}$ and $f=3.96$ for the 8662 Å line;

- $a = -1.31 \cdot 10^{-4}$, $b = 3.9 \cdot 10^{-2}$, $c = 1.36 \cdot 10^{-1}$, $d = 3.56 \cdot 10^{-5}$, $e = 2.17 \cdot 10^{-2}$ and $f=1.86$ for the 8498 Å line;

- $a = -3.55 \cdot 10^{-4}$, $b = 8.3 \cdot 10^{-2}$, $c = 4.41 \cdot 10^{-1}$, $d = 1.20 \cdot 10^{-4}$, $e = 2.04 \cdot 10^{-2}$ and $f=5.12$ for the 8542 Å line.

For all three lines a strong dependence of EW on metallicity is dimostrated by the c coefficient high values.

3. Departure from LTE

In Figure 1 we show the NLTE effect on the EW (left panel) and on the line profiles (right panel) for the 8662 Å line; the results obtained for the 8498 Å and 8542 Å line profiles are very similar to the 8662 Å line case.

For both the three lines the departure from LTE mainly depend on metallicity: it is below 10% when solar or [Fe/H]=−1.0 models are considered, while it strongly enhanced, up to 35%, for [Fe/H]=−2.0 models (see Figure 1 left panel). Furthermore, the sensitivity of NLTE effects to $\log g$ and T_{eff} is appreciable only in low metallicity stars, for these metal-poor stars NLTE effects are higher in lower T_{eff} and lower $\log g$ atmospheres.

Computed LTE and NLTE profiles for 8662 Å line are shown for two models and two R values in Figure 1 (right panel). It is clear that, even in models

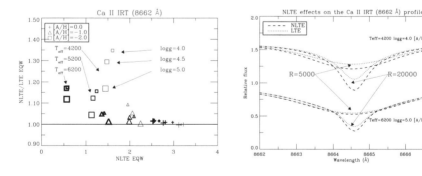

Figure 1. *Left*: EW departure from LTE for the 8662 Å Ca II line. Crosses indicate the solar metallicity models, triangles refer to [Fe/H]=−1.0 and squares to [Fe/H]=−2.0. *Right*: synthetic profiles obtained using LTE (dotted lines) and NLTE (dashed lines) approaches for two different resolutions: R=5 000 and R=20 000. The profiles shown on top have been obtained using model atmospheres with T_{eff}=4200 K, $\log g$=4.0 and [Fe/H]=−2.0, while bottom profiles were obtained from atmospheres with T_{eff}=6200 K, $\log g$=5.0 and [Fe/H]=0.0.

where the EW departure from LTE is negligible (solar metallicity, see Figure 1 left panel), the difference between LTE and NLTE is readily apparent in the line core.

In Figure 1 the EW of the 8662 Å line for the model with [Fe/H]=0.0 departs from LTE by ∼2%, while the CD departs from LTE by more than the 9% at R=20 000 and by ∼4% at R=5 000. For the [Fe/H]=−2.0 model we have an EW departure of ∼14% while CD departure is of ∼24% at R=20 000 and 18% at R=5 000.

This indicates that dealing with line profiles, instead of EW, the NLTE approach is needed, even if, at lower resolutions the NLTE effects on the profile become less evident.

4. Conclusions

The choice of low resolution (R=5 000) will make GAIA observations useful only in analysis based on EW measurements while many informations on line profiles will be lost. On the other hand, this low resolution will allow lower magnitude limits in observations, furthermore, a NLTE approach is not required except for high precision measurements.

When dealing with line profiles, as opposed to EW, a higher resolution (R∼15 000–20 000) and a NLTE approach are required. This is particularly true when investigating the diagnostic power of the CD of the Ca II triplet lines as activity indicators, in this case the activity effects on CD are of the same order of magnitude of the NLTE effects (Busà et al., in preparation) and will not be revealed at lower resolutions.

References

Busà, I., Andretta, V., Gomez, M. T., Terranegra, L. 2001, A&A 373, 993
Carlsson, M. 1986, Uppsala Observatory Internal Report No. 33
Hauschildt, P. H., Allard, F., & Baron, E. 1999, ApJ 512, 377

GAIA spectroscopy and cataclysmic variables

Mark Cropper

Mullard Space Science Laboratory, University College London, Holmbury St Mary, Dorking, Surrey RH5 6NT, United Kingdom

Tom Marsh

Department of Physics and Astronomy, University of Southampton Hampshire SO17 1BJ, United Kingdom

Abstract. The Radial Velocity Spectrometer on GAIA will enlarge the current sample of cataclysmic variables by a factor 3–10 to $V \sim 16$ mag. This sample will be much less affected by the selection effects of current samples. Here we briefly discuss the scientific gains that will accrue in this field from the RVS survey, both for CVs themselves and for the more general understanding of the physics of accretion disks.

1. Introduction

Cataclysmic Variables (CVs) are semi-detached binaries with a white dwarf primary star (see Warner 1995 for a review). CVs are interesting for the following reasons: they are numerous, being common end points of binary evolution; they are significant sources of high energy radiation in the Galaxy, and they are among the best laboratories for the study of accretion physics (which is responsible for $\sim 30\%$ of the energy production in the Universe), particularly regarding the physics and structure of disks.

While CVs have been an extremely active area of research and much progress has been made, there are still significant open areas. These include: (a) space density – what is the completeness of the existing sample of CVs; what are the selection effects; how many are in protracted low mass-transfer states; (b) evolution – what are the underlying processes generating the observed period distribution; what are the end-points of evolution; what is the effect of the common-envelope phase, and (c) disk theory – what is the source of the dissipation in the disk; what is the nature of the disk instability cycle; what is the vertical structure of the disk; what is the effect of the tidal interactions from the secondary, and the impact of the stream.

2. CVs in the GAIA-RVS passband

There is surprisingly little observational data for CVs in the RVS passband (around the CaII near-IR triplet). The major work has been carried out by one of us (Marsh & Dhillon 1997, North 2001), principally on non-magnetic systems.

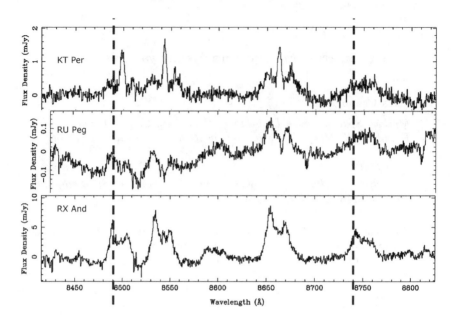

Figure 1. Spectra of three non-magnetic CVs KT Per, RU Peg and RX And in the RVS passband (denoted by the vertical dotted lines). Broad CaII triplet lines from the disk are evident in emission, together with narrower lines originating from the heated secondary star (contributing to the line asymmetry in RX And). Also evident are Paschen lines of H.

Spectra show broad ($\sim 3\,000$ km s^{-1}, 80 Å) double-peaked CaII lines, originating in the cooler outer edges of the disk, with superimposed narrow emission (from the heated face of the secondary star), or, in longer period systems where the secondary is larger and brighter in this band, photospheric absorption features instead of emission. Also in this band are broad Paschen lines, one of which lies under the CaII line at 8662 Å. The spectra are substantially different in the different systems, depending on the orbital period, accretion rate, disk state etc. Examples from three non-magnetic systems are shown in Figure 1.

In the shortest period systems, the secondary star also is a white dwarf, or the degenerate core of main sequence star. These double-degenerate systems are expected to be numerous, but only a few are known. Some show extremely strong lines of NI (uniquely in astrophysical objects). In magnetic CVs, the CaII triplet can be seen strongly in emission in the one or two systems studied but very few data are available.

CVs therefore show strong and distinctive lines in the RVS band. This means that the RVS spectra will be useful for identifying the systems and will contain substantial astrophysical information in the line strengths, shapes and in the velocity variation.

3. The impact of GAIA

3.1. Sensitivity

For an RVS limit of $V = 17.5$ mag, and with four parameters fitted to the spectra to register them (velocity mean, period, amplitude, phase), we expect a practical limiting magnitude of $V \sim 16$. There are currently 140 CVs brighter than V=16 (Downes et al. 2001). Extrapolating from the initial findings of the Sloan survey (Szkody et al. 2002) the incompleteness level is $\sim 30\%$ on the basis of selection by colour only (which has strong selection effects). GAIA will provide spectroscopy, which is very sensitive in the search for signatures of accretion (line emission). This should lead to the discovery of a large number of intrinsically faint systems, with estimates perhaps exceeding ~ 1000 to $V \sim 16$ mag. All of these CVs will have excellent parallaxes and hence luminosities.

3.2. Minimisation of selection effects

GAIA will provide a minimally biased sample of CVs for population and evolution studies. This will result from the combination of absolute luminosities and colours from photometry and astrometry, together with colours and variability from photometry and also spectral signatures (emission lines, radial velocities). It should be noted that spectroscopy from RVS is particularly powerful: many CVs hardly vary (nova-likes) and many systems which would be considered detached based on photometry are evidently accreting only once spectroscopy is obtained. Periods are much easier to determine with radial velocity measurements than by photometry. The spectroscopy will also pick up new classes of unexpected objects such as short period systems with K- rather than M-type secondaries. Clearly these secondaries have been stripped of their outer layers (such as QZ Ser, and the 64 min system 1RXSJ 232953.9+062814; Thorstensen et al. 2002). There is also the likelihood of finding longer period double-degenerate and other 'graveyard' CVs, since evolution models predict large numbers of these. In current surveys they would be indistinguishable from white dwarfs on the basis of their colours.

3.3. The general characterisation of the CV sample

The survey will provide the fraction with CaII emission compared to absorption; the relationship of these with the different classes of CV (magnetic/non-magnetic), luminosity, secondary spectral type, etc. Ultimately, information will be gained on the reason for the different line strengths and other characteristics.

3.4. Disk structure

The CaII triplet lines are more clearly double-peaked than other strong lines in the optical spectrum because they originate in the cooler outer regions of the disk. For a large number of disk CVs, Doppler tomography will be possible. This will produce maps of disk velocities including the effects of tidal distortions and the infalling stream, and the run of calcium emission within the disk. It will also identify other aspects of disk structure: for example the spiral waves that have been seen in some Dwarf Novae (Steeghs, Harlaftis & Horne, 1997). Because of the importance this has for the structure of accretion disks in general, these are currently the subject of considerable discussion as to whether they are shocks or 3-body effects (Ogilvie, 2002). The CaII triplet produces maps of intrinsically higher resolution than the more commonly used Hα, because thermal and Stark broadening is less for this relatively heavy ion.

3.5. Irradiative heating of stellar atmospheres

Many (most) CVs seem to have narrow CaII triplet emission components. The strength of these lines and their ratios are inputs for atmospheric heating models of the secondary. CaII is less saturated than the more usually used Hα, so it is generally easier to model (for example Marsh & Duck, 1996).

3.6. Mass ratios

Mass ratios provide an important test of the population models and inputs for the theory of superhumps which are seen in many of these systems (these can be used to explore the 3-body interactions between the outer edge of the disk and the two stars). The mass ratio is available from the orbital period and velocities of the secondary (K_2). If the inclination is available from reflection modeling, or from eclipses, or directly from the astrometry (a typical binary separation of 1 $R_\odot$ is 50 μas at a typical distance of 100 pc so that the binary orbit and inclination will be obtained directly for a few CVs in the RVS sample) then the masses of the secondary and primary can be calculated. These are fundamental parameters, and the secondary mass is an important input in determining the relationship of the secondary stars in CVs to normal main-sequence stars.

3.7. Physics of the accretion disk

CaII line ratios give information on the optical depth of the line emitting region. Generally normal CVs have optically thick ratios (see Marsh & Dhillon 1997). Intrinsically faint systems (not yet found, but are expected to be the majority) could display optically thin line ratios. The CaII population level calculations are difficult, since they must take into account the vertical structure of the disk, the effects of irradiation, magnetic activity in disk etc: such measurements would however be a future resource for disk outburst models.

References

Downes, R.A., Webbink, R.F., Shara, M.M., Ritter, H., Kolb, U. & Duerbeck, H.W. 2001, PASP 113, 764

Marsh, T.R. & Dhillon, V.S. 1997, MNRAS 292, 385

Marsh, T.R. & Duck, S.R. 1996, MNRAS, 278, 565

North, R. 2001, PhD Thesis, University of Southampton

Ogilvie, G.I. 2002, MNRAS 330, 937

Steeghs, D., Harlaftis, E.T. & Horne, K. 1997, MNRAS 290, L28

Szkody, P. et al. 2002, AJ 123, 430

Thorstensen, J.R., Fenton, W.H., Patterson, J.O., Kemp, J., Krajci, T. & Baraffe, I. 2002, ApJL 567, L49

Warner, B. 1995, Cataclysmic Variables, Cambridge University Press

An extragalactic reference frame for GAIA and SIM using quasars from the Sloan Digital Sky Survey

Eva K. Grebel, Michael Odenkirchen & Coryn A.L. Bailer-Jones

MPI for Astronomy, Königstuhl 17, D-69117 Heidelberg, Germany

Abstract. The quasar survey of the Sloan Digitial Sky Survey is expected to ultimately comprise up to 100 000 spectroscopically confirmed quasars spread over $1/4$ of the sky at 90% completeness to $i' < 19.1$ mag. Not only do these quasars represent a useful extragalactic reference frame for ground-based and space astrometry, but their spectra also allow one to define quasar candidate selection criteria for the multi-color photometric system of the GAIA satellite.

1. Introduction

Knowledge of the absolute motions and orbits of individual stars in our Milky Way, of star clusters, and of nearby dwarf galaxies is the prerequisite for understanding their dynamical history. We already have detailed information about the stellar content and star formation histories of single-age and composite stellar populations in the Milky Way and beyond, and a growing body of detailed chemical stellar abundances is becoming available. But we still lack accurate orbital information for all but the nearest stars. This information is the vital missing piece for constraining the origin, interaction and merger history of dwarf galaxies and the Milky Way, the formation of the Galactic Halo, the kinematics of different Galactic components, etc. The two planned major astrometric space missions, GAIA and the Space Interferometry Mission (SIM), will offer unparalleled capabilities for accurate studies of space motions and kinematics of the Milky Way and its satellites.

2. Quasars as an extragalactic reference frame

Proper motion measurements for field stars, open and globular clusters, stars in tidal streams, and resolved nearby galaxies require a non-moving, unchanging reference frame that should ideally span the entire sky in a closely-spaced grid. The only sources that fulfill these stringent requirements are quasars and point-like AGN/Seyfert galaxies (with the caveat of variability). SIM, a pointed mission, will rely on a primary, moving reference grid of some 4000 stars brighter than 13th magnitude, and a few luminous quasars. In special areas of interest local reference points will need to be established. GAIA as an all-sky survey mission will expected to be able to establish its own extragalactic reference frame by identifying quasars through the multi-color information that it will gather.

Independent of these space missions, for nearby Local Group targets the required time baseline for high-precision proper motion measurements can be reached within a few years provided that (a) the observations are obtained with sufficiently high angular resolution and sensitivity, and (b) the extragalactic reference sources are sufficiently bright and numerous (e.g., Tinney 1999). Indeed there are ongoing efforts to improve proper motion measurements for nearby dwarf galaxies using quasars as astrometric references and exploiting the superior resolution of HST (e.g. Piatek et al. 2002).

Quasars with bright apparent magnitudes are rare, and existing quasar catalogs either do not extend to fainter magnitudes or are very limited in their area coverage. Ongoing attempts to measure proper motions suffer from a lack of sufficiently numerous known background sources, and for the future astrometric missions a well-established, dense grid of such sources extending to the limiting magnitudes of $V \sim 20$ mag of SIM and GAIA would be desirable. The Sloan Digital Sky Survey offers the possibility of establishing such a reference frame.

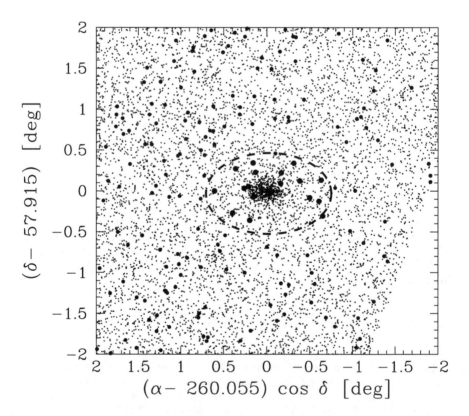

Figure 1. An SDSS field centered on the Milky Way satellite Draco, whose approximate tidal radius is indicated by the dashed line (Odenkirchen et al. 2001). Small dots denote red stars in Draco and in the Milky Way. Large dots represent quasars spectroscopically confirmed by the SDSS. Approximately 17 quasars are found behind Draco.

3. The SDSS quasar survey

The ongoing Sloan Digital Sky Survey (SDSS; York et al. 2000, Stoughton et al. 2002) is expected to cover up to 10 000 deg^2 centered on the North Galactic cap once completed, or one quarter of the sky. The SDSS carries out deep, homogeneous, five-color imaging with a dedicated drift-scan camera and spectroscopic follow-up of mainly galaxies and quasars. Primary science goals of the SDSS are cosmological issues such as large-scale structure, galaxy and quasar census, luminosity functions, and evolution as a function of redshift.

The SDSS photometrically identifies potential quasars and obtains spectra for all low-redshift candidates ranging from $15 < i' < 19.1$ mag, and $i' < 20.2$ mag for high-redshift suspects. Quasar candidates are selected based on their morphology and location in the SDSS color space, or because they match sources from the FIRST radio survey (Richards et al. 2002). This selection also comprises nuclei of Seyfert galaxies as well as low-redshift AGN. There is overlap with the stellar locus of F to M stars for quasars in the redshift range $2.2 < z < 3$ (Stoughton et al. 2002; Richards et al. 2002). Therefore, this region is more sparsely sampled in the spectroscopic target selection process in order to reduce the number of stellar spectra. The overall completeness of the quasar selection is estimated to be better than 90% (Richards et al. 2002). Once completed the SDSS will contain approximately 100 000 spectroscopically confirmed quasars. The astrometric accuracy of the SDSS is ~ 50 mas.

4. Example Draco: quasar density and distribution

We illustrate the usefulness of the SDSS QSO survey for establishing an extragalactic reference frame using the example of Draco, a Milky Way dwarf satellite at a distance of 80 kpc (see Odenkirchen et al. 2001 for a detailed structural study of Draco). 17 spectroscopically confirmed SDSS quasars are located within Draco's tidal radius of $40'$. The resulting quasar density behind Draco is ~ 16 quasars deg^2, or ~ 9 quasars deg^2 within the sensitivity limits of GAIA. The average quasar density of the SDSS to a limiting magnitude of $i' = 19.1$ mag is 13 ± 2 quasars deg^2 (Richards et al. 2002).

5. Using SDSS spectra to define quasar selection criteria for GAIA

The spectroscopic information from the SDSS can be put to further use for GAIA. With a wavelength coverage from 3900 Å to 9100 Å at a spectral resolution of 2000 and a wide redshift range, SDSS quasar spectra can be used to simulate their resulting spectral energy distribution in most of the GAIA passbands. In turn, this information will help in defining photometric quasar candidate selection criteria, which will be the next steps of our work. These criteria will facilitate establishing an all-sky reference frame during the GAIA mission, which can then easily be tested against the one quarter of the sky covered by the SDSS. The more detailed sampling of the spectral energy distribution through GAIA's 11 intermediate-band filters may in addition complement the SDSS quasar census in regions of overlap with the stellar locus, but this remains yet to be tested.

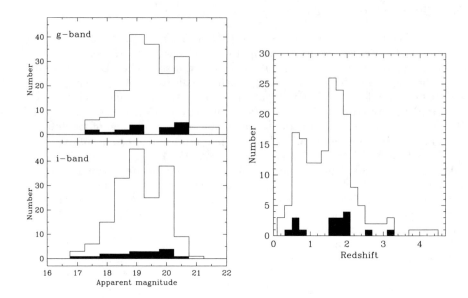

Figure 2. *Left*: distribution of magnitudes of SDSS quasars behind Draco (filled histograms) versus magnitudes of quasars within the 4°× 4° region of Figure 1 around Draco (open histograms). *Right*: redshift distribution of Draco quasars (filled histogram) compared to the redshift distribution in the field around Draco (open histogram).

Acknowledgments. Funding for the creation and distribution of the SDSS Archive has been provided by the Alfred P. Sloan Foundation, the Participating Institutions, the National Aeronautics and Space Administration, the National Science Foundation, the U.S. Department of Energy, the Japanese Monbukagakusho, and the Max Planck Society. The SDSS website is http://www.sdss.org/ All data presented here are public domain data.

References

Odenkirchen, M. et al. 2001, AJ 122, 2538
Piatek, S. et al. 2002, AJ 124, 3198
Richards, G.T. et al. 2002, AJ 123, 2945
Stoughton, C. et al. 2002, AJ 123, 485
Tinney, C.G. 1999, MNRAS 303, 565
York, D.G. et al. 2000, AJ 120, 1579

AGB stars as tracers of star formation histories: implications for GAIA photometry and spectroscopy

Arunas Kučinskas

Lund Observatory, Box 43, SE-221 00 Lund, Sweden

Institute of Theoretical Physics & Astronomy, Goštauto 12, Vilnius 2600, Lithuania

Lennart Lindegren

Lund Observatory, Box 43, SE-221 00 Lund, Sweden

Toshihiko Tanabé

Institute of Astronomy, The University of Tokyo, Tokyo, 181-015, Japan

Vladas Vansevičius

Institute of Physics, Goštauto 12, Vilnius 2600, Lithuania

Abstract. We argue that tracing star formation histories with GAIA using main sequence turn-off (MSTO) point dating will mainly be effective in cases of very mild interstellar extinction ($E_{B-V} < 0.5$). For higher reddenings the MSTO approach will be severely limited both in terms of traceable ages ($t < 0.5$ Gyr at 8 kpc; $E_{B-V} = 1.0$) and/or distances ($d = 2$ kpc if $t \leq 15$ Gyr; $E_{B-V} = 1.0$), since the MSTO will be located at magnitudes too faint for GAIA. AGB stars may alternatively provide precise population ages with GAIA for a wide range of ages and metallicities, with traceable distances of up to 250 kpc at $E_{B-V} = 0.0$ (15 kpc if $E_{B-V} = 2.0$). It is essential however that effective temperatures, metallicities, and reddenings of individual stars are derived with the precision of $\sigma(\log T_{\text{eff}}) \sim 0.01$, $\sigma([M/H]) \sim 0.2$, and $\sigma(E_{B-V}) \sim 0.03$, to obtain $\sigma(\log t) \sim 0.15$. This task is quite challenging for GAIA photometry and spectroscopy, though preliminary tests show that comparable precisions may be achieved with GAIA medium band photometry.

1. Introduction

Out of the numerous methods available for assessing ages of stellar populations only a few will be practically applicable for use with GAIA, because of obvious limitations related to the detection limits of GAIA, etc (e.g. Cacciari 2002). Though the main sequence turn-off (MSTO) point approach is perhaps the most reliable and accurate of all dating methods available today, it will be severely distance-limited when used with GAIA.

The extent of these limitations is illustrated in Figure 1. The left panel shows the maximum age obtainable with GAIA employing MSTO fitting for

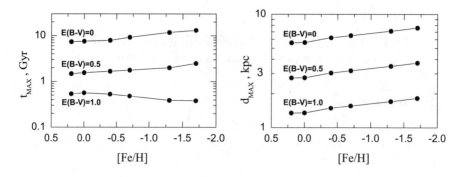

Figure 1. *Left*: maximum age for which MSTO point can be dated at 8 kpc with GAIA, for different [Fe/H] and E_{B-V}. *Right*: maximum distance at which stellar population up to 15 Gyr old can be dated using MSTO with GAIA, for different [Fe/H] and E_{B-V}.

stellar population placed at a distance of 8 kpc (assuming that the MSTO point is located at $V \sim 18.5$ mag, while the detection limit of GAIA is $V = 20$ mag). The right panel shows the maximum distance at which ages up to 15 Gyr may be quantified (in both cases, the MSTO point on the isochrones of Girardi et al. 2000 was dated). It is obvious that in cases of negligible interstellar extinction the MSTO approach may work very efficiently with GAIA, providing reliable ages of stellar populations up to ~ 8 kpc. However, even mild interstellar extinction changes the situation dramatically, shifting the MSTO point below the detection limits of GAIA for many stellar populations within the Galaxy.

2. AGB stars with GAIA: star formation histories

A possible way to complement the MSTO approach would be with tracers that are intrinsically brighter than MSTO stars, which could thus be used for probing stellar populations on larger distance scales. We have shown, that stars on early-AGB (non-thermally pulsing) may be well suited for this purpose, providing reliable ages of stellar populations from the isochrone fitting to AGB sequences on the observed HR diagrams (Kučinskas et al. 2002).

Though similar approaches have been used in the past (e.g. employing AGB sequences in color-magnitude diagrams), the precision in derived ages was low. We propose to work in the M_{bol} vs. T_{eff} plane, using effective temperatures of individual AGB stars derived from the fit of the observed spectral energy distributions (SEDs) with synthetic SEDs. This yields a significant increase in precision of derived population ages, which is a result of the precisely derived T_{eff} of individual AGB stars. Since distances will be known for the majority of Galactic AGB stars with GAIA (which, combined with data from ground- and/or space-based infrared surveys, will give precise M_{bol}), AGB stars will allow stellar populations on large distance scales to be traced with GAIA.

We illustrate the proposed approach employing the star clusters NGC 1783, NGC 2121 and Kron 3 in the Magellanic Clouds. The observed SEDs of AGB

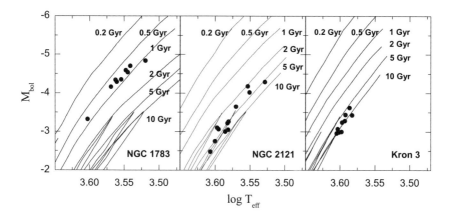

Figure 2. Observed HR diagrams with AGB sequences in star clusters of the Magellanic Clouds: NGC 1783, 2121 (both LMC) and Kron 3 (SMC). Isochrones are from Bertelli et al. (1994).

stars were constructed using $BVRIJHK$ fluxes obtained from the literature. The BaSel 2.2 library of stellar spectra (Lejeune et al. 1998) was used to produce a template of synthetic photometric colors at different $T_{\rm eff}$ and for the metallicities of individual MC clusters. HR diagrams of individual clusters with AGB sequences were constructed using $M_{\rm bol}$ derived from observed integrated fluxes of individual stars and $T_{\rm eff}$ obtained from the SED fitting procedure (Figure 2). AGB ages of individual clusters derived using different sets of isochrones, as well as those obtained from MSTO in previous studies, are given in Table 1.

There is indeed a good agreement between AGB ages obtained using isochrones of Bertelli et al. (1994; B94) or Lejeune & Schaerer (2001; LS01) on one hand, and MSTO estimates on the other. AGB ages obtained with Girardi et al. (2000; G00) isochrones are however considerably older than MSTO estimates (Table 1). This effect tends to increase with decreasing metallicity. It should be stressed, thus, that these intrinsic discrepancies between different sets

Table 1. AGB ages of star clusters in the Magellanic Clouds derived from different sets of isochrones (two metallicities are used for Kron 3). M89 = Mould et al. (1989), M98 = Mighell et al. (1998), R00 = Rich et al. (2000), R01 = Rich et al. (2001).

Cluster	[Fe/H]	AGB ages, this work (Gyr)			MSTO ages	Ref
		B94	G00	LS01		
NGC 1783	-0.4	0.8 ± 0.1	1.1 ± 0.2	–	0.9 ± 0.4	M89
NGC 2121	-0.7	3.8 ± 0.8	6.0 ± 1.6	–	3.2 ± 0.5	R01
Kron 3	-1.3	10.3 ± 2.8	> 15	7.0 ± 1.6	8.0 ± 0.5	R00
	-1.0	4.9 ± 1.1	> 15	–	6.0 ± 1.3	M98

Table 2. Predicted accuracies, versus V magnitude, for T_{eff}, [M/H], E_{B-V} (individual AGB stars, 1X photometric system of Vansevičius & Bridžius 2003), and ages from AGB stars derived with GAIA. d is the maximum distance at which these accuracies may be expected.

V	$\sigma(\log T_{\text{eff}})$	$\sigma([M/H])$	$\sigma(E_{B-V})$	$\sigma(\log t)$	$d_{E_{B-V}=0}$ (kpc)	$d_{E_{B-V}=2}$ (kpc)
18	< 0.01	< 0.2	< 0.03	0.15	100	6
20	0.03	0.3	0.06	0.45	250	15

of isochrones clearly indicate a need for better understanding of stellar evolution on the AGB, to fully exploit the potential of AGB stars for tracing stellar populations in the Galaxy and beyond. Possibilities for tracing star formation histories using AGB stars with GAIA, employing photometric metallicities, gravities and reddenings (GAIA 1X medium band photometric system, Vansevičius & Bridžius 2003) are summarized in Table 2 (the error in age for a single star, $\sigma(\log t)$, is a lower limit since it reflects only the errors in T_{eff} and E_{B-V}). We conclude that, contrary to MSTO dating, AGB stars may provide precise estimates of ages ($t > 0.5$ Gyr) throughout the Galaxy and beyond ($d \lesssim 250$ kpc) for populations within a large range of metallicities, even if interstellar extinction is non-negligible. To achieve this, metallicities to ±0.2 dex, and effective temperatures to ±0.01 dex are highly desirable. A precise knowledge of interstellar reddening ($\sigma(E_{B-V}) \sim 0.03$) is also essential.

Acknowledgments. This work was supported by the Wenner-Gren Foundations.

References

Bertelli, G., Bressan, A., Chiosi, C. et al. 1994, A&AS 106, 275 (B94)
Cacciari, C. 2002, in GAIA: A European Project, O.Bienaymé and C.Turon ed.s, EAS Publ. Ser. 2, pag. 163
Girardi, L., Bressan, A., Bertelli, G. & Chiosi, C. 2000, A&AS 141, 371 (G00)
Kučinskas, A., Vansevičius, V. & Tanabé, T. 2002, Ap&SS 280, 151
Lejeune, T., Cuisinier, F. & Buser, R. 1998, A&AS 130, 65
Lejeune, T. & Schaerer, D. 2001, A&A 366, 538 (LS01)
Marigo, P. & Girardi, L. 2001, A&A 377, 132
Mighell, K.J., Sarajedini, A. & French, R.S. 1998, AJ 116, 2395 (M98)
Rich, R.M., Shara, M.M., Fall, S.M. & Zurek, D. 2000, AJ 119, 197 (R00)
Rich, R.M., Shara, M.M. & Zurek, D. 2001, AJ 122, 842 (R01)
Mould, J., Kristian, J., Nemec, J. et al. 1989, ApJ 339, 84 (M89)
Vansevičius, V. & Bridžius, A. 2003, in GAIA Spectroscopy, Science and Technology, U. Munari ed., ASP Conf. Ser. 298, pag. 41

GAIA Spectroscopy, Science and Technology
ASP Conference Series, Vol. 298, 2003
U. Munari ed.

Photometric properties of theoretical spectral libraries for GAIA photometry

Thibault Lejeune

Observatório Astronómico da Universidade de Coimbra, Portugal

Abstract. Photometric properties of various spectral libraries of synthetic stellar spectra derived from the ATLAS 9 (Castelli et al. 1997), the PHOENIX NextGen (Hauschildt et al. 1999a,b), the NMARCS (Bessell et al. 1998), and the BaSeL (Lejeune et al. 1998, Westera et al. 2002) models are discussed in view of their application to the definition of the GAIA photometric system. Comparisons with empirical UBVRI calibrations show significant discrepancies at low temperatures between the different models, and with respect to the BaSeL 3.1 and the Alonso et al. (1996, 1999) empirical calibrations.

1. Introduction

The GAIA Galactic survey will observe more than 1 billion of stars in our Galaxy, and will obtain photometry in 11 medium-band and 5 large-band filters with the challenging goal to determine the fundamental stellar parameters (T_{eff}, $\log g$, [Fe/H]) across a very wide range of stellar types. In order to test the capabilities of the photometric system and the performances of the classification algorithms, various stellar libraries of synthetic and/or empirical spectra are required. In this study, I compare the photometric properties of the most widely used grids of synthetic spectra (ATLAS, NMARCS, PHOENIX, BaSeL) for cool stars in view of their application to GAIA photometry.

2. Grids of synthetic stellar spectra

The coverage in stellar parameters for each of the grids of models used in this comparative study is given in Table 1. The BaSeL 2.2 hybrid spectral library was constructed from empirical ([Fe/H]=0) and semi-empirical ([Fe/H]$\neq$ 0) color-temperature relations (see Lejeune et al. 1997, 1998 for details), while the new BaSeL 3.1 models are based on purely empirical calibrations, defined for the metallicity range $-2.0 \leq$[Fe/H]≤ 0.0 from a large collection of globular cluster *UBVRIJHKL* photometric data (cf. Westera et al. 2002 for details). Hence, the BaSeL 3.1 calibrations provide the only existing set of empirical metallicity-dependent T_{eff}-*UBVRIJHKL* transformations for $-2.0 \leq$[Fe/H]≤ 0.0 over a large temperature range, from 2000 K to 50 000 K, and are used as reference in this study. Synthetic spectra from the ATLAS 9 atmosphere models have been computed by Castelli et al. (1997) with no overshooting parameter.

Table 1. Parameter coverage of the different grids of models

Models	notes	T_{eff}	$\log g$	[Fe/H]	λ (nm)
BaSeL 2.2[1]	hybrid lib.	2000 - 50000 K	−1.0 - 5.5	−5.0 - 1.0	$9.1 - 1.6\,10^5$
BaSeL 3.1[2]	hybrid lib.	2000 - 50000 K	−1.0 - 5.5	−2.0 - 0.5	$9.1 - 1.6\,10^5$
ATLAS 9[3]	no overs.	3500 - 50000 K	0.0 - 5.0	−2.5 - 0.5	$9.1 - 1.6\,10^5$
PHOENIX	giants[4]	2000 - 7000 K	−0.7 - 0.0	−0.7 - 0.0	$10 - 10^6$
NextGen	dwarfs[5]	1000 - 7000 K	3.5 - 6.0	−4.0 - 0.0	$10 - 10^6$
NMARCS[6]	giants	3600 - 4750 K	−0.5 - 3.5	−0.6 - 0.6	BVRIJHKL
–	dwarfs	2600 - 4000 K	4.5 - 5.0	−2.0 - 0.3	BVRIJHKL

[1] Lejeune et al. 1997, 1998; [2] Westera et al. 2002; [3] Castelli et al. 1997;
[4] Hauschildt et al. 1999a; [5] Hauschildt et al. 1999b; [6] Bessell et al. 1998.

3. Comparisons in the two-color diagrams

In Figure 1, model colors in the range $2500 \le T_{\text{eff}} \le 5500$ K, computed from the different grids of synthetic spectra given in Table 1, are compared in the $(B-V)/(V-I)$ color-color diagram for the same value of the surface gravity ($\log g = 1$) with [Fe/H] $= -0.6, -0.3$ and 0.0. From the figure, it is clear that large differences exist between the different grids, specially in the low temperature regime ($T_{\text{eff}} < 4000$ K). Above 4000 K, NMARCS models provide the best overall agreement with the BaSeL 3.1 empirical calibrations, both in $B-V$ and $V-I$, in particular for the solar metallicity. In contrast, the PHOENIX models appear too blue in $B-V$, while they are in good agreement with the BaSeL 3.1 $V-I$ colors. The $B-V$ model colors from the ATLAS 9 grid are systematically redder below 4500 K. Below 4000 K, very large differences of several tenth of magnitude exist between all the model colors.

Similar comparisons (Lejeune et al., in prep.) for the dwarf models show that large deviations between the models and the BaSeL 3.1 empirical calibrations also exist, and are maximum below 4000 K. For the M dwarfs, model colors computed from the PHOENIX models provide the best agreement with the BaSeL 3.1 empirical calibrations.

4. Temperature-color calibrations

We also compared the theoretical temperature-color calibrations, $T_{\text{eff}}/(U-B)$ and $T_{\text{eff}}/(B-V)$, predicted by each grid of models in the range $3000 \le T_{\text{eff}} \le 6000$ K, with the temperature scales adopted in BaSeL 3.1 and the empirical calibrations of Alonso et al (1996, 1999) derived from the IRFM method. Our results (Lejeune et al., in prep) show that, for $T_{\text{eff}} > 4000$ K, the theoretical scales for all the models agree well with the BaSeL 3.1 empirical relations for [Fe/H] $= 0.0$ and -2.0. Surprisingly, below 4000 K the theoretical calibrations

Properties of spectral libraries for GAIA photometry

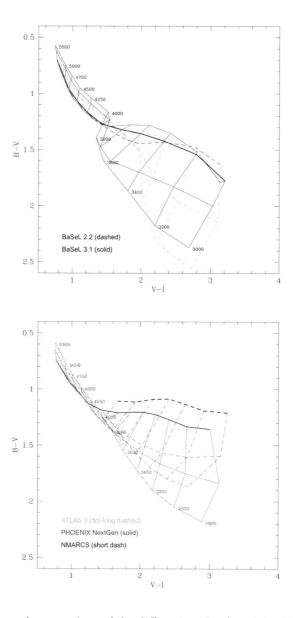

Figure 1. A comparison of the different grids of models with $\log g = 1$ in the $(B-V)/(V-I)$ two-color diagram. *Top* panel: the BaSeL 2.2 and 3.1 model colors. *Bottom* panel: the ATLAS 9 (Castelli et al.), the PHOENIX and the NMARCS models. The lines connect the models with the same metallicity ([Fe/H] = $-0.6, -0, 3$ and 0.0 (thick line), and the models with the same T_{eff} as indicated in each panel. The models show very large differences in the M giant regime (see text).

from ATLAS 9 model atmospheres are found in very good agreement with the empirical values. The $T_{\rm eff}/(B-V)$ relation from the NMARCS model colors for dwarfs deviates significantly (more than 0.5 mag at $T_{\rm eff}$ = 3000 K) from the BaSeL 3.1 empirical calibrations, while the NextGen model calibrations are closer. For the giants, the NMARCS models provide in general a better match to the empirical relations than the NextGen models. Comparisons of the BaSeL 3.1 calibrations with the metallicity-dependent temperature-color empirical relations of Alonso et al. in the temperature range 4000 – 7000 K show a good agreement (less than 0.05 mag) for the dwarf sequences, but a more significant discrepancy (> 0.1 mag on average) for the giant sequence with [Fe/H] = −2.0.

5. Conclusions

UBVRI colors of the most widely used theoretical spectral libraries (ATLAS 9, PHOENIX, NMARCS, and BaSeL) in the temperature range 3000 – 7000 K have been compared with empirical calibrations available for −2.0 ≤[Fe/H]≤ +0.5. We found that the ATLAS 9 model colors from Castelli et al. (1997) agree well or very well with the empirical color-temperature relations at all metallicities and over the range of effective temperatures between 6000 K and 3500 K. The most significant deviations are found for the $T_{\rm eff}/(U-B)$ relation at solar metallicity. Below 4000 K, important deviations are found for the *UBV* colors between both the PHOENIX NextGen and the NMARCS model predictions with respect to the BaSeL 3.1 empirical relations, although some uncertainties still exist in the empirical data at these low temperatures. In a general way, the deviations appear to be less important with the PHOENIX NextGen model colors for dwarfs, and with the NMARCS models for giants.

These differences have to be accounted for, or maybe reduced with some spectral corrections methods, in the selection of models in order to construct a spectral library well suitable to the definition of the GAIA photometric system.

Acknowledgments. This work was supported by the "Fundação para a Ciência e a Tecnologia" through the grant PRAXIS-XXI/BPD/22061/99, and by the project "PESO/P/PRO/15128/1999".

References

Alonso, A., Arribas, S. & Martinez-Roger, C. 1996, A&A 313, 873
Alonso, A., Arribas, S. & Martinez-Roger, C. 1999, A&AS 139, 335
Bessell, M. S., Castelli, F. & Plez, B. 1998, A&A 333, 231
Castelli, F., Gratton, R.G. & Kurucz, R.L. 1997, A&A 318, 841
Hauschildt, P.H., et al., 1999a, ApJ 512, 377
Hauschildt, P.H., et al., 1999b, ApJ 525, 871
Lejeune, T., Cuisinier, F. & Buser, R. 1997, A&AS 125, 229
Lejeune, T., Cuisinier, F. & Buser, R. 1998, A&AS 130, 65
Westera, P., Lejeune, T., Buser, R. et al. 2002, A&A 381, 524

GAIA spectroscopy of symbiotic binaries

Paola M. Marrese[1,2] & Ulisse Munari[1]

1. Padova Astronomical Observatory - INAF, Asiago Station, Via Osservatorio 8, I-36012 Asiago (VI), Italy
2. Department of Astronomy, University of Padova, Vicolo Osservatorio 8, I-35122, Padova, Italy

Abstract. The spectral appearance of symbiotic binaries over the GAIA range is investigated. The majority of symbiotics shows the giant absorption spectrum, while a small fraction is dominated by the emission of the nebula. The potential of GAIA spectroscopic observations (both ground-based and from space) in determining the galactic population of symbiotics and their role as viable precursors to SN Ia is assessed.

1. Introduction

Symbiotic stars are a rather inhomogeneous group of long period, detached, interacting binaries. They show the spectral signatures of a cool giant (CN, TiO and VO bands, neutral metallic lines) together with emission lines tracing high temperatures conditions (e.g. He II 4686 Å). A typical spectrum is shown in Figure 1: at the longest wavelengths the absorption spectrum of the cool giant dominates, while moving to the blue it is veiled and overwhelmed by the nebular continuum excited by the hot compact source. Symbiotics can present photometric variability both periodic and irregular, on a huge range of time scales from flickering to decades long outbursts. In the majority of symbiotics the evolved component (G, K, M and Carbon types) transfers material via stellar wind onto the compact companion (a white dwarf with typically $L \sim 10^3$ $L_\odot$ and $T_{\rm eff} \sim 10^5$ K). The material lost by the cool star forms a circumstellar nebula ($N_{\rm e} > 10^6$ cm^{-3} and $T_{\rm e} \sim 17\,000$ K) and a good fraction of it is accreted by the WD. The accretion rate ($\dot{M}_{\rm acc}$) determines how the accreted material is processed (Kenyon 1986): (a) if $\dot{M}_{\rm acc} \sim 10^{-7} M_\odot$ yr^{-1} a stable shell burning can take place; (b) if $\dot{M}_{\rm acc}$ is higher than that, the WD envelope expands to supergiant dimensions, and (c) at lower $\dot{M}_{\rm acc}$ the material is quietly accumulated on the WD until conditions for hydrogen burning are reached, and depending on the degree of electron degeneracy, the consequent outburst can be violent with mass ejection, or mildly progress without mass-ejection (the latter being the most frequent case).

2. Symbiotic stars as precursors to SN Ia

Given the importance of SN Ia as standard candles to determine the distance scale of the Universe, it is disappointing that their progenitors are still vaguely identified. The knowledge of the precursors is of basic importance to determine their homogeneity and the similarity of evolutionary channels that can

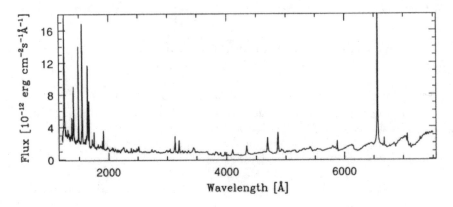

Figure 1. Ultraviolet and optical spectrum of the symbiotic star AG Peg in quiescence (from Mikolajewska 1997).

lead to their explosion at all cosmological ages, thus increasing the confidence in their use. The role of symbiotics as possible SN Ia precursors has been widely discussed in the literature (Iben & Tutukov 1984, 1996, Wheeler 1990, Munari & Renzini 1992, Kenyon et al. 1993, Munari 1994, Hachisu et al. 1996, 1999).

There are different ways to decide if symbiotics can be viable precursors to SN Ia, among them: (a) if their white dwarfs can grow in mass to the Chandrasekhar limit; (b) the analysis of their population properties to understand if their number in the Galaxy is large enough to account for the observed Galactic SN Ia rate, and (c) if they belong to the old spheroidal population that dominates in elliptical galaxies where only SN Ia explode. The total number of symbiotics in the Galaxy (extrapolating from the ∼250 known objects) is still quite uncertain, with estimates ranging from 3 000 (Allen 1984) to 30 000 (Kenyon et al. 1993) to 300 000 (Munari & Renzini 1992). This is basically due to the uncertainty in the galactic population to which the different subclasses of symbiotics belong to.

3. GAIA impact on symbiotic stars knowledge

The evolutionary status of symbiotics can be best provided by the study of their cool component, which is better investigated in the far-red and near-IR region where it is intrinsically more luminous, less contaminated by the veiling effect of the blue nebular continuum (which affects the visual region) and less affected by the dust emission (which may influence the IR region). We have accomplished an high S/N, high resolution ($R{\sim}18\,000$), spectroscopic survey of symbiotic binaries in the GAIA wavelength region with the 1.82 m Asiago telescope + Echelle + CCD. The survey so far includes 35 objects at $\delta > -25°$ brighter than V∼12 mag. The Ca II wavelength region of symbiotics was investigated, at lower resolution, also by Schulte-Ladbeck (1988, 16 objects, resolution 12 Å), Huang et al. (1994, 6 objects, $R = 2\,000$), and Zhu et al. (1999, 12 objects, $R = 2\,000$).

A representative sample of the Asiago observed spectra can be found in Figure 2, where three different types of symbiotic stars are shown: (a) V1016 Cyg

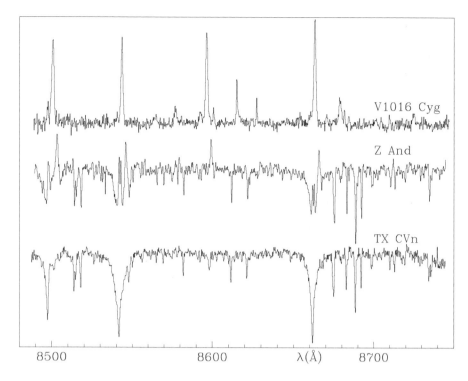

Figure 2. Spectra of a sample of symbiotic stars in the GAIA wavelength range. The contribution of the nebular component relative to the cool giant is increasing from TX CVn to V1016 Cyg.

(which contains a Mira variable, like ∼20% of known symbiotics) shows an overwhelming contribution of the nebular component; (b) the spectrum of Z And (representative of those systems usually named *classical* symbiotics, ∼40% of the total) shows the presence of the nebular emission confined to Ca II and Paschen lines; while (c) in the case of TX CVn−like systems (∼40% of the total) we find a pure cool giant spectrum. Data presented in Katz (2003) and Munari (2003b) can be used to identify spectral features in Figure 2. For those systems showing only the absorption lines, GAIA−like spectroscopic observations allow: (i) to derive a spectral classification, (ii) to perform an atmospheric analysis ($T_{\rm eff}$, $\log g$, chemical abundances) and (iii) to derive the radial velocity of the cool component. For the systems showing only the nebular emissions, GAIA−like spectroscopy provides (α) diagnostic tracers of the physical properties of the nebular regions and (β) their motion and kinematical location in the binary system. The symbiotics showing both absorption and emission lines will be the most important, as for them the characteristic and kinematical tracing of both cool and nebular component can be derived. Based on orbital period distribution in the Belczynski et al. (2000) catalog, the 5 years mission lifetime will allow GAIA to derive reliable spectroscopic orbital periods for ∼60% of symbiotics. For the systems with longer periods, epoch velocities will differ from the systemic one by less than a few km s^{-1} and therefore will still be useful to trace their galactic kinematics and population. In addition, GAIA will provide an all sky survey and will couple spectroscopy with accurate astrometry and multi−epoch, multiband photometry: thousands of new symbiotic stars will be discovered and their properties accurately derived (cf. Munari 2003a).

It will be well possible to: (*a*) determine the galactic population to which symbiotics belong, (*b*) check the conclusions by Whitelock & Munari (1992) and Munari (1994) who suggest that the majority of symbiotics belongs to the Bulge−Thick Disk population (while the systems containing Mira variables are younger Thin Disk objects) and (*c*) decide if symbiotics can be viable precursors to SN Ia. This will be accomplished by: (*i*) the determination of symbiotics position and motion in the Galaxy (i.e. spheroidal or flattened spatial distribution, participation to the Galactic rotation, velocity dispersion); (*ii*) the analysis of their photometric properties (absolute magnitudes and intrinsic colors); (*iii*) the photospheric chemical analysis and (*iv*) their local density and thus the estimate of their total number in the Galaxy.

References

Allen, D.A. 1984, PASAu 5, 369

Belczynski, K., Mikolajewska, J., Munari, U. et al. 2000, A&AS 146, 407

Hachisu, I., Kato, M. & Nomoto, K. 1996, ApJL 470, L97

Hachisu, I., Kato, M. & Nomoto, K. 1999, ApJ 552, 487

Huang, C.C., Friedjung, M. & Zhou, Z.X. 1994, A&AS 106, 413

Iben, I.Jr. & Tutukov, A.V. 1984, ApJS 54, 335

Iben, I.Jr. & Tutukov, A.V. 1996, ApJS 105, 145

Katz, D. 2003, in GAIA Spectroscopy, Science and Technology, U. Munari ed., ASP Conf. Ser. 298, pag. 65

Kenyon, S.J. 1986, The symbiotic stars, Cambridge University Press

Kenyon, S.J., Livio, M., Mikolajewska, J. & Tout, C.A. 1993, ApJ 407, L81

Mikolajewska, J. 1997, in Physical Processes in Symbiotic Binaries and Related Systems, J. Mikolajewska ed., Copernicus Found. for Polish Astron., pag. 4

Munari, U. & Renzini, A. 1992, ApJ 397, L87

Munari, U. 1994, Mem SAIT 65, 157

Munari, U. 2003a, in Symbiotic Stars probing stellar evolution, R.L.M. Corradi, J. Mikolajewska and & T.J.Mahoney ed.s, ASP Conf. Ser., in press

Munari, U. 2003b, in GAIA Spectroscopy, Science and Technology, U. Munari ed., ASP Conf. Ser. 298, pag. 51

Schulte-Ladbeck, R.E. 1988, A&A 189, 97

Wheeler, J.C. 1990, NYASA 617, 8

Whitelock, P.A. & Munari, U. 1992, A&A 255, 171

Zhu, Z.X., Friedjung, M., Zhao, G. et al. 1999, A&AS 140, 69

Expanding the Asiago library of real spectra for GAIA

Paola M. Marrese[1,2], Ulisse Munari[1], Federico Boschi[1] & Lina Tomasella[1]

1. Padova Astronomical Observatory - INAF, Asiago Station, Via Osservatorio 8, I-36012 Asiago (VI), Italy
2. Department of Astronomy, University of Padova, Vicolo Osservatorio 8, I-35122, Padova, Italy

Abstract. The Asiago library of real spectral in the GAIA wavelength range so far contains high resolution (R~17 000), high S/N spectra of a total of 223 MK standard and reference stars, 165 of them covering F, G, K and M spectral types. The Asiago library explores the high end while Cenarro et al. (2001) the low end of the range of possible spectral resolutions considered for the GAIA spectrograph (from 20 000 to 5 000).

1. Introduction

The primary scientific goal of the GAIA mission is the determination of the structure, composition, evolution and formation history of the Galaxy. To accomplish this, data from astrometry, photometry and spectroscopy will be combined, and each target star will be characterized in terms of its position and motion in the Galaxy and its atmospheric parameters (T_{eff}, $\log g$, $Z/Z_\odot$, individual chemical abundances). To assess the performance of the GAIA spectral observations, feed input data to simulations of spectra recording on the spectrograph focal plane, and training of data reduction and analysis pipelines, an appropriate set of spectra of real stars is required.

2. The Asiago library of GAIA–like spectra

High resolution (R~17 000) and high S/N ($\geq$100) GAIA–like spectra were obtained (with the Asiago 1.82 m telescope + Echelle + CCD) for a consistent sample of MK standard and reference stars of the cooler spectral types.

This survey is intended to integrate and enrich the GAIA atlas mapping the MK system presented by Munari & Tomasella (1999, 73 cool objects + 58 hotter ones), keeping the same resolution and instrumental setup for homogeneity purposes. Given the predominance of cool spectral types among the GAIA targets, we observed additional 92 cool stars. The new atlas includes dwarfs, giants and supergiants ranging from F2 to M7, spread over a wide range of metallicities ($-2.59 \leq$[Fe/H]$\leq +0.38$).

The choice of the targets for this supplementary survey was driven (and limited) by (a) the observability from northern latitudes ($\delta > -25°$), (b) convenient object brightness (V<10 mag), (c) the existence of a standard star for a

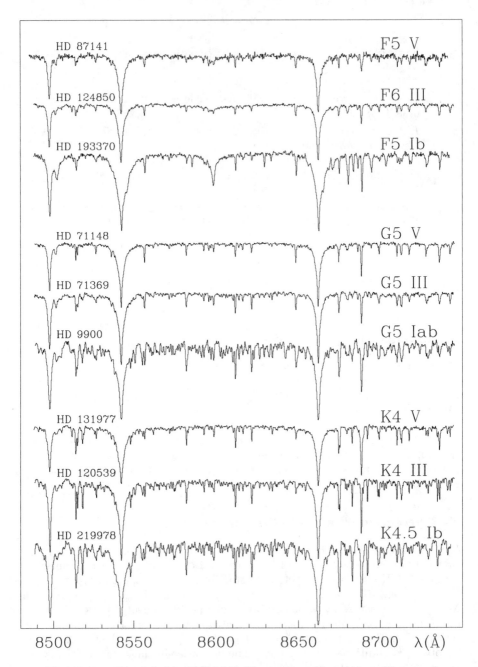

Figure 1. Ground−based GAIA−like spectra from Asiago library of MK standard and reference stars to illustrate gravity effects at F5, G5 and K4. The metallicity of all stars falls within $-0.2 \leq [Fe/H] \leq +0.2$). The spectra are shifted to null radial velocity.

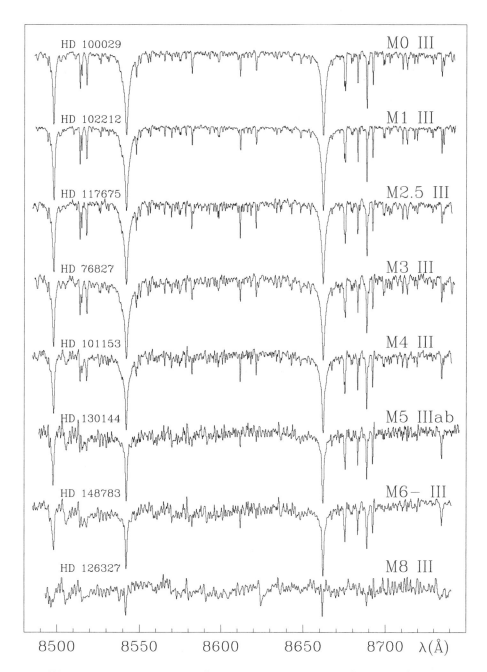

Figure 2. Temperature effects on M giant spectra from the Asiago GAIA library. The spectra are shifted to null radial velocity.

given MK type and (*d*) possibly the existence of a spectroscopic determination of [Fe/H]. Further details on the target stars, inclusion criteria and data reduction will be found in Marrese et al. (2003, in press), with spectra available at CDS. Summing up from Munari & Tomasella (1999) and the present work, the Asiago library contains a total of 165 cool MK template stars observed over the GAIA wavelength region. Table 1 shows the Asiago library coverage of the cool spectral types at different metallicities, while examples of the observed spectra are presented in Figures 1 and 2, where the effects of temperature and gravity on the spectral features are illustrated.

	V	IV	III	II	I		V	IV	III	II	I
F0	×	×	×	△	×	G9	×		△		
F2	×O	×			×	K0	×		O△	△	△□
F3	×O		△			K1			△		
F4	O		△			K2			△□		△
F5	△	×	× △		× △	K3	× △		O△□	△	O
F6	△		△			K4			×O△		△
F7	×				×	K5	△		O△	×	O
F8	O△	△	×		O	K6			△		
F9	△					K7			△	×	
G0	△	△	×O		△	M0	×		×O		× △
G1		×			O△	M1	×		×		×O
G2		×			× △	M2	△		× □	× △	× △
G3						M3	×		× △□	×	×
G4						M4	×		× △		×
G5	× △	O	O△	×	O△	M5	×		×		×
G6	×		O			M6	×		× △		
G7	×			□		M7				×	
G8	△		O△		× △	M8	×				

Table 1. Mapping of cool MK spectral types in the Asiago library of GAIA−like spectra. *crosses*: stars without a [Fe/H] determination; *circles*: [Fe/H]≤−0.2; *triangles*: −0.2≤[Fe/H]≤+0.2; *squares*: [Fe/H]≥+0.2. Each symbol may refer to more than one object.

3. Conclusions

The adopted resolution (R∼17 000) places the Asiago library at the upper limit of the range which was considered for the GAIA RVS instrument, and the spectra can be easily degraded to match the RVS final one (i.e. R∼11 500). This library, together with Cenarro et al. (2001) one (706 stars, R∼6 000), will support (*i*) focal plane simulations, feeding to the optical train and comparison with recovered spectra, (*ii*) neural network training to characterize the properties of the sources (Bailer−Jones 2003), as well as (*iii*) testing of model atmospheres and synthetic spectra.

References

Bailer-Jones, C.A.L. 2003, in GAIA Spectroscopy, Science and Technology, U. Munari ed., ASP Conf. Ser. 298, pag. 199
Cenarro, A.J., Cardiel, N., Gorgas, J., et al. 2001, MNRAS 326, 959
Munari, U. & Tomasella, L. 1999, A&A 137, 521

GAIA and the boon to extra-solar planet studies

Eugene F. Milone

RAO, The University of Calgary, Physics & Astronomy Dept., 2500 Univ. Dr. NW, Calgary, AB, T2N 1N4, Canada

Michael D. Williams

RAO, The University of Calgary, Physics & Astronomy Dept., 2500 Univ. Dr. NW, Calgary, AB, T2N 1N4, Canada

Abstract. GAIA will be a boon to extra-solar planetary detections and the subsequent analysis will provide basic and fundamental data on the detected planets. We discuss the methods and the analytical tools currently on hand for accomplishing this work.

1. Introduction

The GAIA satellite offers an excellent opportunity for the detection and characterization of extra-solar planetary systems. The main method for detecting these systems will be astrometry. Although the velocity resolution of GAIA's spectrograph may not be good enough to detect the effects of planetary companions, the spectrograph will be useful for classifying the central star and determining the star's dynamical associations. Due to the length of time between successive observations, GAIA's photometric data will not be useful for detailed study of transit events, however they may be useful for detecting extra-solar planets. Photometric detection of extra-solar planets is especially useful for systems that are too far away to be detected astrometrically. GAIA's great contribution to extra-solar planetary research comes from its multi-instrument approach and the large and diverse group of stars that will be sampled.

2. Stellar properties and planetary formation

The stellar mass can not be directly measured, since the motions of the planet can not be measured. The mass of the star has to be determined indirectly using stellar evolution models. GAIA's ability to combine parallax measurements and multi-band photometry to determine stellar parameters such as spectral type, effective temperature, luminosity class, chemical abundances, surface gravity, distance, and interstellar reddening, allows an estimate of the stellar mass to be made (which is dependent on the stellar structure model used).

Another method of estimating the stellar mass was used by Mazeh et al. (2000) to determine the mass of HD 209458a. They compared HIRES spectra to Kurucz stellar atmosphere model spectra and determined the stellar parameters

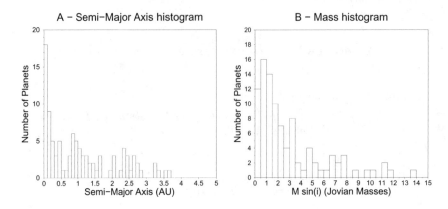

Figure 1. The distribution of orbital semi-major axis (*left*) and $M \sin i$ (*right*).

by finding the best fit model. It may be possible to apply this approach to data from the GAIA spectrograph.

The more than 70 extra-solar planets discovered to date have raised more questions about planet formation than they have answered. Some of the properties of these systems are plotted in Figures 1 and 2. The systems that have been found so far are very different from our solar system. Figure 1a shows that the majority of planets discovered thus far are much closer to their stars than the giant planets in our solar system, but the statistics are skewed toward massive planets with small orbital semi-major axis due to the sensitivity bias in radial velocity searches. GAIA and other astrometric searches are excellent complements to radial velocity searches since they are biased toward systems with longer periods. GAIA's mission duration of five years means that it will have a good probability of detecting planets with orbital periods up to 5 years. GAIA has the opportunity to detect planetary systems similar to the Sun-Jupiter system.

The peak of the distribution extra-solar planetary masses (in this case they are projected masses) seems to be around 1 Jovian mass (Figure 1b). The distribution of masses is not continuous from binary star companions down to planetary sized companions. This break has been called the 'brown dwarf desert', and it may be an indication that what we are calling extra-solar planet systems may have a different formation process than binary stars or star-brown dwarf systems. Brown dwarfs are not rare; they may have space densities as high 0.1 pc^3 and are often found in low mass ratio (i.e. close to unity) binary systems (Basri 2000), but brown dwarfs are rarely detected in high mass ratio (i.e. far from unity) binary systems. At this point in time a good observational method for discriminating between planets (sub-stellar objects that formed from a circum-stellar disk) and brown dwarfs (sub-stellar objects that formed in a similar fashion to stars) has not been found. Black (1997) suggested that the eccentricity of the object's orbit may be used to distinguish between planets and brown dwarfs. He plotted the eccentricity as a function of the log of the period and found that pre-main sequence stars followed a linear trend. The giant planets in our solar systems follow a different trend, they all have very small eccentricities. Black

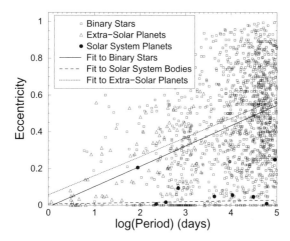

Figure 2. Black's (1997) proposed critria for distinguishing planets from brown dwarfs. The extra-solar planets follow the same trend as the binary star system. There is no clear trend for stars or extra-solar planets with orbits larger then the circularization limits.

found that the orbits of most of the planetary systems, followed the binary star trend as opposed to the giant planet trend. Figure 2 is an updated version of Black's plot. GAIA can be a testbed for criteria to differentiate between brown dwarfs and planets. GAIA should provide a large sample of sub-stellar objects, many with orbital periods as long as 5 years. Planets with longer orbital periods may not be as affected by orbital evolution.

Just as interesting as search programs that have found multitudes of planets are some that have failed to find any. Gilliland et al. (2000) conducted a search for planetary transits in the globular cluster 47 Tuc. Although the search had a photometric precision required to detect planetary transits, it detected none, indicating that the frequency of planets in the globular cluster is at least a factor of ten less than that of the solar neighborhood. The lack of planets in 47 Tuc may be due either to the lower metallicity or to the higher density of stars in the cluster. GAIA's multi-instrument approach may shed some light on this mystery. GAIA will observe stars in a variety of interstellar environments: galactic field stars, open cluster stars, and population II stars in the galactic disk. GAIA's proper motion observations combined with distance determinations and radial velocity data will be able to determine cluster associations of stars. Spectroscopy can determine the metallicity of the parent stars. From the sample of planetary systems found by GAIA it will be possible to examine the effects of stellar density and metallicity on the frequency of planetary companions.

3. Photometry of transits

GAIA's main means of detecting extra-solar planets will be its micro-arcsecond astrometry, but it is still important to consider the photometric detection of

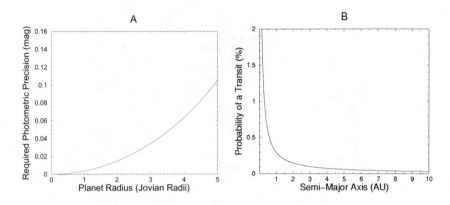

Figure 3. Some of the factors that will effect the probability of detecting extra-solar planetary transits.

transits. GAIA's astrometric search is limited to stars that are closer than about 200 pc, it should be possible to detect transit of extra-solar planets at a much greater distance. The required photometric precision as a function of planetary radius for a three sigma detection is plotted in Figure 3a. Although transits are short (~3 hours) compared to the length of time between observations (~ 1 month) it is still possible that over the 5 year lifetime of the mission GAIA will detect many transits. The probability of a transit occurring decreases greatly with increasing semi-major axis (Figure 3b), so short period transits are the most likely to occur, they are also the mostly to be detected by GAIA. Analysis of Hipparcos data showed that it had detected the planetary transit of HD 209458 (Robichon & Arenou 2000), so it is worth the effort of search light curves produced by GAIA for planetary transits.

4. Conclusion

GAIA offers a unique opportunity for extra-solar planet research. GAIA will observe a large sample of stars. From this large sample it may be possible to examine the effects of the interstellar environment on planet formation.

References

Basri, G. 2000, ARA&A 38, 485
Black, D.C. 1997, ApJL 490, L171
Gilliland, R.L. et al. 2000, ApJL 545, L47
Mazeh, T. et al. 2000, ApJL 532, L55
Robichon, N. & Arenou, F. 2000, A&A 355, 295

Evaluating the GAIA performance on photometry of near-contact and contact binaries

Panagiotis G. Niarchos & Vassilios N. Manimanis

Department of Astrophysics, Astronomy and Mechanics, University of Athens, GR 157 84 Zographos, Athens, Greece

Abstract. The physical parameters of two eclipsing binaries (one near-contact and one contact) are derived by using ground based photometric observations and Hipparcos/Tycho photometric data, which mimic the photometric observations that should be obtained by GAIA. The results are compared and the achievable precision of the basic stellar parameters derived by GAIA photometry is discussed.

1. Introduction

GAIA is designed to obtain extremely precise astrometry (in the micro-arcsec regime), multi-band photometry and medium/high resolution spectroscopy for a large sample of stars. The main scientific objectives of the GAIA mission and the expected benefits for astrophysical research are given by Gilmore et al. (1998) and Perryman et al. (2001), while the goals of GAIA spectroscopy and photometry are discussed by Munari (1999a,b). The GAIA large-scale photometric survey will have significant intrinsic scientific value for the study of variable stars of nearly all types, including detached eclipsing binaries, near contact or contact binaries and pulsating stars . It is expected that about 1×10^6 eclipsing binaries (EBs) with $V \leq 16$ mag will be discovered. The number of photometric points in the five-year mission lifetime is estimated to be 100 to 150 and the observing fashion will be quite similar to Hipparcos operational mode. Even if for only 1% of the observed EBs reliable physical parameters could be derived, this would be a great contribution to stellar astrophysics and a giant leap in comparison with what has been obtained so far from ground based observations (cf. Andersen 1991).

2. Selection of systems

There is an intrinsic limit on the number of photometric observations to the GAIA operational mode (100 to 150 points). How these observations can be compared with the state-of-the-art ground based observations? What is the accuracy to which eclipsing binaries can be investigated using GAIA data alone? The aim of the present investigation is to give an answer to the questions above regarding the semi-detached and contact binaries. Such an investigation for detached binaries has been carried out by Munari et al. (2001). GAIA is expected to collect a number of photometric points per band per star similar to that of

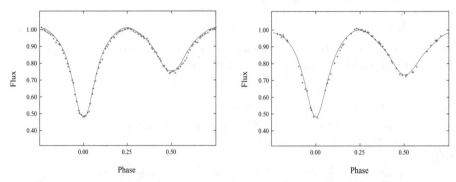

Figure 1. Light curves of the semi-detached binary V1010 Ophiuchi. *Left*: ground-based observations of Leung (1974). *Right*: Hipparcos H_P data.

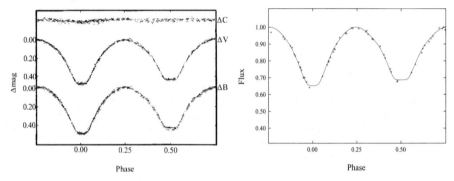

Figure 2. Light curves of the contact binary system V566 Ophiuchi. *Left*: ground-based observations of Bookmyer (1976). *Right*: Hipparcos H_P data.

Hipparcos. For the present investigation we have than used the following criteria for the selection of the systems:
(a) the systems have been observed by the Hipparcos/Tycho mission; (b) ground-based photometric observations of high quality exist, and (c) accurate spectroscopic mass ratios (determined from radial velocity measurements using modern techniques) are available.

The systems selected according to the above criteria are: the near-contact (semi-detached system) V1010 Oph and the contact (W UMa type) system V566 Oph. The V light curves from ground-based observations and those obtained from Hipparcos H_P observations are analyzed and the derived elements are compared.

3. Method of analysis and results

The Wilson-Devinney DC and LC programs have been used for the light curve analysis. The free parameters were: the phase of conjunction ϕ_0, the inclination

Table 1. Basic parameters from light curve solutions.

	V1010 Oph*		V566 Oph	
	ground based	Hipparcos	ground based[†]	Hipparcos
i (°)	85.86(68)	81.06(95)	80.30(24)	81.92(1.04)
T_1 (K)	8200*	8200*	7000*	7000*
T_2 (K)	5652(33)	5748(73)	6881(12)	6761(39)
Ω_1	2.7755*	2.7755*	2.2575(34)	2.2620(76)
Ω_2	2.7721(280)	2.7772(252)	2.2575(34)	2.2620(76)
$q = m_2/m_1$	0.4485*	0.4485*	0.2369*	0.2369*
$L_1/(L_1+L_2)$	0.8964(36)	0.8890(114)	0.7879(25)	0.8018(50)
r_1 (pole)	0.4317(18)	0.4356(27)	0.4891(10)	0.4880(18)
r_1 (side)	0.4613(22)	0.4663(36)	0.5342(16)	0.5326(25)
r_1 (back)	0.4909(27)	0.4975(47)	0.5628(21)	0.5609(33)
r_2 (pole)	0.3036(95)	0.3065(56)	0.2608(13)	0.2596(21)
r_2 (side)	0.3189(117)	0.3225(70)	0.2738(16)	0.2723(24)
r_2 (back)	0.3628(148)	0.3693(128)	0.3235(39)	0.3204(50)

*: assumed
[†]: from Van Hamme & Wilson (1985)
★: one spot on star 2: latitude 0°, longitude 72°, size 20°, T.F. 0.65

i, the temperature of the secondary component T_2, the fractional luminosity of the primary L_1 and the potential Ω (according to the mode used), while the fixed (adopted) ones were: the temperature of the primary T_1 (from the spectral type), the gravity darkening coefficient g (0.32 for convective envelopes, 1.0 for radiative), the albedo A (0.5 for convective, 1.0 for radiative), and limb darkening coefficient × (from tables by Claret et al. 1995 and Diaz-Cordoves et al. 1995). The spectroscopic mass ratio q was used as fixed parameter and the

Table 2. Absolute elements in solar units.

	M_1	M_2	R_1	R_2	L_1	L_2
			V1010 Oph			
ground based	1.87(12)	0.90(5)	2.08(6)	1.48(8)	10.7(7)	3.47(35)
Hipparcos	1.88(14)	0.89(7)	2.08(1)	1.46(1)	11.7(8)	3.07(8)
difference (%)	0.5	1.1	0	1.4	9.3	12
			V566 Oph			
ground based	1.40(3)	0.33(1)	1.47(1)	0.79(1)	4.57(4)	1.26(1)
Hipparcos	1.54(11)	0.36(4)	1.518(3)	0.819(4)	4.99(24)	1.26(2)
difference (%)	10	9.1	3.5	3.7	9.2	0

third light was zero in all cases. The basic elements derived from the light curve analysis are given in Table 1. Parameter uncertainties in Tables 1 and 2 are given in parentheses and refer to the last quoted digits. The elements of Table 1 were combined with the existing spectroscopic data to compute the physical parameters (absolute elements) of the systems, which are given in Table 2. In the same table the percentage difference between the absolute elements derived from ground-based observations and those from GAIA like observations are also given.

4. Conclusions

The main results of this investigation are:
(i) there are special advantages of GAIA mission (with combined astrometric, spectroscopic and photometric observations) compared to the classical ground based approach to eclipsing binary studies;
(ii) for the near-contact systems, the derived absolute elements from GAIA-like observations differ from those of ground based observations within the limits of the combined errors; and
(iii) for the contact systems, although these differences are larger than in the case of near-contact systems, they are mostly within the limits of the combined errors.

A final analysis of the expected GAIA performance on the photometry of eclipsing binaries could be drawn when enough stars will be investigated to cover both various spectral types as well as various kinds of interaction (detached, semi-detached and contact binaries).

References

Andersen, J. 1991, A&A Rev 3, 91
Bookmyer, B.B. 1976, PASP 88, 473
Claret, A., Diaz-Cordoves, J., & Gimenez, A. 1995, A&AS 114, 247
Diaz-Cordoves, J., Claret, A., & Gimenez, A. 1995, A&AS 110, 329
Gilmore, G., Perryman, M., Lindegren, L. et al. 1998, SPIE 3350, 541
Leung, K.-C. 1974, AJ 79, 852
Munari, U. 1999a, Baltic Astr. 8, 73
Munari, U. 1999b, Baltic Astr. 8, 123
Munari, U., Tomov, T., Zwitter, T. et al. 2001, A&A 378, 477
Perryman, M.A.C., De Boer, K.S., Gilmore, G. et al. 2001, A&A 369, 339
Van Hamme, W., & Wilson, R.E. 1985, A&A 152, 25

Hydrogen-to-helium ratio in WR stars from GAIA spectroscopy

Andrzej Niedzielski

Torun Centre of Astronomy, N. Copernicus University, Poland

Tiit Nugis

Tartu Astrophysical Observatory, 61602 Tõravere, Estonia

Abstract. We present possible application of GAIA spectroscopic observations for chemical composition determinations in nitrogen sequence Wolf-Rayet stars. The narrow spectral window of GAIA contains Paschen hydrogen and HeII lines of the $n - 6$ series which are well suited for this kind of analysis in both single and binary WR stars as it was pointed out already by Nugis & Niedzielski (1995).

1. Introduction: chemical composition of WR stars

Wolf-Rayet (WR) stars were discovered as a class of peculiar stars by C. J. E. Wolf and G. Rayet in 1867. Contrary to the spectra of most stars which are dominated by narrow absorption lines, the spectra of WR stars show broad and bright emission lines. Because of their peculiar spectral characteristics and high luminosities, it is easy to identify them by spectroscopic observations, even at large distances.

One of the most important characteristics of the WR stars is chemical composition. Reliable estimates of the hydrogen-to-helium ratio in the outer layers (winds) of WR stars are very important for clarifying the evolutionary status of these stars. Different studies generally agree that WN-star winds show compositions which correspond to the nuclear-processed material of hydrogen burning through CNO cycle. In some portion of these stars (mostly in WNL stars) we may see the H/He ratios corresponding to the mixing of the processed and unprocessed layers. The winds of WC and WO stars show nuclear-processed material corresponding to the He-burning phase and WC/WN stars show the mixture of materials corresponding to hydrogen and helium burning zones. For some stars the chemical composition estimates by different methods are quite strongly differing and for most of the binaries chemical composition estimates are lacking at all.

Ambarzumian (1933) was the first one to estimate quantitatively helium abundance in WR star envelopes. He compared intensities of HeII 4686 and HeII 4860+HI 4861 lines and used a recombination theory. Ambarzumian found that helium is more abundant as compared to hydrogen. Rublev (1972) and Smith (1973) were first to determine H/He ratios in several WN stars using the Pickering decrement method (Castor & van Blerkom 1970). They found that hydrogen is indeed strongly deficient as compared to solar value. Nugis (1975) presented first results of H/He determination in WC stars and found this ratio to be very close to 0. The extensive survey of Conti, Leep & Perry (1983) confirmed hydrogen depletion in 60 WN stars, while the survey of Torres et al (1986) confirmed the lack of hydrogen in optical spectra of 80 WC stars. The decrease of hydrogen content in WN stars during their evolution is therefore well established fact. Standard models of stellar evolution predict abrupt transition from hydrogen depleted to hydrogen-free WN atmospheres (Langer et al. 1994) while models of rotating massive stars (Maeder & Meynet 2000) allow for gradual change. Systematic survey of hydrogen abundances in WN stars may prove the importance of rotation in the evolution of WR stars.

2. GAIA vs chemical composition of WR stars

The near infrared spectral region of WR stars has been studied fragmentally and with insufficient spectral resolution. Only few WR stars have been observed in the 8000–10 000 Å range (Vreux et al. 1983, Vreux et al. 1989, Vreux et al. 1990). The spectral resolution of these observations was however low ($\Delta\lambda \approx 2$ Å). Yet it seems to be a very important region in different aspects. In the case of WN stars the study of HI and HeII lines in this range ought to give us the most reliable estimates of the H/He ratios (Nugis & Niedzielski 1995).

The GAIA spectroscopic survey is expected to cover the range 8480 –8740 Å. In this range lie some diffuse interstellar bands, the strongest of which is at 8620 Å. Equivalent width of this diffuse interstellar band correlates well with the strength of IS reddening (Munari 2000, 2002) and can be used for more correct determination of IS reddening for many WR stars.

In the GAIA spectroscopic range there are located hydrogen Paschen series lines P12 - P16. These lines are blended with the He I lines with the same principal quantum numbers and with the He II lines of the $n - 6$ series with $n = 24, 26, 28, 30$ and 32. Some of the odd n lines of He II $n - 6$ series are unblended and the difference of even n and odd n members of He II $n - 6$ series gives us information on H/He ratios in WN stars. Nugis & Niedzielski (1995) concluded that for WN stars the best estimates of the hydrogen-to-helium ratios can be obtained from the neighboring HeII $n - 6$ series lines ($n \geq 12$).

This study showed that theoretical fluxes of HeII + HI lines are quite seriously changed as compared to the case when the mutual influence of HeII and HI is neglected in radiation transfer treatments (especially in the case of WN5 - 9 stars). The level populations of HI are not strongly influenced by the suppression of stellar continuum radiation (due to absorption by blending HeII atoms) and by line radiation of surrounding HeII atoms, but line radiation of HeII is quite strongly suppressed (especially in the Pickering series) due to absorption caused by HI atoms lying in front of HeII atoms (in relation to a distant observer). This suppression is quite substantial in the case of WN7-9 stars and quite remarkable also in WN5-6 stars. The HeII $n - 6$ series lines are only slightly influenced by this effect and they are better suited for determining the hydrogen-to-helium ratios in late type WN stars (HI second principal quantum number state is strongly overpopulated as compared to the third state). It was

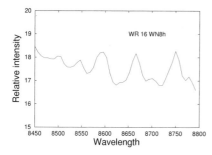 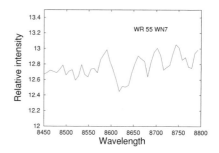

Figure 1. Examples of WR spectra with large and small amount of hydrogen within GAIA spectral window. Left panel shows spectrum of WR 16 (=HD 86161) with H/He=1.8, right one presents WR 55 (=HD 117866) with H/He=0.44 (Nugis & Niedzielski 1995). Note the intensity of HeII line at 8703 compared to neighboring H+He blends. Presented spectra come from Vreux et al. 1989.

also found that the lines of HeII Pickering series are (differently from the He II $n - 6$ series) quite strongly blended with the lines of other elements and this makes them not suitable for reliable determination of H/He ratios.

Besides the effects mentioned above, it is important to take into account that with the increase of n the interval between neighbouring lines decreases, and starting from some value of n it becomes smaller than the Doppler shift corresponding to the terminal wind velocity ($\Delta\lambda = \lambda v_\infty/c$). This means that the line He II $n - 6$ with high enough n in the approaching wind part may absorb the emission of receding wind part from the line He II $(n + 1) - 6$. High members of the HeII $n - 6$ series are optically thick in WNL star winds (Nugis & Niedzielski 1995) and therefore the effect of neighbour-absorption has to be accounted for carefully.

Figure 1 shows the spectra of two WNL in the range 8450-8800 Å. It can be seen that this neighbour-absorption effect becomes important starting from the line He II 8626 Å $(27 - 6)$. The blue side of that line is suppressed due to blending with the diffuse IS band at 8620 Å and the red side is suppressed by the line 8701 Å $(26 - 6)$ absorption. We can use the lines $24 - 6$, $25 - 6$ and $26 - 6$ for finding H/He ratios.

Nugis & Niedzielski (1995) derived empirical relations that allow to obtain H/He ratios from the equivalent widths of high members of He II $n - 6$ and H I Paschen series lines. The study of Nugis & Lamers (2002) of optically thick winds of WR stars shows that the velocity laws of their winds are much more gradual as compared to standard models (velocity law indexes β are about 5-6 instead of unity in standard models) and therefore new models are needed to be used for scaling the relationships for determination of H/He ratios.

The GAIA spectroscopic survey can give us very valuable good quality spectral information for deriving H/He ratios for many WN stars both in our Galaxy and in more distant galaxies.

Another important aspect of the GAIA spectral window is the reliable estimates of H/He ratios for WR binary stars. The optical spectra can not be used for this purpose because Balmer and Pickering series lines are severely blended with the absorption lines of the companion. Therefore the near IR spectral range offers the unique possibility to determine H/He ratio for many WR binaries for the first time.

Acknowledgments. AN thanks J.M. Vreux for making available the unpublished NIR spectra of WR stars. This work was supported by the Estonian Science Foundation grant No. 5003 and by N. Copernicus University grant 365A.

References

Ambarzumian, V. 1933, Pulkovo Obs. Circ. No. 7, 11
Conti, P.S., Leep, E.M. & Perry, D.N. 1983, ApJ 268, 228
Castor, J.I.& van Blerkom, D. 1970, ApJ 161, 485
Langer, N., Hamann, W.-R., Lennon, M., Najarro, F., Pauldrach, A.W.A. & Puls, J. 1994, A&A 290, 819
Maeder, A. & Meynet, G. 2000, ARA&A 38, 143
Munari, U. 2000, in Molecules in Space and in the Laboratory, I.Porceddu and S.Aiello ed.s, Soc. Ital. Fisica, pag. 179 (astro-ph/0010271)
Munari, U. 2002, in GAIA: A European Space Project, O. Bienaymé and C. Turon ed.s, EDP Sciences, pag. 39
Nugis, T. 1975, in Variable Stars and Stellar Evolution, IAU Symp. 67, V.E. Sherwood and L. Plaut ed.s, Reidel, pag. 291
Nugis, T. & Niedzielski, A. 1995, A&A 300, 237
Nugis, T.,& Lamers, H.J.G.L.M. 2002, A&A 389, 162
Rublev, S.V. 1972, Izv. Spets. Astrofiz. Obs. 4, 3
Smith, L.F. 1973, in Wolf-Rayet and High-Temperature Stars, IAU Symp. 49, M. Bappu & J. Sahade ed.s, Reidel, pag. 15
Torres, A.V., Conti, P.S. & Massey, P. 1986, ApJ 300, 379
Vreux, J.M., Dennefeld, M. & Andrillat Y. 1983, A&AS 54, 437
Vreux, J.M., Dennefeld, M., Andrillat, Y. & Rochowicz K. 1989, A&AS 81, 353
Vreux, J.M., Andrillat, Y. & Biemont, E. 1990, A&A 238, 207

The extended tidal tails of Palomar 5: tracers of the Galactic potential

Michael Odenkirchen, Eva K. Grebel, Hans-Walter Rix

MPI for Astronomy, Königstuhl 17, D-69117 Heidelberg, Germany

Walter Dehnen

Ast. Inst. Potsdam, An der Sternwarte 16, D-14482 Potsdam, Germany

Heidi Jo Newberg

Rensselaer Polytechnic Institute, 110 Eight St., Troy, NY 12180, USA

Constance M. Rockosi

University of Washington, Box 951580, Seattle, WA 98195, USA

Brian Yanny

Fermilab, P.O. Box 500, Batavia, IL 60510, USA

Abstract. We detected extended, curved stellar tidal tails emanating from the sparse, disrupting Halo globular cluster Palomar 5, which cover $10°$ on the sky. These streams allow us to infer the orbit of Palomar 5 and to ultimately constrain the Galactic potential at its location.

1. Palomar 5: a globular cluster torn apart by the Milky Way

Palomar 5 is an extraordinarily sparse globular cluster in the outer Halo of the Milky Way, at a distance of 23 kpc from the Sun (Figure 1). Its peculiar properties (very low mass, large core, relatively flat luminosity function) fostered the idea that this cluster might be a likely victim of disruptive Galactic tides.

Using deep multi-color photometry from the Sloan Digital Sky Survey (SDSS; York et al. 2000, Gunn et al. 1998) we found unambiguous, direct evidence for the suspected tidal disruption of Palomar 5 (Odenkirchen et al. 2001; Rockosi et al. 2002): For the first time, two massive tails of stellar debris with well-defined S-shape geometry were detected, emanating in opposite directions from the cluster.

As the SDSS is scanning more and more of the sky we have now extended our search over an area of ~ 87 deg^2. Contaminating objects were removed by eliminating extended sources and by applying an optimized smooth color-magnitude-dependent weighting function. This optimized weighting enhances the density contrast between cluster and field stars by almost a factor of 20 and provides a least-squares solution for the spatial distribution of the cluster

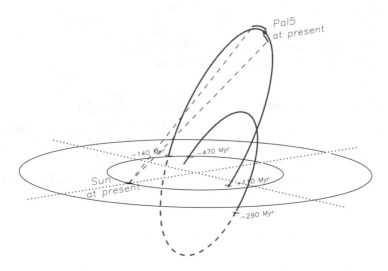

Figure 1. The present location of Palomar 5 in the Milky Way and its inferred plunging Galactic orbit (fat solid and dashed line). The approximate times of past and future disk passages are indicated.

population. The resulting surface density map of Palomar 5 stars is shown in Figure 2.

2. A narrow, curved, 10 deg stream of debris

We find that the tidal tails extend over an arc of at least 10° on the sky and form a narrow stream with a FWHM of only 18'. This corresponds to a projected length of $\simeq 4$ kpc in space, and a projected FWHM of 120 pc. The northern tail is visible out to 6.5° from the cluster. The southern tail is traced over 3.5° but probably continues beyond the border of the currently available field (Figure 2). The stellar mass in the tails adds up to 1.2 times the mass of stars in the cluster, i.e., the tails contain more mass than what is left in the cluster. Palomar 5 thus presents a text-book example of a tidally disrupting globular cluster. It is so far the only known stellar system besides the Sagittarius dwarf galaxy that demonstrates the formation of a Halo stream within the Milky Way.

The tails have a clumpy structure (Figure 2). This implies that the mass loss has been episodic, and suggests that it was triggered by disk and/or Bulge shocks. Indeed Palomar 5 passes through the Galactic disk at intervals of a few 100 Myr (Figure 1). In Figure 3 we present the radial profile of the stellar surface density (i.e. the azimuthally averaged surface density as a function of distance from the cluster center) from the core of the cluster out to the current end points of the tails. The profile shows a characteristic break near the cluster's tidal radius at about 16'. Inside this radius, the profile decreases approximately like r^{-3}. Beyond this limit the profile is flatter and approximately follows an $r^{-1.5}$ power law. The overall decrease thus differs from a simple r^{-1} power law that would result from a constant linear density along the tails.

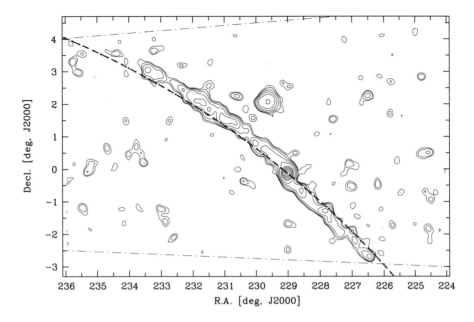

Figure 2. Contour map of the surface density of Palomar 5 stars derived by weighted star counts of SDSS point sources. The fat dashed line shows the best-fit local orbit of the cluster. The curved dash-dotted lines indicate the border of the current SDSS scan region.

3. Clues on the cluster's orbit and mass loss rate

Location and curvature of the tails are direct tracers of the cluster's Galactic motion and hence provide unique information about the orbit of Palomar 5. The best-fit local orbit is shown as dashed line in Figure 2. The direction of the cluster's motion is determined by its orientation with respect to the Galactic center: The southern, leading tail and the northern, trailing tail indicate that Palomar 5 is on a prograde orbit. Using a standard three-component model for the Galactic potential we infer that Palomar 5 is observed close to the tails' maximum distance from the Galactic disk and has recently passed through apogalacticon (implying that the tidal stream is currently relatively dense). In about 100 Myr the cluster will cross the disk at a distance of only 7 kpc from the Galactic center (see Figure 1). This will produce a strong tidal shock that might lead to complete disruption.

The amount by which the tails are offset from the orbit of the cluster is directly related to the velocity at which the tidal debris drifts away from the cluster. The observed mean offset (about 75 pc in projection), the parameters of our model orbit, and the total amount of stellar mass seen in the tails lead to an estimate of the mean mass loss rate of about $5\,\mathrm{M}_\odot\,\mathrm{Myr}^{-1}$. Assuming this rate to be more or less constant (as suggested by numerical simulations) we conclude that 10 Gyr ago Palomar 5 may have had a mass of about $5 \cdot 10^4 M_\odot$.

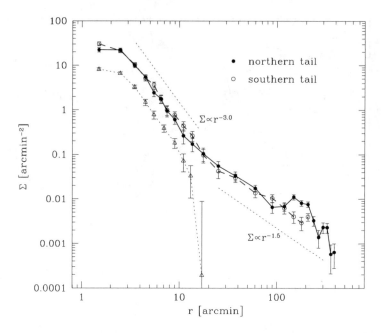

Figure 3. Radial profile of the stellar surface density in the cluster and tails from number counts in sectors of concentric rings. For comparison the leftmost data points show the cluster profile from counts in sectors perpendicular to the tails (arbitrarily shifted by -1 in log Σ).

This is about ten times as much as it has today, but still considerably less than the mass of an average present-day Galactic globular cluster.

The current data allow us to predict the tangential velocity of Palomar 5 as a function of the parameters of the Galactic potential. GAIA and SIM will allow us to accurately measure the velocities and proper motions of stars in the cluster and in the extended tails, fully characterizing the kinematics of the stream independent of any galactic model. In return, these kinematics impose strong constraints on the Galactic potential at the location of the cluster.

Acknowledgments. Funding for the creation and distribution of the SDSS Archive has been provided by the Alfred P. Sloan Foundation, the Participating Institutions, the National Aeronautics and Space Administration, the National Science Foundation, the U.S. Department of Energy, the Japanese Monbukagakusho, and the Max Planck Society. The SDSS website is http://www.sdss.org/

References

Gunn, J.E. et al. 1998, AJ 116, 3040
Odenkirchen, M. et al. 2001, ApJL 548, L165
Rockosi, C.M. et al. 2002, AJ 124, 349
York, D.G. et al. 2000, AJ 120, 1579

Missing spectroscopy of θ Tucanae

Margit Paparó

Konkoly Observatory of the Hungarian Academy of Sciences, P.O. Box 67, H-1525 Budapest XII, Hungary

Abstract. In the course of a frequency analyses of θ Tucanae, the "well-behaving" single star turned out to be a binary system. Additional spectroscopy confirmed it as a double line spectroscopic binary with a mass ratio 0.896. According to the composite spectra, the UV excess and infrared excess, the companion seems to be a late F or early G type star. However, a delicate modelling of the binary system where the primary is a δ Scuti type star needs the exact spectral type of the companion. The possibility for searching similar binary systems in GAIA database is discussed.

1. Introduction

1.1. Regular frequency distribution

In an international photometric campaign, 246 hours observation spread over 42 nights at three sites (South-Africa, Chile and Australia) were obtained in 1993 for θ Tucanae, a δ Scuti type variable star. About 2300 new Strömgren y, 2000 b and 1500 v and u photoelectric observations allowed us to get a stable mathematical representation of the light variation back to 20 years with 10 excited pulsational modes. The schematic amplitude spectrum of θ Tucanae (Figure 1) reveals a high level regular arrangement of modes (Paparó et al. 1996).

1.2. Quanted values for amplitude ratios and phase differences

Two non-adiabatic parameters, A_b/A_y amplitude ratio and $\phi_{b-y} - \phi_b$ phase difference, were derived as new, useful criterion for possible mode identification, which is at present a serious unsolved problem for δ Scuti stars (Figure 2). The modes are obviously arranged, according to two parameters, on levels in groups. The systematic arrangement of modes may hint light on the excitation mechanism. Those modes seem to survive and have distinguished values for both the amplitude ratio and phase difference. The modes having different values seem to be damped in the excitation mechanism. The two parameters may be connected to the radial order (n) and horizontal quantum number (l) of the excited modes. At the present stage we only know which theoretical modes must share the same behaviour, however, a pure theoretical calibration is needed (Paparó et al. 2000).

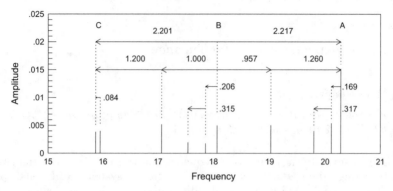

Figure 1. The schematic amplitude spectrum of θ Tucanae. High level regular arrangement of modes is shown.

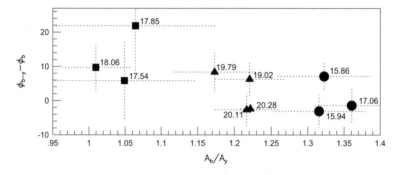

Figure 2. Groups and levels according to A_b/A_y and $\phi_{b-y} - \phi_b$ are shown. This is a discriminative plane for mode identification.

1.3. A δ Scuti star in binary system

The high precision photometry allowed us the detailed investigation of the variation of the mean light in θ Tucanae. A periodic behaviour has been derived back to 15 years (Figure 3). Alternative deep and shallow minima are nicely fitted by two periods (0.28151 and 0.14206 cycle day^{-1}) with nearly 1 : 2 ratio. A prediction of the binary nature has been concluded (Paparó et al., 1996) and a lately spectroscopic investigation confirmed (De Mey et al. 1998). The double lined spectra revealed the radial velocity curve for both companion with a 0.896 mass ratio. A resonant effect between one of the characteristic orbital period (connected to the ellipsoidal shape of the secondary) and the dominant pulsational mode has been obtained (Paparó et al. 2002).

1.4. Target for modelling

A δ Scuti star with extremely regular behaviour in a binary system is, no doubt, a delicate target for modelling from both pulsational and evolutional points of view. However, some information is still missing.

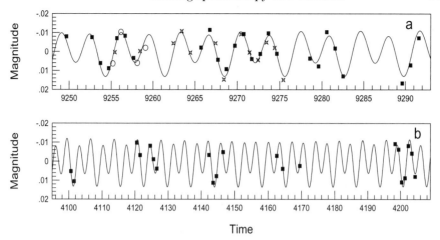

Figure 3. The periodic variation of θ Tucanae's mean light level (*top*) for the campaign and (*bottom*) back to 15 years. Different symbols mark different sites. Time is marked by HJD - 2440000.

2. What is known for the secondary?

2.1. Composite spectrum

According to the catalogues the spectral type of θ Tucanae is A4m or A7IV. A refined classification, based on lines of calcium, hydrogen and metallic lines, resulted in kA7hA7mF0(IV) spectral type by Gray & Garrison (1989). The new spectral type could be interpreted as a composite spectrum of a binary system with a F type component.

2.2. Chromospheric activity

A search for chromospheric activity among the A and F stars has been carried out by Simon & Landsman (1991), based on IUE spectra from measurements of CII 1335 Å. The earliest star of any luminosity class in the sample to show CII emission is θ Tucanae. The unusual ultraviolet behaviour of θ Tucanae could be a sign of a late type companion showing chromospheric activity.

2.3. Infrared excess: 12 μm flux

θ Tucanae is an IRAS source. It was detected at 12 μm but not at 25, 60 or 100 μm. The 12-μm flux given by King (1990) is 0.37 Jy. The star has 0.31 mag infrared excess. Taking the distance as 88 pc and accepting $m-M = 4.8$ distance modulus the infrared absolute magnitude is $M_{12} = 0.31$ mag. A comparison of the infrared absolute magnitude to a model of the infrared sky (Wainscoat et al. 1992) suggests that a component could have a F8 - G2 III spectral type.

2.4. Luminosity ratio for components in infrared

We can estimate the brightness of the companion using the relation between the bolometric and infrared luminosity for δ Scuti stars. To get the observed infrared flux we need a secondary beside θ Tucanae which is 2 mag brighter at 12 μm than the primary.

2.5. Radial velocity curve for both component

As we mentioned before, a spectroscopic investigation has got the radial velocity curve of both component, however, the precise spectral type of the secondary has not been derived. The double lined spectra suggest similar luminosity for both component.

2.6. The secondary's spectral type is needed

The present paper is aimed to call the attention to spectroscopic investigation of θ Tucanae for the sake of the secondary's spectral type.

2.7. A search for similar binary systems in GAIA data

The discovery of θ Tucanae's binary nature was obtained by killing off (averaging) the pulsational light variability of the system. The mean light level variation was investigated separately. The situation is similar to that of GAIA data where the scarce measurements of a given star will kill off the short period variability. However, there is hope to use the mean light level variation for searching for binary systems with variable star as a main component. The example of θ Tucanae focuses the attention on the spectroscopic and photometric requirements of GAIA measurements to make them proper for searching for binary systems with pulsating component.

Acknowledgments. Research grant of Hungarian Astronomical Foundation is acknowledged

References

De Mey, K., Daems, K. & Sterken, C., 1998, A&A 336, 527
Gray, R.O. & Garrison, R.F. 1989, ApJS 70, 623
King, J.R., 1990, PASP 102, 658
Paparó, M., Sterken, C., Spoon, H.W.W. & Birch, P.V. 1996, A&A 315, 400
Paparó, M. & Sterken, C., 2000, A&A 362, 245
Paparó, M., Shibahashi, H. & Sterken, C. 2002, in ASP Conf. Ser. 259, 94
Simon, T. & Landsman, W. 1991, ApJ 380, 200
Wainscoat, R.J., Cohen, M., Volk, K., Walker, H.J. & Schwartz, D.E., 1992, ApJS 83, 111

GAIA spectroscopy of Carbon stars

Yakiv V. Pavlenko
Main Astronomical Observatory of the National Academy of Sciences, Golosiiv woods, 03680, Kyiv-127, Ukraine

Paola M. Marrese & Ulisse Munari
Astronomical Observatory of Padova - INAF, Asiago station, 36012 Asiago (VI), Italy

Abstract. GAIA spectra of a grid of Carbon stars have been obtained with the Asiago 1.82m telescope and computed via synthetic modeling. The spectral appearance of Carbon stars over the GAIA wavelength range is best described as a forest of closely spaced strong absorption lines (nearly all due to the CN molecule) that perform stupendously in terms of radial velocity accuracy, at the level of 0.1 km s^{-1} on a single GAIA well exposed spectrum at 17 000 resolution. CaII triplet lines can still be recognized down to the coolest Carbon types, turning into emission in some carbon Miras. The completely different absorption pattern for ^{12}CN and ^{13}CN allows measurement of the ^{12}C/^{13}C ratio. Smooth and marked spectral transitions are observed for Carbon spectra over the GAIA wavelength range, offering good prospects for classification.

1. Introduction

Classical Carbon stars are evolved objects at the tip of the AGB branch, and as such they are bright and visible over great distances in the Galaxy. The GAIA 8480-8740 Å wavelength range is placed toward the peak of the Carbon star spectral energy distribution, favouring their observation. The large and homogeneous observational database that GAIA will collect on Carbon star in the Galaxy and nearby dwarf satellites is obviously expected to deliver a fresh and a deeper view into the realm of this highly evolved type of stars.

Spectral surveys previously conducted over the GAIA wavelength range as well as grids of computed synthetic spectra have only marginally considered Carbon stars, whose appearance over the GAIA spectral interval is basically unknown.

To fill this gap, we have observed over the GAIA wavelength range a well distributed grid of Carbon stars mapping the Keenan and Morgan (1941) classification scheme (still a good and compact description of the spectrum) and selected from the Alksnis et al. (2001) catalogue, and computed a preliminary grid of synthetic spectra. This contribution outlines briefly the main results.

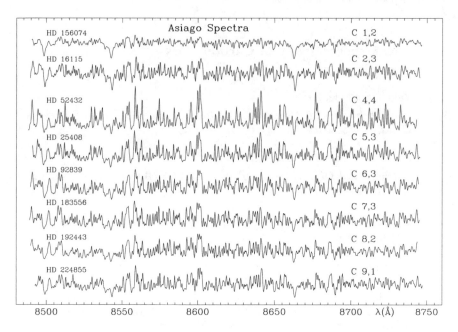

Figure 1. Spectra of some of the observed Carbon stars arranged along the classification sequence. Note the decrease in CaII visibility.

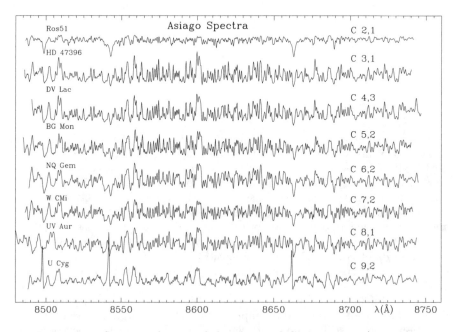

Figure 2. Spectra of some of the observed Carbon stars (continued from Figure 1). Note the CaII emission in U Cyg.

Table 1. List of Carbon stars for which high S/N and 20 000 resolution spectra over the GAIA wavelength range have been obtained with the Asiago 1.82 m + Echelle + CCD spectrograph. The majority of them is displayed in Figures 1 and 2.

			α_{J2000}	δ_{J2000}	V
HD 156074		C1,2	17 13 31.2	+42 06 22	7.6
Ros51		C2,1	13 55 44.2	−18 14 40	7.8
C* 888		C2,2	07 46 26.5	−08 19 19	11.2
HD 16115		C2,3	02 35 06.5	−09 26 34	8.2
HD 47396	NY Gem	C3,1	06 39 17.8	+22 36 18	12.5
BD +21.64		C3,3	00 34 15.8	+22 23 36	9.7
	V738 Mon	C3,4	06 43 12.5	−08 45 30	9.9
	DV Lac	C4,3	22 45 04.2	+56 37 18	9.9
HD 52432	V614 Mon	C4,4	07 01 01.9	−03 15 09	7.3
HD 1546	VX And	C4,5	00 19 54.0	+44 42 34	8.5
	BG Mon	C5,2	06 56 22.8	+07 04 40	9.2
HD 25408	UV Cam	C5,3	04 05 53.9	+61 47 40	7.6
	VW Gem	C5,4	06 42 08.6	+31 27 18	8.2
HD 59643	NQ Gem	C6,2	07 31 54.5	+24 30 13	7.7
HD 92839	VY UMa	C6,3	10 45 04.0	+67 24 41	6.0
HD 42272	TU Gem	C6,4	06 10 53.1	+26 00 53	7.4
	V Hya	C6,5	10 51 37.3	−21 15 00	9.7
HD 63353	W CMi	C7,2	07 48 45.5	+05 23 35	9.0
HD 183556	UX Dra	C7,3	09 21 35.5	+76 33 35	6.3
	TT Tau	C7,4	04 51 31.3	+28 31 37	7.9
HD 34842	UV Aur	C8,1	05 21 48.9	+32 30 43	9.6
HD 192443	RS Cyg	C8,2	20 13 23.7	+38 43 44	7.7
HD 224855	WZ Cas	C9,1	00 01 15.9	+60 21 19	7.1
HD 193680	U Cyg	C9,2	20 19 36.6	+47 53 39	6.7

2. Observed spectra

The observed spectra have been secured with the Asiago Echelle+CCD spectrograph at a resolution ∼ 18 000 (close to the upper limit for the GAIA spectrograph) and S/N always in excess of 200 (thus all visible features are real and not noise artifacts). A list of the observed Carbon stars is given in Table 1, while Figures 1 and 2 show some of the spectra arranged into temperature sequences.

The features progress with the temperature index is quite evident, suggesting that a detailed classification scheme can be devised for Carbon stars over the GAIA wavelength range.

The collected spectra have been cross-correlated with templates from the synthetic grid (see below) to estimate the accuracy that can be achieved in the

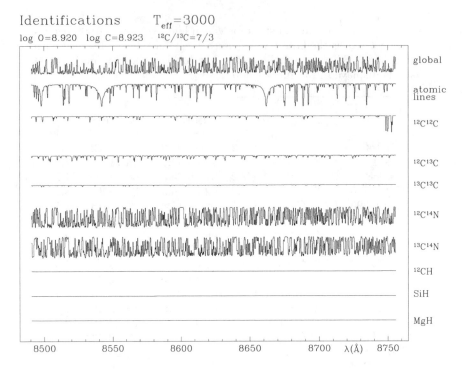

Figure 3. Synthetic spectra of diatomic molecules over the GAIA wavelength region.

radial velocities. As expected, the accuracy turned out to be impressively good: all spectra displayed in Figures 1 and 2 can be cross-correlated to 0.1 km s^{-1} error. This is evidently due to *all* pixels being part of strong lines, thus *all* pixels carrying a strong radial velocity information.

3. Synthetic spectra

We computed a grid of stellar model atmospheres with 2500$\leq T_{\rm eff} \leq$3500 K and 0$\leq \log g \leq$1 by SAM12 program (Pavlenko 2002, http://www.mao.kiev.ua/staff/yp/TOP-mod.htm). SAM12 is a modification of ATLAS12 (Kurucz 1999, http://kurucz.harvard.edu/). Opacity sampling treatment was used to account for atomic and molecular absorption. Atomic and molecular line data were taken from Kurucz (1993), Goorvitch (1994) and/or VALD (Kupka et al. 1999). Chemical equilibrium was computed for the case of carbon-rich plasma, i.e C/O>1. Molecular constants of chemical equilibrium were taken from Tsuji (1973).

A grid of synthetic Carbon spectra has been computed for the indicated atmospheric parameters by WITA6 program (Pavlenko 1999). Computations were carried out for the same opacity sources and abundances as in model atmosphere computations. We adopted a micro-turbulent velocity V_t =3.5 km s^{-1},

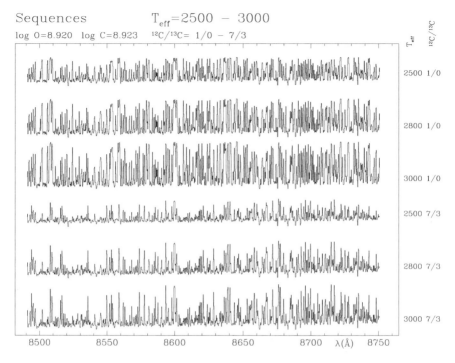

Figure 4. A dependence of synthetic spectra of Carbon stars on input parameters.

Voigt profiles were taken for all lines, and damping constants were from VALD (Kupka et al. 1999) and Kurucz (1993) databases, or computed in the framework of classical Unsold (1949) approach.

4. Results

We carried out an identifications of all molecular features observed in the region. To do it, we computed synthetic spectra taking into account only line lists of a given species. Results are showed in Figure 3. The main contributors to line opacity are the CN molecule and the atomic lines.

A sample of synthetic spectra computed by taking into account absorption by all species are showed in Fig. 4. In general, C-giant spectra show a dependence on a larger number of input parameters than M stars at the same temperature. The impact of temperature and $^{12}C/^{13}C$ on computed spectra is showed in Figure 4. A more general approach should consider the effects on the computed spectra of different micro-turbulent velocities, [Fe/H] and [He/H], stellar winds and sphericity effects. Detailed consideration of these effects lies beyond the scope of this contribution. Extensive studies of carbon stars are carried out by various groups (cf. Wallerstein & Knapp 1998, Guliermo & Wallerstein 2000, and references therein). Taking into account refined input parameters for synthetic spectra computations, different parameterization's algorithms (see Jones et al. 2002, Pavlenko & Jones 2002) can be used to derive more precisely the fundamental parameters of Carbon stars.

Our analysis of computed and observed spectra of Carbon stars over the GAIA wavelength range indicates that (a) the CN molecule is by far the strongest shaper of the Carbon stars appearance in the GAIA wavelength range, (b) other molecules (in particular C_2) play a negligible role, (c) the spectral pattern of the CN molecule is completely different for the ^{12}CN and ^{13}CN variants, allowing a determination of the ^{12}C/^{13}C ratio from GAIA spectra, (d) CaII lines in absorption are recognizable over the whole Carbon star sequence, (e) CaII lines can appear in emission in some carbon Miras (perhaps only at selected pulsation phases), and (f) the Carbon star perform stupendously in term of accuracy of radial velocities.

References

Alksnis, A., Balklavs, A., Dzervitis, U. et al. 2001, Balt.Astron. 10, 1

Jones, H. R.A., Pavlenko, Y.V, Viti, S. & Tennyson, J. 2002, MNRAS 330, 675

Goorvitch, D. 1994, ApJS 95, 535

Guliermo, G. & Wallerstein, G. 2000, AJ 119, 1839

Keenan, P.C. & Morgan, W.W. 1941, ApJ 94, 501

Kupka, F., Piskunov, N., Ryabchikova, T.A., Stempels, H.C. & Weiss, W.W. 1999, A&AS 138, 119

Kurucz, R.L. 1993a, Kurucz's CD-ROM Series, CfA Harvard, Cambridge

Pavlenko, Y.V. 1999, Astr. Reports 43, 94

Pavlenko, Ya. V. & Jones, H.R.A. 2002, A&A 396, 967

Tsuji, T. 1973, A&A 23, 411

Unsold, A. 1949, Physik der Sternatmospharen, Springer

Wallerstein, G. & Knapp, G.R. 1998, ARA&A 36, 369

Coarse estimation of physical parameters of eclipsing binaries by means of automative scripting

Andrej Prša

Department of Mathematics and Physics, University of Ljubljana, Jadranska 19, 1000 Ljubljana, Slovenia

Abstract. Because of GAIA's estimated harvest of $\sim 10^5$ eclipsing binaries (Munari et al. 2001) automative procedures for extracting physical parameters from observations must be introduced. We present preliminary results of automative scripting applied to five eclipsing binaries, for which photometric and radial velocity observations were taken from literature. Although the results are encouraging, extensive testing on a wider sample has to be performed.

1. Introduction

Eclipsing binaries have solid grounds in today's astrophysics due to relatively simple geometry modeling that allows astronomers to extract physical parameters such as masses and radii from observed light curves and radial velocity curves in absolute units, which is very difficult (if at all possible) for other kinds of celestial objects. Modeling approaches are developing rapidly with increasing computing power of PCs. Most widely used modeling algorithms are WD (Wilson & Devinney 1971), WINK (Wood 1971), FOTEL (Hadrava 1986) and others. However, what most of these models lack is the level of automation required for scanning missions. GAIA will retrieve data for $\sim 10^5$ eclipsing binaries to 15 mag (ESA-SCI-2000-4) and fully automative approaches are absolutely necessary.

2. Automative approach

The fact that most of the existing modeling software lacks automation isn't a simple oversight, but is a consequence of few inevitable problems one faces while seeking a solution (R.Wilson, notes accompanying WD98 code, hereafter W98): (*1*) extraction of physical parameters of eclipsing binaries is a highly non-linear inverse problem and is as such subjected to modeling degeneracies; (*2*) solution convergence may be slow or even fail because of shallow, wide minima in parameter space, numerical algorithms' inefficiency and non-linearity along with strong parameters' correlations, and (*3*) the choice of parameters to be adjusted changes from iteration to iteration based on common sense and it is difficult to assess in advance how a certain change in parameter values changes the credibility of the entire solution.

Bearing this in mind we use Wilson–Devinney model for basis and propose an automative approach, which is subject to the following main principles: (*a*)

Table 1. List of chosen test stars. RVs are for both stars.

	type	available data	references
BH Vir	Detached	BV, $ubvy$, RVs	Clement et al. (1997)
GK Dra	Detached, δ-Sct	BV, RVs	Zwitter (2003)
TY Boo	Over-contact	BV, RVs	Milone et al. (1991)
UV Leo	Detached	BV, RVs	Frederic & Etzel (1996)
V505 Per	Detached	$H_P V_T B_T$, RVs	Munari et al. (2001)

the underlaying formalism of solution-seeking shouldn't be changed in any way; (*b*) there is more to astrophysics than parameter estimation, so any automative processing should be limited only to obtain coarse solutions, (*c*) the successfulness should be tested on wide and diverse types of eclipsing binaries, and (*d*) the time cost of modeling analysis should be considerably reduced.

We refer to this process as *automative scripting*. Based on photometric and spectroscopic observations, with the assumption that the orbital period P, the epoch HJD_0 and the type of an eclipsing binary (detached, semi-detached, contact or over-contact) are known, we follow the predefined pattern of what modeling actions to perform in a particular order. Such a pattern is a *script*. Our goal is to create such a script (or a set of scripts), that would lead us to a coarse modeling solution without any interactive approach.

3. Preliminary test results

We have performed some preliminary testing of automative scripting on dozen different eclipsing binaries taken from literature. Figure 1 shows the results of automative scripting performed on five stars given in Table 1. The measurements consist of 100–1600 data points per filter in photometry and 10–100 data points in radial velocities. A typical time scale for the complete process of automative scripting on a 1.4 GHz processor is ∼35 seconds. Table 2 contains basic extracted parameters, which may be compared against literature values in Table 3.

Few remarks should be made: (*i*) the solutions were reached without any human intervention, (*ii*) the solution degeneracy is reduced by using the method of multiple subsets (W98) instead of exclusive use of differential corrections, (*iii*) further degeneracy elimination is possible because of the presence of radial velocities, which along with photometric data uniquely determine the semi-major axis and both components' masses. Because of GAIA's RVS capabilities we expect tremendous improvements to the field of eclipsing binaries.

Note the discrepancy between calculated and adopted potentials Ω_1 and Ω_2 in Tables 2 and 3: this is a consequence of the degeneracy between the inclination and stellar radii ($\propto \Omega^{-1}$), the direct example of how blindly trusting automatically retrieved solutions may be dangerous. Since we can recognize this problem, we may avoid it by using more adequate scripts, but there may be many caveats we must still anticipate. The results demonstrated in the previous section

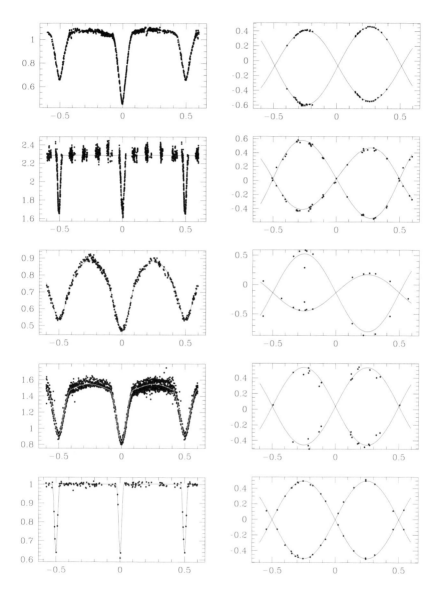

Figure 1. Light curves (*left*) and radial velocity curves (*right*) of the five stars given in Table 1: BH Vir, GK Dra, TY Boo, UV Leo and V505 Per, respectively from top to bottom. The independent variable is in all cases phase and the dependent variable is relative flux for light curves and normalized radial velocities to $2\pi a/P$ for radial velocity curves. Measurements are marked with filled circles and the synthetic solution from Table 2 with a solid line.

Table 2. Computed stellar parameters using automative scripting. Standard deviation in units of the last decimal place (W98).

	$a[R_\odot]$	q	T_2/T_1	$i[°]$	Ω_1, Ω_2	$v_\gamma[\text{km s}^{-1}]$
BH Vir	4.7(3)	0.90(0)	0.92(0)	85.9(1)	5.2, 4.7(0)	−0.2(8)
GK Dra	28.9(4)	1.26(0)	1.01(0)	83.9(2)	10.5, 10.5(1)	1.7(12)
TY Boo	2.0(2)	2.15(1)	1.00(0)	76.4(1)	4.0, 4.0(4)	−43.8(15)
UV Leo	4.0(3)	1.00(1)	0.97(1)	84.1(1)	5.1, 4.4(0)	12.8(6)
V505 Per	15.1(1)	0.98(1)	0.99(0)	87.0(2)	11.5, 11.2(2)	−0.7(6)

Table 3. Literature parameters from references given in Table 1.

	$a[R_\odot]$	q	T_2/T_1	$i[°]$	Ω_1	Ω_2	$v_\gamma[\text{km s}^{-1}]$
BH Vir	4.7	0.89	0.91	87.5	4.7	4.8	0.0
GK Dra	29.0	1.26	0.98	85.8	14.1	12.2	3.8
TY Boo	2.32	2.13	1.00	77.5	5.4	5.4	−38.7
UV Leo	3.91	0.96	0.98	82.6	4.6	4.3	13.0
V505 Per	15.0	0.98	0.98	87.8	11.5	11.2	0.0

are encouraging; however, one has to keep a fair amount of scientific scepticism about the adopted automation philosophy. This study was performed with an extremely undersampled statistics. A thorough study over large numbers of diverse eclipsing binaries is compulsory to gain confidence in the automation procedure. At this time we are still uncertain about the universality of the developed algorithm. We may be facing systematic errors due to conformity of our script: it *may* invoke the existence of a preferred subspace within the parameter space, a dire consequence that would compromise all retrieved solutions.

References

Clement, R. et al. 1997, A&AS 124, 499
Frederik, M.C., & Etzel, P.B. 1996, AJ 111, 2081
Hadrava, P. & Kadrnoska, J. 1986, Cel. Mech. 39, 267
Milone, E.F., Groisman, G., Fry, D.J.I. & Bradstreet, D.M. 1991, ApJ 370, 677
Munari, U. et al. 2001, A&A 378, 477
Wilson, R.E. & Devinney, E.J. 1971, ApJ 166, 605
Wood, D.B. 1971, AJ 76, 701
Zwitter, T. 2003, in GAIA Spectroscopy, Science and Technology, U. Munari ed., ASP Conf Ser. 298, pag. 329

GAIA spectroscopy of active solar-type stars

Silvia Ragaini, Vincenzo Andretta, Maria Teresa Gomez, Luciano Terranegra

INAF – Osservatorio Astronomico di Capodimonte, Naples, Italy

Innocenza Busà & Isabella Pagano

INAF – Osservatorio Astrofisico di Catania, Catania, Italy

Abstract. We present a preliminary study of the Ca II infrared triplet on sample models of active solar-type stars, with the aim of verifying the feasibility of using GAIA spectroscopy for studies and surveys of chromospheric and photospheric activity in cool stars. GAIA spectra have been simulated using synthetic profiles broadened with two possible instrumental profiles and using different values of the S/N ratio.

1. Introduction

Spectroscopic observations offer powerful tools for the study of the structure of the stellar atmospheres and their activity. Synthetic spectra, generated by means of numerical modelling, can be compared with observed spectral features, providing a diagnosis of physical properties (thermodynamical and dynamical) of the stellar plasma. Traditional diagnostics, such as the Ca II H & K lines in cool stars, suffer some observational limitations related to their position in the blue spectral range: low stellar flux in cool stars, low CCD sensitivity and line crowding. There are, however, other spectral lines in the visible and near infrared, that can usefully complement the information provided by traditional diagnostics and that are easily observable also in relatively faint stars. Some of these other diagnostics, among which there are the Ca II near-infrared triplet and the resonance lines of the alkali lines, have received only a relatively limited attention.

The spectral interval covering the Ca II triplet is free from line crowding due to atmospheric, interstellar and stellar absorption lines; the extended wings of the Ca II triplet absorption lines probe a wide range of photosheric layers, and are thus sensitive to fundamental stellar atmospheric parameters: T_{eff}, $\log g$, [Fe/H]. Their cores, on the other hand, are so opaque that they are formed in the uppermost atmospheric layers (chromospheres) and their central depths have also been shown to provide good indicators of chromospheric activity (e. g. Foing et al. 1989 and reference therein).

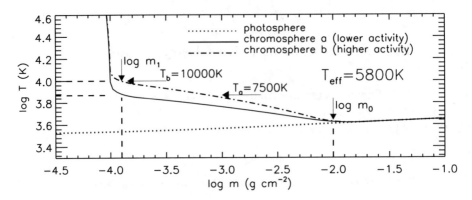

Figure 1. The model photosphere (dotted line) corresponding to $T_{\rm eff}$=5800 K, is shown here together with two model chromospheres (dash-dotted and solid lines, respectively). The arrows at T_a=7500 K and T_b=10 000 K mark the end temperature (the beginning of the transition region, TR) of the model chromospheres. In both model chromospheres the position of the temperature minimum has been placed at $\log m_0 = -2$, and the start of the TR at $\log m_1 = -4$

2. The calculations

The method adopted here to investigate the usefulness of GAIA spectra for studies of activity in cool stars is based on simulations of observed spectra. The synthetic spectra have been obtained by means of radiative transfer calculations, using the multi-level NLTE code MULTI (Carlsson 1986). The starting model photospheres for these calculations are from existing grids (NextGen, Hauschildt et al., 1999), with stellar parameters characteristic of late-type dwarf stars. In particular, we have considered two photospheres with $T_{\rm eff}$ = 5800 K and 4600 K, both with $\log g = 4.5$ and solar abundances ([Fe/H]=0.0). We have added to these photospheres two model chromospheres, each characterized by the same position of the temperature minimum, and by different temperature chromospheric gradients (Figure 1).

The resulting line profiles of the Ca II 8542 Å line are shown in Figure 2. In order to evaluate the effect of the instrumental resolution of the GAIA spectrometer, which at this moment is still to be fine tuned, we have broadened these synthetic spectra with instrumental profiles corresponding to low ($R = 5\,000$) and medium ($R = 20\,000$) resolution.

In the case of low spectral resolution (panels a and b of Figure 2), the line profile reversal in the core, a typical chromospheric signature more evident in the models with higher chromospheric gradient, is completely washed out. In realistic spectra, with a finite value of the S/N ratio, one would expect that such a signature would be difficult to measure, except perhaps for the brighest stars.

We have then investigated the effect of noise on the chromospheric signature in the core of these lines. In order to do so, we have added to the spectra shown in Figure 2 some statistical (Poissonian) noise, for different values of the S/N ratio: 10, 30 and 100. We have repeated this procedure 60 times, obtaining a

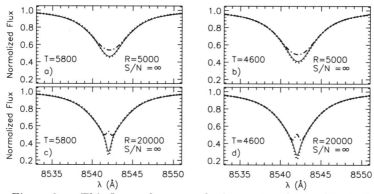

Figure 2. This figure shows synthetic spectra obtained including instrumental broadening for two different resolutions: $R=5\,000$ (panels a,b) and $R=20\,000$ (panels c,d). The pairs of panels a,c and b,d correspond to $T_{\text{eff}} = 5800$ K and 4600 K, respectively. Line-style coding as in Figure 1.

set of independent synthetic, noisy spectra. From each chromospheric profile, we have then subtracted the theoretical photospheric profile, obtaining, at each value of the S/N ratio, a set of 'measured' line excesses.

In Figure 3 we show the mean values of these chromospheric excesses, together with their standard deviation, as function of the S/N ratio. The results shown in that figure confirm quantitatively the impression from the profiles of Figure 2: at low resolution ($R=5\,000$), only strong emission can be easily detected, while for less active chromospheres very high signal-to-noise ratios (S/N> 100) would be required. The medium-resolution case ($R=20\,000$) is instead more promising, allowing a determination of chromospheric activity even for moderate activity levels and relatively low S/N ratios.

3. Conclusions

We have shown in this paper how it is possible with GAIA spectra to estimate the activity level of solar-type stars by measuring the flux excess, relative to an unperturbed photosphere, in the Ca II triplet. From the fig.3, it is clear that even with a modest S/N ratio it is possible to quantify the emission in the core of the Ca II triplet, in stars of medium or high activity levels. For low activity stars, higher S/N ratios will be required. Therefore, GAIA will be able to provide a large and homogeneous sample of Ca II triplet excesses. Above certain levels of activity, which we estimate to be corresponding to about $R_{HK} \sim -4.8$, we expect the sample to be nearly complete. A more precise evaluation of this level, will depend both on the estimated relation Ca II triplet vs. R_{HK} (Busà et al., in preparation), and on the specific S/N ratio for a given magnitude GAIA will be able to attain. From a different point of view, the fact that Ca II triplet excesses in bright stars with medium/high activity levels can be measured reliably with only relatively few measurements, will also open the possibility of a survey of time dependence of activity during the lifetime of the

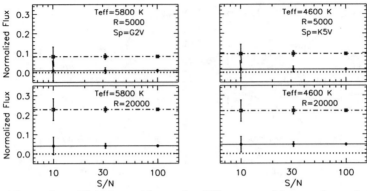

Figure 3. This figure shows the difference, with error bars, between the central depression of the chromospheric noised spectra and the photospheric ones (excess emission), versus the S/N ratio. The horizontal lines represent the value of the excess emission in spectra with S/N=∞.

mission. Finally, we emphasize that the Ca II triplet excesses are defined here from theoretical photospheric profiles, which can be computed reliably given the stellar parameters (T_{eff}, $\log g$, [Fe/H]) provided by the same GAIA spectra: in other words, these excesses can be self-consistently derived from GAIA spectra.

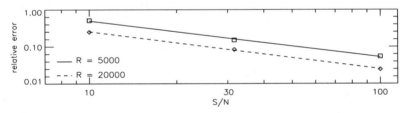

Figure 4. Dependence of the logarithm of the relative error on the central depression, as function of the logarithm of the S/N ratio. The data are fitted with a line of slope -1, or, equivalently: error $\propto$ $(S/N)^{-1}$. Both plots refer to $T_{\text{eff}} = 4600$ K, at the two resolutions: $R=5\,000$ and $R=20\,000$.

References

Carlsson, M. 1986, Uppsala Observatory, Internal Report N. 33
Foing, B.M. et al. 1989, A&AS 80, 189
Hauschildt, P. H., Allard, F., & Baron, E. 1999, ApJ 512, 377

CaII triplet monitoring of stars with the R CrB type variability

Alexander E. Rosenbush

MAO of the NAS of Ukraine, Zabolotnogo Str.27, Kyiv, 03680 Ukraine

Abstract. We present some results of the spectral high ($R \lesssim 35\,000$) resolution monitoring for 4 years of the CaII triplet lines in stars with the R Coronae Borealis type variability, which are objects with a high mass loss rate. The CaII lines of all program stars have broad and complex profiles. At the light maximum the central intensity of CaII 8542 Å line in R CrB itself can change from 0.1 to 0.2 during one day without visible variations in the line wings. Here an emission is always observed in the line core. Its intensity is not affected considerably by the formation of a dust shell obscuring the photospheric flux of the central star during a light minimum. The shift of emissions corresponds to the outflow velocity of several km s^{-1} for a layer of the stellar atmosphere where these emissions are formed. SU Tau and SV Sge, stars with the R CrB type variability, and XX Cam and HD 182040, non-variable HdC stars, have similar profiles with small distinctions. Arguments that the outflow velocity in the stellar photosphere is non-zero for stars with the R CrB type variability and that it may reach 10 km s^{-1} are presented. This fact is important for the unprecedently high precision of the proper motion data of the GAIA mission because this value shall be included in the radial velocity of star as an unknown component.

1. Introduction

Stars with the R CrB type variability have a very high mass loss rate of 10^{-6} M$_\odot$ yr^{-1} (cf. Rosenbush 1996). The star pulsation sometimes induces conditions for the condensation of a screening shell in addition to the well-known permanent shell, and we observe a visual light minimum (cf. Rosenbush 2000, 2001a, 2001b). The weakening of photospheric flux gives us the unique possibility to observe the radiation of permanent shell, in particular, numerous emission lines. At the light maximum they are observed as emissions only in cores of some strong absorption lines, for instance, ScII 4246 Å (Keenan & Greenstein 1963). The goal of our monitoring of the CaII triplet and NaI doublet lines in R CrB itself and other stars with similar variability was to study the variations of these lines profiles in the quiet and active states of a star.

The unprecedented precision of proper motion data and the mass-measuring of radial velocities of stars by the GAIA mission raises an important issue connected with mass loss by stars. It is obvious that the matter is lost from any layer of stellar atmosphere including also the photosphere. Obviously the photo-

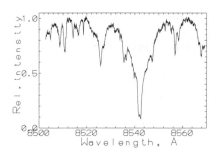 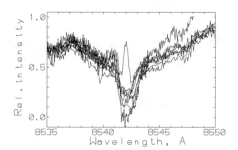

Figure 1. *Left*: the 8542 Å spectral region of the Ca II line in R CrB at the light maximum. *Right*: the range of variations of the Ca II 8542 Å line profile at the ligth maximum. The profile with the sharp emission at the minimum is shown.

spheric lines will be shifted by an amount corresponding to the outflow velocity. This velocity will be included as an unknown component in the radial velocity of star, then it transfers to the kinematic characteristics of star and can distort the kinematic picture of the Galaxy.

2. Calcium line profile and its variations

Spectra of R CrB and other stars were obtained on the coude spectrograph of the 2.6 m telescope of the Crimean Astrophysical Observatory, Ukraine (Rosenbush 2000, 2001a).

At the light maximum the Ca II triplet presents as broad lines with a possible asymmetry in the line wings (Figure 1a). All three calcium lines are identical and have a 1-5 km s^{-1} blue-shifted emission in the line core. Figure 1b demonstrates a range of the 8542 Å line profile variations at the light maximum and displays one of line profiles at the 1998 minimum. The deepest profile corresponds to the unique state of a slightly dimming brightness of R CrB in 2000 January-May. The next profile above it also is deep and corresponds to the state of the star before its 1998 minimum. The other five profiles characterize the star at the quiet state. The deeper blue side of line core is more typical for them, but all have a lesser line depth. We do not know whether this asymetry is related either to the red-shifted sharp emission or to the more intense blue-shifted absorption component, detailed calculations and further observations are needed. For the present we note that the line profiles at the light minimum presented in Figure 2a in the relative-absolute scale indicate clearly a stable emission flux in the line core. An interesting change of the line profile over one day was observed immediately before the 1998 minimum. If the first two profiles of the 8542 Å line we observed at the light maximum do not distinguish one from the other (MJD 51033 and 51038, where MJD is the modified Julian date MJD=JD−2400000), whereas the third profile (MJD 51039) obtained several days before the beginning of the 1998 minimum had a considerable difference: the central line intensity changed from 0.1 up to 0.2 and the intensity in the line wings slightly decreased. This variation does not fall beyond the observed range of profile variations, therefore this observed intensity variation in the line center just over one day was real. The FWHM of emission during the minimum was about 14 km s^{-1} and the average shift was about $-0.4(\pm 0.7)$ km s^{-1}. This emission can be clearly identified with the sharp emission observed in the line core at the maximum. The permanent asymmetry of absorption line engages our attention.

The observations of R CrB in the Na I D lines refer to the phase of the minimal brightness. The lines had a complicated shape: in addition to the photospheric line a sharp emission appeared in its core as well as a very broad blue shifted absorption. The blue shift of the sharp emission was about 10 km s^{-1}. The broad absorptions had the equivalent width of about 0.6 Å for D2 and 0.4 Å for D1; the shift corresponded to the outflow velocity of matter of about 135 km s^{-1}. It should be noted that the Ca II triplet lines did not show the corresponding circumstellar absorption. As the observations of other stars show, the calcium lines also had no interstellar components. We can say that the above-mentioned asymmetry of the absorption calcium line profile in R CrB at the quiet state is typical for HD 182040 and XX Cam, cool carbon stars with the hydrogen deficit without the R CrB type variability (Figure 2b). In other stars with the R CrB type variability, for instance, SV Sge near a light maximum, the asymmetry is less pronounced. In SU Tau at the recovery phase 0.6 magnitudes below the light maximum the line profile is nearly symmetric with a possible emission in the core. But the latter data is related just to a single observation. We add that the asymmetry of line cores also may be the consequence of a differences of stellar pulsation parameters in different layers of stellar atmosphere.

3. The radial velocity of star and the velocity of matter outflow away from its atmosphere

Contrary to the above-mentioned small blue shift of sharp emission in the core of the Ca II lines, the sharp emission in the core of the Na I lines which appeared after the weakening of the photospheric flux had a blue shift of 8-10 km s^{-1} (Rosenbush 2001a). The photospheric line itself in this star also had a blue shift of 2-4 km s^{-1} relative to the iron and carbon lines in this spectral region. The similar shifts are observed in other stars with the similar variability: -12 km s^{-1} in SV Sge, according to our observations, -10 km s^{-1} in S Aps and RZ Nor (Skuljan & Cottrell 1999), V482 Cyg (Kameswara Rao & Lammbert 1993). These authors do not exclude the circumstellar nature of this shift. In R CrB itself, which has a high galactic latitude, a weak interstellar line at -42 km s^{-1} is observed, and other stronger lines are not expected therefore the shift of photospheric lines is more like of stellar nature. We are inclined to believe that this shift in varying degrees is also characteristic of other photospheric lines. Indirect arguments in favour of different matter outflow velocity in different layers of the atmosphere are presented by RY Sgr. It is known that the amplitude of the radial velocity pulsation for low-excitation lines is higher than for high-excitation lines (Lawson et al. 1991). It is also known from the theoretical study of stellar pulsation (Fadeyev 1984) that the periods of pulsations of different layers are equal but their amplitude is greater in outer layers. The study of

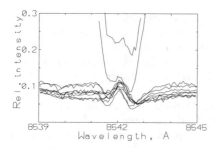 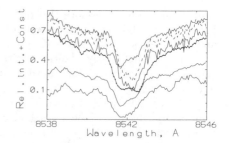

Figure 2. *Left*: the relative-absolute spectral energy distibutions in the region of the Ca II 8542 Å line (the upper line is the unique profile for MJD 51039). *Right*: the profiles of the Ca II line (from bottom to top) in SV Sge (near the light maximum, with a shift by −0.2 in intensity relative to R CrB), HD 182040 (−0.1), R CrB (two dates, 51033 solid line with dots, and 52332 solid line), SU Tau (0.6 mag below the maximum, +0.1, the dashed line), XX Cam (+0.2).

the RY Sgr pulsation during the 1967 minimum led Alexander et al. (1972) to the conclusion that the radial velocity of the star defined by emission lines is more negative approximately by 45 km s^{-1} than for absorption lines. We are inclined to believe the distinction of 10 km s^{-1} between the average velocity of the pulsation in 1970 and in 1988 (Rosenbush 1996) to be a consequence of a variation in the mass loss rate. We think that this value, 10 km s^{-1}, may be identified with the velocity of matter outflow in the photosphere. If all these velocities are aligned in order from the inner atmosphere to the outer one, we obtain the following. The matter in the photospheric layer flows with a velocity of about 10 km s^{-1}. In the layer of sharp emission origin, which according to our study (Rosenbush 2000) has a radius of about 20 stellar radii, the outflow velocity reaches 25 km s^{-1} in RY Sgr. In R CrB this value is about 10 km s^{-1}.

References

Alexander, J.B., et al. 1972, MNRAS 158, 305
Fadeiev, Iu.A. 1984, Ap&SS 100, 329
Keenan, P.C., & Greenstein, J.L. 1963, Contr.Perkins Obs., Ser.II, N. 13, 197
Lawson, W.A., Cottrell, P.L., & Clark, M. 1991, MNRAS 251, 687
Kameswara Rao, N. & Lambert, D.L. 1993, PASP 105, 574
Rosenbush, A.E. 1996, Ap. 39, 78
Rosenbush, A.E. 2000, Ap. 43, 435
Rosenbush, A.E. 2001a, Ap. 44, 78
Rosenbush, A.E. 2001b, Kinem.Phys.Cel.Bod.Suppl. 3, 399 (astro-ph/0104341)
Skuljan, Lj., & Cottrell, P.L. 1999, MNRAS 302, 341.

Spectroscopy with FLAMES: applications to GAIA RVS

Frédéric Royer

Observatoire de Genève, 51 chemin des Maillettes, CH-1290 Sauverny

Abstract. The reduction of GAIA RVS instrument data can use applicable solutions defined for other spectrographs. The Data Reduction Software of the GIRAFFE spectrograph, part of the FLAMES facility at VLT, is developed in Geneva. Some reduction recipes can be adapted to GAIA RVS in terms of localization, extraction and calibration. On the other hand, the Argus mode of the instrument is described as a way to simulate GAIA observations for crowded fields.

1. Description of FLAMES and GIRAFFE

The FLAMES (Fiber Large Array Multi-Element Spectrograph) facility is mounted on Kueyen, the second Unit Telescope of the VLT. This instrument, soon available to the community (by Spring 2003), is described by Pasquini et al. (2000) and details can be found at http://www.eso.org/instruments/flames

VLT Unit Telescope #2 (Kueyen) feeds GIRAFFE, at the Nasmyth focus, through a set of optical fibers either deployed on individual objects (MEDUSA mode, 132 fibers) or grouped in for integral field observation (ARGUS/IFU modes, 320 fibers in all). The spectrograph produces on the 2k×4k detector a set of 137/320 spectra. Two spectral resolutions with set-ups covering all the wavelength range 360 – 940 nm are available; high resolution 15 000 – 45 000 (22 set-ups) and low resolution 5000 – 13 000 (8 set-ups). The GIRAFFE spectrograph offers three different observing modes:

1. In the MEDUSA mode 132 fibers are distributed in a field of $25'$ diameter. Some of these have to be attributed to sky measurement.

2. In the IFU (Integral Field Units) mode, 15 fiber bundles are distributed in the same $25'$ field, each bundle containing 20 fibers corresponding to an area of $3'' \times 2''$ on the sky (spatial sampling of $0''\!.52$).
   In this mode the PSF on the detector is much more difficult to disentangle because it is less well sampled (the fiber size being smaller) and the spectra undergo significant cross-talk (being close-packed).

3. ARGUS, a single integral field unit containing 300 fibers covering a rectangular area of $11''\!.5 \times 7''\!.3$ (spatial sampling of $0''\!.52$) or $6''\!.6 \times 4''\!.2$ (spatial sampling of $0''\!.3$). As for the IFU mode, 15 sky fibers are available distributed anywhere around the Argus field (the microlens aperture on the sky is $0''\!.52 \times 0''\!.52$).

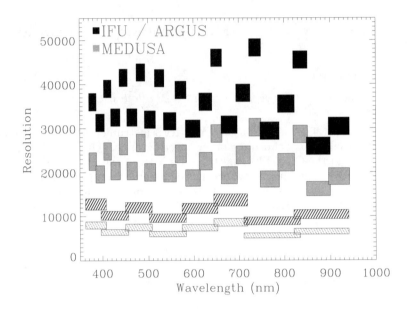

Figure 1. Every setups of GIRAFFE are displayed in the wavelength–resolution plane. Black and grey respectively stand for IFU/ARGUS and MEDUSA fibers, whereas plain and shaded boxes correspond to high and low resolution respectively.

The resolution is diplayed as a function of the wavelength in Fig. 1, for the whole set of available configurations. The better spectral resolution in the IFU/ARGUS mode is due to the better sampling on the sky (the smaller diameter of fibers).

2. Integral field observation and crowding

In the focal plane of the GAIA RVS instrument, the three dimensions α, δ (spatial coordinates) and λ (wavelength) are projected on two dimensions. An increasing crowding will rise from this projection effect. Integral field observation allows the measurement of $(\alpha, \delta, \lambda)$ as a datacube, and can be useful to simulate GAIA RVS observations from real data. The ARGUS mode of GIRAFFE covers a rectangular area, with two available spatial sampling: $0\rlap{.}''52$ or $0\rlap{.}''3$. By using the spectral window 8484 – 9001 Å (on 4096 pixels, with $R \sim 25\,900$), covering the wavelength range of GAIA, the resulting extracted spectra can be injected in a simulation to study the same region as seen by GAIA.

The spatial resolution of GAIA RVS is $1\rlap{.}''0 \times 1\rlap{.}''5$ and the spectral resolution will be lower than 20 000, so that GIRAFFE ARGUS data can degraded in both spaces to simulate GAIA RVS data.

3. Data reduction software

The Data Reduction Software (DRS) of GIRAFFE, developed at Observatoire de Genève, Institut d'astronomie de l'Université de Lausanne and Observatoire de Paris (Blecha et al. 2000) is divided in two parts:

- the *BaseLine Data Reduction Software* (BLDRS), which aims at removing the instrumental signature from the observed data and calibrating the spectra in physical units.

- and the *Ancillary Data Analysis Software* (ADAS), dedicated to the analysis of the output spectra of the BLDRS.

3.1. Data reduction recipes

The BaseLine Data Reduction software does, among other things, the localization of the spectra on the detector and performs the extraction.

Localization The localization process carried out using the flat-field calibration frames is performed in two steps. First a localization mask is derived from scratch using threshold detection. And then the PSF perpendicular to the dispersion if fitted inside the localization lanes, along the centroids.

Extraction This function applies to any preprocessed science or calibration frame. The extraction of the spectra is carried out inside the localization lanes, and three options are offered: the first and the second ones can be applied in the MEDUSA mode because contamination should be negligible in this mode, while the third one should be adopted for the ARGUS/IFU. In this latter option, each spectral bin is described as a linear combination of the PSF multiplying the spectral elements which are to be extracted. This is Horne (1986) optimal extraction generalized to the case where the spectra are not strictly independent due to the cross-talk, and it is well adapted to any situation (including the case where the inter-spectra contamination is severe).

These localization and extraction processes have to be tested on real IFU/ARGUS data to see at which contamination level they are reliable, and then see if they are applicable to GAIA RVS.

4. Radial velocities

Computing the radial velocities of the observed targets is part of the ADAS package. The method and the CORAVEL templates used are described in Royer et al. (2002).

The accuracy of the measured radial velocity has been assessed in the GIRAFFE spectral range limited to the wavelength interval 8480 – 8740 Å, i.e. the GAIA RVS bandpass.

The correlation was done with classical binary masks of the CORAVEL type, optimized for each set-up, taking into account the nominal resolution of the GIRAFFE spectrum ($R \sim 16\,300$). The Solar Flux Atlas of Kurucz et al.

(1984) has been used to simulate observed data, with different signal-to-noise ratio, (SNR) and to create the mask. Thus there is no mismatch between the masks and the spectra.

Four different binary masks have been created, with different tolerances on the width of spectral lines to be retained. For *Mask1*, the tolerance is the lowest, there are 24 holes between 8480 and 8740 Å. In the other one, *Mask2*, there are 33 holes. And for both tolerances, a mask if built with and without the calcium triplet lines ('+Ca') among the holes. The results in the accuracy of the radial velocities are listed in Table 1, for the Sun.

Table 1. The expected accuracy in radial velocity (in km s^{-1}), for the solar spectrum, as a function of S/N and the templates.

Template	signal-to-noise ratio							
	1	2	5	10	20	40	100	200
Mask1	—	10.544	2.088	0.974	0.432	0.196	0.091	0.043
Mask2	31.734	6.997	2.429	1.134	0.517	0.271	0.098	0.052
Mask1+Ca	9.651	3.884	1.494	0.826	0.354	0.157	0.074	0.034
Mask2+Ca	12.034	5.289	1.823	0.914	0.402	0.210	0.078	0.041

One can notice that taking into account wider lines in the correlation improves the resulting accuracy for low signal-to-noise data, but spoils it in the case of high signal-to-noise. In the case of the Sun, using the calcium triplet for the radial velocities of low SNR increases significantly the precision of the radial velocity, but for high SNR the gain factor is much lower. The optimzation of the templates should be investigated for other spectral types.

These results were computed from simulated spectra, but GIRAFFE will soon start producing data, which will be used to carry out further tests.

References

Blecha, A., Cayatte, V., North, P. et al. 2000, in Optical and Infrared Telescope Instrumentation and Detectors, SPIE 4008, 467

Horne, K. 1986, PASP 98, 609

Kurucz, R. L., Furenlid, I. & Brault, J. 1984, National Solar Observatory Atlas Sunspot, National Solar Observatory, New Mexico

Pasquini, L., Avila, G., Allaert, E., Ballester, P., Biereichel, P., Buzzoni, B., Cavadore, C. et al. 2000, in Optical and Infrared Telescope Instrumentation and Detectors, SPIE 4008, 129

Royer, F., Blecha, A., North, P., Simond, G., Baratchart, S., Cayatte, V., Chemin, L. & Palsa, R. 2002, in Astronomical Data Analysis II, SPIE 4847, in press

An analysis of the precision of stellar radial velocities obtained using the HERCULES spectrograph

Jovan Skuljan, John B. Hearnshaw & Stuart I. Barnes

Department of Physics and Astronomy, University of Canterbury, Private Bag 4800, Christchurch 8020, New Zealand

Abstract. High-precision stellar radial velocities have been measured at Mt John University Observatory using the new HERCULES spectrograph. An analysis has been made to examine the overall stability and final precision of radial velocities based on observations obtained during the past several months. The relative velocities are measured by cross-correlation between one-dimensional spectra in logarithmic wavelength space. Some delicate reduction steps are performed in order to achieve the highest precision possible. A typical precision of a few metres per second is easily obtained during one observing night. Stellar velocities are somewhat affected by insufficient scrambling by the optical fibre. Some minor modifications to the spectrograph and the reduction procedure are currently under consideration and they are expected to bring the overall precision of radial velocities to a level of several metres per second.

1. Introduction

A new high-resolution spectrograph called HERCULES has been installed at Mt John University Observatory, to be used on the 1 m telescope (Hearnshaw et al. 2002, in prep., hereafter Hetal2002). A 1024 × 1024 SITe CCD camera is currently used as a detector in four different spectral regions.

Over the past twelve months a large number of stellar spectra has been collected and a dedicated reduction software package has been developed. Some preliminary analysis of the results has been made in order to examine the overall stability and final precision of stellar velocities obtained with this spectrograph.

2. Spectrograph description

The High Efficiency and Resolution Canterbury University Large Echelle Spectrograph (HERCULES) is a fibre-fed Echelle spectrograph that was designed and built at the University of Canterbury and has been in operation at Mt John University Observatory since April 2001 (Hetal2002). HERCULES receives light from the f/13.5 Cassegrain focus of the 1 m McLellan telescope. Resolving powers of $R = 41\,000$, $70\,000$ and $80\,000$ are available, depending on the optical fibre used (100-μm with no slit, 100-μm with a 50-μm slit and 50-μm with no slit). An R2 200 × 400-mm Echelle grating provides dispersion, while a large BK7 prism is used for cross-dispersion in double pass. The spectrograph has no moving parts except for the positioning and focusing of the CCD. The wavelength coverage is designed to be 380–880 nm in a single exposure on a 50 × 50 mm CCD chip. The maximum efficiency of the whole system (fibre, spectrograph and detector) is about 18% in 2 arc second seeing. The whole instrument is installed in a large vacuum tank at 2–4 torr. The tank is in a thermally isolated and insulated environment.

Some basic characteristics of the spectrograph are presented in Table 1.

Table 1. HERCULES specifications

parameter	value
Echelle grating size	204 × 408 mm
Groove spacing	31.6 per mm
Collimator	$D_{coll} = 210$ mm, $f_{coll} = 783$ mm (parabolic)
Cross-disperser	BK7 prism
Prism apex angle	$\alpha = 49.5°$
Prism triangular face	$a = 258$ mm, $h = 276$ mm
Prism width	$b = 255$ mm
Camera type	folded Schmidt
Camera primary mirror	$D_{cam} = 500$ mm, $f_{cam} = 973$ mm (spherical)

3. Reduction procedure

A dedicated computer program called HRSP (Hercules Reduction Software Package) has been developed in order to achieve the maximum efficiency in the reduction of HERCULES CCD images. The program has been written in C and has been optimized for HERCULES Echelle spectra.

A standard Echelle reduction procedure is used, including: the background and cosmic ray subtraction, order extraction, flat-fielding, continuum normalization and wavelength calibration. Relative radial velocities are obtained by cross-correlation between two spectra of the same star. Every order is cross-correlated separately and the arithmetic mean velocity is calculated. Spectra are prepared for cross-correlation by first subtracting the mean flux and then by applying a cosine-bell window to the edges (Simkin 1974).

4. Blue sky velocities

Blue sky spectra are used for monitoring the stability of radial velocity determinations. There are many advantages of using the blue sky as a calibration source and these include the fact that the sky is uniformly bright, which eliminates any possible problems caused by poor fibre scrambling. Also, the radial velocity of the Sun is known to high accuracy.

A typical short-term precision of HERCULES sky spectra is about 3–4 m s^{-1}, which is close to the photon-noise limit. This has enabled us to detect

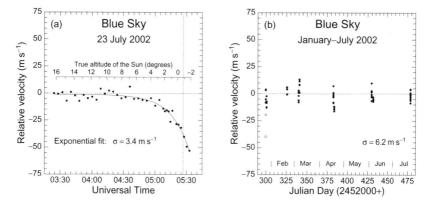

Figure 1. Left: radial velocities of the blue sky and the effect of differential atmospheric extinction. A vertical dashed line at 5:27 UT is used to mark the sunset. Right: long-term radial velocities of the blue sky. Open circles represent the observations with residuals larger than $2.5\,\sigma$.

some subtle effects, such as the slowly increasing blue shift towards the sunset (Figure 1a). This is caused by the differential extinction in the Earth's atmosphere, combined with the Sun's rotation. The upper solar limb appears slightly brighter than the lower one and this produces an apparent shift in the average radial velocity of as much as $40-50$ m s^{-1}, when the Sun is on the horizon.

Our long-term precision is somewhat lower (~ 6 m s^{-1}), as seen in Figure 1b. One of the reasons is that some of the observations were made close to the sunset, so that they were affected by the differential extinction effect described above. It is essential that the observations of the blue sky are performed at least one hour before the sunset in order to eliminate this effect.

5.   Stellar velocities

The precision of stellar velocities is somewhat lower when compared to the blue sky. The measurements are affected by insufficient scrambling of the optical fibre. As a result, the radial velocity depends on the actual position of the stellar image (the seeing) with respect to the optical fibre. An experiment has been made in which a bright star (α Cen) was observed using very short exposure times (of a few seconds), while the sidereal rate of the telescope was made slightly worse, so that the star drifted very slowly across the fibre. No guiding or repositioning of the telescope was done while a series of exposures was taken. The result of this experiment is shown in Figure 2a. The radial velocities show relative differences of as much as 70 m s^{-1} when the star moves from one side of the optical fibre to the other.

The effect of poor scrambling becomes especially prominent at very short exposure times, when an instantaneous stellar image is captured on the fibre. The precision of radial-velocities of very bright stars is only about $10-15$ m s^{-1}. The results for fainter stars are somewhat better, due to longer exposures. When a

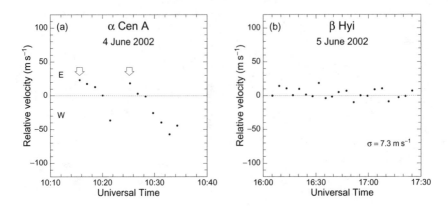

Figure 2. *Left*: the effect of poor scrambling by the optical fibre. The star drifts across the fibre while a sequence of observations is made. The telescope is repositioned only at the beginning of each sequence (as indicated by arrows). *Right*: radial velocity of β Hyi over a short period of time. The star has been re-centred on the fibre before each exposure. Exposure time: 30–40 s.

star is exposed for a few minutes, both the atmospheric scintillation and frequent repositioning of the telescope during the guiding process produce an averaging effect, so that all parts of the optical fibre entrance get equally illuminated. A precision of $5-10$ m s^{-1} is obtained for moderately bright stars, as shown in Figure 2b. Finally, a typical precision for a sixth-magnitude star (exposure times of several minutes) is about 5 m s^{-1}.

6. Conclusion

Observations collected so far demonstrate that the radial velocity precision of HERCULES spectra is somewhat limited by the scrambling properties of the optical fibre. Blue sky spectra give a short-term precision of $3-4$ m s^{-1}, which is close to the photon-noise limit. The long-term precision can be brought to the same level by careful reduction and by avoiding observing the sky while the Sun is low. A similar high precision in stellar radial velocities can be achieved by improving the telescope positioning and guiding and, possibly, by introducing a double scrambler.

References

Simkin, S.M. 1974, A&A 31, 129

Line indices as age, metallicity and abundance ratio indicators

Rosaria Tantalo

Department of Astronomy, University of Padova, Vicolo dell'Osservatorio 2, 35122 Padova, Italy

Cesare Chiosi

INAF, Viale del Parco Mellini 84, 00136 Roma, Italy

Abstract. We investigate the ability of the line strength indices to assess the metallicity, age and abundance ratios [α/Fe] of elliptical galaxies. The analysis is made adopting new sets of isochrones with solar-scaled and α-enhanced mixtures.

1. Introduction

The integrated line strength indices (Faber et al. 1977) of single stellar populations (SSP) and/or large assemblies of stars, are known to depend on the metallicity, age, and the degree of enhancement in α-elements of their stellar content.

The indices of individual stars are calculated with the aid of the *fitting functions* (Worthey, Faber & González 1992; Worthey & Ottaviani 1997) which correlate the intensity of an index to T_{eff}, $\log g$ and [Fe/H] of the star.

Worthey et al. (1994) analyzed the sensitivity of each index to the age and metallicity of SSPs, and found that H_β, H_γ etc. are more sensitive to the age, whereas Mg_2, Mg_b, $\langle Fe \rangle$ and others are more sensitive to the metallicity. Therefore, those indices seem to have the potential of partially resolving the *age-metallicity degeneracy*.

Over the years, an extensive use of the two indices diagnostics has been made to infer the age and the metallicity of early type galaxies (Kuntschner & Davies 1998, Tantalo et al. 1998, Trager et al. 2000) neglecting the dependence on the degree of enhancement in α-elements. Only recently, the effect of this begins to be considered.

2. SSPs index with enhancement

In presence of enhancement in α-elements one has to modify the relationship between the total metallicity Z and the iron content [Fe/H] by suitably defining the enhancement factor Γ (cf. Tantalo et al. 1998; Tantalo & Chiosi 2002, in prep.). Splitting the metallicity Z in the sum of two terms:

$$Z = \sum_j X_j + X_{Fe}$$

we may define the so-called *enhancement degree* Γ in the following way:

$$[\text{Fe/H}] = [\text{Z/H}] - \Gamma$$

which is equivalent to $A[\text{E/Fe}]$ of (Trager et al. 2000, hereafter TFWG20) where A is a suitable factor which depends on the pattern of abundances that are enhanced with respect to the solar partition.

Assumed a certain degree of enhancement Γ and a total metallicity Z, the abundances of all elements but [Fe/H] can be arbitrarily varied provided their sum is equal to Γ. The results will of course depend on the exact way this partition is made. In the following we will examinate results obtained for plausible Γs and partitions.

3. The Tripicco & Bell calibration

Assumed a partition of enhanced abundances, the indices of the SSPs are first obtained using the Worthey's *fitting functions* at zero enhancement and then re-scaled to the particular set of abundances. To this aim we adopt the method developed by TFWG20 which stems from the study of Tripicco & Bell (1995, hereafter TB95) who first introduced the concept of *response functions*. These are obtained by changing one at a time the abundance of chemical elements in synthetic spectra and calculating the corresponding variations in the indices. The correction for any specific index is given by:

$$\delta I = \frac{\Delta I}{I_0} = \left\{ \prod_i [1 + R_{0.3}(X_i)]^{([X_i/H]/0.3)} \right\} - 1$$

where $R_{0.3}(X_i)$ is the response function for element i at $[X_i/H]=+0.3$ dex. This is given by $(\delta R_i \star \sigma_0)$ where δR_i is the value for each element in unit of standard errors σ_0 as tabulated in TB95.

4. Description of different cases

4.1. Case A: $C^0 O^+$ model by TFWG20

To compare our results with those of TFWG20 we have derived the indices for solar-scaled stellar models (isochrones) and different degree of enhancement (to a first approximation the effect of enhancement on stellar models is neglected). To this aim we adopt the isochrones by Bertelli et al. (1994) and the pattern of enhanced abundances as in the model $C^0 O^+$ by TFWG20. The following groups of elements are considered: (a) *enhanced elements*[1]: N, Mg, Na, Si, Ti

---

[1] These are scaled up by the same factor

included in the E notation as in TFWG20, (b) *depressed elements*[2] (i.e. Fe-peak Group): Fe, Ca, Cr included in the D notation, and (c) *fixed elements*, which means solar-scaled.

This allow us to test the effect of a simple enhancement scheme on standard stellar models/isochrones.

4.2. Case B: Salasnich et al. 2000 mixture

The new stellar models by Salasnich et al. (2000, hereafter S20) are used to calculate isochrones and three sets of SSPs indices adopting the TB95 calibration for three different degree of enhancement $\Gamma=0.00$ dex, $\Gamma=0.3557$ dex and $\Gamma=0.50$ dex. This allow us to combine the effect of enhancement both on the stellar models/isochrones and on the indices in a self consistent way.

4.3. Case C: Bertelli et al. 1994 and S20-mixture

Finally, we calculate the SSPs indices using the old isochrones by Bertelli et al. (1994), and the mixture of chemical abundances as in S20, namely solar, $\Gamma=0.3557$ dex and $\Gamma=0.50$ dex. This allow us to test the effect of a complex enhancement scheme on standard stellar models/isochrones.

5. Results from the minimum-distance method

We apply the above SSPs indices to the *Pristine IDS* sample by S.C. Trager (in his 1997 Ph.D thesis) using the minimum-distance method described by TFWG20. In our analysis we consider six different indices (H_β, Mg_2, Mg_b, $\langle Fe \rangle$, NaD and $C_2\,4668$ Å) and all possible combinations in groups of three. A given group of indices is considered to provide good results if the observational and theoretical ones coincide within an uncertainty of 10%. We find that for each of Case A, B, and C above there is a number of groups (varying from case to case) which meet the requirement. For the Salasnich mixture (Case B) we find four combinations of indices that are useful to derive age, metallicity and Γ for each galaxies in the sample. The results are shown in left panel of Figure 1. It is soon evident that a large spread exist among the determination of the same parameter provided by different index-triplet.

In order to understand the reason for this, we isolate the good index-triplet common to all Cases A, B and C and use this to derive age, metallicity and Γ.

Limited to the age parameter, the results are shown in the right panel of Figure 1 where we compare the ages obtained for each galaxy according to Case A, B, and C. It can be see that the major source of disagreement resides in the enhancement factor and the way it is taken into account (top-right and bottom-left).

Finally a last remark is worth noticing: a comparable large scatter in the age (and other parameters as well) would result by changing the index-triplet in use, among the combination with the same quality.

---

[2] These are scaled down by the same factor

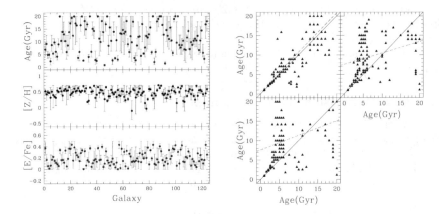

Figure 1. *Left*: age, metallicity and enhancement factor obtained from Case B using different index-triplets (namely H_β-Mg_b-Mg_2, H_β-Mg_b-$C_2\,4668$ Å, Mg_b-Mg_2-$C_2\,4668$ Å and Mg_b-NaD-$C_2\,4668$ Å). The dots show the mean values for each galaxy whereas the errorbars indicate the maximum and minimum estimate. *Right*: comparison between the ages obtained from different sets of isochrones and different enhancement patterns adopting the same index-triplet (H_β-Mg_b-$C_2\,4668$ Å). *Top-left*: relation between different isochrones (Case A vs. C). The one to one relation and the linear best fit of the results are indicated by solid and dotted lines, respectively. *Top-right*: Case C vs. B. *Bottom-left*: Case A vs. B.

As main conclusion of this analysis, the many uncertainties still affecting index calibration, real pattern of abundances in the so-called α-enhanced mixture, and the apparent equal sensitivity of different indices to age, metallicity and Γ, does not allow us to safely use the index diagnostics to infer these important parameters for a stellar population.

References

Bertelli, G., Bressan, A., Chiosi, C. et al. 1994, A&AS 106, 275
Faber, S.M., Burstein, D. & Dressler, A. 1977, AJ 82, 941
Kuntschner, H. & Davies, R.L. 1998, MNRAS 295, L29
Salasnich, B., Girardi, L., Weiss, A. & Chiosi, C. 2000, A&A 361, 1023
Tantalo, R., Chiosi, C. & Bressan, A. 1998, A&A 333, 419
Trager, S.C., Faber, S.M., Worthey, G. & González, J.J. 2000, AJ 119, 1645
Tripicco, M.J. & Bell, R.A. 1995, AJ 110, 3035
Worthey, G. & Ottaviani, D. 1997, ApJS 111, 377
Worthey, G., Faber, S.M. & Gonzalez, J.J. 1992, ApJ 398, 69
Worthey, G., Faber, S.M., González, J.J. & Burstein, D. 1994, ApJS 94, 687

Comments on atomic data for the GAIA spectral region

Glenn M. Wahlgren & Sveneric Johansson

Atomic Astrophysics, Lund Observatory
Box 43, SE-22100, Lund, Sweden

Abstract. Maximizing the scientific return from spectra obtained from the GAIA radial velocity spectrometer will require an intimate knowledge of the spectral lines that may be observable. Of particular concern is line blending, which may influence both the derived radial velocities and elemental abundances at the proposed spectral resolutions. We have investigated the available data for spectral lines in the proposed GAIA wavelength interval and provide comments on its nature.

1. Introduction

We have investigated the status of atomic line data useful for interpreting spectra to be obtained by the GAIA radial velocity spectrometer (RVS). While the spectral resolution of the RVS is yet to be finalized, the working hypothesis here is that the resolving power may be in the range $R = \lambda/\Delta\lambda = 10\,000$ to $20\,000$.

The atomic line data considered here are restricted to wavelengths, energy levels, oscillator strengths and line structure arising from hyperfine, isotopic and Zeeman broadening mechanisms. Atomic data are available from a variety of sources. However, rather than review all of these sources we have chosen to consider only one, the VALD database (Vienna Atomic Line Database, http://www.astro.unvie.ac.at/~vald). VALD includes data from many sources, both experimental and theoretical, and has become a useful resource to astronomers attempting to locate atomic line data. Among the contributors to the VALD database, the largest by number of lines is the line lists of Kurucz (1993). Currently, the VALD database does not include all relevant published atomic data from the past several years.

2. The presence of an element

The GAIA spectral interval, 8480 – 8740 Å, implies that a spectral transition possesses a wavenumber in the interval 11 800 – 11 400 cm^{-1}. Even if the magnitude of these wavenumbers is relatively small, corresponding to a transition energy of approximately 1.4 eV, the lines in this wavelength regime can involve transitions from the ground state or reflect transitions between rather highly-excited states. Due to the atomic structure, some alkali atoms and especially rare earth elements have low excitation transitions in the near-infrared region. For medium excitation (2-4 eV) most lines come from the iron-group elements, where the extended parent structure makes the low configurations span a large

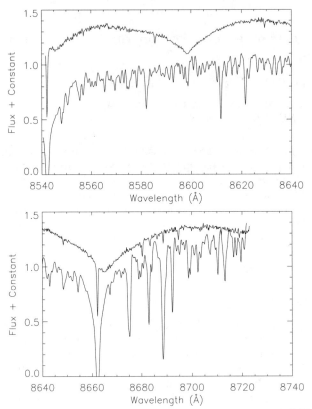

Figure 1. A comparison of spectra for the proposed GAIA wavelength region. In both panels the upper spectrum is of the warm star HR 3383 (A1 Vm) and the lower spectrum is of the cool, luminous star α Ori (M2 Iab). Atomic and molecular line opacity is a prominent feature for cool stars at these wavelengths.

energy range and even overlap. This is especially true for elements with many valence electrons, such as iron. High-excitation lines of iron-group elements (5-6 eV in first spectra and 10-12 eV in second spectra) fall in this wavelength range, as well as high-excitation (8-12 eV) lines in spectra of light p-shell elements like C I, N I, O I and rare gas atoms, all of them having their resonance lines in the far UV region. In this sense, essentially all elements are capable of transitions in the 8480 – 8740 Å region, and their detected presence will depend upon the excitation potential, the oscillator strength and the elemental abundance. The stellar effective temperature also plays a significant role through the dominant ionization state. Figure 1 compares spectra from cool and warm stars, where the dominant ionization state is either neutral or singly-ionized, respectively, for iron-group elements.

Although VALD does contain lines from the lowest four atomic/ionic states, it is dominated by lines from neutral elements in the GAIA spectral interval.

The current state of available atomic line data can be briefly summarized by the following regarding the presence of the various elements.

- *Light elements* (Z ≤ 20): for only C, N, O, F, Ar, and Ca are lines present from the neutral (first spectrum), singly- (second spectrum), and doubly-ionized (third spectrum) states. Neutral element lines are present for all other elements except Be (no spectral lines) and Al (second and third spectra).

- *Fe-group elements* (21 ≤ Z ≤ 29): all elements with the exception of Sc and Cu present lines from the three lowest spectra. Both Sc and Cu are not represented by third spectrum lines.

- *Heavy elements* (Z ≥ 30): of the 30 heavy elements with line data in this region, 17 are represented by only the neutral state, with the remainder having either only lines from the second spectrum (6 elements) or both of the first two spectra (7 elements). No element has lines from the third spectrum.

3. Oscillator strengths

Oscillator strengths, of f-values, for spectral lines in the GAIA wavelength interval possess a wide range of uncertainty. For the GAIA interval, of the 450 transitions in the NIST (http://physics.nist.gov/PhysRefData/contents.html) critically compiled database, approximately 30% have experimentally determined transition probabilities, to a highly variable level of accuracy. Experimentally, the transition probability, A, is determined from a combination of measured energy level lifetime, τ, and branching fractions, BF, using the relation $A = BF/\tau$. The f-value is then determined analytically from its relationship to the transition probability. Both τ and BF are best determined experimentally for neutral and singly-ionized species, and are usually limited in excitation energy by experimental realities.

The majority of f-value data originates from databases for theoretically calculated data, the most extensive of which may arguably be the semi-empirically calculated values of Kurucz (1993). In his database experimental data replace the calculated data when available. The accuracy of the individual f-values is impossible to estimate. More sophisticated calculations have been performed on specific lines or problems, showing good agreement with experimental data available. Such calculated data are included in the evaluated compilations from NIST, which also contain error estimates for all lines and are mainly based on experimental data. Most experimental f-value data concern low-excitation lines, but efforts to reach higher excitation lines in iron-group elements are made in the FERRUM project (Johansson et al. 2002). This project shows that the theoretical calculations utilizing the orthogonal operator technique (Raassen & Uylings 1998) provide the best theoretical f-values for iron-group elements.

The general problem with the reliability of calculated f-values for complex spectra is the confusion of energy level assignments due to mixing and configuration interaction among levels at high excitation. This leads to an increase in uncertainty. GAIA spectroscopy involves many atomic transitions from such high-excitation states.

4. Line structure

Spectral line structure is known to influence analyses for stellar properties such as elemental abundances and magnetic field strengths through the atomic properties hyperfine structure, isotopic shift and Zeeman broadening. These line structure mechanisms have been characterized for many atomic species, and are actually determined for individual energy levels. The detectability of direct line structure effects in GAIA spectra is small due to the rather low spectral resolution and potentially great line blending for cooler stars. However, its effects can be manifested as systematic abundance errors. Line structure is generally not included in the VALD or Kurucz line lists and must be sought out from the published literature. Compilations of references for line structure data are available from NIST. A listing of references for the first three spectra of lanthanide elements has been produced by Wahlgren (2002). A website of atomic data is also available from D. Verner (http://www.pa.uky.edu/~verner/atom.html).

5. Concluding remarks

Many elements are represented by spectral lines in the proposed GAIA spectral region of 8480 – 8740 Å. However, the atomic line data are incomplete and of variable quality. Many species are represented by few and/or weak lines that may give the mistaken impression that adequate data are available. As mostly neutral atoms are represented in the current spectral line databases it would appear that the atomic data situation is best for the coolest stars, and less so for stars for which ions are the dominant species. Among the heavier elements (rare earths and heavier) the uncertainties in f-values are particularly large. Experimental f-values in this region are particularly lacking.

The ominous tone of these comments can be lifted by noting that the current situation is correctable. Modern experimental and theoretical techniques can be applied to supply the necessary atomic data. As the present timeline for the GAIA mission places its launch ten years into the future there is ample time for obtaining required atomic data. However, timing is still a critical issue since there are relatively few researchers equipped to provide such data and the time necessary for conducting accurate laboratory analyses can be months to years. Prioritizations as to which species are most in need of data should be made as soon as possible if data are required for pre-launch planning exercises.

References

Johansson, S. et al. 2002, Physica Scripta 100, 71
Kurucz, R.L. 1993, Kurucz's CD-ROM N. 23, CfA Harvard, Cambridge
Raassen, A.J.J., & Uylings, P.H.M. 1998, A&A 340, 300
Wahlgren, G.M. 2002, Physica Scripta 100, 22

GAIA Spectroscopy, Science and Technology
ASP Conference Series, Vol. 298, 2003
U. Munari ed.

Derivation of stellar parameters from DIVA spectral data

Philip G. Willemsen & Torsten A. Kaempf

Sternwarte der Universität Bonn, Auf dem Hügel 71, D-53121 Bonn

Abstract. The DIVA satellite will obtain spectral data for about a third of the 40 million stars seen in the survey. The overlapping spectral orders in the cross-scan direction will provide a DISPI (DISspersed Position coded Intensity). The information on all stellar parameters is contained in the DISPI so that deconvolution is not neccessary. Several methods have been tested to derive $T_{\rm eff}$, $\log g$ and [M/H] from the DISPI. This poster focuses on analyses with artificial neural networks. The relatively modest spectral resolution of the DISPIs allows to discriminate well in temperature, moderately well in gravity and crudely in metallicity. By including Balmer jump information from the DIVA UV telescope the classifications for gravity and temperature can be improved.

1. Introduction

The *Deutsches Instrument für Vielkanalphotometrie und Astrometrie*, or DIVA, is like its predecessor Hipparcos and its follow-up GAIA an astrometric satellite. During its scheduled two year operation time it will scan the sky some 60 times, allowing measurements of position, parallax and proper motion. The survey will be complete down to $V \simeq 15.5$ mag, covering 40 million stars. In addition a dispersion grating will give spectral data for 13 million stars down to $V \simeq 13.5$ mag. A secondary UV instrument, equipped with two filters long- and shortward of the Balmer jump, will measure fluxes for roughly the same amount of objects. Launch of the satellite is scheduled for 2005-2006, while the final catalogue will be available four years later.

The studies presented in this poster focused on the spectrophotometric (SC) data. Using information on the transmission of the grating's diffraction orders and the CCD's quantum efficiency, a program by Scholz (1998) at the AIP in Potsdam simulated the dispersed images. Input spectra were taken from the BaSeL library (Westera et al. 2002). An example of a spectrum and its SC counterpart can be seen in Figure 1. The diffraction orders (at resolutions 12, 6, and 4 nm per eff. pixel for the first, second and third orders, respectively) are overlapping and thus we do not get pure spectra. In order to avoid confusion, we chose to call these one-dimensional position related intensity functions DISPIs. Nevertheless, it contains the full information of a spectrum with similar resolution, albeit in another form. Our goal was to find the best analysis method to derive the stellar parameters $T_{\rm eff}$, $\log g$ and [M/H] from these DISPIs.

Before starting to analyze DISPIs, the following steps to simulate DIVA data had to be done. First, the simulation software calculates noise-free two-

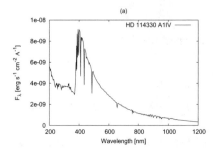 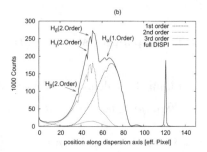

Figure 1. A spectrum of HD 114330, an A1IV star, scaled to $V = 10$ mag (*left*) and its corresponding noise-free DISPI (*right*). Some of the Balmer line features in the DISPI were marked with the order they originate from.

dimensional dispersed images from an input spectrum. Then noise is added and a scaling applied, i.e. to select a magnitude. A sky level is subtracted from each data point to give the final dispersed image. In order to obtain a one-dimensional DISPI, a rectangular aperture is used to sum the TDI-rows in dispersion direction. The width of this aperture used for extraction was chosen to give the best S/N. Also, some studies were made to find the minimal length for the DISPI, containing all essential information (Willemsen et al. 2001). This restriction was made to not surpass the download bandwidth capacity. Finally, the DISPIs are area-normalized, since we wanted to classify them only from their shape.

All 8300 spectra of the latest BaSeL library release were transformed. Each DISPI was then also scaled to different magnitudes. Our available parameter space therefore was roughly $2\,000 \leq T_{\text{eff}} \leq 50\,000$, $-1.02 \leq \log g \leq 5.5$, $-5.0 \leq [\text{M/H}] \leq +1.0$, and $8 \leq V \leq 12$ mag.

Concerning the UV telescope, a program was written, that calculates the fluxes in both filters. This is done by first applying idealized rectangular transmission curves to the spectra. Next, for the foreseen exposure time of 22.6 seconds, photon counts were obtained and noise was applied (the same CCDs as in the main instrument are used).

For the simulation of end-of-mission measurements, the S/N was predicted for one hundred stacked single exposures. A relation was found that gives the magnitude of a single frame with the same S/N as the stacked frame. This is important in order to understand how precise the classification of the final (stacked) mission data will be.

2. Analyses with artificial neural networks (ANNs)

Feedforward neural networks have proven useful in a number of scientific disciplines for interpolating multidimensional data, and thus providing a nonlinear mapping between an input domain (DISPI + UV) and an output domain (stellar parameters). The software used in this work was provided by C. Bailer-Jones.

Before any classification can be done, a network has to be trained, meaning the tuning of internal weights, that map the DISPI plus UV data to the desired stellar parameters. This is done by giving the network a large sample of both input and output data, called training set. From these the weights are determined through several thousand iterations, until the errors of calculated and given output are minimized. Afterwards, the network can be used to classify the remaining N DISPIs. For this work, we randomly chose half of our data as training set, the other half was the validation set. The latter was used to calculate the error of classification via

$$A = \frac{1}{N} \cdot \sum_{p=1}^{N} |C(p) - T(p)| \quad (1)$$

where p denotes the pth pattern and T is the target (or "true") value for this parameter. The quantity $C(p)$ is the average classification output of several networks (different starting values for the weights can give different results).

We devided the DISPI sample into several smaller ones with different temperature ranges. We chose seven different ranges, with a broad distinction between low- (L_1: $2\,000 \leq T_{\text{eff}} < 4\,000$ K; L_2: $4\,000 \leq T_{\text{eff}} < 6\,000$ K; L_3: $6\,000 \leq T_{\text{eff}} < 8\,000$ K), mid- (M_1: $8\,000 \leq T_{\text{eff}} < 10\,000$ K; M_2: $10\,000 \leq T_{\text{eff}} < 12\,000$ K) and high temperatures (H_1: $12\,000 \leq T_{\text{eff}} < 20\,000$ K; H_2: $20\,000 \leq T_{\text{eff}} < 50\,000$ K).

We discovered that the separation into such small temperature regions often yielded better results. This is understandable as the classification of $\log g$ and [M/H] depends upon the presence of spectral features in a DISPI which is also closely related to the temperature of a star. However, it should be noted that the chosen temperature ranges are not based on the intrinsic physical properties of a star. The chosen distinction was mainly motivated to see what can be learned from DISPIs in different temperature regimes in principle. The broad range for the mid-temperature samples (*M*-samples) was especially chosen to account for the special situation when the Balmer jump and the H-lines change their meaning as indicators for temperature and surface gravity. Under real conditions one would have to employ a broad classifier to separate DISPIs into smaller (overlapping) temperature ranges. This should be possible by using colour information from the star's corresponding undispersed image (Evans 2001).

3. Results and conclusions

Figure 2 shows the classification accuracy of the stellar parameters that can be obtained with our ANNs as a function of DISPI brightness. Overall, T_{eff} can be retrieved to very acceptable accuracy, which is understandable as temperature determines the basic shape of spectral energy distributions. Classification of $\log g$ is less precise, as it is a weaker signal compared to temperature. Also, in the L_2 sample the errors are slightly larger than in the others (~ 0.5 dex). This is because in the intermediate temperature range neither molecular lines nor the Balmer jump provide more accurate gravity information. [M/H] is the most difficult parameter to extract. This was to be expected, since such low spectral resolution strongly dilutes details of spectral line information, and thus of metal abundances, except perhaps in structures such as the G-band, etc.

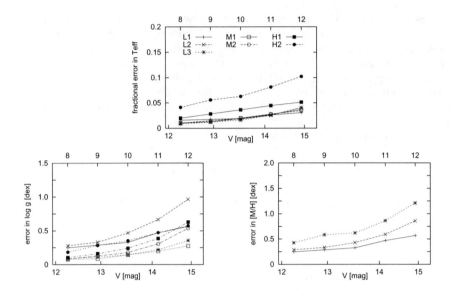

Figure 2. The classification error (Eq.1) for T_{eff}, $\log g$ and [M/H] as a function of V magnitude for all temperature sets. The lower scale refers to end-of-mission data in which 100 DISPIs of objects with the magnitude given on the upper scale are stacked. For [M/H] only the results for the temperature ranges L_1 to L_3 are presented, as for higher temperatures the metallicity information is scanty in DISPIs.

Compared to first analyses without extra information by the UV ratio, the results are equal or better. Especially the T_{eff} classification improved considerably and $\log g$ can be found more precisely for low temperature DISPIs.

The DISPI concept shows that astrophysical parameters can be derived in one go; there is no need for a multiple filter system.

A full paper on our results is in preparation for submission to A&A.

References

Evans, D. W. 2001, DIVA-TD0233-01, DIVA tech. notes archive[1]
Scholz, R.-D. 1998, DIVA-TD0196-01, DIVA tech. notes archive[1]
Westera, P., Lejeune, T., Buser, R., Cuisinier, F., & Bruzual, G. 2002, A&A, 381, 524
Willemsen, P. G., Kaempf, T. A., Bailer-Jones, C. A. L., & de Boer, K. S. 2001, DIVA-TD0271-01, DIVA tech. notes archive[1]

---

[1] http://www.ari.uni-heidelberg.de/diva/diva.html

Crowding in the GAIA spectrograph focal plane

Tomaž Zwitter

University of Ljubljana, Dept. of Physics, Jadranska 19, 1000 Ljubljana, Slovenia

Arne A. Henden

Universities Space Research Association/U.S. Naval Observatory, Flagstaff AZ 86002-1149, USA

Abstract. Superpositions of stellar tracings are present in every slitless spectrograph. The probability for such overlaps in the GAIA RVS spectrograph focal plane is estimated using photometric observations of 66 stellar fields, mostly close to the Galactic plane. It is shown that overlaps of bright stars ($V < 17$ mag) are common near the Galactic plane, and no spectrum is free from superpositions of faint star tracings. Most overlappers are of spectral type K.

1. Introduction

GAIA RVS is a slitless spectrograph, so some degree of crowding due to spectral tracing overlaps in the focal plane is to be expected. Overlapping spectra increase the effective background signal and make it highly non uniform. In an accompanying paper (Zwitter 2003) we show that superpositions of spectral lines of background stars can be removed by careful modelling. Here we use photometry of actual stellar fields in 66 directions in the Galaxy to assess the probability for an overlap, as well as the typical spectral type of overlappers.

2. Observations

We use B, V, and I_C photometry of 66 fields around a sample of symbiotic stars (Henden & Munari 2000, 2001). These fields were observed with the USNO Flagstaff Station, 1.0m telescope and two CCD detectors. The exposures were designed to permit accurate photometry of the symbiotic star, and so were of varying depth depending on the brightness of the target star. However, all fields are relatively complete to $V \sim 17$ with stellar detections reaching $V = 20.5$ mag. Only photometric, good seeing nights were used, with extinction and transformation coefficients determined from nightly measures of Landolt standard stars. Aperture photometry was used, with a minimum of three observations on separate photometric nights per field. Typical zeropoint errors are around 0.01 mag. More complete description of the observations and techniques can be found in the referenced papers. These fields can be used to assess stellar density and dis-

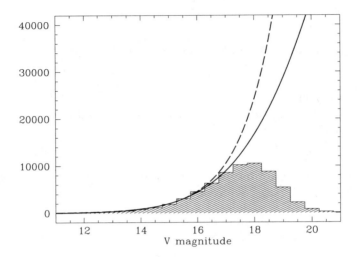

Figure 1. Luminosity function from 66 stellar fields with 76 000 stars. Histogram represents magnitude distribution of V-mag measurements, solid line is the assumed true distribution (eq. 1), while dashed line is a potential fit ($dN/dV \propto 2.3^V$) to the bright stars ($V < 16$ mag).

tribution of spectral types down to the GAIA faintness limit. A total of 75 959 stars were analyzed.

The luminosity function for $V > 15$ mag is frequently described by a power law. Here we adopt a heuristic law

$$dN/dV \propto 2.3^{V-0.05(V-15)^2} \qquad (1)$$

which gives a better fit to the data (see Fig. 1). The relation is compatible with star counts predictions by the Galaxy model (ESA-SCI-2000-4, Torra et al. 1999) using Hakkila et al. (1997) extinction law.

Average star density of $V < 17$ mag stars equals 15 100 stars deg^{-2} close to the Galactic plane ($|b| < 20^o$) and 1900 stars deg^{-2} away from it ($|b| > 20°$). This is somewhat larger than the corresponding values (6100 and 1200 stars deg^{-2}) from the Galaxy model (ESA-SCI-2000-4). This is probably a statistical anomaly; symbiotic stars tend to lie close to the galactic plane and selection effects may emphasize regions of higher star density. Even so only 10% of directions close to the Galactic plane ($|b| < 20°$) reach the density of 40 000 ($V < 17$ mag) stars deg^{-2}. Spectral types of field stars cluster around an early K type at the bright end, reaching a mid-K for the faintest targets (Figure 2).

3. Crowding and the sampling law

The sampling law of the GAIA satellite guaranties that the arrangement of stars in the focal plane will be different for each of the ~ 100 transits of a given star, providing that the spin and precession periods of the satellite are

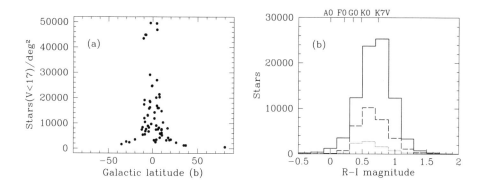

Figure 2. *Left*: observed star density as a function of Galactic coordinates. *Right*: distribution of observed colours for different magnitude classes: $14 < V < 16$ (dotted line), $16 < V < 18$ (dashed line), and $18 < V < 20$ (solid line). The latter is corrected for incompletness. Labels on top mark colours of unreddened main sequence stars.

kept incommensurate. Two stars that badly overlap in one passage have non-overlapping tracings on the next pass. This is a simple consequence of the fact that the length of the spectral tracing in the dispersion direction is much larger than its width. There will be unfortunate cases, for example close optical doubles, with tracings overlapping in a significant fraction of observations. But such stars are rare and so of no interest to us here. Below we discuss the results on spectral overlaps for typical randomly positioned stars in the focal plane.

Two spectra overlap with a probability p if the length of the spectrum is larger than the free length

$$L = (ns)^{-1} \ln[(1-p)^{-1}] \qquad (2)$$

where the star density equals n stars deg^{-2} and the width of the stellar tracing is s arcsec ($s \sim 4.5$ arcsec for the Astrium design of the GAIA spectrograph). The results are presented in Table 1.

4. Conclusions

We shall comment on the result for a resolution of 10 000 assuming a spectral width of 4.5 arcsec. Results for other resolutions and spectral widths can be easily judged from Table 1.

The wavelength interval of GAIA spectra covers 250 Å around $\lambda_c = 8615$ Å. The spectral length for $R = 10\,000$ is 580 pixels, assuming 2 pixels per resolution element. Since each pixel covers 1 arcsec on the sky in the dispersion direction the spectral overlap with another $V < 17$ mag star will occur if the distance between spectral heads is smaller than 580 arcsec. One can see from Table 1 that this happens in some 21 out of 100 spectra if the star density is 1200

Table 1. Severity of crowding following Eq. 2.

star density n (stars deg^{-2})	average free length (arcsec)	probability (p) that the distance between spectral heads is smaller than L (arcsec)				
		$L=1$	$L=10$	$L=100$	$L=500$	$L=1000$
600	4800	0.000	0.002	0.021	0.099	0.188
1 200	2400	0.000	0.004	0.041	0.188	0.341
3 000	960	0.001	0.010	0.099	0.406	0.647
6 000	480	0.002	0.021	0.188	0.647	0.875
20 000	144	0.007	0.067	0.501	0.969	0.999
50 000	57	0.017	0.159	0.824	1.000	1.000
100 000	28	0.034	0.293	0.969	1.000	1.000

stars deg^{-2}, a typical value for field stars with $V < 17$ mag at high Galactic latitudes. Density of fainter stars at the same position is much higher (cf. Eq. 1), so overlaps with $V \sim 18$ mag stars are common even at high latitudes. If one considers overlapping stars down to $V = 20$ mag the probability for an overlap increases to 97%. If the resolution were smaller than $R = 10\,000$ the length of the spectral tracing covering the GAIA spectral wavelength interval would be shorter. Still it is clear that lower spectral resolutions cannot make spectra free from spectral overlaps of faint stars even at high Galactic latitudes. The fraction of spectra overlapped with $V < 17$ mag background stars increases to 70% at the star density of 6 000 ($V < 17$ mag) stars deg^{-2} (i.e. a value typical close to the Galactic plane).

Spectral overlap is never complete. Even at extreme star densities of 50 000 stars deg^{-2} only 16% of the spectra would have an overlapping $V < 17$ mag spectrum starting within 10 arcsec (= 10 pixels) behind its head.

Spectral tracings in the focal planes of the GAIA spectrograph will overlap at all star densities and at all resolutions. The policy cannot and should not be to keep only the spectra that are free from overlaps. Should this be the case one would throw away most of the collected GAIA spectra.

References

Hakkila, J., Myers, J.M., Stidham, B.J. & Hartmann, D.H. 1997, AJ 114, 2043
Henden, A. & Munari, U. 2000, A&AS 143, 343
Henden, A. & Munari, U. 2001, A&A 372, 145
Torra, J., Chen, B., Figueras, F., Jordi, C. & Luri, X. 1999, Balt. Astr. 8, 171
Zwitter, T. 2003, in GAIA Spectroscopy, Science and Technology, U. Munari ed., ASP Conf. Ser. 298, pag. 493

Information recovery from overlapping GAIA spectra

Tomaž Zwitter

University of Ljubljana, Dept. of Physics, Jadranska 19, 1000 Ljubljana, Slovenia

Abstract. The RVS instrument aboard GAIA is a slitless spectrograph, so some spectral tracing overlap is inevitable. We show that this background can be accurately modeled and subtracted. Radial velocity accuracy is not degraded significantly unless the star density is higher than the value typical for the Galactic plane.

1. Introduction

GAIA RVS is a slitless spectrograph, so some degree of spectrum overlap is always present (Zwitter & Henden 2003, hereafter ZH03). The overlap is proportional to resolution, but even low resolutions ($R \sim 5000$) do not avoid it. On the other hand most astrophysical information other than radial velocity is washed away at resolutions lower than $R = 10\,000$ (e.g. Thevenin 2003). Overlapped stars are different during each transit, so some spectra of the same object suffer from bad overlaps, others do not. For this purpose it is essential that sidereal spin and precession periods of the satellite are not commeasurable.

Information recovery from overlapped spectra can use information from other GAIA instruments. In particular, the astrometric position measurements tell which stars are overlapping at each of the transits of a given field over the focal plane. Also the distance between spectra heads (that could be treated as a rough 'velocity' shift) is accurately known from the star mappers in the spectro plane.

Photometric classification of stellar spectra allows to choose correct spectral templates for all overlapping spectra. Expected errors will be small, of the order of 100 K in temperature, 0.2 in $\log g$ and similar in metallicity. This information will build up during the mission, so reduction reiterations will be needed.

2. Radial velocity measurement

Radial velocity supplies the sixth component in the position-velocity space, so it is the prime reason for existence of a spectrograph aboard GAIA. We ran a set of simulations to recover radial velocity. The results allowing only for zodiacal light background and read-out noise were published in Zwitter (2002), up-to-date results are discussed in Munari et al. (2003).

Zodiacal light and read-out noise are not the only source of background. Spectra of other stars (partly) overlap our spectrum, depending on star density and spectral resolution (ZH03). It is important that we know positions, magnitudes and rough spectral types of overlapping background stars. We assume that:

- stellar positions are accurately known. Star mappers as well as astrometry easily justify this assumption;

- stellar magnitudes are accurately known. Typical errors will be ~ 0.001 mag for bright stars and ≤ 0.02 mag for stars of $V=18$ mag (ESA-SCI-2000-4). This is well below the spectroscopic shot noise, so this assumption is justified. Contemporaneous star mapper flux measurements can supply the required information in case of variable stars;

- stellar types are roughly known. A mismatch of 250 K in temperature and 0.5 in $\log g$ and [Fe/H] was assumed. This is some 3× larger than expected typical final-mission stellar classification errors from photometry (Jordi 2003). So this assumption is justified even for mid-mission analysis. Simulated star types clustered around K1 V, i.e. a typical spectral type of background stars, and their luminosity function followed results of ZH03.

- Stellar radial velocities are roughly known. The assumed errors were

$$\sigma(RV) = v_\circ \; 2.51^{V-14.0} \qquad v_\circ = 0.2\,\mathrm{km\ s^{-1}} \tag{1}$$

where $\sigma(RV)$ is the standard deviation of the difference between assumed and true radial velocities and V is the visual magnitude of the star. This is compatible to mission averaged results for a K1 V star (Zwitter 2002). It turns out that radial velocity spread of background stars is not critical for the final results, so this assumption is justified.

Figure 1 shows examples of background spectra. Vertical axis is spectral flux in units of continuum flux from a K1 V star of magnitude $V = 14$. Thus the level of 0.01 corresponds to a contribution of a $V = 19$ background star. Randomly chosen examples of spectra at different resolutions (R) and star densities (n) are plotted. Star density $n = 1200$ ($V < 17$ mag) stars deg^{-2} is common at high Galactic latitudes ($|b| > 20°$), the value of 6100 is typical for the Galactic plane, while $n = 40\,000$ ($V < 17$ mag) stars deg^{-2} is representative of rather high stellar densities, encountered in $\leq 10\%$ of cases at $|b| < 20°$ (ZH03). Number of overlapping background spectra and their flux level generally increases with resolution.

One should not conclude that these rather bright and jumpy background spectra jeopardize derivation of radial velocities. The background can be very well modeled from information that is available (positions, magnitudes, rough spectral types and velocities of overlapping stars - see above). Note that this info yields spectral tracings (thin lines in Figure 1) that are *very* similar to those of real stars (thick lines).

Figure 2 reports final results for different star densities. Note that errors on derived radial velocity do not increase significantly, except for the faintest targets ($V > 17$ mag) and the highest star densities ($n = 40\,000$ ($V < 17$ mag) stars deg^{-2}). Even those errors could be reduced by filtering spectra so that only regions of spectral lines are cross-correlated. Curves are a bit noisy due to a limited Monte Carlo computing time, but the main result is clear: spectral

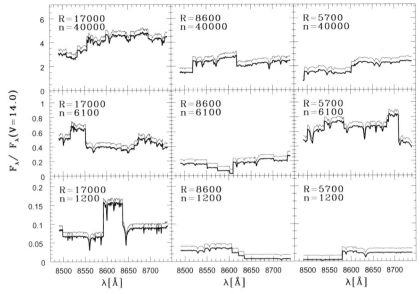

Figure 1. Examples of background spectra of different resolutions (R) and star densities (n). Thick lines: actual spectra. Thin lines: their reconstruction from available information. A small vertical shift was applied to the latter for better legibility.

lines of background stars do not spoil radial velocity analysis. The background stars merely increase the level of the background, thus increasing its shot noise.

3. Recovery of other astrophysically relevant information

GAIA spectrograph will be able to measure other information than radial velocities: abundances of individual elements, rotational velocity, temperature, $\log g$, and metallicity. All these will be recovered from detailed spectral profiles of individual lines. For this purpose: (a) spectral line profiles should be measured with high-enough accuracy; (b) the profile should not be spoiled by notable spectral lines from overlapping background stars.

The first condition effectively limits the analysis to bright targets, $V < 15$ mag, and the second to environments of moderate star density, $n < 6000$ ($V < 17$ mag) stars deg^{-2}.

4. Conclusions

Radial velocities can be recovered even for faint targets. Because the spectral shape of the background can be modeled accurately, overlapping stars degrade the RV accuracy mainly by increasing the background shot noise. It was found that overlapping spectra generally do not reduce the radial velocity accuracy substantially, even if observing at high resolution ($R \sim 10\,000$) and close to the

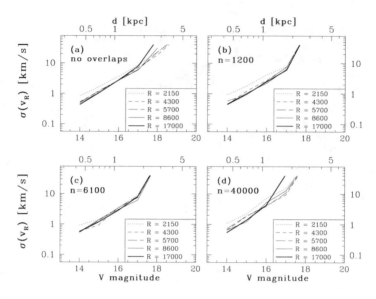

Figure 2. Radial velocity errors for a mission averaged spectrum of a K1 V star at different resolutions R. a: no spectral overlaps, b: observations at star densities characteristic for high Galactic latitudes, c: for the Galactic plane, and d: for a high density environment.

Galactic plane. Degrading becomes substantial if the S/N per 1 Å bin of the non-overlapped spectrum is < 5. This happens at $V \sim 17.5$ mag for mission averaged spectra and at $V \sim 15$ mag for single transits. Recovery of other astrophysical information is more difficult and mostly limited to bright targets in not too dense environments. For accurate modeling of the background the quality of stellar models as well as an accurate knowledge of flux throughput of the instrument (and filters used to bracket the spectral range) is of utmost importance.

References

Jordi, C. 2003, in GAIA Spectroscopy, Science and Technology, U. Munari ed., ASP Conf. Series 298, pag. 209

Munari, U., Zwitter, T., Katz, D. & Cropper, M. 2003, in GAIA Spectroscopy, Science and Technology, U. Munari ed., ASP Conf. Series 298, pag. 275

Thévenin, F. 2003, in GAIA Spectroscopy, Science and Technology, U. Munari ed., ASP Conf. Series 298, pag. 291

Zwitter, T. 2002, A&A 386, 748

Zwitter, T. & Henden, A. 2003, in GAIA Spectroscopy, Science and Technology, U. Munari ed., ASP Conf. Series 298, pag. 489

INDICES

Subject index

abundance stratification	173
accretion disk	407
accretion stream	409
adaptive optics	389
α−process elements	174, 245, 385, 477
AGB stars	257, 270, 415
AGN	181, 411
angular,	
acceleration	32
momentum	59
anisotropy,	
orbital	131
parameter	134
anticenter, galactic	247
ADSD database	221
Am stars	59
Ap stars	59
APS measuring machine	145
APT robotic telescope	372
ARVI instrument	65
ASAS survey	372
Asiago spectral library	427
asteroid,	
masses	6
photometry	367
astrometric instrument,	
attitude control	17
basic angle	16
CCD gating	17
focal plane	16
object detection	17
optical design	15
performance	18
windowing	17
astrometry,	25
centroiding	27
parameters propagation	28
modeling	30
reference frame	411

relativistic model	25, 159
ATLAS model atmospheres	173, 419, 454
atomic line data	174, 399, 481
attitude control	22
automatic analysis,	
eclipsing binary	315, 457
lines	399
spectra	57, 176
wav. coincidence stat.	399
Baade-Wesselink method	248
Baade window	52
bar, galactic	248, 261
BaSel library	417, 419, 485
Be stars	59
binary stars,	
astrometric	54, 342, 354, 410
atmosphere representation	323
components evolution	308
contact	435
detached	316
distances from	309, 313
double-contact	316
eclipsing	54, 303, 313, 323, 329, 435, 447, 457
formation	340, 351
heating	410
mass-luminosity relation	313, 316, 357
mass ratio	339, 351, 410
overcontact	316, 326, 435
period distribution	339, 347
period search	331, 348
pulsating component in	447
reflection	315, 336, 410
resolved	4, 339
selection effects	339, 355
semi-major axis distribution	340, 351
semi-detached	316
spotted	59, 437
synchronization	60, 319, 359
BSC catalog	370
brown dwarfs	59, 310, 341, 432
Bulge	248, 261, 385
BHB stars	142
carbon stars,	
dwarf	379
model spectra	191, 451
real spectra	451
CAMC survey	371
cataclysmic variables	407

CDS Strasbourg	106, 430
central star of planetary nebulae	295
Cepheids,	
binarity	240
cluster, in	241
discovery	238, 239
double-mode	239
first overtone	239, 247
fundamental mode	239, 249
general properties	237
instability strip, region	237
instability wedge	238
limiting mag	56
LMC, SMC, in	239, 246
metallicity gradient	247
P-L relation	237, 249
P-L-Color relation	238, 246
Population II	245
pulsation curve	56
Chandrasekhar limit	424
chromaticity effect	380
chromospheric activity	61, 230, 406, 449, 461
classification,	57, 199
Carbon stars, of	451
hierarchical scheme	203
global model	203
local model	201
parallel scheme	203
two-dimensional	292
WR stars, of	295, 419
cold electronics	82
convection models,	174
non-local	177
convective envelope	59
CORAVEL masks	471
core-mass luminosity relation	262
cosine-bell windowing	474
crowding	53, 70, 110, 489
dark matter,	
baryonic	129
cold	130, 153, 384
disk	130
haloes	129
hot	129
scale length	130
velocity dispersion	130, 135
data,	
access	97
analysis	10, 97

ground processing	97	
handling	22, 77, 97	
lossless compression	78	
lossy compression	79	
on-board processing	71	
degeneracy, mass and anisotropy	131	
δ Sct stars	245, 447	
DENIS survey	372	
detectors	17, 69	
diffuse interst. band 8620 Å	61, 240, 440	
Disk	248, 376	
distance scale	237, 246, 258, 314	
DISPI spectra	485	
DIVA astrometric mission	374, 382, 485	
Doppler tomography	409	
DUCAC survey	373	
dust,		
condensation temperature	270	
formation	184	
settling	186	
dynamical time	131	
dwarf novae	409	
dwarf spheroidal satellite galaxies,	129-136, 411	
triaxial models	135	
eclipsing binaries,		
accuracy of orbits	305, 317, 335, 437, 457	
accuracy of physical param.	306, 324, 335, 359, 437, 457	
Asiago program	329	
atmosphere modeling	325	
automatic data treatment	315, 338, 457	
interacting	410	
intrinsic variability	336	
parameters ramping	326	
overcontact systems	310, 314, 326, 435	
W UMa type	435	
enhancement, degree, chemical	478	
EROS microlensing exp.	239, 246	
equation of state	182	
equatorial disk	59	
expert systems	315	
Faber-Jackson relation	253	
FAME astrometric mission	374	
FERRUM project	483	
fiber,		
ARGUS integral field	469	
Autofib2 positioner	389	
Echidna positioner	382	
fed instruments	53	
FLAMES facility	469	

HERCULES facility	473
IFU integral field	469
MEDUSA mode	469
scrambling	475
FIRST radio survey	413
fundamental physics	6
galactic tides	443
galaxies,	4
elliptical	424, 477
GALEX ultraviolet mission	374
gamma-ray bursts	366
GCPD database	371
GEMINI telescope	375
general relativity,	159
light deflection	164
light trajectories	160
Schwarzschild model	165
global astrometry	6, 14
globular clusters,	54, 250
luminosity function	253
mass-loss rate	445
Miras in	259
tidal tails	443
grazing incidence	6
ground-based follow-ups	350, 376
kinematics	
galactic	52
internal	53
GSC-II	106, 138, 371, 375, 382
GASS simulator	115
GDASS	98
GIBIS simulator	115
GIRAFFE spectrograph	469
GIS solution	101
GranTeCan	390
GSPC2 catalog	371
H^- opacity	292
Halo,	
age	376
dark	376
gas infall	248
inner	142
metallicity	51
outer	142
parameter	134
prograde motion	142
retrograde motion	143
streams	52, 444
tracers	142

HDF North, South	371
heating models	410
Henden database	371
HgMn stars	59
Hipparcos,	
operation mode	7
catalog	382
compared to GAIA	26
input catalog	371
photometry	334
support radial velocities	391
HIRES spectrograph	431
hot bottom burning	262
Hertzsprung bump	56
HR diagram	307, 417
HST	412
Hubble constant	246
ICRS reference system	106
ING telescopes	387
interferometric diameter	184
interstellar lines	61
iron abundance	176
irradiation	408
jet ejection	230
Kurucz's model atmospheres	173, 215, 292, 419, 454
L3CCD,	84
in photon-counting mode	82
Lagrangian comoving frame	181
launch date	24
light,	
gravitational bending	6
propagation delay	39
limb darkening	54
line,	
automatic analysis	402
blanketing	181, 403
broadening	59, 182, 410
central depth analysis	404, 463
emission core	233, 403, 463, 466
equivalent width	404
formation	182, 455, 465
indexes	477
intensity parameter	399
isotopic shift	484
hyperfine structure	484
lists	173, 179, 189, 399
$\log gf$	399, 481
number of	56
optical depth	410

profile	182, 230, 463, 466
wav. coincidence stat.	399
Zeeman effect	481
LINEAR asteroid survey	371
lithium	262
Local Group	248, 412
local standard of rest	143
LONEOS database	371
LOTIS imaging GRB	372
LOTOSS survey	371
low signal detection	83
LSST asteroid survey	252, 373
magnetic,	
activity	410
field	59, 484
radiative transfer	174
stars	230
MACHO microlensing exp.	239, 246, 371, 376
MAGIC telescope	390
MAMA measuring machine	145
MARCS spectral models	191, 419
2MASS survey	372
median combine	83
microturbolence velocity	176, 454
minimum distance method	57, 201, 293, 479
Mira variables,	245, 257
C-rich	56, 261, 453
diameter	264
galactic kinematics	259, 260
globular clusters, in	259
initial masses	259
instability strip	246
limiting mag	56
line splitting	264
O-rich	258
period-luminosity relation	262
pulsation curves	55
S-type	261
short period groups	258
symbiotic	425
mixing length	173
MK classification system	57, 427
model,	
atmosphere	191
grid	215, 419, 451
spectrum	191
molecular spectra	454

motion,
    across-scan 15
    prograde 445
    tangential 52
    uniform rectilinear 33
multiple object spectroscopy (MOS) 382
NEAT asteroid survey 371
nebular spectrum 425
neural network 201, 486
NextGen model grid 187, 403, 419, 461
NIST database 483
NLTT catalog 376
NTLE 175, 181, 292, 403, 461
non-stellar objects 106
north galactic pole (NGP) 137, 149, 377, 413
novae 253, 366
OB associations 54
OH/IR sources 257
oscillating blue stragglers 245
OGLE microlensing exp. 239, 246, 310, 371
opacity,
    distribution function 173, 191
    sampling 178, 182, 191
open clusters 53, 60
optical depth 410
orbit acquisition 21
overshooting 173
P Cyg profile 230
Pan-STARRS asteroid survey 252, 373
parallactic effect 35
parallax, 35, 40
    time variability 37
parameterization, 199
    calibration of algorithms 206
parametrized post-Newtonian (PPN) 6
pattern matching approach 201
payload configuration 18
payload temperature 20
peculiar stars, 61, 227
    chemically 228
Period-Luminosity relation (PL),
    Cepheids 237, 238, 246, 249
    Miras 262
    RR Lyr 141, 248
    RV Tau stars 269
perspective acceleration 7, 31, 38, 53, 66
PDS measuring machine 145
phase space 51, 275
PHOENIX model atmosphere 181, 419

photographic surveys	145, 365, 375
photometric,	
broad bands	210
catalogs	366
GAIA system	41, 41, 210, 335, 413, 419
Johnson-Cousin system	370
medium bands	210
performance	209
SLOAN system	370
test targets	42
Planck mission	21
plane parallel atmosphere	173
PLANET microlensing exp.	246
planetary nebulae,	168
central stars	295
distances to	169
extragalactic distance indicator	54, 170
planets, extra-Solar	5, 431
Plummer,	
density	132
radius	134
PMM measuring machine	145
point spread function	75, 86
POSS-I, -II	138, 365, 375, 377
post-AGB phase	268
power demand	22
proper motion,	34, 412
reduced	377
pulsating stars	55, 228, 237, 245, 257, 267, 448, 465
mode identification	448
quasars	4, 373, 411
QUEST survey	373
R CrB type variability	465
radial velocity,	
accuracy	54, 275, 454, 472, 475
barycentric	55, 275, 466
IAU standards	275
performance	73
time variability	34, 466
RAPTOR survey	371
RAVE project	275, 381
reddening	47, 67, 211, 416
relativity,	
general	159
special	181
rotational,	
accuracy of velocities	285
broadening	174
induced mixing	285

magnetic fields	286
ROTSE survey	371
RR Lyrae,	142, 245, 248
limiting mag	56
period-luminosity relation	141, 248
pulsation curve	55
RS CVn stars	310
RVS instrument,	65, 278, 293
Consortium	393
working group	393
RV Tau stars,	268
period-luminosity relation	269
SAM12 model atmosphere	454
satellite configuration,	20
envelope	23
instrument model	117
lifetime	23
mass	23
precession	14
scanning law	105, 346
simulation	113, 121
spin axis	14
temperature	13
SB9 orbit catalog	346
seeing	387
service module	21
Seyfert galaxies	411
shock fronts	264, 271, 410
silicon carbide (SiC)	20
SIM mission	25, 131, 411, 446
simulation	102, 113, 119
SLOAN survey	250, 370, 373, 409, 411, 443
S/N of GAIA spectra	278
SN type Ia	253, 423
Sobolev approximation	181
solar,	
array	22
motion	143
system, objects	5
Soyuz-Fregat launcher	21
spectral,	
analysis	57
databases	221, 451
dispersion, orientation	67
dispersion, range	72
energy distribution	416
synthetic grids	57, 176, 186, 215, 419, 451
spectro-photometric instrument	18

SRd stars	268
Stark broadening	410
static model	190
stellar,	
age-metallicity degeneracy	477
bar	385
density	105, 490
density profile	340
evolutionary tracks	308
IMF	351, 357, 361
isochrones	417, 477
formation history	4, 151, 415
formation rate	152, 351
mass-loss	61, 230, 467
mass-luminosity relation	313, 316, 357
pulsations	228, 237, 245, 257, 267, 448
radii	359
rotation	59, 230, 285, 359, 440
scale height	341
shell	465
spots	230
streams	384
systems	130
warp	385
wind	59, 230
superCOSMOS measuring machine	145
superhumps	410
surface gravity	174
supernovae	4, 366, 385
symbiotic stars	423
TASS survey	372
technology program	23
telemetric flow	103, 105
thermal,	
line broadening	410
control, passive	20
pulses	270
Thick Disk,	149
age	376
metal weak tail	140
origin	385
velocity gradient	155
Thin Disk,	149
age	376
velocity ellipsoid	156, 386
vertical tilt	156
tidal energy dissipation	286
tilt mechanism	71
transverse velocity	72

Tully-Fisher relation	253
turn-off	415
Tycho-2 catalog	369, 382
UCAC survey	372
UKST telescope	382
UML language	101, 115, 122
USNO-B catalog	138, 369
VALD database	174, 192, 400, 454, 482
variable stars, intrinsic	5, 227, 237, 245, 257, 261, 267, 295, 337, 407, 423, 439, 447, 461, 465
virial velocity	129
VISTA telescope	373
VLT telescopes	469
VPH gratings	383
VST telescope	373
wavelength calibration,	
accuracy	87, 277
with standard stars	120, 279
with absorption cells	85
with fiber Bragg gratings (FBGs)	93
wavelength grid	182
weak-lined stars	268
Wesenheit magnitudes	246
white dwarfs,	4, 60, 241, 344, 376, 409, 424
cooling and age	376
wind terminal velocity	298, 441
WBVR catalog	370
WHT telescope	389
WITA model spectra	454
WR stars,	295, 439
chemical abundances	439, 454
in the Galaxy	296
in LMC, SMC	296
WC sequence	298
WN sequence	297
WN, WC, WO subdivision	295
zodiacal light	278

Object index

And	Z	425	Dra	UX	453	Ori	α	190, 482
	RX	408		BY	59, 233	Peg	51	88
	VX	453		GK	330, 336, 458		RU	408
	WY	272		dSph	131, 412		AG	424
Ant	T	242	For	dSph	250		OO	308, 331, 336
Aps	S	467		Cluster	253	Per	UY	242
Aql	η	56	Gem	ξ	56		KT	408
Ari	β	174		TU	453		V505	308, 331, 336, 458
	UX	233		VW	453		V570	308, 331, 336
Aur	UV Aur	453		AD	56		OB-1	263
	AB	231		NY	453	Phe	AI	308
Boo	α	194		NQ	453	Sge	γ	185
	TY	458	Her	AC	272		S	56
Cam	UV	453		stream	260		SV	467
	XX	467	Hya	R	259, 262		GY	242
	CK	238		V	453	Sgr	U	56, 242
Car	S	260		W	264		W	56
	VY	242		HS	308		RY	467
	XY	56		mov. group	259		WZ	242
	XX	56	Hyi	β	476		BB	242
	GT	242	Lac	DV	453		dSph	250, 384, 444
Cas	SU	242		V411	238	Sco	KQ	242
	WZ	453	Leo	R	264	Sct	R	272
	XY	56		UV	331, 336, 458		RU	242
	CEa	242		I dSph	253		EV	242
	CEb	242	Lib	33	177		V367	242
	CF	242	Lyn	CN	55	Ser	QZ	409
	DL	242	Lyr	α	184	Tau	17	86
Cen	α A	475		RR	142		T	232
	V	242	Men	UX	308		SU	467
Cep	δ	56, 242, 252	Mon	T	242		SZ	242
	V380	231		U	271		TT	453
Cet	τ	86		BG	453		V781	336
	o	263		CV	242	Tuc	θ	447
CMi	α	174		V614	453		47	385, 433
	W	453		V738	453	UMa	VY	453
CVn	RS	233		V838	232, 367		AW	317
	TX	425	Mus	TU	325, 327	UMi	α	238
Com	Cluster	252	Nor	S	56, 242		dSph	131, 250
CrB	R	466		RZ	467	Vel	RZ	242
Cyg	U	453		TW	56, 242		SW	242
	RS	453		QZ	242		γ^2	296
	SU	242		V340	242	Vir	BH Vir	458
	V482	467	Oph	Y	238		Cluster	253
	V1016	425		RS	231	Vul	S	242
	V1726	242		XX	232		SV	242
	P	232		V566	437		PU	231
	N-1 2001	231		V1010	437	BD	+21.64	453

511

HD	9900	428	HD	184738	298	NGC	4874	253
HD	16115	453	HD	193370	428	NGC	4881	253
HD	71148	428	HD	209458	431	NGC	4889	253
HD	71369	428	HD	219978	428	OMHR	58793	379
HD	76827	429	HR	1099	233	VB	10	180
HD	87141	428	HR	3383	482	WR	8	299
HD	100029	429	HR	5941	232	WR	14	297
HD	101153	429	HR	6118	231	WR	16	441
HD	102212	429	HR	7635	185	WR	17	297
HD	114330	486	IC	1613	262	WR	23	297
HD	117675	429	IC	4051	253	WR	31	296
HD	120539	428	Kron	3	417	WR	48	299
HD	123657	193	M	31	253	WR	52	297
HD	124850	428	M	33	263	WR	55	441
HD	126327	429	M	101	263	WR	57	297
HD	130144	429	NGC	1783	417	WR	66	296
HD	131977	428	NGC	1866	243	WR	92	297
HD	141477	193	NGC	2031	243	Barnard'	star	53
HD	148783	429	NGC	2121	417	Ros	51	453
HD	156074	453	NGC	2136	243	C∗	888	453
HD	182040	467	NGC	3379	253	Palomar	5	443

1RXS J232953.9+062814	409
GSC2U J131147.2+292348	379
2MASSW J0036159+182110	186
XTE J0421+560	229

Author index

Allard, F.	179	Feast, M.	257
Andretta, V.	403, 461	Figueras, F.	97, 209
Ansari, S.G.	97, 399		
Arenou, F.	339	Gilmore, G.	127
Aufdenberg, J.	179	Gomboc, A.	285
		Gomez, M.T.	403, 461
Babusiaux, C.	113	Grebel, E.K.	411, 443
Bailer-Jones, C.A.L.	199, 411		
Barnes, S.I.	473	Halbwachs, J.-L.	339
Baron, E.	179	Hauschildt, P.	179
Bertelli, G.	153	Hearnshaw, J.B.	473
Bienaymé, O.	147	Henden, A.A.	365, 489
Bijaoui, A.	291	Hodgkin, S.T.	375
Bono, G.	245		
Boschi, F.	427	Jancart, S.	345
Bridžius, A.	41	Johansson, S.	481
Bucciarelli, B.	159	Jordi, C.	97, 209
Busà, I.	403, 461		
		Kaempf, T.A.	485
Cacciari, C.	137	Katz, D.	65, 105, 119, 275, 291
Carollo, D.	375		
Carrasco, J.M.	209	Kinman, T.	137
Castelli, F.	215	Kovaleva, D.	357
Chiosi, C.	153, 477	Kučinskas, A.	415
Corradi, R.L.M.	168, 387		
Cropper, M.	75, 275, 407	Lattanzi, M.G.	137, 159, 375
Crosta, M.T.	159	Lejeune, T.	419
		Lennon, D.J.	387
de Felice, F.	159	Lindegren, L.	415
Dehnen, W.	443	Luri, X.	97, 113
Desidera, S.	85		
Drimmel, R.	137	Magrini, L.	168
		Malkov, O.	357
Eggenberger, A	339	Manimanis, V.N.	435

Marrese, P.M.	423, 427, 451	Schweitzer, A.	179
Marsh, T.	407	Siebert, A.	147
Masana, E.	97	Skuljan, J.	473
Mayor, M.	339	Smart, R.L.	137, 375
McLean, B.J.	375	Söderhjelm, S.	351
Mignard, F.	25	Sordo, R.	221
Milone, E.F.	303, 431	Soubiran, C.	147
Morin, D.	105	Spagna, A.	137, 375
Munari, U.	51, 85, 93, 215, 221, 227, 275, 423, 427, 451	Steinmetz, M.	381
		Stütz, Ch.	173
		Szabados, L.	237
Nesvacil, N.	173		
Newberg, H.J.	443	Tanabé, T.	415
Niarchos, P.G.	435	Tantalo, R.	477
Niedzielski, A.	295, 439	Terranegra, L.	403, 461
Nugis, T.	439	Thévenin, F.	291
		Tomasella, L.	427
Ochsenbein, F.	105	Torra, J.	97, 209
Odenkirchen, M.	411, 443		
		Udry, S.	339
Pace, O.	13		
Pagano, I.	403, 461	Vallenari, A.	153
Paparó, M.	447	Van Hamme, W.	323
Pasetto, S.	153	Vansevičius, V.	41, 415
Pavlenko, Ya.V	451	Vecchiato, A.	159
Perinotto, M	168	Viala, Y.-P.	105
Pernechele, C.	93		
Perryman, M.A.C	3, 391	Wahlgren, G.M.	267, 481
Pletz, B.	189	Weiss, W.W.	173
Pourbaix, D.	345	Willemsen, P.G.	485
Prša, A.	457	Williams, M.D.	431
		Wilson, R.E.	313, 323
Ragaini, S.	461	Wyithe, S.B.	313
Rix, H.-W.	443		
Rockosi, C.M.	443	Yanny, B.	443
Rodonò, M.	403		
Rosenbush, A.E.	465	Zwitter, T.	215, 275, 329, 489, 493
Royer, F.	469		
Rutten, R.G.M.	387		

A LISTING OF THE VOLUMES

Published
by

THE ASTRONOMICAL SOCIETY OF THE PACIFIC
(ASP)

An international, nonprofit, scientific and educational organization
founded in 1889

All book orders or inquiries concerning

ASTRONOMICAL SOCIETY OF THE PACIFIC
CONFERENCE SERIES
(ASP - CS)

and

INTERNATIONAL ASTRONOMICAL UNION VOLUMES
(IAU)

should be directed to the:

Astronomical Society of the Pacific Conference Series
390 Ashton Avenue
San Francisco   CA   94112-1722   USA

Phone:	800-335-2624	(Within USA)
Phone:	415-337-2126	
Fax:	415-337-5205	

E-mail:	service@astrosociety.org
Web Site:	http://www.astrosociety.org

Complete lists of proceedings of past IAU Meetings are maintained at the IAU Web site at the URL: http://www.iau.org/publicat.html

Volumes 32 - 189 in the IAU Symposia Series may be ordered from:

Kluwer Academic Publishers
P. O. Box 117
NL 3300 AA Dordrecht
The Netherlands

Kluwer@wKap.com

ASP CONFERENCE SERIES VOLUMES
Published by the Astronomical Society of the Pacific

PUBLISHED: 1988 (* asterisk means OUT OF STOCK)

Vol. CS -1   PROGRESS AND OPPORTUNITIES IN SOUTHERN HEMISPHERE
              OPTICAL ASTRONOMY: CTIO 25TH Anniversary Symposium
              eds. V. M. Blanco and M. M. Phillips
              ISBN 0-937707-18-X

Vol. CS-2   PROCEEDINGS OF A WORKSHOP ON OPTICAL SURVEYS FOR QUASARS
              eds. Patrick S. Osmer, Alain C. Porter, Richard F. Green, and Craig B. Foltz
              ISBN 0-937707-19-8

Vol. CS-3   FIBER OPTICS IN ASTRONOMY
              ed. Samuel C. Barden
              ISBN 0-937707-20-1

Vol. CS-4   THE EXTRAGALACTIC DISTANCE SCALE:
              Proceedings of the ASP 100th Anniversary Symposium
              eds. Sidney van den Bergh and Christopher J. Pritchet
              ISBN 0-937707-21-X

Vol. CS-5   THE MINNESOTA LECTURES ON CLUSTERS OF GALAXIES
              AND LARGE-SCALE STRUCTURE
              ed. John M. Dickey
              ISBN 0-937707-22-8

PUBLISHED: 1989

Vol. CS-6 * SYNTHESIS IMAGING IN RADIO ASTRONOMY: A Collection of Lectures
              from the Third NRAO Synthesis Imaging Summer School
              eds. Richard A. Perley, Frederic R. Schwab, and Alan H. Bridle
              ISBN 0-937707-23-6

PUBLISHED: 1990

Vol. CS-7   PROPERTIES OF HOT LUMINOUS STARS: Boulder-Munich Workshop
              ed. Catharine D. Garmany
              ISBN 0-937707-24-4

Vol. CS-8 * CCDs IN ASTRONOMY
              ed. George H. Jacoby
              ISBN 0-937707-25-2

Vol. CS-9   COOL STARS, STELLAR SYSTEMS, AND THE SUN: Sixth Cambridge Workshop
              ed. George Wallerstein
              ISBN 0-937707-27-9

Vol. CS-10 * EVOLUTION OF THE UNIVERSE OF GALAXIES:
              Edwin Hubble Centennial Symposium
              ed. Richard G. Kron
              ISBN 0-937707-28-7

Vol. CS-11   CONFRONTATION BETWEEN STELLAR PULSATION AND EVOLUTION
              eds. Carla Cacciari and Gisella Clementini
              ISBN 0-937707-30-9

Vol. CS-12   THE EVOLUTION OF THE INTERSTELLAR MEDIUM
              ed. Leo Blitz
              ISBN 0-937707-31-7

PUBLISHED: 1991

Vol. CS-13   THE FORMATION AND EVOLUTION OF STAR CLUSTERS
              ed. Kenneth Janes
              ISBN 0-937707-32-5

ASP CONFERENCE SERIES VOLUMES
Published by the Astronomical Society of the Pacific

PUBLISHED: 1991 (* asterisk means OUT OF STOCK)

Vol. CS-14 ASTROPHYSICS WITH INFRARED ARRAYS
ed. Richard Elston
ISBN 0-937707-33-3

Vol. CS-15 LARGE-SCALE STRUCTURES AND PECULIAR MOTIONS IN THE UNIVERSE
eds. David W. Latham and L. A. Nicolaci da Costa
ISBN 0-937707-34-1

Vol. CS-16 Proceedings of the 3rd Haystack Observatory Conference on ATOMS, IONS, AND MOLECULES: NEW RESULTS IN SPECTRAL LINE ASTROPHYSICS
eds. Aubrey D. Haschick and Paul T. P. Ho
ISBN 0-937707-35-X

Vol. CS-17 LIGHT POLLUTION, RADIO INTERFERENCE, AND SPACE DEBRIS
ed. David L. Crawford
ISBN 0-937707-36-8

Vol. CS-18 THE INTERPRETATION OF MODERN SYNTHESIS OBSERVATIONS OF SPIRAL GALAXIES
eds. Nebojsa Duric and Patrick C. Crane
ISBN 0-937707-37-6

Vol. CS-19 RADIO INTERFEROMETRY: THEORY, TECHNIQUES, AND APPLICATIONS, IAU Colloquium 131
eds. T. J. Cornwell and R. A. Perley
ISBN 0-937707-38-4

Vol. CS-20 FRONTIERS OF STELLAR EVOLUTION:
50th Anniversary McDonald Observatory (1939-1989)
ed. David L. Lambert
ISBN 0-937707-39-2

Vol. CS-21 THE SPACE DISTRIBUTION OF QUASARS
ed . David Crampton
ISBN 0-937707-40-6

PUBLISHED: 1992

Vol. CS-22 NONISOTROPIC AND VARIABLE OUTFLOWS FROM STARS
eds. Laurent Drissen, Claus Leitherer, and Antonella Nota
ISBN 0-937707-41-4

Vol CS-23 * ASTRONOMICAL CCD OBSERVING AND REDUCTION TECHNIQUES
ed. Steve B. Howell
ISBN 0-937707-42-4

Vol. CS-24 COSMOLOGY AND LARGE-SCALE STRUCTURE IN THE UNIVERSE
ed. Reinaldo R. de Carvalho
ISBN 0-937707-43-0

Vol. CS-25 ASTRONOMICAL DATA ANALYSIS, SOFTWARE AND SYSTEMS I - (ADASS I)
eds. Diana M. Worrall, Chris Biemesderfer, and Jeannette Barnes
ISBN 0-937707-44-9

Vol. CS-26 COOL STARS, STELLAR SYSTEMS, AND THE SUN:
Seventh Cambridge Workshop
eds. Mark S. Giampapa and Jay A. Bookbinder
ISBN 0-937707-45-7

Vol. CS-27 THE SOLAR CYCLE: Proceedings of the
National Solar Observatory/Sacramento Peak 12th Summer Workshop
ed. Karen L. Harvey
ISBN 0-937707-46-5

ASP CONFERENCE SERIES VOLUMES
Published by the Astronomical Society of the Pacific

PUBLISHED: 1992 (asterisk means OUT OF STOCK)

Vol. CS-28 AUTOMATED TELESCOPES FOR PHOTOMETRY AND IMAGING
eds. Saul J. Adelman, Robert J. Dukes, Jr., and Carol J. Adelman
ISBN 0-937707-47-3

Vol. CS-29 Viña del Mar Workshop on CATACLYSMIC VARIABLE STARS
ed. Nikolaus Vogt
ISBN 0-937707-48-1

Vol. CS-30 VARIABLE STARS AND GALAXIES
ed. Brian Warner
ISBN 0-937707-49-X

Vol. CS-31 RELATIONSHIPS BETWEEN ACTIVE GALACTIC NUCLEI
AND STARBURST GALAXIES
ed. Alexei V. Filippenko
ISBN 0-937707-50-3

Vol. CS-32 COMPLEMENTARY APPROACHES TO DOUBLE
AND MULTIPLE STAR RESEARCH, IAU Colloquium 135
eds. Harold A. McAlister and William I. Hartkopf
ISBN 0-937707-51-1

Vol. CS-33 * RESEARCH AMATEUR ASTRONOMY
ed. Stephen J. Edberg
ISBN 0-937707-52-X

Vol. CS-34 ROBOTIC TELESCOPES IN THE 1990's
ed. Alexei V. Filippenko
ISBN 0-937707-53-8

PUBLISHED: 1993

Vol. CS-35 * MASSIVE STARS: THEIR LIVES IN THE INTERSTELLAR MEDIUM
eds. Joseph P. Cassinelli and Edward B. Churchwell
ISBN 0-937707-54-6

Vol. CS-36 PLANETS AROUND PULSARS
ed. J. A. Phillips, S. E. Thorsett, and S. R. Kulkarni
ISBN 0-937707-55-4

Vol. CS-37 FIBER OPTICS IN ASTRONOMY II
ed. Peter M. Gray
ISBN 0-937707-56-2

Vol. CS-38 NEW FRONTIERS IN BINARY STAR RESEARCH: Pacific Rim Colloquium
eds. K. C. Leung and I.-S. Nha
ISBN 0-937707-57-0

Vol. CS-39 THE MINNESOTA LECTURES ON THE STRUCTURE
AND DYNAMICS OF THE MILKY WAY
ed. Roberta M. Humphreys
ISBN 0-937707-58-9

Vol. CS-40 INSIDE THE STARS, IAU Colloquium 137
eds. Werner W. Weiss and Annie Baglin
ISBN 0-937707-59-7

Vol. CS-41 ASTRONOMICAL INFRARED SPECTROSCOPY:
FUTURE OBSERVATIONAL DIRECTIONS
ed. Sun Kwok
ISBN 0-937707-60-0

ASP CONFERENCE SERIES VOLUMES
Published by the Astronomical Society of the Pacific

PUBLISHED: 1993 (* asterisk means OUT OF STOCK)

Vol. CS-42   GONG 1992: SEISMIC INVESTIGATION OF THE SUN AND STARS
               ed. Timothy M. Brown
               ISBN 0-937707-61-9

Vol. CS-43   SKY SURVEYS: PROTOSTARS TO PROTOGALAXIES
               ed. B. T. Soifer
               ISBN 0-937707-62-7

Vol. CS-44   PECULIAR VERSUS NORMAL PHENOMENA IN A-TYPE AND RELATED STARS,
               IAU Colloquium 138
               eds. M. M. Dworetsky, F. Castelli, and R. Faraggiana
               ISBN 0-937707-63-5

Vol. CS-45   LUMINOUS HIGH-LATITUDE STARS
               ed. Dimitar D. Sasselov
               ISBN 0-937707-64-3

Vol. CS-46   THE MAGNETIC AND VELOCITY FIELDS OF SOLAR ACTIVE REGIONS,
               IAU Colloquium 141
               eds. Harold Zirin, Guoxiang Ai, and Haimin Wang
               ISBN 0-937707-65-1

Vol. CS-47   THIRD DECENNIAL US-USSR CONFERENCE ON SETI --
               Santa Cruz, California, USA
               ed. G. Seth Shostak
               ISBN 0-937707-66-X

Vol. CS-48   THE GLOBULAR CLUSTER-GALAXY CONNECTION
               eds. Graeme H. Smith and Jean P. Brodie
               ISBN 0-937707-67-8

Vol. CS-49   GALAXY EVOLUTION: THE MILKY WAY PERSPECTIVE
               ed. Steven R. Majewski
               ISBN 0-937707-68-6

Vol. CS-50   STRUCTURE AND DYNAMICS OF GLOBULAR CLUSTERS
               eds. S. G. Djorgovski and G. Meylan
               ISBN 0-937707-69-4

Vol. CS-51   OBSERVATIONAL COSMOLOGY
               eds. Guido Chincarini, Angela Iovino, Tommaso Maccacaro, and Dario Maccagni
               ISBN 0-937707-70-8

Vol. CS-52   ASTRONOMICAL DATA ANALYSIS SOFTWARE AND SYSTEMS II - (ADASS II)
               eds. R. J. Hanisch, R. J. V. Brissenden, and Jeannette Barnes
               ISBN 0-937707-71-6

Vol. CS-53   BLUE STRAGGLERS
               ed. Rex A. Saffer
               ISBN 0-937707-72-4

PUBLISHED: 1994

Vol. CS-54 *   THE FIRST STROMLO SYMPOSIUM: THE PHYSICS OF ACTIVE GALAXIES
               eds. Geoffrey V. Bicknell, Michael A. Dopita, and Peter J. Quinn
               ISBN 0-937707-73-2

Vol. CS-55   OPTICAL ASTRONOMY FROM THE EARTH AND MOON
               eds. Diane M. Pyper and Ronald J. Angione
               ISBN 0-937707-74-0

Vol. CS-56   INTERACTING BINARY STARS
               ed. Allen W. Shafter
               ISBN 0-937707-75-9

ASP CONFERENCE SERIES VOLUMES
Published by the Astronomical Society of the Pacific

PUBLISHED: 1994 (* asterisk means OUT OF STOCK)

Vol. CS-57　　STELLAR AND CIRCUMSTELLAR ASTROPHYSICS
　　　　　　　eds. George Wallerstein and Alberto Noriega-Crespo
　　　　　　　ISBN 0-937707-76-7

Vol. CS-58 *　THE FIRST SYMPOSIUM ON THE INFRARED CIRRUS
　　　　　　　AND DIFFUSE INTERSTELLAR CLOUDS
　　　　　　　eds. Roc M. Cutri and William B. Latter
　　　　　　　ISBN 0-937707-77-5

Vol. CS-59　　ASTRONOMY WITH MILLIMETER AND SUBMILLIMETER WAVE
　　　　　　　INTERFEROMETRY,
　　　　　　　IAU Colloquium 140
　　　　　　　eds. M. Ishiguro and Wm. J. Welch
　　　　　　　ISBN 0-937707-78-3

Vol. CS-60　　THE MK PROCESS AT 50 YEARS: A POWERFUL TOOL FOR ASTROPHYSICAL
　　　　　　　INSIGHT, A Workshop of the Vatican Observatory --Tucson, Arizona, USA
　　　　　　　eds. C. J. Corbally, R. O. Gray, and R. F. Garrison
　　　　　　　ISBN 0-937707-79-1

Vol. CS-61　　ASTRONOMICAL DATA ANALYSIS SOFTWARE AND SYSTEMS III - (ADASS III)
　　　　　　　eds. Dennis R. Crabtree, R. J. Hanisch, and Jeannette Barnes
　　　　　　　ISBN 0-937707-80-5

Vol. CS-62　　THE NATURE AND EVOLUTIONARY STATUS OF HERBIG Ae/Be STARS
　　　　　　　eds. Pik Sin Thé, Mario R. Pérez, and Ed P. J. van den Heuvel
　　　　　　　ISBN 0-9837707-81-3

Vol. CS-63　　SEVENTY-FIVE YEARS OF HIRAYAMA ASTEROID FAMILIES:
　　　　　　　THE ROLE OF COLLISIONS IN THE SOLAR SYSTEM HISTORY
　　　　　　　eds. Yoshihide Kozai, Richard P. Binzel, and Tomohiro Hirayama
　　　　　　　ISBN 0-937707-82-1

Vol. CS-64 *　COOL STARS, STELLAR SYSTEMS, AND THE SUN:
　　　　　　　Eighth Cambridge Workshop
　　　　　　　ed. Jean-Pierre Caillault
　　　　　　　ISBN 0-937707-83-X

Vol. CS-65 *　CLOUDS, CORES, AND LOW MASS STARS:
　　　　　　　The Fourth Haystack Observatory Conference
　　　　　　　eds. Dan P. Clemens and Richard Barvainis
　　　　　　　ISBN 0-937707-84-8

Vol. CS-66 *　PHYSICS OF THE GASEOUS AND STELLAR DISKS OF THE GALAXY
　　　　　　　ed. Ivan R. King
　　　　　　　ISBN 0-937707-85-6

Vol. CS-67　　UNVEILING LARGE-SCALE STRUCTURES BEHIND THE MILKY WAY
　　　　　　　eds. C. Balkowski and R. C. Kraan-Korteweg
　　　　　　　ISBN 0-937707-86-4

Vol. CS-68 *　SOLAR ACTIVE REGION EVOLUTION:
　　　　　　　COMPARING MODELS WITH OBSERVATIONS
　　　　　　　eds. K. S. Balasubramaniam and George W. Simon
　　　　　　　ISBN 0-937707-87-2

Vol. CS-69　　REVERBERATION MAPPING OF THE BROAD-LINE REGION
　　　　　　　IN ACTIVE GALACTIC NUCLEI
　　　　　　　eds. P. M. Gondhalekar, K. Horne, and B. M. Peterson
　　　　　　　ISBN 0-937707-88-0

Vol. CS-70 *　GROUPS OF GALAXIES
　　　　　　　eds. Otto-G. Richter and Kirk Borne
　　　　　　　ISBN 0-937707-89-9

ASP CONFERENCE SERIES VOLUMES
Published by the Astronomical Society of the Pacific

PUBLISHED: 1995 (* asterisk means OUT OF STOCK)

Vol. CS-71	TRIDIMENSIONAL OPTICAL SPECTROSCOPIC METHODS IN ASTROPHYSICS, IAU Colloquium 149 eds. Georges Comte and Michel Marcelin ISBN 0-937707-90-2
Vol. CS-72	MILLISECOND PULSARS: A DECADE OF SURPRISE eds. A. S Fruchter, M. Tavani, and D. C. Backer ISBN 0-937707-91-0
Vol. CS-73	AIRBORNE ASTRONOMY SYMPOSIUM ON THE GALACTIC ECOSYSTEM: FROM GAS TO STARS TO DUST eds. Michael R. Haas, Jacqueline A. Davidson, and Edwin F. Erickson ISBN 0-937707-92-9
Vol. CS-74	PROGRESS IN THE SEARCH FOR EXTRATERRESTRIAL LIFE: 1993 Bioastronomy Symposium ed. G. Seth Shostak ISBN 0-937707-93-7
Vol. CS-75	MULTI-FEED SYSTEMS FOR RADIO TELESCOPES eds. Darrel T. Emerson and John M. Payne ISBN 0-937707-94-5
Vol. CS-76	GONG '94: HELIO- AND ASTERO-SEISMOLOGY FROM THE EARTH AND SPACE eds. Roger K. Ulrich, Edward J. Rhodes, Jr., and Werner Däppen ISBN 0-937707-95-3
Vol. CS-77	ASTRONOMICAL DATA ANALYSIS SOFTWARE AND SYSTEMS IV - (ADASS IV) eds. R. A. Shaw, H. E. Payne, and J. J. E. Hayes ISBN 0-937707-96-1
Vol. CS-78	ASTROPHYSICAL APPLICATIONS OF POWERFUL NEW DATABASES: Joint Discussion No. 16 of the 22nd General Assembly of the IAU eds. S. J. Adelman and W. L. Wiese ISBN 0-937707-97-X
Vol. CS-79 *	ROBOTIC TELESCOPES: CURRENT CAPABILITIES, PRESENT DEVELOPMENTS, AND FUTURE PROSPECTS FOR AUTOMATED ASTRONOMY eds. Gregory W. Henry and Joel A. Eaton ISBN 0-937707-98-8
Vol. CS-80 *	THE PHYSICS OF THE INTERSTELLAR MEDIUM AND INTERGALACTIC MEDIUM eds. A. Ferrara, C. F. McKee, C. Heiles, and P. R. Shapiro ISBN 0-937707-99-6
Vol. CS-81	LABORATORY AND ASTRONOMICAL HIGH RESOLUTION SPECTRA eds. A. J. Sauval, R. Blomme, and N. Grevesse ISBN 1-886733-01-5
Vol. CS-82 *	VERY LONG BASELINE INTERFEROMETRY AND THE VLBA eds. J. A. Zensus, P. J. Diamond, and P. J. Napier ISBN 1-886733-02-3
Vol. CS-83 *	ASTROPHYSICAL APPLICATIONS OF STELLAR PULSATION, IAU Colloquium 155 eds. R. S. Stobie and P. A. Whitelock ISBN 1-886733-03-1
ATLAS	INFRARED ATLAS OF THE ARCTURUS SPECTRUM, 0.9 - 5.3 μm eds. Kenneth Hinkle, Lloyd Wallace, and William Livingston ISBN: 1-886733-04-X

ASP CONFERENCE SERIES VOLUMES
Published by the Astronomical Society of the Pacific

PUBLISHED: 1995 (* asterisk means OUT OF STOCK)

Vol. CS-84 THE FUTURE UTILIZATION OF SCHMIDT TELESCOPES, IAU Colloquium 148
eds. Jessica Chapman, Russell Cannon, Sandra Harrison, and Bambang Hidayat
ISBN 1-886733-05-8

Vol. CS-85 * CAPE WORKSHOP ON MAGNETIC CATACLYSMIC VARIABLES
eds. D. A. H. Buckley and B. Warner
ISBN 1-886733-06-6

Vol. CS-86 FRESH VIEWS OF ELLIPTICAL GALAXIES
eds. Alberto Buzzoni, Alvio Renzini, and Alfonso Serrano
ISBN 1-886733-07-4

PUBLISHED: 1996

Vol. CS-87 NEW OBSERVING MODES FOR THE NEXT CENTURY
eds. Todd Boroson, John Davies, and Ian Robson
ISBN 1-886733-08-2

Vol. CS-88 * CLUSTERS, LENSING, AND THE FUTURE OF THE UNIVERSE
eds. Virginia Trimble and Andreas Reisenegger
ISBN 1-886733-09-0

Vol. CS-89 ASTRONOMY EDUCATION: CURRENT DEVELOPMENTS,
FUTURE COORDINATION
ed. John R. Percy
ISBN 1-886733-10-4

Vol. CS-90 THE ORIGINS, EVOLUTION, AND DESTINIES OF BINARY STARS
IN CLUSTERS
eds. E. F. Milone and J. -C. Mermilliod
ISBN 1-886733-11-2

Vol. CS-91 BARRED GALAXIES, IAU Colloquium 157
eds. R. Buta, D. A. Crocker, and B. G. Elmegreen
ISBN 1-886733-12-0

Vol. CS-92 * FORMATION OF THE GALACTIC HALO .... INSIDE AND OUT
eds. Heather L. Morrison and Ata Sarajedini
ISBN 1-886733-13-9

Vol. CS-93 RADIO EMISSION FROM THE STARS AND THE SUN
eds. A. R. Taylor and J. M. Paredes
ISBN 1-886733-14-7

Vol. CS-94 MAPPING, MEASURING, AND MODELING THE UNIVERSE
eds. Peter Coles, Vicent J. Martinez, and Maria-Jesus Pons-Borderia
ISBN 1-886733-15-5

Vol. CS-95 SOLAR DRIVERS OF INTERPLANETARY AND TERRESTRIAL DISTURBANCES:
Proceedings of 16th International Workshop National Solar
Observatory/Sacramento Peak
eds. K. S. Balasubramaniam, Stephen L. Keil, and Raymond N. Smartt
ISBN 1-886733-16-3

Vol. CS-96 HYDROGEN-DEFICIENT STARS
eds. C. S. Jeffery and U. Heber
ISBN 1-886733-17-1

Vol. CS-97 POLARIMETRY OF THE INTERSTELLAR MEDIUM
eds. W. G. Roberge and D. C. B. Whittet
ISBN 1-886733-18-X

ASP CONFERENCE SERIES VOLUMES
Published by the Astronomical Society of the Pacific

PUBLISHED: 1996 (* asterisk means OUT OF STOCK)

Vol. CS-98	FROM STARS TO GALAXIES: THE IMPACT OF STELLAR PHYSICS ON GALAXY EVOLUTION eds. Claus Leitherer, Uta Fritze-von Alvensleben, and John Huchra ISBN 1-886733-19-8
Vol. CS-99	COSMIC ABUNDANCES: Proceedings of the 6th Annual October Astrophysics Conference eds. Stephen S. Holt and George Sonneborn ISBN 1-886733-20-1
Vol. CS-100	ENERGY TRANSPORT IN RADIO GALAXIES AND QUASARS eds. P. E. Hardee, A. H. Bridle, and J. A. Zensus ISBN 1-886733-21-X
Vol. CS-101	ASTRONOMICAL DATA ANALYSIS SOFTWARE AND SYSTEMS V – (ADASS V) eds. George H. Jacoby and Jeannette Barnes ISBN 1080-7926
Vol. CS-102	THE GALACTIC CENTER, 4th ESO/CTIO Workshop ed. Roland Gredel ISBN 1-886733-22-8
Vol. CS-103	THE PHYSICS OF LINERS IN VIEW OF RECENT OBSERVATIONS eds. M. Eracleous, A. Koratkar, C. Leitherer, and L. Ho ISBN 1-886733-23-6
Vol. CS-104	PHYSICS, CHEMISTRY, AND DYNAMICS OF INTERPLANETARY DUST, IAU Colloquium 150 eds. Bo Å. S. Gustafson and Martha S. Hanner ISBN 1-886733-24-4
Vol. CS-105	PULSARS: PROBLEMS AND PROGRESS, IAU Colloquium 160 ed. S. Johnston, M. A. Walker, and M. Bailes ISBN 1-886733-25-2
Vol. CS-106	THE MINNESOTA LECTURES ON EXTRAGALACTIC NEUTRAL HYDROGEN ed. Evan D. Skillman ISBN 1-886733-26-0
Vol. CS-107	COMPLETING THE INVENTORY OF THE SOLAR SYSTEM: A Symposium held in conjunction with the 106th Annual Meeting of the ASP eds. Terrence W. Rettig and Joseph M. Hahn ISBN 1-886733-27-9
Vol. CS-108	M.A.S.S. -- MODEL ATMOSPHERES AND SPECTRUM SYNTHESIS: 5th Vienna - Workshop eds. Saul J. Adelman, Friedrich Kupka, and Werner W. Weiss ISBN 1-886733-28-7
Vol. CS-109	COOL STARS, STELLAR SYSTEMS, AND THE SUN: Ninth Cambridge Workshop eds. Roberto Pallavicini and Andrea K. Dupree ISBN 1-886733-29-5
Vol. CS-110	BLAZAR CONTINUUM VARIABILITY eds. H. R. Miller, J. R. Webb, and J. C. Noble ISBN 1-886733-30-9
Vol. CS-111	MAGNETIC RECONNECTION IN THE SOLAR ATMOSPHERE: Proceedings of a Yohkoh Conference eds. R. D. Bentley and J. T. Mariska ISBN 1-886733-31-7

ASP CONFERENCE SERIES VOLUMES
Published by the Astronomical Society of the Pacific

PUBLISHED: 1996 (* asterisk means OUT OF STOCK)

Vol. CS-112　　THE HISTORY OF THE MILKY WAY AND ITS SATELLITE SYSTEM
　　　　　　　　eds. Andreas Burkert, Dieter H. Hartmann, and Steven R. Majewski
　　　　　　　　ISBN 1-886733-32-5

PUBLISHED: 1997

Vol. CS-113　　EMISSION LINES IN ACTIVE GALAXIES: NEW METHODS AND TECHNIQUES,
　　　　　　　　IAU Colloquium 159
　　　　　　　　eds. B. M. Peterson, F.-Z. Cheng, and A. S. Wilson
　　　　　　　　ISBN 1-886733-33-3

Vol. CS-114　　YOUNG GALAXIES AND QSO ABSORPTION-LINE SYSTEMS
　　　　　　　　eds. Sueli M. Viegas, Ruth Gruenwald, and Reinaldo R. de Carvalho
　　　　　　　　ISBN 1-886733-34-1

Vol. CS-115　　GALACTIC CLUSTER COOLING FLOWS
　　　　　　　　ed. Noam Soker
　　　　　　　　ISBN 1-886733-35-X

Vol. CS-116　　THE SECOND STROMLO SYMPOSIUM:
　　　　　　　　THE NATURE OF ELLIPTICAL GALAXIES
　　　　　　　　eds. M. Arnaboldi, G. S. Da Costa, and P. Saha
　　　　　　　　ISBN 1-886733-36-8

Vol. CS-117　　DARK AND VISIBLE MATTER IN GALAXIES
　　　　　　　　eds. Massimo Persic and Paolo Salucci
　　　　　　　　ISBN-1-886733-37-6

Vol. CS-118　　FIRST ADVANCES IN SOLAR PHYSICS EUROCONFERENCE:
　　　　　　　　ADVANCES IN THE PHYSICS OF SUNSPOTS
　　　　　　　　eds. B. Schmieder. J. C. del Toro Iniesta, and M. Vázquez
　　　　　　　　ISBN 1-886733-38-4

Vol. CS-119　　PLANETS BEYOND THE SOLAR SYSTEM
　　　　　　　　AND THE NEXT GENERATION OF SPACE MISSIONS
　　　　　　　　ed. David R. Soderblom
　　　　　　　　ISBN 1-886733-39-2

Vol. CS-120　　LUMINOUS BLUE VARIABLES: MASSIVE STARS IN TRANSITION
　　　　　　　　eds. Antonella Nota and Henny J. G. L. M. Lamers
　　　　　　　　ISBN 1-886733-40-6

Vol. CS-121　　ACCRETION PHENOMENA AND RELATED OUTFLOWS, IAU Colloquium 163
　　　　　　　　eds. D. T. Wickramasinghe, G. V. Bicknell, and L. Ferrario
　　　　　　　　ISBN 1-886733-41-4

Vol. CS-122　　FROM STARDUST TO PLANETESIMALS:
　　　　　　　　Symposium held as part of the 108th Annual Meeting of the ASP
　　　　　　　　eds. Yvonne J. Pendleton and A. G. G. M. Tielens
　　　　　　　　ISBN 1-886733-42-2

Vol. CS-123　　THE 12th 'KINGSTON MEETING': COMPUTATIONAL ASTROPHYSICS
　　　　　　　　eds. David A. Clarke and Michael J. West
　　　　　　　　ISBN 1-886733-43-0

Vol. CS-124　　DIFFUSE INFRARED RADIATION AND THE IRTS
　　　　　　　　eds. Haruyuki Okuda, Toshio Matsumoto, and Thomas Roellig
　　　　　　　　ISBN 1-886733-44-9

Vol. CS-125　　ASTRONOMICAL DATA ANALYSIS SOFTWARE AND SYSTEMS VI
　　　　　　　　eds. Gareth Hunt and H. E. Payne
　　　　　　　　ISBN 1-886733-45-7

ASP CONFERENCE SERIES VOLUMES
Published by the Astronomical Society of the Pacific

PUBLISHED: 1997 (* asterisk means OUT OF STOCK)

Vol. CS-126   FROM QUANTUM FLUCTUATIONS TO COSMOLOGICAL STRUCTURES
eds. David Valls-Gabaud, Martin A. Hendry, Paolo Molaro, and Khalil Chamcham
ISBN 1-886733-46-5

Vol. CS-127   PROPER MOTIONS AND GALACTIC ASTRONOMY
ed. Roberta M. Humphreys
ISBN 1-886733-47-3

Vol. CS-128   MASS EJECTION FROM AGN (Active Galactic Nuclei)
eds. N. Arav, I. Shlosman, and R. J. Weymann
ISBN 1-886733-48-1

Vol. CS-129   THE GEORGE GAMOW SYMPOSIUM
eds. E. Harper, W. C. Parke, and G. D. Anderson
ISBN 1-886733-49-X

Vol. CS-130   THE THIRD PACIFIC RIM CONFERENCE ON
RECENT DEVELOPMENT ON BINARY STAR RESEARCH
eds. Kam-Ching Leung
ISBN 1-886733-50-3

PUBLISHED: 1998

Vol. CS-131   BOULDER-MUNICH II: PROPERTIES OF HOT, LUMINOUS STARS
ed. Ian D. Howarth
ISBN 1-886733-51-1

Vol. CS-132   STAR FORMATION WITH THE INFRARED SPACE OBSERVATORY (ISO)
eds. João L. Yun and René Liseau
ISBN 1-886733-52-X

Vol. CS-133   SCIENCE WITH THE NGST (Next Generation Space Telescope)
eds. Eric P. Smith and Anuradha Koratkar
ISBN 1-886733-53-8

Vol. CS-134   BROWN DWARFS AND EXTRASOLAR PLANETS
eds. Rafael Rebolo, Eduardo L. Martin, and Maria Rosa Zapatero Osorio
ISBN 1-886733-54-6

Vol. CS-135   A HALF CENTURY OF STELLAR PULSATION INTERPRETATIONS:
A TRIBUTE TO ARTHUR N. COX
eds. P. A. Bradley and J. A. Guzik
ISBN 1-886733-55-4

Vol. CS-136   GALACTIC HALOS: A UC SANTA CRUZ WORKSHOP
ed. Dennis Zaritsky
ISBN 1-886733-56-2

Vol. CS-137   WILD STARS IN THE OLD WEST: PROCEEDINGS OF THE 13th NORTH
AMERICAN WORKSHOP ON CATACLYSMIC VARIABLES
AND RELATED OBJECTS
eds. S. Howell, E. Kuulkers, and C. Woodward
ISBN 1-886733-57-0

Vol. CS-138   1997 PACIFIC RIM CONFERENCE ON STELLAR ASTROPHYSICS
eds. Kwing Lam Chan, K. S. Cheng, and H. P. Singh
ISBN 1-886733-58-9

Vol. CS-139   PRESERVING THE ASTRONOMICAL WINDOWS:
Proceedings of Joint Discussion No. 5 of the 23rd General Assembly of the IAU
eds. Syuzo Isobe and Tomohiro Hirayama
ISBN 1-886733-59-7

ASP CONFERENCE SERIES VOLUMES
Published by the Astronomical Society of the Pacific

PUBLISHED: 1998 (* asterisk means OUT OF STOCK)

Vol. CS-140   SYNOPTIC SOLAR PHYSICS --18th NSO/Sacramento Peak Summer Workshop
eds. K. S. Balasubramaniam, J. W. Harvey, and D. M. Rabin
ISBN 1-886733-60-0

Vol. CS-141   ASTROPHYSICS FROM ANTARCTICA:
A Symposium held as a part of the 109th Annual Meeting of the ASP
eds. Giles Novak and Randall H. Landsberg
ISBN 1-886733-61-9

Vol. CS-142   THE STELLAR INITIAL MASS FUNCTION: 38th Herstmonceux Conference
eds. Gerry Gilmore and Debbie Howell
ISBN 1-886733-62-7

Vol. CS-143 * THE SCIENTIFIC IMPACT OF THE GODDARD HIGH RESOLUTION
SPECTROGRAPH (GHRS)
eds. John C. Brandt, Thomas B. Ake III, and Carolyn Collins Petersen
ISBN 1-886733-63-5

Vol. CS-144   RADIO EMISSION FROM GALACTIC AND EXTRAGALACTIC COMPACT
SOURCES, IAU Colloquium 164
eds. J. Anton Zensus, G. B. Taylor, and J. M. Wrobel
ISBN 1-886733-64-3

Vol. CS-145   ASTRONOMICAL DATA ANALYSIS SOFTWARE AND SYSTEMS VII – (ADASS VII)
eds. Rudolf Albrecht, Richard N. Hook, and Howard A. Bushouse
ISBN 1-886733-65-1

Vol. CS-146   THE YOUNG UNIVERSE GALAXY FORMATION
AND EVOLUTION AT INTERMEDIATE AND HIGH REDSHIFT
eds. S. D'Odorico, A. Fontana, and E. Giallongo
ISBN 1-886733-66-X

Vol. CS-147   ABUNDANCE PROFILES: DIAGNOSTIC TOOLS FOR GALAXY HISTORY
eds. Daniel Friedli, Mike Edmunds, Carmelle Robert, and Laurent Drissen
ISBN 1-886733-67-8

Vol. CS-148   ORIGINS
eds. Charles E. Woodward, J. Michael Shull, and Harley A. Thronson, Jr.
ISBN 1-886733-68-6

Vol. CS-149   SOLAR SYSTEM FORMATION AND EVOLUTION
eds. D. Lazzaro, R. Vieira Martins, S. Ferraz-Mello, J. Fernández, and C. Beaugé
ISBN 1-886733-69-4

Vol. CS-150   NEW PERSPECTIVES ON SOLAR PROMINENCES, IAU Colloquium 167
eds. David Webb, David Rust, and Brigitte Schmieder
ISBN 1-886733-70-8

Vol. CS-151   COSMIC MICROWAVE BACKGROUND
AND LARGE SCALE STRUCTURES OF THE UNIVERSE
eds. Yong-Ik Byun and Kin-Wang Ng
ISBN 1-886733-71-6

Vol. CS-152   FIBER OPTICS IN ASTRONOMY III
eds. S. Arribas, E. Mediavilla, and F. Watson
ISBN 1-886733-72-4

Vol. CS-153   LIBRARY AND INFORMATION SERVICES IN ASTRONOMY III -- (LISA III)
eds. Uta Grothkopf, Heinz Andernach, Sarah Stevens-Rayburn,
and Monique Gomez
ISBN 1-886733-73-2

ASP CONFERENCE SERIES VOLUMES
Published by the Astronomical Society of the Pacific

PUBLISHED: 1998 (* asterisk means OUT OF STOCK)

Vol. CS-154   COOL STARS, STELLAR SYSTEMS AND THE SUN: Tenth Cambridge Workshop
eds. Robert A. Donahue and Jay A. Bookbinder
ISBN 1-886733-74-0

Vol. CS-155   SECOND ADVANCES IN SOLAR PHYSICS EUROCONFERENCE:
THREE-DIMENSIONAL STRUCTURE OF SOLAR ACTIVE REGIONS
eds. Costas E. Alissandrakis and Brigitte Schmieder
ISBN 1-886733-75-9

PUBLISHED: 1999

Vol. CS-156   HIGHLY REDSHIFTED RADIO LINES
eds. C. L. Carilli, S. J. E. Radford, K. M. Menten, and G. I. Langston
ISBN 1-886733-76-7

Vol. CS-157   ANNAPOLIS WORKSHOP ON MAGNETIC CATACLYSMIC VARIABLES
eds. Coel Hellier and Koji Mukai
ISBN 1-886733-77-5

Vol. CS-158   SOLAR AND STELLAR ACTIVITY: SIMILARITIES AND DIFFERENCES
eds. C. J. Butler and J. G. Doyle
ISBN 1-886733-78-3

Vol. CS-159   BL LAC PHENOMENON
eds. Leo O. Takalo and Aimo Sillanpää
ISBN 1-886733-79-1

Vol. CS-160   ASTROPHYSICAL DISCS: An EC Summer School
eds. J. A. Sellwood and Jeremy Goodman
ISBN 1-886733-80-5

Vol. CS-161   HIGH ENERGY PROCESSES IN ACCRETING BLACK HOLES
eds. Juri Poutanen and Roland Svensson
ISBN 1-886733-81-3

Vol. CS-162   QUASARS AND COSMOLOGY
eds. Gary Ferland and Jack Baldwin
ISBN 1-886733-83-X

Vol. CS-163   STAR FORMATION IN EARLY-TYPE GALAXIES
eds. Jordi Cepa and Patricia Carral
ISBN 1-886733-84-8

Vol. CS-164   ULTRAVIOLET–OPTICAL SPACE ASTRONOMY BEYOND HST
eds. Jon A. Morse, J. Michael Shull, and Anne L. Kinney
ISBN 1-886733-85-6

Vol. CS-165   THE THIRD STROMLO SYMPOSIUM: THE GALACTIC HALO
eds. Brad K. Gibson, Tim S. Axelrod, and Mary E. Putman
ISBN 1-886733-86-4

Vol. CS-166   STROMLO WORKSHOP ON HIGH-VELOCITY CLOUDS
eds. Brad K. Gibson and Mary E. Putman
ISBN 1-886733-87-2

Vol. CS-167   HARMONIZING COSMIC DISTANCE SCALES IN A POST-HIPPARCOS ERA
eds. Daniel Egret and André Heck
ISBN 1-886733-88-0

Vol. CS-168   NEW PERSPECTIVES ON THE INTERSTELLAR MEDIUM
eds. A. R. Taylor, T. L. Landecker, and G. Joncas
ISBN 1-886733-89-9

ASP CONFERENCE SERIES VOLUMES
Published by the Astronomical Society of the Pacific

PUBLISHED: 1999 (* asterisk means OUT OF STOCK)

Vol. CS-169 11th EUROPEAN WORKSHOP ON WHITE DWARFS
eds. J.-E. Solheim and E. G. Meištas
ISBN 1-886733-91-0

Vol. CS-170 THE LOW SURFACE BRIGHTNESS UNIVERSE, IAU Colloquium 171
eds. J. I. Davies, C. Impey, and S. Phillipps
ISBN 1-886733-92-9

Vol. CS-171 LiBeB, COSMIC RAYS, AND RELATED X- AND GAMMA-RAYS
eds. Reuven Ramaty, Elisabeth Vangioni-Flam, Michel Cassé, and Keith Olive
ISBN 1-886733-93-7

Vol. CS-172 ASTRONOMICAL DATA ANALYSIS SOFTWARE AND SYSTEMS VIII
eds. David M. Mehringer, Raymond L. Plante, and Douglas A. Roberts
ISBN 1-886733-94-5

Vol. CS-173 THEORY AND TESTS OF CONVECTION IN STELLAR STRUCTURE:
First Granada Workshop
ed. Álvaro Giménez, Edward F. Guinan, and Benjamín Montesinos
ISBN 1-886733-95-3

Vol. CS-174 CATCHING THE PERFECT WAVE: ADAPTIVE OPTICS AND
INTERFEROMETRY IN THE 21st CENTURY,
A Symposium held as a part of the 110th Annual Meeting of the ASP
eds. Sergio R. Restaino, William Junor, and Nebojsa Duric
ISBN 1-886733-96-1

Vol. CS-175 STRUCTURE AND KINEMATICS OF QUASAR BROAD LINE REGIONS
eds. C. M. Gaskell, W. N. Brandt, M. Dietrich, D. Dultzin-Hacyan,
and M. Eracleous
ISBN 1-886733-97-X

Vol. CS-176 OBSERVATIONAL COSMOLOGY: THE DEVELOPMENT OF GALAXY SYSTEMS
eds. Giuliano Giuricin, Marino Mezzetti, and Paolo Salucci
ISBN 1-58381-000-5

Vol. CS-177 ASTROPHYSICS WITH INFRARED SURVEYS: A Prelude to SIRTF
eds. Michael D. Bicay, Chas A. Beichman, Roc M. Cutri, and Barry F. Madore
ISBN 1-58381-001-3

Vol. CS-178 STELLAR DYNAMOS: NONLINEARITY AND CHAOTIC FLOWS
eds. Manuel Núñez and Antonio Ferriz-Mas
ISBN 1-58381-002-1

Vol. CS-179 ETA CARINAE AT THE MILLENNIUM
eds. Jon A. Morse, Roberta M. Humphreys, and Augusto Damineli
ISBN 1-58381-003-X

Vol. CS-180 SYNTHESIS IMAGING IN RADIO ASTRONOMY II
eds. G. B. Taylor, C. L. Carilli, and R. A. Perley
ISBN 1-58381-005-6

Vol. CS-181 MICROWAVE FOREGROUNDS
eds. Angelica de Oliveira-Costa and Max Tegmark
ISBN 1-58381-006-4

Vol. CS-182 GALAXY DYNAMICS: A Rutgers Symposium
eds. David Merritt, J. A. Sellwood, and Monica Valluri
ISBN 1-58381-007-2

Vol. CS-183 HIGH RESOLUTION SOLAR PHYSICS: THEORY, OBSERVATIONS,
AND TECHNIQUES
eds. T. R. Rimmele, K. S. Balasubramaniam, and R. R. Radick
ISBN 1-58381-009-9

ASP CONFERENCE SERIES VOLUMES
Published by the Astronomical Society of the Pacific

PUBLISHED: 1999 (* asterisk means OUT OF STOCK)

Vol. CS-184	THIRD ADVANCES IN SOLAR PHYSICS EUROCONFERENCE: MAGNETIC FIELDS AND OSCILLATIONS eds. B. Schmieder, A. Hofmann, and J. Staude ISBN 1-58381-010-2
Vol. CS-185	PRECISE STELLAR RADIAL VELOCITIES, IAU Colloquium 170 eds. J. B. Hearnshaw and C. D. Scarfe ISBN 1-58381-011-0
Vol. CS-186	THE CENTRAL PARSECS OF THE GALAXY eds. Heino Falcke, Angela Cotera, Wolfgang J. Duschl, Fulvio Melia, and Marcia J. Rieke ISBN 1-58381-012-9
Vol. CS-187	THE EVOLUTION OF GALAXIES ON COSMOLOGICAL TIMESCALES eds. J. E. Beckman and T. J. Mahoney ISBN 1-58381-013-7
Vol. CS-188	OPTICAL AND INFRARED SPECTROSCOPY OF CIRCUMSTELLAR MATTER eds. Eike W. Guenther, Bringfried Stecklum, and Sylvio Klose ISBN 1-58381-014-5
Vol. CS-189	CCD PRECISION PHOTOMETRY WORKSHOP eds. Eric R. Craine, Roy A. Tucker, and Jeannette Barnes ISBN 1-58381-015-3
Vol. CS-190	GAMMA-RAY BURSTS: THE FIRST THREE MINUTES eds. Juri Poutanen and Roland Svensson ISBN 1-58381-016-1
Vol. CS-191	PHOTOMETRIC REDSHIFTS AND HIGH REDSHIFT GALAXIES eds. Ray J. Weymann, Lisa J. Storrie-Lombardi, Marcin Sawicki, and Robert J. Brunner ISBN 1-58381-017-X
Vol. CS-192	SPECTROPHOTOMETRIC DATING OF STARS AND GALAXIES ed. I. Hubeny, S. R. Heap, and R. H. Cornett ISBN 1-58381-018-8
Vol. CS-193	THE HY-REDSHIFT UNIVERSE: GALAXY FORMATION AND EVOLUTION AT HIGH REDSHIFT eds. Andrew J. Bunker and Wil J. M. van Breugel ISBN 1-58381-019-6
Vol. CS-194	WORKING ON THE FRINGE: OPTICAL AND IR INTERFEROMETRY FROM GROUND AND SPACE eds. Stephen Unwin and Robert Stachnik ISBN 1-58381-020-X

PUBLISHED: 2000

Vol. CS-195	IMAGING THE UNIVERSE IN THREE DIMENSIONS: Astrophysics with Advanced Multi-Wavelength Imaging Devices eds. W. van Breugel and J. Bland-Hawthorn ISBN 1-58381-022-6
Vol. CS-196	THERMAL EMISSION SPECTROSCOPY AND ANALYSIS OF DUST, DISKS, AND REGOLITHS eds. Michael L. Sitko, Ann L. Sprague, and David K. Lynch ISBN: 1-58381-023-4
Vol. CS-197	XVth IAP MEETING DYNAMICS OF GALAXIES: FROM THE EARLY UNIVERSE TO THE PRESENT eds. F. Combes, G. A. Mamon, and V. Charmandaris ISBN: 1-58381-24-2

ASP CONFERENCE SERIES VOLUMES
Published by the Astronomical Society of the Pacific

PUBLISHED: 2000 (* asterisk means OUT OF STOCK)

Vol. CS-198   EUROCONFERENCE ON "STELLAR CLUSTERS AND ASSOCIATIONS: CONVECTION, ROTATION, AND DYNAMOS"
eds. R. Pallavicini, G. Micela, and S. Sciortino
ISBN: 1-58381-25-0

Vol. CS-199   ASYMMETRICAL PLANETARY NEBULAE II: FROM ORIGINS TO MICROSTRUCTURES
eds. J. H. Kastner, N. Soker, and S. Rappaport
ISBN: 1-58381-026-9

Vol. CS-200   CLUSTERING AT HIGH REDSHIFT
eds. A. Mazure, O. Le Fèvre, and V. Le Brun
ISBN: 1-58381-027-7

Vol. CS-201   COSMIC FLOWS 1999: TOWARDS AN UNDERSTANDING OF LARGE-SCALE STRUCTURES
eds. Stéphane Courteau, Michael A. Strauss, and Jeffrey A. Willick
ISBN: 1-58381-028-5

Vol. CS-202 *   PULSAR ASTRONOMY – 2000 AND BEYOND, IAU Colloquium 177
eds. M. Kramer, N. Wex, and R. Wielebinski
ISBN: 1-58381-029-3

Vol. CS-203   THE IMPACT OF LARGE-SCALE SURVEYS ON PULSATING STAR RESEARCH, IAU Colloquium 176
eds. L. Szabados and D. W. Kurtz
ISBN: 1-58381-030-7

Vol. CS-204   THERMAL AND IONIZATION ASPECTS OF FLOWS FROM HOT STARS: OBSERVATIONS AND THEORY
eds. Henny J. G. L. M. Lamers and Arved Sapar
ISBN: 1-58381-031-5

Vol. CS-205   THE LAST TOTAL SOLAR ECLIPSE OF THE MILLENNIUM IN TURKEY
eds. W. C. Livingston and A. Özgüç
ISBN: 1-58381-032-3

Vol. CS-206   HIGH ENERGY SOLAR PHYSICS – *ANTICIPATING HESSI*
eds. Reuven Ramaty and Natalie Mandzhavidze
ISBN: 1-58381-033-1

Vol. CS-207   NGST SCIENCE AND TECHNOLOGY EXPOSITION
eds. Eric P. Smith and Knox S. Long
ISBN: 1-58381-036-6

ATLAS   VISIBLE AND NEAR INFRARED ATLAS OF THE ARCTURUS SPECTRUM 3727-9300 Å
eds. Kenneth Hinkle, Lloyd Wallace, Jeff Valenti, and Dianne Harmer
ISBN: 1-58381-037-4

Vol. CS-208   POLAR MOTION: HISTORICAL AND SCIENTIFIC PROBLEMS, IAU Colloquium 178
eds. Steven Dick, Dennis McCarthy, and Brian Luzum
ISBN: 1-58381-039-0

Vol. CS-209   SMALL GALAXY GROUPS, IAU Colloquium 174
eds. Mauri J. Valtonen and Chris Flynn
ISBN: 1-58381-040-4

Vol. CS-210   DELTA SCUTI AND RELATED STARS: Reference Handbook and Proceedings of the 6th Vienna Workshop in Astrophysics
eds. Michel Breger and Michael Houston Montgomery
ISBN: 1-58381-043-9

ASP CONFERENCE SERIES VOLUMES
Published by the Astronomical Society of the Pacific

PUBLISHED: 2000 (* asterisk means OUT OF STOCK)

Vol. CS-211   MASSIVE STELLAR CLUSTERS
eds. Ariane Lançon and Christian M. Boily
ISBN: 1-58381-042-0

Vol. CS-212   FROM GIANT PLANETS TO COOL STARS
eds. Caitlin A. Griffith and Mark S. Marley
ISBN: 1-58381-041-2

Vol. CS-213   BIOASTRONOMY `99: A NEW ERA IN BIOASTRONOMY
eds. Guillermo A. Lemarchand and Karen J. Meech
ISBN: 1-58381-044-7

Vol. CS-214   THE Be PHENOMENON IN EARLY-TYPE STARS, IAU Colloquium 175
eds. Myron A. Smith, Huib F. Henrichs and Juan Fabregat
ISBN: 1-58381-045-5

Vol. CS-215   COSMIC EVOLUTION AND GALAXY FORMATION:
STRUCTURE, INTERACTIONS AND FEEDBACK
The 3rd Guillermo Haro Astrophysics Conference
eds. José Franco, Elena Terlevich, Omar López-Cruz, and Itziar Aretxaga
ISBN: 1-58381-046-3

Vol. CS-216   ASTRONOMICAL DATA ANALYSIS SOFTWARE AND SYSTEMS IX
eds. Nadine Manset, Christian Veillet, and Dennis Crabtree
ISBN: 1-58381-047-1   ISSN: 1080-7926

Vol. CS-217   IMAGING AT RADIO THROUGH SUBMILLIMETER WAVELENGTHS
eds. Jeffrey G. Mangum and Simon J. E. Radford
ISBN: 1-58381-049-8

Vol. CS-218   MAPPING THE HIDDEN UNIVERSE: THE UNIVERSE BEHIND THE MILKY WAY
THE UNIVERSE IN HI
eds. Renée C. Kraan-Korteweg, Patricia A. Henning, and Heinz Andernach
ISBN: 1-58381-050-1

Vol. CS-219   DISKS, PLANETESIMALS, AND PLANETS
eds. F. Garzón, C. Eiroa, D. de Winter, and T. J. Mahoney
ISBN: 1-58381-051-X

Vol. CS-220   AMATEUR - PROFESSIONAL PARTNERSHIPS IN ASTRONOMY:
The 111th Annual Meeting of the ASP
eds. John R. Percy and Joseph B. Wilson
ISBN: 1-58381-052-8

Vol. CS-221   STARS, GAS AND DUST IN GALAXIES: EXPLORING THE LINKS
eds. Danielle Alloin, Knut Olsen, and Gaspar Galaz
ISBN: 1-58381-053-6

PUBLISHED: 2001

Vol. CS-222   THE PHYSICS OF GALAXY FORMATION
eds. M. Umemura and H. Susa
ISBN: 1-58381-054-4

Vol. CS-223   COOL STARS, STELLAR SYSTEMS AND THE SUN:
Eleventh Cambridge Workshop
eds. Ramón J. García López, Rafael Rebolo, and María Zapatero Osorio
ISBN: 1-58381-056-0

Vol. CS-224   PROBING THE PHYSICS OF ACTIVE GALACTIC NUCLEI
BY MULTIWAVELENGTH MONITORING
eds. Bradley M. Peterson, Ronald S. Polidan, and Richard W. Pogge
ISBN: 1-58381-055-2

ASP CONFERENCE SERIES VOLUMES
Published by the Astronomical Society of the Pacific

PUBLISHED: 2001 (* asterisk means OUT OF STOCK)

Vol. CS-225 VIRTUAL OBSERVATORIES OF THE FUTURE
eds. Robert J. Brunner, S. George Djorgovski, and Alex S. Szalay
ISBN: 1-58381-057-9

Vol. CS-226 12th EUROPEAN CONFERENCE ON WHITE DWARFS
eds. J. L. Provencal, H. L. Shipman, J. MacDonald, and S. Goodchild
ISBN: 1-58381-058-7

Vol. CS-227 BLAZAR DEMOGRAPHICS AND PHYSICS
eds. Paolo Padovani and C. Megan Urry
ISBN: 1-58381-059-5

Vol. CS-228 DYNAMICS OF STAR CLUSTERS AND THE MILKY WAY
eds. S. Deiters, B. Fuchs, A. Just, R. Spurzem, and R. Wielen
ISBN: 1-58381-060-9

Vol. CS-229 EVOLUTION OF BINARY AND MULTIPLE STAR SYSTEMS
A Meeting in Celebration of Peter Eggleton's 60th Birthday
eds. Ph. Podsiadlowski, S. Rappaport, A. R. King, F. D'Antona, and L. Burderi
IBSN: 1-58381-061-7

Vol. CS-230 GALAXY DISKS AND DISK GALAXIES
eds. Jose G. Funes, S. J. and Enrico Maria Corsini
ISBN: 1-58381-063-3

Vol. CS-231 TETONS 4: GALACTIC STRUCTURE, STARS, AND
THE INTERSTELLAR MEDIUM
eds. Charles E. Woodward, Michael D. Bicay, and J. Michael Shull
ISBN: 1-58381-064-1

Vol. CS-232 THE NEW ERA OF WIDE FIELD ASTRONOMY
eds. Roger Clowes, Andrew Adamson, and Gordon Bromage
ISBN: 1-58381-065-X

Vol. CS-233 P CYGNI 2000: 400 YEARS OF PROGRESS
eds. Mart de Groot and Christiaan Sterken
ISBN: 1-58381-070-6

Vol. CS-234 X-RAY ASTRONOMY 2000
eds. R. Giacconi, S. Serio, and L. Stella
ISBN: 1-58381-071-4

Vol. CS-235 SCIENCE WITH THE ATACAMA LARGE MILLIMETER ARRAY (ALMA)
ed. Alwyn Wootten
ISBN: 1-58381-072-2

Vol. CS-236 ADVANCED SOLAR POLARIMETRY: THEORY, OBSERVATION, AND
INSTRUMENTATION, The 20th Sacramento Peak Summer Workshop
ed. M. Sigwarth
ISBN: 1-58381-073-0

Vol. CS-237 GRAVITATIONAL LENSING: RECENT PROGRESS AND FUTURE GOALS
eds. Tereasa G. Brainerd and Christopher S. Kochanek
ISBN: 1-58381-074-9

Vol. CS-238 ASTRONOMICAL DATA ANALYSIS SOFTWARE AND SYSTEMS X
eds. F. R. Harnden, Jr., Francis A. Primini, and Harry E. Payne
ISBN: 1-58381-075-7

Vol. CS-239 MICROLENSING 2000: A NEW ERA OF MICROLENSING ASTROPHYSICS
ed. John Menzies and Penny D. Sackett
ISBN: 1-58381-076-5

ASP CONFERENCE SERIES VOLUMES
Published by the Astronomical Society of the Pacific

PUBLISHED: 2001 (* asterisk means OUT OF STOCK)

Vol. CS-240	GAS AND GALAXY EVOLUTION, A Conference in Honor of the 20th Anniversary of the VLA eds. J. E. Hibbard, M. P. Rupen, and J. H. van Gorkom ISBN: 1-58381-077-3
Vol. CS-241	CS-241 THE 7TH TAIPEI ASTROPHYSICS WORKSHOP ON COSMIC RAYS IN THE UNIVERSE ed. Chung-Ming Ko ISBN: 1-58381-079-X
Vol. CS-242	ETA CARINAE AND OTHER MYSTERIOUS STARS: THE HIDDEN OPPORTUNITIES OF EMISSION SPECTROSCOPY eds. Theodore R. Gull, Sveneric Johannson, and Kris Davidson ISBN: 1-58381-080-3
Vol. CS-243	FROM DARKNESS TO LIGHT: ORIGIN AND EVOLUTION OF YOUNG STELLAR CLUSTERS eds. Thierry Montmerle and Philippe André ISBN: 1-58381-081-1
Vol. CS-244	YOUNG STARS NEAR EARTH: PROGRESS AND PROSPECTS eds. Ray Jayawardhana and Thomas P. Greene ISBN: 1-58381-082-X
Vol. CS-245	ASTROPHYSICAL AGES AND TIME SCALES eds. Ted von Hippel, Chris Simpson, and Nadine Manset ISBN: 1-58381-083-8
Vol. CS-246	SMALL TELESCOPE ASTRONOMY ON GLOBAL SCALES, IAU Colloquium 183 eds. Wen-Ping Chen, Claudia Lemme, and Bohdan Paczyński ISBN: 1-58381-084-6
Vol. CS-247	SPECTROSCOPIC CHALLENGES OF PHOTOIONIZED PLASMAS eds. Gary Ferland and Daniel Wolf Savin ISBN: 1-58381-085-4
Vol. CS-248	MAGNETIC FIELDS ACROSS THE HERTZSPRUNG-RUSSELL DIAGRAM eds. G. Mathys, S. K. Solanki, and D. T. Wickramasinghe ISBN: 1-58381-088-9
Vol. CS-249	THE CENTRAL KILOPARSEC OF STARBURSTS AND AGN: THE LA PALMA CONNECTION eds. J. H. Knapen, J. E. Beckman, I. Shlosman, and T. J. Mahoney ISBN: 1-58381-089-7
Vol. CS-250	PARTICLES AND FIELDS IN RADIO GALAXIES CONFERENCE eds. Robert A. Laing and Katherine M. Blundell ISBN: 1-58381-090-0
Vol. CS-251	NEW CENTURY OF X-RAY ASTRONOMY eds. H. Inoue and H. Kunieda ISBN: 1-58381-091-9
Vol. CS-252	HISTORICAL DEVELOPMENT OF MODERN COSMOLOGY eds. Vicent J. Martínez, Virginia Trimble, and María Jesús Pons-Bordería ISBN: 1-58381-092-7

PUBLISHED: 2002

Vol. CS-253	CHEMICAL ENRICHMENT OF INTRACLUSTER AND INTERGALACTIC MEDIUM eds. Roberto Fusco-Femiano and Francesca Matteucci ISBN: 1-58381-093-5

ASP CONFERENCE SERIES VOLUMES
Published by the Astronomical Society of the Pacific

PUBLISHED: 2002 (* asterisk means OUT OF STOCK)

Vol. CS-254   EXTRAGALACTIC GAS AT LOW REDSHIFT
eds. John S. Mulchaey and John T. Stocke
ISBN: 1-58381-094-3

Vol. CS-255   MASS OUTFLOW IN ACTIVE GALACTIC NUCLEI: NEW PERSPECTIVES
eds. D. M. Crenshaw, S. B. Kraemer, and I. M. George
ISBN: 1-58381-095-1

Vol. CS-256   OBSERVATIONAL ASPECTS OF PULSATING B AND A STARS
eds. Christiaan Sterken and Donald W. Kurtz
ISBN: 1-58381-096-X

Vol. CS-257   AMiBA 2001: HIGH-Z CLUSTERS, MISSING BARYONS, AND CMB POLARIZATION
eds. Lin-Wen Chen, Chung-Pei Ma, Kin-Wang Ng, and Ue-Li Pen
ISBN: 1-58381-097-8

Vol. CS-258   ISSUES IN UNIFICATION OF ACTIVE GALACTIC NUCLEI
eds. Roberto Maiolino, Alessandro Marconi, and Neil Nagar
ISBN: 1-58381-098-6

Vol. CS-259   RADIAL AND NONRADIAL PULSATIONS AS PROBES OF STELLAR PHYSICS, IAU Colloquium 185
eds. Conny Aerts, Timothy R. Bedding, and Jørgen Christensen-Dalsgaard
ISBN: 1-58381-099-4

Vol. CS-260   INTERACTING WINDS FROM MASSIVE STARS
eds. Anthony F. J. Moffat and Nicole St-Louis
ISBN: 1-58381-100-1

Vol. CS-261   THE PHYSICS OF CATACLYSMIC VARIABLES AND RELATED OBJECTS
eds. B. T. Gänsicke, K. Beuermann, and K. Reinsch
ISBN: 1-58381-101-X

Vol. CS-262   THE HIGH ENERGY UNIVERSE AT SHARP FOCUS: CHANDRA SCIENCE, held in conjunction with the 113th Annual Meeting of the ASP
eds. Eric M. Schlegel and Saeqa Dil Vrtilek
ISBN: 1-58381-102-8

Vol. CS-263   STELLAR COLLISIONS, MERGERS AND THEIR CONSEQUENCES
ed. Michael M. Shara
ISBN: 1-58381-103-6

Vol. CS-264   CONTINUING THE CHALLENGE OF EUV ASTRONOMY: CURRENT ANALYSIS AND PROSPECTS FOR THE FUTURE
eds. Steve B. Howell, Jean Dupuis, Daniel Golombek, Frederick M. Walter, and Jennifer Cullison
ISBN: 1-58381-104-4

Vol. CS-265   ω CENTAURI, A UNIQUE WINDOW INTO ASTROPHYSICS
eds. Floor van Leeuwen, Joanne D. Hughes, and Giampaolo Piotto
ISBN: 1-58381-105-2

Vol. CS-266   ASTRONOMICAL SITE EVALUATION IN THE VISIBLE AND RADIO RANGE, IAU Technical Workshop
eds. J. Vernin, Z. Benkhaldoun, and C. Muñoz-Tuñón
ISBN: 1-58381-106-0

Vol. CS-267   HOT STAR WORKSHOP III: THE EARLIEST STAGES OF MASSIVE STAR BIRTH
ed. Paul A. Crowther
ISBN: 1-58381-107-9

Vol. CS-268   TRACING COSMIC EVOLUTION WITH GALAXY CLUSTERS
eds. Stefano Borgani, Marino Mezzetti, and Riccardo Valdarnini
ISBN: 1-58381-108-7

ASP CONFERENCE SERIES VOLUMES
Published by the Astronomical Society of the Pacific

PUBLISHED: 2002 (* asterisk means OUT OF STOCK)

Vol. CS-269   THE EVOLVING SUN AND ITS INFLUENCE ON PLANETARY ENVIRONMENTS
eds. Benjamín Montesinos, Álvaro Giménez, and Edward F. Guinan
ISBN: 1-58381-109-5

Vol. CS-270   ASTRONOMICAL INSTRUMENTATION AND THE BIRTH AND GROWTH OF ASTROPHYSICS: A Symposium held in honor of Robert G. Tull
eds. Frank N. Bash and Christopher Sneden
ISBN: 1-58381-110-9

Vol. CS-271   NEUTRON STARS IN SUPERNOVA REMNANTS
eds. Patrick O. Slane and Bryan M. Gaensler
ISBN: 1-58381-111-7

Vol. CS-272   THE FUTURE OF SOLAR SYSTEM EXPLORATION, 2003-2013
Community Contributions to the NRC Solar System Exploration Decadal Survey
ed. Mark V. Sykes
ISBN: 1-58381-113-3

Vol. CS-273   THE DYNAMICS, STRUCTURE AND HISTORY OF GALAXIES
eds. G. S. Da Costa and H. Jerjen
ISBN: 1-58381-114-1

Vol. CS-274   OBSERVED HR DIAGRAMS AND STELLAR EVOLUTION
eds. Thibault Lejeune and João Fernandes
ISBN: 1-58381-116-8

Vol. CS-275   DISKS OF GALAXIES: KINEMATICS, DYNAMICS AND PERTURBATIONS
eds. E. Athanassoula, A. Bosma, and R. Mujica
ISBN: 1-58381-117-6

Vol. CS-276   SEEING THROUGH THE DUST:
THE DETECTION OF HI AND THE EXPLORATION OF THE ISM IN GALAXIES
eds. A. R. Taylor, T. L. Landecker, and A. G. Willis
ISBN: 1-58381-118-4

Vol. CS 277   STELLAR CORONAE IN THE CHANDRA AND XMM-NEWTON ERA
eds. Fabio Favata and Jeremy J. Drake
ISBN: 1-58381-119-2

Vol. CS 278   NAIC–NRAO SCHOOL ON SINGLE-DISH ASTRONOMY:
TECHNIQUES AND APPLICATIONS
eds. Snezana Stanimirovic, Daniel Altschuler, Paul Goldsmith, and Chris Salter
ISBN: 1-58381-120-6

Vol. CS 279   EXOTIC STARS AS CHALLENGES TO EVOLUTION, IAU Colloquium 187
eds. Christopher A. Tout and Walter Van Hamme
ISBN: 1-58381-122-2

Vol. CS 280   NEXT GENERATION WIDE-FIELD MULTI-OBJECT SPECTROSCOPY
eds. Michael J. I. Brown and Arjun Dey
ISBN: 1-58381-123-0

Vol. CS 281   ASTRONOMICAL DATA ANALYSIS SOFTWARE AND SYSTEM XI
eds. David A. Bohlender, Daniel Durand, and Thomas H. Handley
ISBN: 1-58381-124-9   ISSN: 1080-7926

Vol. CS 282   GALAXIES: THE THIRD DIMENSION
eds. Margarita Rosado, Luc Binette, and Lorena Arias
ISBN: 1-58381-125-7

Vol. CS 283   A NEW ERA IN COSMOLOGY
eds. Nigel Metcalfe and Tom Shanks
ISBN: 1-58381-126-5

ASP CONFERENCE SERIES VOLUMES
Published by the Astronomical Society of the Pacific

PUBLISHED: 2002 (* asterisk means OUT OF STOCK)

Vol. CS 284   AGN SURVEYS
eds. R. F. Green, E. Ye. Khachikian, and D. B. Sanders
ISBN: 1-58381-127-3

Vol. CS 285   MODES OF STAR FORMATION AND THE ORIGIN OF FIELD POPULATIONS
eds. Eva K. Grebel and Walfgang Brandner
ISBN: 1-58381-128-1

PUBLISHED: 2003

Vol. CS 286   CURRENT THEORETICAL MODESL AND HIGH RESOLUTION SOLAR OBSERVATIONS: PREPARING FOR ATST
eds. Alexei A. Pevtsov and Han Uitenbroek
ISBN: 1-58381-129-X

Vol. CS 287   GALACTIC STAR FORMATION ACROSS THE STELLAR MASS SPECTRUM
eds. J.M. De Buizer and N.S. van der Bliek
ISBN:1-58381-130-3

Vol. CS 288   STELLAR ATMOSPHERE MODELING
eds. I. Hubeny, D. Mihalas and K. Werner
ISBN: 1-58381-131-1

Vol. CS 289   THE PROCEEDINGS OF THE IAU 8TH ASIAN-PACIFIC REGIONAL MEETING, VOLUME 1
eds. Satoru Ikeuchi, John Hearnshaw and Tomoyuki Hanawa
ISBN: 1-58381-134-6

Vol. CS 290   ACTIVE GALACTIC NUCLEI: FROM CENTRAL ENGINE TO HOST GALAXY
eds. S. Collin, F. Combes and I. Shlosman
ISBN: 1-58381-135-4

Vol. CS-291   HUBBLE'S SCIENCE LEGACY:
FUTURE OPTICAL/ULTRAVIOLET ASTRONOMY FROM SPACE
eds. Kenneth R. Sembach, J. Chris Blades, Garth D. Illingworth and Robert C. Kennicutt, Jr.
ISBN: 1-58381-136-2

Vol. CS-292   INTERPLAY OF PERIODIC, CYCLIC AND STOCHASTIC VARIABILITY IN SELECTED AREAS OF THE H-R DIAGRAM
ed. Christiaan Sterken
ISBN: 1-58381-138-9

Vol. CS-293   3D STELLAR EVOLUTION
eds. S. Turcotte, S. C. Keller and R. M. Cavallo
ISBN: 1-58381-140-0

Vol. CS-294   SCIENTIFIC FRONTIERS IN RESEARCH ON EXTRASOLAR PLANETS
eds. Drake Deming and Sara Seager
ISBN: 1-58381-141-9

Vol. CS-295   ASTRONOMICAL DATA ANALYSIS SOFTWARE AND SYSTEMS XII
eds. Harry E. Payne, Robert I. Jedrzejewski and Richard N. Hook
ISBN: 1-58381-142-7

Vol. CS-296   NEW HORIZONS IN GLOBULAR CLUSTER ASTRONOMY
eds. Giampaolo Piotto, Georges Meylan, S. George Djorgovski and Marco Riello
ISBN: 1-58381-143-5

Vol. CS-297   STAR FORMATION THROUGH TIME
eds. Enrique Perez, Rosa M. Gonzalez Delgado and Guillormo Tonorio Tagle
ISBN: 1-58381-144-3

ASP CONFERENCE SERIES VOLUMES
Published by the Astronomical Society of the Pacific

PUBLISHED: 2003 (* asterisk means OUT OF STOCK)

Vol. CS-298   GAIA SPECTROSCOPY: SCIENCE AND TECHNOLOGY
ed. Ulisse Munari
ISBN: 1-58381-145-1

A LISTING OF IAU VOLUMES MAY BE FOUND ON THE NEXT PAGE

INTERNATIONAL ASTRONOMICAL UNION (IAU) VOLUMES
Published by the Astronomical Society of the Pacific

PUBLISHED: 1999 (* asterisk means OUT OF STOCK)

Vol. No. 190   NEW VIEWS OF THE MAGELLANIC CLOUDS
eds. You-Hua Chu, Nicholas B. Suntzeff, James E. Hesser, and David A. Bohlender
ISBN: 1-58381-021-8

Vol. No. 191   ASYMPTOTIC GIANT BRANCH STARS
eds. T. Le Bertre, A. Lèbre, and C. Waelkens
ISBN: 1-886733-90-2

Vol. No. 192   THE STELLAR CONTENT OF LOCAL GROUP GALAXIES
eds. Patricia Whitelock and Russell Cannon
ISBN: 1-886733-82-1

Vol. No. 193   WOLF-RAYET PHENOMENA IN MASSIVE STARS AND STARBURST GALAXIES
eds. Karel A. van der Hucht, Gloria Koenigsberger, and Philippe R. J. Eenens
ISBN: 1-58381-004-8

Vol. No. 194   ACTIVE GALACTIC NUCLEI AND RELATED PHENOMENA
eds. Yervant Terzian, Daniel Weedman, and Edward Khachikian
ISBN: 1-58381-008-0

PUBLISHED: 2000

Vol. XXIVA   TRANSACTIONS OF THE INTERNATIONAL ASTRONOMICAL UNION
REPORTS ON ASTRONOMY 1996-1999
ed. Johannes Andersen
ISBN: 1-58381-035-8

Vol. No. 195   HIGHLY ENERGETIC PHYSICAL PROCESSES AND MECHANISMS FOR EMISSION FROM ASTROPHYSICAL PLASMAS
eds. P. C. H. Martens, S. Tsuruta, and M. A. Weber
ISBN: 1-58381-038-2

Vol. No. 197   ASTROCHEMISTRY: FROM MOLECULAR CLOUDS TO PLANETARY SYSTEMS
eds. Y. C. Minh and E. F. van Dishoeck
ISBN: 1-58381-034-X

Vol. No. 198   THE LIGHT ELEMENTS AND THEIR EVOLUTION
eds. L. da Silva, M. Spite, and J. R. de Medeiros
ISBN: 1-58381-048-X

PUBLISHED: 2001

IAU SPS   ASTRONOMY FOR DEVELOPING COUNTRIES
Special Session of the XXIV General Assembly of the IAU
ed. Alan H. Batten
ISBN: 1-58381-067-6

Vol. No. 196   PRESERVING THE ASTRONOMICAL SKY
eds. R. J. Cohen and W. T. Sullivan, III
ISBN: 1-58381-078-1

Vol. No. 200   THE FORMATION OF BINARY STARS
eds. Hans Zinnecker and Robert D. Mathieu
ISBN: 1-58381-068-4

Vol. No. 203   RECENT INSIGHTS INTO THE PHYSICS OF THE SUN AND HELIOSPHERE: HIGHLIGHTS FROM SOHO AND OTHER SPACE MISSIONS
eds. Pål Brekke, Bernhard Fleck, and Joseph B. Gurman
ISBN: 1-58381-069-2

Vol. No. 204   THE EXTRAGALACTIC INFRARED BACKGROUND AND ITS COSMOLOGICAL IMPLICATIONS
eds. Martin Harwit and Michael G. Hauser
ISBN: 1-58381-062-5

INTERNATIONAL ASTRONOMICAL UNION (IAU) VOLUMES
Published by the Astronomical Society of the Pacific

PUBLISHED: 2001 (* asterisk means OUT OF STOCK)

Vol. No. 205 GALAXIES AND THEIR CONSTITUENTS
 AT THE HIGHEST ANGULAR RESOLUTIONS
 eds. Richard T. Schilizzi, Stuart N. Vogel, Francesco Paresce, and Martin S. Elvis
 ISBN: 1-58381-066-8

Vol. XXIVB TRANSACTIONS OF THE INTERNATIONAL ASTRONOMICAL UNION
 REPORTS ON ASTRONOMY
 ed. Hans Rickman
 ISBN: 1-58381-087-0

PUBLISHED: 2002

Vol. No. 12 HIGHLIGHTS OF ASTRONOMY
 ed. Hans Rickman
 ISBN: 1-58381-086-2

Vol. No. 199 THE UNIVERSE AT LOW RADIO FREQUENCIES
 eds. A. Pramesh Rao, G. Swarup, and Gopal-Krishna
 ISBN: 58381-121-4

Vol. No. 206 COSMIC MASERS: FROM PROTOSTARS TO BLACKHOLES
 eds. Victor Migenes and Mark J. Reid
 ISBN: 1-58381-112-5

Vol. No. 207 EXTRAGALACTIC STAR CLUSTERS
 eds. Doug Geisler, Eva K. Grebel, and Dante Minniti
 ISBN: 1-58381-115-X

PUBLISHED: 2003

Vol. XXVA TRANSACTIONS OF THE INTERNATIONAL ASTRONOMICAL UNION
 REPORTS ON ASTRONOMY 1999-2002
 ed. Hans Rickman
 ISBN: 1-58381-137-0

Vol. No. 208 ASTROPHYSICAL SUPERCOMPUTING USING PARTICLE SIMULATIONS
 eds. Junichiro Makino and Piet Hut
 ISBN: 1-58381-139-7

Vol. No. 209 PLANETARY NEBULAE: THEIR EVOLUTION AND ROLE IN THE UNIVERSE
 eds. Sun Kwok, Michael Dopita and Ralph Sutherland
 ISBN: 1-58381-148-6

Vol. No. 211 BROWN DWARFS
 ed. Eduardo Martín
 ISBN: 1-58381-132-X

Vol. No. 212 A MASSIVE STAR ODYSSEY: FROM MAIN SEQUENCE TO SUPERNOVA
 eds. Karel A. van der Hucht, Artemio Herrero and César Esteban
 ISBN: 1-58381-133-8

Ordering information is available at the beginning of the listing